The Role of Toll-Like Receptors (TLRs) in Infection and Inflammation 2.0

The Role of Toll-Like Receptors (TLRs) in Infection and Inflammation 2.0

Ralf Kircheis
Oliver Planz

MDPI

Basel • Beijing • Wuhan • Barcelona • Belgrade • Novi Sad • Cluj • Manchester

Editors
Ralf Kircheis
Research & Development
Advanced Medical Solutions
Saal a.d. Donau
Germany

Oliver Planz
Institute of Cell Biology
and Immunology
Eberhard Karls University
Tuebingen
Germany

Editorial Office
MDPI AG
Grosspeteranlage 5
4052 Basel, Switzerland

This is a reprint of articles from the Special Issue published online in the open access journal *International Journal of Molecular Sciences* (ISSN 1422-0067) (available at: www.mdpi.com/journal/ijms/special_issues/N6BP5CK662).

For citation purposes, cite each article independently as indicated on the article page online and as indicated below:

Lastname, A.A.; Lastname, B.B. Article Title. *Journal Name* **Year**, *Volume Number*, Page Range.

ISBN 978-3-7258-2344-4 (Hbk)
ISBN 978-3-7258-2343-7 (PDF)
doi.org/10.3390/books978-3-7258-2343-7

Contents

About the Editors

Ralf Kircheis

Dr. Ralf Kircheis has a background in biomedicine. He has outstanding experience in a broad area of medical sciences, including immunology, oncology, gene therapy, drug and gene delivery, and drug development. Ralf Kircheis is the author and editor of multiple research articles, review articles, book chapters, and patent applications.

Oliver Planz

Professor Oliver Planz is a leading internationally recognized scientist in the areas of virology and innate immunity. He is an author and editor of multiple research articles, review articles, book chapters, and patent applications.

Preface

Toll-like receptors (TLRs) are key players in the innate immune system in host' first-line defense against pathogens. TLRs recognize pathogen-associated molecular patterns (PAMPs) from bacteria, viruses, and other microorganisms, and self-derived damage-associated molecular patterns (DAMPs) released from dying or lytic cells, as well as chromatin-associated molecular patterns (CAMPs), thus playing a critical role not only in immune surveillance but also in disease pathogenesis. TLR activation triggers the production of proinflammatory cytokines and type I interferons, which are key mediators of the host immune response against bacterial, viral, and fungal infections. However, dysregulation and excessive activation can be detrimental, leading to hyperinflammation, sepsis, and loss of tissue integrity. TLRs are involved in the pathogenesis of acute bacterial and viral infections, as well as chronic infectious and non-infectious inflammatory diseases.

This Special Issue 2.0 continues the conceptional line started in the previous Special Issue, focusing on the involvement of Toll-like receptors (TLRs) and their related signaling pathways in viral infection and inflammation, and emphasizing the therapeutic potential of TLR system modulation for the treatment of acute viral infections and their long-term sequelae and to provide potential targets for treating chronic viral and inflammatory diseases. This Special Issue provides an excellent collection of contemporary reviews and research articles covering a broad variety of topics, ranging from the protective activity of TLRs in viral infections to detrimental disturbances of the host immune system. The most intriguing theme discussed in this second Special Issue is the intercorrelation between acute viral infections and triggered delayed effects, including neuroinflammatory and neurodegenerative long-term sequelae. Notably, new data increasingly show that neurological pathways, such as neurogenesis, neuroinflammation, and cognitive and behavioural aspects, may be influenced by innate, TLR-mediated pathways. In this context, the connection between infection-triggered acute disbalance in TLR activation and resulting neurological sequelae following bacterial (sepsis) or viral infections, including long COVID-associated neurological sequelae, may provide new targets for future treatment options.

The two editors of this Special Issue, Dr. Ralf Kircheis and Professor Oliver Planz, have many years of outstanding experience in the areas of immunology, oncology, virology, drug delivery, gene therapy, and drug development. This Special Issue may be of particular interest for researchers and students working in the areas of innate immunology, viral and bacterial infections, and acute and chronic inflammation, but also for all other medical researchers and practitioners who are interested in the molecular and cellular mechanisms of the innate immune system.

Ralf Kircheis and Oliver Planz
Editors

International Journal of *Molecular Sciences*

MDPI

Editorial

Special Issue "The Role of Toll-Like Receptors (TLRs) in Infection and Inflammation 2.0"

Ralf Kircheis [1,*] and Oliver Planz [2]

1 Syntacoll GmbH, 93342 Saal an der Donau, Germany
2 Institute of Cell Biology and Immunology, Eberhard Karls University Tuebingen, 72076 Tuebingen, Germany; oliver.planz@uni-tuebingen.de
* Correspondence: ralf.kircheis@admedsol.com; Tel.: +49-15116790606

Toll-like receptors (TLRs) are key players in the innate immune system, in host' first-line defense against pathogens [1,2]. TLRs recognize pathogen-associated molecular patterns (PAMPs) from bacteria, viruses, and other microorganisms, and self-derived damage-associated molecular patterns (DAMPs) released from dying or lytic cells [3–9]. as well as chromatin-associated molecular patterns (CAMPs) [10], thus playing a critical role not only in immune surveillance but also in disease pathogenesis [3–12]. The TLR activation triggers the production of proinflammatory cytokines and type I interferons, which are key mediators of the host immune response against bacterial, viral, and fungal infections [13–18]. However, dysregulation and excessive activation can be detrimental, leading to hyperinflammation, sepsis, and loss of tissue integrity. TLRs are involved in the pathogenesis of acute bacterial and viral infections, as well as chronic infectious and non-infectious inflammatory diseases [19–24].

This Special Issue 2.0 continues the conceptional line started in the previous Special Issue [25], focusing on the involvement of Toll-like receptors (TLRs) and their related signaling pathways in viral infection and inflammation, and emphasizing the therapeutic potential of TLR system modulation for the treatment of acute viral infections and their long-term sequelae and to provide potential targets for treating chronic viral and inflammatory diseases. This Special Issue provides an excellent collection of contemporary reviews and research articles covering a broad variety of topics, ranging from the protective activity of TLRs in viral infections to detrimental disturbances of the host immune system. The most intriguing theme discussed in this second Special Issue is the intercorrelation between acute viral infections and triggered delayed effects, including neuroinflammatory and neurodegenerative long-term sequelae.

The first review [26] provides an excellent overview on TLRs, triggered signaling cascades, and their function in immune surveillance and disease pathogenesis (see Figure 1). The authors discuss recent advancements in TLR research and the use of specific agonists and antagonists with application in immunotherapy and vaccine development. The authors illustrate the activation of specific TLRs, e.g., TLR7 and TLR9, in antiviral therapies or as adjuvants in cancer vaccines, as well as the targeting of TLR2 and TLR4 with antagonists to mitigate hyper-inflammatory responses in sepsis and viral infections, including COVID-19 and HSV-1 infection. Finally, the review describes the role of the RAGE/TLR system in diabetic encephalopathy, cancer, and cardiovascular diseases. Furthermore, the connection between innate and adaptive immunity and their impact on autoimmune diseases and neuroimmune diseases are discussed.

Finally, the work provides an outlook to future directions, such as the integration of computational modeling and personalized medicine approaches [26]. Computational modeling could represent a breakthrough in revolutionizing TLR research by providing tools for predicting TLR signaling pathways and analyzing the impact of potential therapeutics by simulating TLR modulation in silico before proceeding with in vivo studies [27–32].

Citation: Kircheis, R.; Planz, O. Special Issue "The Role of Toll-Like Receptors (TLRs) in Infection and Inflammation 2.0". *Int. J. Mol. Sci.* **2024**, *25*, 9709. https://doi.org/10.3390/ijms25179709

Received: 30 August 2024
Accepted: 5 September 2024
Published: 7 September 2024

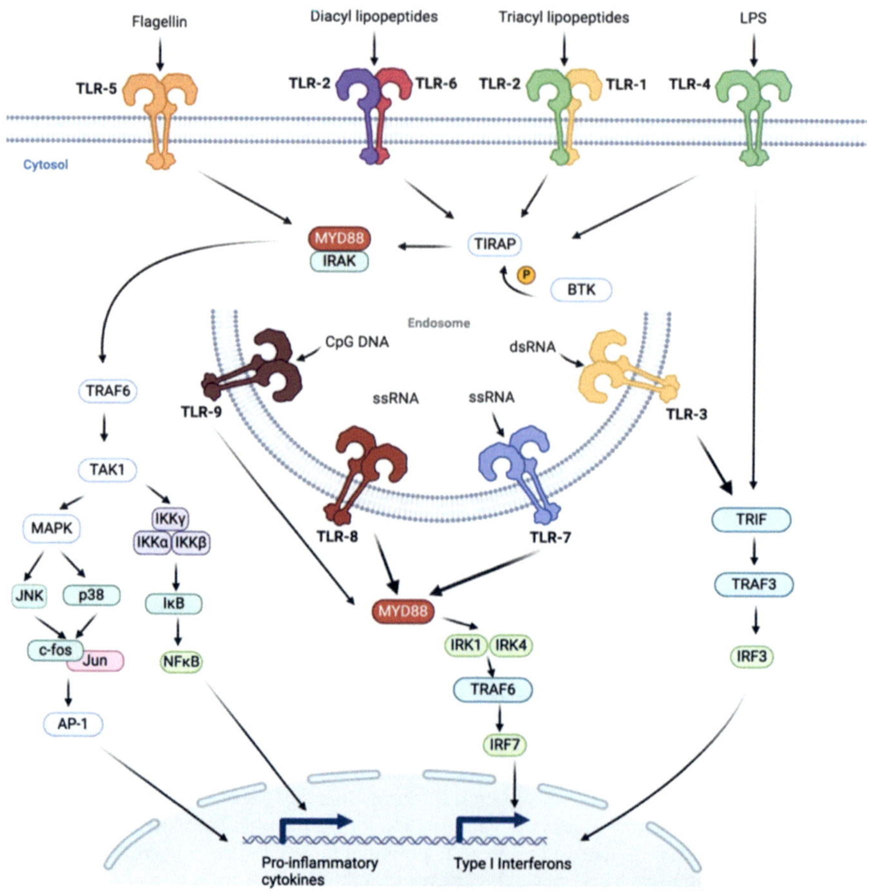

Figure 1. Overview of Toll-like receptor (TLR) signaling pathways showing cell surface TLRs (TLR-1, TLR-2, TLR-4, TLR-5, and TLR-6), their ligands (flagellin, diacyl lipopeptides, triacyl lipopeptides, LPS), and the endosomal TLRs (TLR-3, TLR-7, TLR-8, and TLR-9) recognizing CpG DNA and RNA molecules. The subsequent signaling cascades involving adaptor proteins such as MYD88, TRIF, and TRAF6 are shown below. These lead to the activation of NF-κB and IRFs and result in the expression of proinflammatory cytokines and type I interferons. Adopted from [26].

The second review [33] provides a comprehensive overview on the Toll-like receptor response to Human Immunodeficiency Virus Type 1 (HIV-1) and co-infection with Hepatitis B or C Virus. HIV-1 is known to modulate the host TLR response with impact on viral persistence or disease resolution, e.g., HIV co-infection is associated with a higher persistence of HBV co-infection, higher HBV DNA levels, and increased cirrhosis and liver-related mortality, as well as a higher risk of hepatocellular carcinoma (HCC) [34–36]. Similarly, HIV co-infection can also aggravate the clinical course of HCV, resulting in higher HCV persistence, higher viral loads, enhanced liver fibrosis [37–41], and the development of HCC [42,43]. Increased expression of TLR6, TLR7, and TLR8 or TLR2, TLR3, and TLR4 mRNA has been observed in chronic untreated HIV-1 infections and in patients with advanced disease, respectively, demonstrating the involvement of TLRs and their signaling during HIV infection. TLR7 polymorphisms were shown to modulate the progression of HIV-1 infection. A decreased secretion of Th1 cytokines such as IL-2 and antiviral interferons, together with an increased secretion of Th2 cytokines, i.e., IL-4 and IL-10, and proinflammatory cytokines TNF-α,

IL-1, IL-6, and IL-8 have been found in HIV-1 infection, correlating with disturbances in TLR activation. HIV regulatory proteins, HIV-1 Tat and Nef, have been demonstrated to modulate cytokine release and TLR activation, leading to enhanced viral replication [44,45]. Furthermore, HIV-1 accessory proteins Vif, Vpr, and Vpu were shown to inhibit cellular restriction factors and the TLR sensing of HIV-1 [46–48]. Activation of TLR signaling during HIV infection and the effect on viral replication are shown in Figure 2.

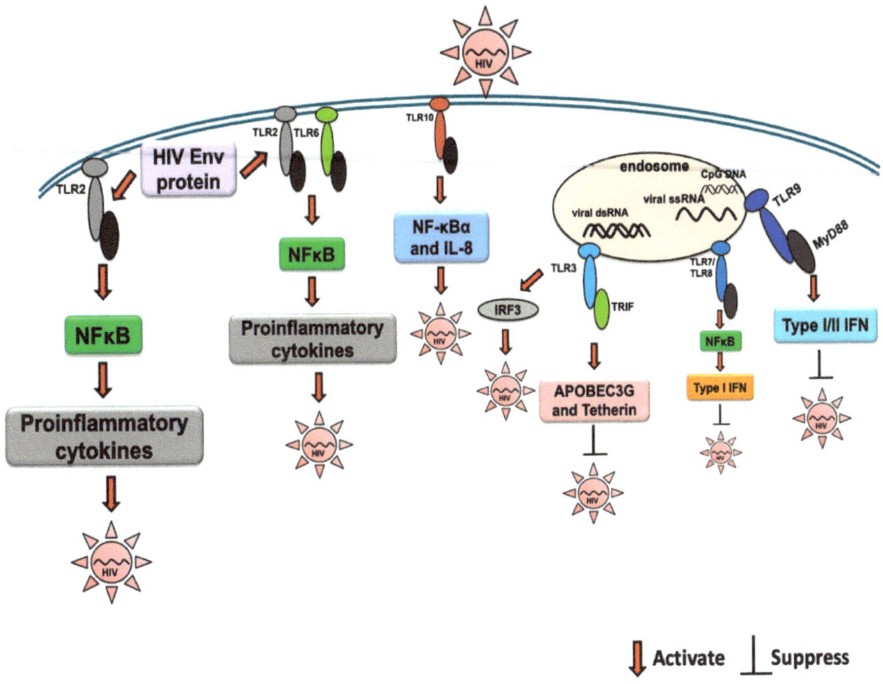

Figure 2. Activation of TLR signaling during HIV infection and the effect on viral replication are depicted. The activation and suppression of signaling are indicated with arrows. Adopted from [33].

Finally, the potential of TLR agonists as latency-reverting agents and immune stimulators and their use in new strategies to cure HIV are discussed. Thus far, a persisting HIV-1 reservoir in long-lived memory CD4+ T cells, with a transcriptionally silent provirus that remains undetected by the host immune system, is seen as the highest challenge to a cure for HIV-1 [49]. Although ART can suppress HIV-1 replication to undetectable levels, it cannot eliminate latent persistent viral reservoirs. For a complete cure of HIV-1, new strategies must be developed that eliminate latently infected cells persisting in people with HIV on ART [50,51]. In this regard, latency-reversing agents that can induce viral reactivation, leading to immune cell recognition and the elimination of latently infected cells, will have a key role [52,53]. TLR agonists as key modulators of the host immune response could act as latency-reversing agents in HIV-1 and HCV/HBV co-infected patients [54].

The third review discusses the latest knowledge on the involvement of TLR in SARS-CoV-2 infection and COVID-19 [55]. TLR activation has been shown to contribute to viral clearance and disease resolution [56,57]. This is also illustrated by a three times higher probability of male patients for severe clinical course and death compared to females, and this may be related to the X chromosome-linked TLR7 gene with a higher and more stable TLR7-driven IFN-α production in female dendritic cells protecting from progression to severe disease [58–60].

However, TLRs represent a double-edged sword that can elicit dysregulated immune signaling exacerbating dysregulated immune response in patients with severe COVID-19. TLR2, TLR3, TLR4, and TLR7 have been associated with COVID-19 severity [61]. Different components of SARS-CoV-2 were shown to activate different TLRs (see Figure 3). The E protein was shown to activate TLR2/TLR1 and TLR2/TLR6, and the S protein has been demonstrated to activate TLR4 correlating with excessive NF-κB activation, cytokine release, and immune dysregulation [62–65]. Furthermore, SARS-CoV-2 was suggested to bind and activate TLR4, leading to increased ACE2 expression, correlating with enhanced viral entry and hyperinflammation [66]. TLR7/8 and TLR3 are activated by the viral ssRNA and the dsRNA intermediate, respectively. The TLR3 receptor has been widely described as a host defense factor against different viruses [67,68]. An in silico study has indicated the NSP10 protein of SARS-CoV-2 to activate TLR3 [69]. The authors discuss the role of TLRs in the pathogenesis of COVID-19 with the examples of TLR7 and TLR3 rare variants, such as the L412F polymorphism in TLR3 that negatively regulates anti-SARS-CoV-2 immune responses [70]. The authors also discuss the interaction of TLR2 and TLR4 with SARS-CoV-2 proteins and involvement of TLR2 in NET formation by SARS-CoV-2 [71].

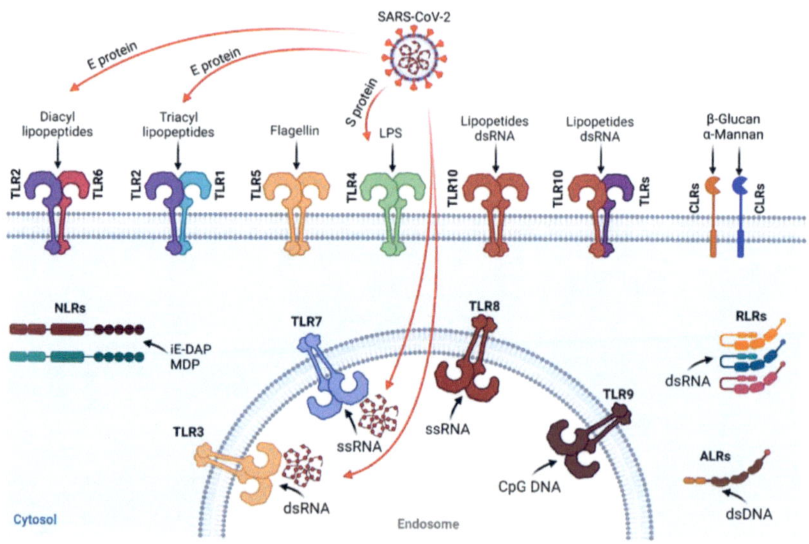

Figure 3. TLR activation by structural proteins of SARS-CoV-2. Adopted from [55].

Overall, in addition to its pivotal role, for an effective antiviral response, excessive TLR-triggered hyperinflammation can alter the equilibrium that drives disease severity [55].

Furthermore, Pedicillo et al. [72] demonstrate the association between TLR-4-triggered inflammation and macrophage imbalance in lung inflammatory infiltrates of lung tissue autopsies from lethal COVID-19 cases. The persistence of the spike protein was shown to correlate with TLR4 upregulation and increase macrophage infiltration. Importantly, a macrophage shift characterized by a downregulation of GAL-3(+) alveolar macrophages, which are crucial for resolving inflammation and promoting tissue repair, paralleled by an increase in CD163(+) myeloid-derived monocyte-macrophages, was shown. The data indicate that TLR-4 expression and activation induce persistently active inflammation, with inefficient resolution, and pathological macrophage shifts, as part of the pathophysiological mechanisms of lethal COVID-19 [72].

The critical role of TLR4 in COVID-19 pathophysiology is further supported by in silico studies analyzing the S protein-induced dimerization of TLR4/MD-2 complexes of the SARS-CoV-2 Omicron variant in comparison to wild-type virus and earlier SARS-CoV-2 variants [73]. This work addresses the question of why the pathogenicity of Omicron variants is significantly lower than that of wild-type virus and earlier Variants of Concern (VoCs). Despite a binding activity to TLR4 that is compatible to that of the wild-typer virus, a lower potency of the Omicron spike protein to trigger the dimerization of TLR4/MD-2 complexes was found. This can explain Omicron's lower pathogenicity compared to wild-type virus and VOCs to a large extent. Furthermore, in silico data indicate the tendency of a decreasing TLR4 dimerization potency starting earlier, during SARS-CoV-2 evolution, somewhere after the transition from wild-type or Alpha variant to Gamma and Delta variants; however, this became most pronounced with the appearance of Omicron variants [73]. Figure 4 illustrates the consequences of TLR4 hyperactivation by the spike protein, particularly of wild-type SARS-CoV-2, compared to the lower activation of the Omicron variants.

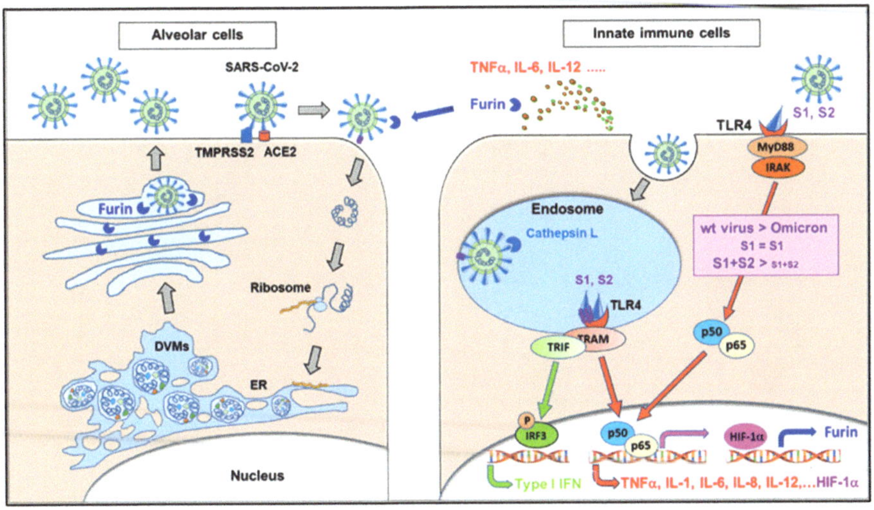

Figure 4. The SARS-CoV-2 binds to ACE2 followed by proteolytic cleavage by TMPRRS2 and fusion with the host cell membrane followed by the uptake of viral RNA into the host cell. The transcription of the viral RNA and translation of viral proteins occurs in double-membrane vesicles (DMVs). The viral components assemble into virus particles that leave the cells via the Golgi apparatus, where the spike protein undergoes proteolytic cleavage by furin. This process preferably occurs in virus-producer cells with high TMPRSS2 expression, e.g., alveolar cells (left side). Alternatively, the virus can be taken up via clathrin-coated pits into endosomes, followed by proteolytic cleavage by cathepsin L. The endosomal uptake is predominant in TMPRSS2-negative, cathepsin L-rich cells, such as innate immune cells (right side). The spike protein of the SARS-CoV-2 acts as a TLR4 agonist, resulting in the dimerization of the TLR4/MD-2 complex triggering downstream signaling, e.g., the NF-κB (p50/p65) pathway. The activation of the NF-κB pathway triggers HIF-1α activation and the expression of cytokines, such as TNFα, IL-1, IL-6, and IL-12. Notably, HIF-1α and IL-12 have been found to enhance furin expression. In contrast to the highly effective TLR4/MD-2 dimerization by the wild-type spike protein trimer, the amino acid substitutions in the Omicron spike protein interfere with the potency of the spike protein for TLR4/MD-2 dimerization, leading to less NF-κB signaling and lower cytokine expression. Adopted from [73,74].

Besides TLR4, another widely expressed and extensively studied pattern recognition receptor is TLR2. TLR2 is unique in that it forms heterodimers with TLR1, TLR4, TLR6, or TLR10, which allows it to recognize a very broad range of pathogens. The study by Colleselli et al. [75] in this Special Issue provides an overview of TLR2, its homo- and heterodimers, and of the pro- and anti-inflammatory properties of TLR2. The review provides an interesting outline of TLR2-associated infectious diseases, such as sepsis, COVID-19, and neurodegenerative diseases, such as Parkinson's disease and Alzheimer's disease (see Figure 5) [76–81].

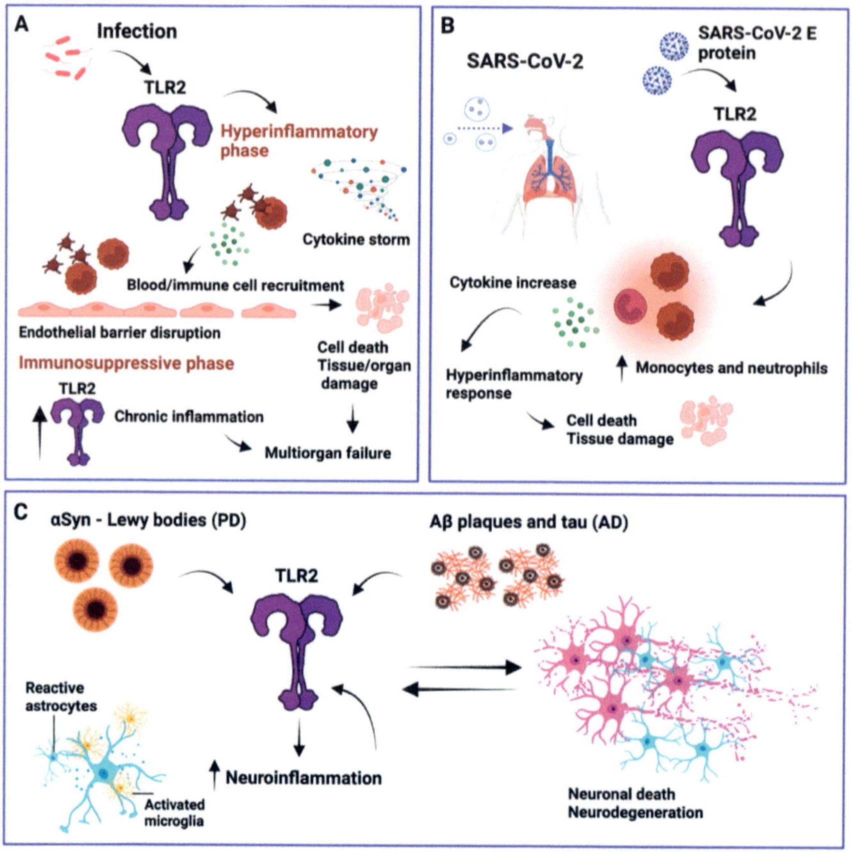

Figure 5. Overview of TLR2 involvement in different infectious and non-infectious inflammatory diseases. (**A**) Sepsis can be triggered by TLR2, which further activates platelets and other immune cells, leading to endothelial barrier disruption, cell death, and tissue and organ damage. In addition, TLR2 has been shown to be elevated during the immunosuppressive phase of sepsis, which can lead to multiorgan failure. (**B**) Severe acute respiratory syndrome coronavirus 2 (SARS-CoV-2). TLR2 can recognize the SARS-CoV-2 envelope (E) protein, leading to an increase in inflammatory monocytes and neutrophils, inducing a hyperinflammatory response that can cause cell death and tissue damage. (**C**) Parkinson's disease (PD) and Alzheimer's disease (AD). TLR2 can recognize α-synuclein (αSyn), amyloid-β (Aβ) plaques, and tau. This promotes neuroinflammation characterized by reactive astrocytes and activated microglia, which further increase TLR2 levels, leading to a feedback loop of neuronal cell death, further TLR2 upregulation, and neuroinflammation. Adopted from [75].

The article by Abarca-Merlin et al. [82] provides an excellent overview on the involvement of TLRs in immunity and neurogenesis and on Toll-like receptors as regulators in the nervous system. TLRs are expressed in resident immune cells as well as in neurons and glial cells of the nervous system (see Figure 6). There is further support for the role of TLRs not only in the immune response but also in physiological and pathophysiological processes of the nervous system, such as neurogenesis, behavior, cognition, infection, neuroinflammation, and neurodegeneration [82–84].

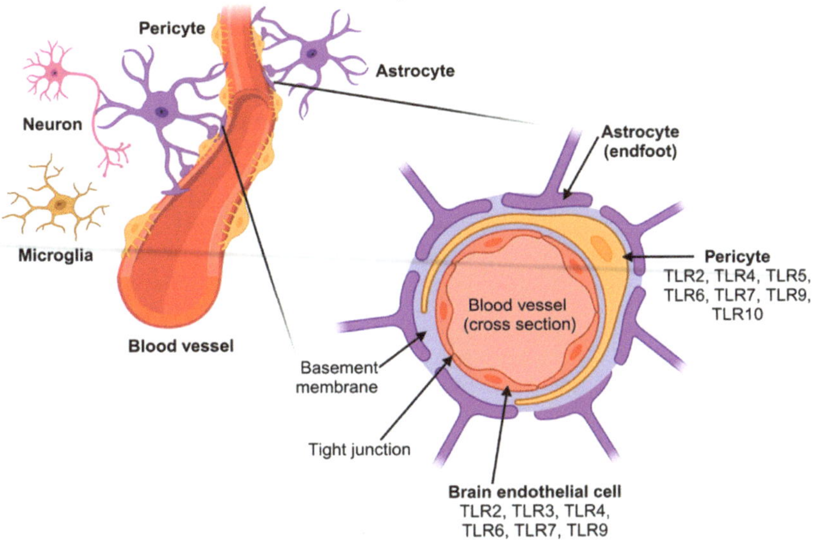

Figure 6. Schematic representation of the blood–brain barrier (BBB) located in the central part of the neurovascular unit (NVU). A cross-section of a blood vessel of the blood–brain barrier is depicted, showing the three main cell types that compose it (endothelial cells, pericytes, and astrocytes), as well as other cell types of the neurovascular unit (e.g., microglia and neurons) and their respective, expressed TLRs. The main function of the neurovascular unit is the formation of the blood–brain barrier and neurovascular coupling. Adopted from [82].

The next work of Dubik et al. [85] on the involvement of TLR in neuroinflammation studied the synergistic targeting of TLR7 and NOD2 for therapeutic intervention in multiple sclerosis. Neuroinflammation regulation is critical for maintaining central nervous system (CNS) homeostasis and can be particularly relevant for the treatment of autoimmune diseases, including multiple sclerosis (MS). Earlier studies highlighted the significance of innate signaling in triggering anti-inflammatory mechanisms, playing a protective role in experimental autoimmune encephalomyelitis (EAE), which represents a predictive model for a MS-like disease. This study investigated the effects of targeting two innate receptors, TLR7 and NOD2, simultaneously to prevent EAE progression. The simultaneous intrathecal administration of NOD2 and TLR7 agonists led to the synergistic induction of Type I IFN and effectively suppressed EAE. The suppression of EAE correlated with a significant decrease in the infiltration of monocytes, granulocytes, and natural killer cells; reduced demyelination; and the downregulation of IL-1β, CCL2, and IFN-γ gene expression in the spinal cord. These results show the potential of targeting the innate TLR-related pathways for alleviating neuroinflammation associated with MS [85].

Hernandez et al. [86]'s article covers neurocognitive impairments in association with LPS-induced acute respiratory distress syndrome (ARDS) and the modulation of the Lung–Brain Axis of communication in wild-type vs. Fat-1 mice. Although ARDS primarily affects the lungs, many ARDS patients also develop neurocognitive impairments. To investigate the connection between the lung and brain during ARDS and the therapeutic potential of specialized pro-resolving mediators (SPMs), fat-1 mice were crossbred with Resolvin E1 (RvE1) receptor knockout mice. ARDS was induced through the intratracheal application of lipopolysaccharide. Protein and mRNA analyses revealed that LPS induced lung inflammation as well as increased inflammatory activation in the hypothalamus.

The authors demonstrated that immune cell trafficking to the brain largely contributed to immune-to-brain communication during ARDS rather than cytokines. Deficiency in RvE1 receptors and enhanced omega-3 polyunsaturated fatty acid levels (fat-1 mice) were shown to affect lung–brain interaction during ARDS by altering the profiles of inflammatory and lipid mediators and glial activity markers [86].

The second group of articles focus on the role of TLRs in metabolic and cardiovascular diseases. Understanding the complex interactions between metabolism and the immune system may help us identify key immunomodulatory factors as therapeutic targets in obesity and cardiovascular diseases. The work of Höpfiner et al. [87] investigated the regulation of Cathelicidin Antimicrobial Peptide (CAMP) gene expression induced by TNFα and cfDNA in adipocytes. CAMP is a regulator of innate immunity expressed in adipocytes. CAMP, therefore, might act as an adipokine in adipose inflammation. TNFα, cell-free nucleic acids (cfDNAs), and TLR9 are all parts of the innate immune system and are functionally active in adipose tissue. Adipocyte inflammation was induced in vitro by TNFα and cfDNA stimulation. TNFα significantly induced CAMP gene expression in mature adipocytes, which was reduced through the inhibition of PI3K signaling. These findings suggest a regulatory role of inflammatory mediators, such as TNFα and cfDNA, in adipocytic CAMP expression as a novel putative molecular mechanism in adipose tissue innate immunity [87].

Mylonas et al. [88] studied the role of MYD88 and proinflammatory chemokines in aortic atheromatosis in the context of novel statin treatment. Their study investigated how statins mitigate proatherogenic inflammation. Statins, such as Rosuvastatin significantly reduced MYD88, CCL4, CCL20, CCR2, TNF-α, IFN-β, IL-1b, IL-2, IL-4, IL-8, and IL-10, both in the thoracic and abdominal aorta. Another statin, Fluvastatin, downregulated MYD88, CCR2, IFN-β, IFN-γ, IL-1b, IL-2, IL-4, and IL-10 in both aortic segments. In conclusion, statin therapy can control the proatherogenic inflammation in hyperlipidemic animals [88].

Finally, a very interesting work by Loaiza et al. studied the impact of extracellular histones and absence of Toll-like receptors on cardiac functional and electrical disturbances in mouse hearts [89]. Sepsis is frequently accompanied by cardiac functional and electrical disturbances. As a causative factor, extracellular histones, e.g., released from activated neutrophils, were shown to significantly contribute to cardiac dysfunction, as demonstrated by Echo-Doppler measurements. This study investigated the roles of extracellular histones and their interactions with TLRs in cardiac dysfunction through in vivo Echo-Doppler studies on mice perfused with extracellular histones in wild-type vs. TLR2, TLR3, or TLR4 knockout (KO) mice. Histone perfusion caused defects in contractility and relaxation, with TLR2 and TLR3 KO mice being partially protected. Specifically, TLR2 KO mice exhibited the greatest rescue effect. In contrast, TLR4 KO exacerbated cardiac dysfunction. Among individual histones, H1 induced the most pronounced abnormalities in cardiac function, cardiomyocyte apoptosis, and LDH release. The study indicates the interaction of histones with TLRs as potential therapeutic targets for septic cardiomyopathy [89].

Overall, this Special Issue 2.0 provides a diverse collection of original articles and review articles covering different aspects of Toll-like receptors and their roles in infection and inflammation. The various articles illustrate well the two sides of TLRs as drivers for the pathogenesis of acute bacterial and viral infections, including COVID-19; as drivers of multiple non-infectious diseases of various tissue origin; as potent players for antiviral, antibacterial immunity; and as therapeutic targets to treat chronical non-infectious inflammatory diseases [90–96]. Notably, new data increasingly show that neurological pathways, such as neurogenesis, neuroinflammation, and cognitive and behavioral aspects may be influenced by innate, TLR-mediated pathways. The SARS-CoV-2 envelope protein was shown to trigger depression-like behaviors and dyssomnia via TLR2-mediated neuroinflammation in mice [97], and the SARS-CoV-2 spike protein was shown to induce TLR4-mediated long-term cognitive dysfunction recapitulating post-COVID-19 syndrome in mice [98]. In this context, the connection between infection-triggered acute disbalance in TLR activation and resulting neurological sequelae following bacterial (sepsis) or viral infections, including long COVID-associated neurological sequelae [99–102], may provide new targets for future treatment options [103].

TLRs are drivers for both the immune response and the pathogenesis of acute bacterial and viral infections, but they are also involved in a broad variety of non-infectious diseases of neurological, metabolic, and cardio-vascular origin. The interconnection of infection-triggered TLR disbalances and resulting long-term neurological, cardio-vascular, and metabolic sequelae may represent promising targets for the development of new therapies.

Author Contributions: Conceptualization, R.K. and O.P.; methodology, R.K.; validation, O.P.; formal analysis, R.K. and O.P.; investigation, R.K.; data curation, R.K.; writing—original draft preparation, R.K.; writing—review and editing, O.P. and R.K.; visualization, R.K. All authors have read and agreed to the published version of the manuscript.

Conflicts of Interest: Author Ralf Kircheis was employed by the company Syntacoll GmbH. The author declared that the research was conducted in the absence of any commercial or financial relationships that could be construed as a potential conflict of interest.

References

1. Mogensen, T.H. Pathogen recognition and inflammatory signaling in innate immune defenses. *Clin. Microbiol. Rev.* **2009**, *22*, 240–273. [CrossRef] [PubMed]
2. Taghavi, M.; Khosravi, A.; Mortaz, E.; Nikaein, D.; Athari, S.S. Role of pathogen-associated molecular patterns (PAMPS) in immune responses to fungal infections. *Eur. J. Pharmacol.* **2017**, *808*, 8–13. [CrossRef] [PubMed]
3. Greenhalgh, D.G.; Green, T.L.; Lim, D.; Cho, K. Bacterial Pathogen–Associated Molecular Patterns Upregulate Human Glucocorticoid Receptor Expression in Peripheral Blood Mononuclear Cells. *Shock* **2022**, *58*, 393–399. [CrossRef]
4. Eppensteiner, J.; Kwun, J.; Scheuermann, U.; Barbas, A.; Limkakeng, A.T.; Kuchibhatla, M.; Elster, E.A.; Kirk, A.D.; Lee, J. Damage- and pathogen-associated molecular patterns play differential roles in late mortality after critical illness. *JCI Insight.* **2019**, *4*, e127925. [CrossRef]
5. Denk, S.; Perl, M.; Huber-Lang, M. Damage-and pathogen-associated molecular patterns and alarmins: Keys to sepsis? *Eur. Surg. Res.* **2012**, *48*, 171–179. [CrossRef]
6. Zindel, J.; Kubes, P. DAMPs, PAMPs, and LAMPs in Immunity and Sterile Inflammation. *Annu. Rev. Pathol. Mech. Dis.* **2020**, *15*, 493–518. [CrossRef]
7. Denning, N.-L.; Aziz, M.; Gurien, S.D.; Wang, P. DAMPs and NETs in Sepsis. *Front. Immunol.* **2019**, *10*, 2536. [CrossRef]
8. De Lorenzo, G.; Ferrari, S.; Cervone, F.; Okun, E. Extracellular DAMPs in Plants and Mammals: Immunity, Tissue Damage and Repair. *Trends Immunol.* **2018**, *39*, 937–950. ISSN 1471-4906. [CrossRef]
9. Mueller, C. Danger-associated molecular patterns and inflammatory bowel disease: Is there a connection? *Dig. Dis.* **2012**, *30*, 40–46. [CrossRef]
10. Nofi, C.P.; Wang, P.; Aziz, M. Chromatin-Associated Molecular Patterns (CAMPs) in sepsis. *Cell Death Dis.* **2022**, *13*, 700. [CrossRef]
11. Kawai, T.; Akira, S. The role of pattern-recognition receptors in innate immunity: Update on Toll-like receptors. *Nat. Immunol.* **2010**, *11*, 373–384. [CrossRef]
12. Li, D.; Wu, M. Pattern recognition receptors in health and diseases. *Signal Transduct. Target. Ther.* **2021**, *6*, 291. [CrossRef] [PubMed]
13. Akira, S.; Takeda, K.; Kaisho, T. Toll-like receptors: Critical proteins linking innate and acquired immunity. *Nat. Immunol.* **2001**, *2*, 675–680. [CrossRef]
14. Medzhitov, R. Toll-like receptors and innate immunity. *Nat. Rev. Immunol.* **2001**, *1*, 135–145. [CrossRef]
15. Kawasaki, T.; Kawai, T. Toll-like receptor signaling pathways. *Front. Immunol.* **2014**, *5*, 461. [CrossRef] [PubMed]
16. Fitzgerald, K.A.; Kagan, J.C. Toll-like Receptors and the Control of Immunity. *Cell* **2022**, *180*, 1044–1066. [CrossRef]
17. Perales-Linares, R.; Navas-Martin, S. Toll-like receptor 3 in viral pathogenesis: Friend or foe? *Immunology* **2013**, *140*, 153–167. [CrossRef]
18. Zheng, W.; Xu, Q.; Zhang, Y.; E, X.; Gao, W.; Zhang, M.; Zhai, W.; Rajkumar, R.S.; Liu, Z. Toll-like receptor-mediated innate immunity against herpesviridae infection: A current perspective on viral infection signaling pathways. *Virol. J.* **2020**, *17*, 192. [CrossRef]
19. Kumar, V. Toll-like receptors in sepsis-associated cytokine storm and their endogenous negative regulators as future immunomodulatory targets. *Int. Immunopharmacol.* **2020**, *89*, 107087. ISSN 1567-5769. [CrossRef]
20. Naqvi, I.; Giroux, N.; Olson, L.; Morrison, S.A.; Llanga, T.; Akinade, T.O.; Zhu, Y.; Zhong, Y.; Bose, S.; Arvai, S.; et al. DAMPs/PAMPs induce monocytic TLR activation and tolerance in COVID-19 patients; nucleic acid binding scavengers can counteract such TLR agonists. *Biomaterials* **2022**, *283*, 121393. [CrossRef]
21. Gao, H.; Leaver, S.K.; Burke-Gaffney, A.; Finney, S.J. Severe sepsis and Toll-like receptors. *Semin. Immunopathol.* **2008**, *30*, 29–40. [CrossRef] [PubMed]

22. Opal, S.M.; E Huber, C. Bench-to-bedside review: Toll-like receptors and their role in septic shock. *Crit. Care* **2002**, *6*, 125–136. [CrossRef]
23. Cook, D.N.; Pisetsky, D.S.; A Schwartz, D. Toll-like receptors in the pathogenesis of human disease. *Nat. Immunol.* **2004**, *5*, 975–979. [CrossRef] [PubMed]
24. Kiziltas, S. Toll-like receptors in pathophysiology of liver diseases. *World J. Hepatol.* **2016**, *8*, 1354–1369. [CrossRef] [PubMed]
25. Kircheis, R.; Planz, O. The Role of Toll-like Receptors (TLRs) and Their Related Signaling Pathways in Viral Infection and Inflammation. *Int. J. Mol. Sci.* **2023**, *24*, 6701. [CrossRef]
26. Chen, Y.-H.; Wu, K.-H.; Wu, H.-P. Unraveling the Complexities of Toll-like Receptors: From Molecular Mechanisms to Clinical Applications. *Int. J. Mol. Sci.* **2024**, *25*, 5037. [CrossRef]
27. Billod, J.-M.; Lacetera, A.; Guzmán-Caldentey, J.; Martín-Santamaría, S. Computational Approaches to Toll-like Receptor 4 Modulation. *Molecules* **2016**, *21*, 994. [CrossRef]
28. Murgueitio, M.S.; Rakers, C.; Frank, A.; Wolber, G. Balancing Inflammation: Computational Design of Small-Molecule Toll-like Receptor Modulators. *Trends Pharmacol. Sci.* **2016**, *38*, 155–168. [CrossRef]
29. Klett, J.; Reeves, J.; Oberhauser, N.; Perez-Regidor, L.; Martin-Santamaria, S. Modulation of toll-like receptor 4. Insights from x-ray crystallography and molecular modeling. *Curr. Top. Med. Chem.* **2014**, *14*, 2672–2683. [CrossRef]
30. Arciero, J.; Ermentrout, G.B.; Siggers, R.; Afrazi, A.; Hackam, D.; Vodovotz, Y.; Rubin, J. Modeling the interactions of bacteria and Toll-like receptor-mediated inflammation in necrotizing enterocolitis. *J. Theor. Biol.* **2013**, *321*, 83–99. [CrossRef]
31. Farzan, M.; Mirzaei, Y.; Aiman, S.; Azadegan-Dehkordi, F.; Bagheri, N. Immunoinformatics-based multi-epitope vaccine design for the re-emerging monkeypox virus. *Int. Immunopharmacol.* **2023**, *123*, 110725. [CrossRef]
32. Brooshghalan, S.E.; Sabahi, M.; Ebadi, S.A.; Sadeghian, Z.; Nayebi, A.M.; Haddadi, R. Silibinin chronic treatment in a rat model of Parkinson disease: A comprehensive in-vivo evaluation and in silico molecular modeling. *Eur. J. Pharmacol.* **2023**, *941*, 175517. [CrossRef] [PubMed]
33. Kayesh, M.E.H.; Kohara, M.; Tsukiyama-Kohara, K. Toll-like Receptor Response to Human Immunodeficiency Virus Type 1 or Co-Infection with Hepatitis B or C Virus: An Overview. *Int. J. Mol. Sci.* **2023**, *24*, 9624. [CrossRef] [PubMed]
34. Kim, H.N.; Newcomb, C.W.; Carbonari, D.M.; Roy, J.A.; Torgersen, J.; Althoff, K.N.; Kitahata, M.M.; Reddy, K.R.; Lim, J.K.; Silverberg, M.J.; et al. Risk of HCC With Hepatitis B Viremia Among HIV/HBV-Coinfected Persons in North America. *Hepatology* **2021**, *74*, 1190–1202. [CrossRef]
35. Xiao, J.; Zhang, Y.; Wu, J.; Chen, X.; Zou, W. HIV/HBV coinfection: Understanding the complex interactions and their impact on spontaneous HBV clearance, chronic liver damage, cirrhosis, and hepatocellular carcinoma. *Aids Rev.* **2024**, *26*, 032–040. [CrossRef]
36. Fattovich, G.; Stroffolini, T.; Zagni, I.; Donato, F. Hepatocellular carcinoma in cirrhosis: Incidence and risk factors. *Gastroenterology* **2004**, *127* (Suppl. 1), S35–S50. [CrossRef]
37. Puoti, M.; Prestini, K.; Putzolu, V.; Zanini, B.; Baiguera, C.; Antonini, M.G.; Pagani, P.; Airoldi, M.; Carosi, G. HIV/HCV co-infection: Natural history. *Curr. Opin. HIV AIDS* **2003**, *17*, 144–146.
38. Maggiorella, M.T.; Sernicola, L.; Picconi, O.; Pizzi, E.; Belli, R.; Fulgenzi, D.; Rovetto, C.; Bruni, R.; Costantino, A.; Taffon, S.; et al. Epidemiological and molecular characterization of HBV and HCV infections in HIV-1-infected inmate population in Italy: A 2017–2019 multicenter cross-sectional study. *Sci. Rep.* **2023**, *13*, 14908. [CrossRef]
39. Abravanel, F.; Raymond, S.; Pambrun, E.; Winnock, M.; Bonnard, P.; Sogni, P.; Trimoulet, P.; Dabis, F.; Salmon-Ceron, D.; Izopet, J.; et al. HIV-1 tropism and liver fibrosis in HIV–HCV co-infected patients. *PLoS ONE* **2012**, *7*, e50289. [CrossRef]
40. Kim, A.Y.; Chung, R.T. Coinfection with HIV-1 and HCV—A one-two punch. *Gastroenterology* **2009**, *137*, 795–798. [CrossRef]
41. Saracino, A.; Bruno, G.; Scudeller, L.; Punzi, G.; Lagioia, A.; Ladisa, N.; Monno, L.; Angarano, G. Does HIV-1 co-receptor tropism correlate with fibrosis progression in HIV/HCV co-infected patients? *J. Clin. Virol.* **2014**, *59*, 167–171. [CrossRef]
42. Hu, J.; Liu, K.; Luo, J. HIV-HBV and HIV-HCV Coinfection and Liver Cancer Development. *Cancer Treat. Res.* **2019**, *177*, 231–250. [CrossRef] [PubMed]
43. Nunnari, G.; Berretta, M.; Pinzone, M.R.; Di Rosa, M.; Berretta, S.; Cunsolo, G.; Malaguarnera, M.; Cosentino, S.; De Paoli, P.; Schnell, J.M.; et al. Hepatocellular carcinoma in HIV positive patients. *Eur. Rev. Med. Pharmacol. Sci.* **2012**, *16*, 1257–1270. [PubMed]
44. El-Hage, N.; Podhaizer, E.M.; Sturgill, J.; Hauser, K.F. Toll-like receptor expression and activation in astroglia: Differential regulation by HIV-1 Tat, gp120, and morphine. *Immunol. Investig.* **2011**, *40*, 498–522. [CrossRef]
45. Qin, X.; Yao, J.; Yang, F.; Nie, J.; Wang, Y.; Liu, P.C. Human immunodeficiency virus type 1 Nef in human monocyte-like cell line THP-1 expands treg cells via toll-like receptor 2. *J. Cell. Biochem.* **2011**, *112*, 3515–3524. [CrossRef]
46. Sauter, D.; Kirchhoff, F. HIV replication: A game of hide and sense. *Curr. Opin. HIV AIDS* **2016**, *11*, 173–181. [CrossRef] [PubMed]
47. Mehrbod, P.; Ande, S.R.; Alizadeh, J.; Rahimizadeh, S.; Shariati, A.; Malek, H.; Hashemi, M.; Glover, K.K.M.; Sher, A.A.; Coombs, K.M.; et al. The roles of apoptosis, autophagy and unfolded protein response in arbovirus, influenza virus, and HIV infections. *Virulence* **2019**, *10*, 376–413. [CrossRef]
48. Hoshino, S.; Konishi, M.; Mori, M.; Shimura, M.; Nishitani, C.; Kuroki, Y.; Koyanagi, Y.; Kano, S.; Itabe, H.; Ishizaka, Y. HIV-1 Vpr induces TLR4/MyD88-mediated IL-6 production and reactivates viral production from latency. *J. Leukoc. Biol.* **2010**, *87*, 1133–1143. [CrossRef]

49. Chen, J.; Zhou, T.; Zhang, Y.; Luo, S.; Chen, H.; Chen, D.; Li, C.; Li, W. The reservoir of latent HIV. *Front. Cell. Infect. Microbiol.* **2022**, *12*, 945956. [CrossRef]
50. Kreider, E.F.; Bar, K.J. HIV-1 Reservoir Persistence and Decay: Implications for Cure Strategies. *Curr. HIV/AIDS Rep.* **2022**, *19*, 194–206. [CrossRef]
51. Li, S.; Moog, C.; Zhang, T.; Su, B. HIV reservoir: Antiviral immune responses and immune interventions for curing HIV infection. *Chin. Med. J.* **2022**, *135*, 2667–2676. [CrossRef] [PubMed]
52. Board, N.L.; Moskovljevic, M.; Wu, F.; Siliciano, R.F.; Siliciano, J.D. Engaging innate immunity in HIV-1 cure strategies. *Nat. Rev. Immunol.* **2022**, *22*, 499–512. [CrossRef] [PubMed]
53. Bricker, K.M.; Chahroudi, A.; Mavigner, M. New Latency Reversing Agents for HIV-1 Cure: Insights from Nonhuman Primate Models. *Viruses* **2021**, *13*, 1560. [CrossRef] [PubMed]
54. Martinsen, J.T.; Gunst, J.D.; Højen, J.F.; Tolstrup, M.; Søgaard, O.S. The Use of Toll-like Receptor Agonists in HIV-1 Cure Strategies. *Front. Immunol.* **2020**, *11*, 1112. [CrossRef]
55. Mantovani, S.; Oliviero, B.; Varchetta, S.; Renieri, A.; Mondelli, M.U. TLRs: Innate Immune Sentries against SARS-CoV-2 Infection. *Int. J. Mol. Sci.* **2023**, *24*, 8065. [CrossRef]
56. Kayesh, M.E.H.; Kohara, M.; Tsukiyama-Kohara, K. An Overview of Recent Insights into the Response of TLR to SARS-CoV-2 Infection and the Potential of TLR Agonists as SARS-CoV-2 Vaccine Adjuvants. *Viruses* **2021**, *13*, 2302. [CrossRef]
57. Isazadeh, A.; Heris, J.A.; Shahabi, P.; Mohammadinasab, R.; Shomali, N.; Nasiri, H.; Valedkarimi, Z.; Khosroshahi, A.J.; Hajazimian, S.; Akbari, M.; et al. Pattern-recognition receptors (PRRs) in SARS-CoV-2. *Life Sci.* **2023**, *329*, 121940. [CrossRef]
58. Ghimire, R.; Shrestha, R.; Amaradhi, R.; Patton, T.; Whitley, C.; Chanda, D.; Liu, L.; Ganesh, T.; More, S.; Channappanavar, R. Toll-like receptor 7 (TLR7)-mediated antiviral response protects mice from lethal SARS-CoV-2 infection. *bioRxiv* **2023**. [CrossRef]
59. Cords, L.; Woost, R.; Kummer, S.; Brehm, T.T.; Kluge, S.; Schmiedel, S.; Jordan, S.; Lohse, A.W.; Altfeld, M.; Addo, M.M.; et al. Frequency of IRF5+ dendritic cells is associated with the TLR7-induced inflammatory cytokine response in SARS-CoV-2 infection. *Cytokine* **2023**, *162*, 156109. [CrossRef]
60. Conti, P.; Younes, A. Coronavirus COV-19/SARS-CoV-2 affects women less than men: Clinical response to viral infection. *J. Biol. Regul. Homeost. Agents* **2020**, *34*, 339–343. [CrossRef]
61. Khanmohammadi, S.; Rezaei, N. Role of Toll-like receptors in the pathogenesis of COVID-19. *J. Med. Virol.* **2021**, *93*, 2735–2739. [CrossRef] [PubMed]
62. Zheng, M.; Karki, R.; Williams, E.P.; Yang, D.; Fitzpatrick, E.; Vogel, P.; Jonsson, C.B.; Kanneganti, T.-D. TLR2 senses the SARS-CoV-2 envelope protein to produce inflammatory cytokines. *Nat. Immunol.* **2021**, *22*, 829–838. [CrossRef]
63. Khan, S.; Shafiei, M.S.; Longoria, C.; Schoggins, J.W.; Savani, R.C.; Zaki, H.; States, U. SARS-CoV-2 spike protein induces inflammation via TLR2-dependent activation of the NF-κB pathway. *eLife* **2021**, *10*, e68563. [CrossRef]
64. Sariol, A.; Perlman, S. SARS-CoV-2 takes its Toll. *Nat. Immunol.* **2021**, *22*, 801–802. [CrossRef] [PubMed] [PubMed Central]
65. Sahanic, S.; Hilbe, R.; Dünser, C.; Tymoszuk, P.; Löffler-Ragg, J.; Rieder, D.; Trajanoski, Z.; Krogsdam, A.; Demetz, E.; Yurchenko, M.; et al. SARS-CoV-2 activates the TLR4/MyD88 pathway in human macrophages: A possible correlation with strong pro-inflammatory responses in severe COVID-19. *Heliyon* **2023**, *9*, e21893. [CrossRef] [PubMed]
66. Aboudounya, M.M.; Heads, R.J. COVID-19 and Toll-like Receptor 4 (TLR4): SARS-CoV-2 May Bind and Activate TLR4 to Increase ACE2 Expression, Facilitating Entry and Causing Hyperinflammation. *Mediat. Inflamm.* **2021**, *10*, 8874339. [CrossRef]
67. Zhang, S.-Y.; Herman, M.; Ciancanelli, M.J.; de Diego, R.P.; Sancho-Shimizu, V.; Abel, L.; Casanova, J.-L. TLR3 immunity to infection in mice and humans. *Curr. Opin. Immunol.* **2013**, *25*, 19–33. [CrossRef]
68. Wang, P.; Liu, J.B.; Wang, X.; Meng, F.Z.; Xiao, Q.H.; Liu, L.; Zhu, J.; Hu, W.H.; Ho, W.Z. Activation of Toll-like receptor 3 inhibits HIV infection of human iPSC-derived microglia. *J. Med. Virol.* **2023**, *95*, e29217. [CrossRef]
69. Choudhury, A.; Das, N.C.; Patra, R.; Mukherjee, S. In silico analyses on the comparative sensing of SARS-CoV-2 mRNA by the intracellular TLRs of humans. *J. Med. Virol.* **2021**, *93*, 2476–2486. [CrossRef]
70. Croci, S.; Venneri, M.A.; Mantovani, S.; Fallerini, C.; Benetti, E.; Picchiotti, N.; Campolo, F.; Imperatore, F.; Palmieri, M.; Daga, S.; et al. The polymorphism L412F in *TLR3* inhibits autophagy and is a marker of severe COVID-19 in males. *Autophagy* **2022**, *18*, 1662–1672. [CrossRef]
71. Sung, P.-S.; Yang, S.-P.; Peng, Y.-C.; Sun, C.-P.; Tao, M.-H.; Hsieh, S.-L. CLEC5A and TLR2 are critical in SARS-CoV-2-induced NET formation and lung inflammation. *J. Biomed. Sci.* **2022**, *29*, 52. [CrossRef] [PubMed]
72. Pedicillo, M.C.; De Stefano, I.S.; Zamparese, R.; Barile, R.; Meccariello, M.; Agostinone, A.; Villani, G.; Colangelo, T.; Serviddio, G.; Cassano, T.; et al. The Role of Toll-like Receptor-4 in Macrophage Imbalance in Lethal COVID-19 Lung Disease, and Its Correlation with Galectin-3. *Int. J. Mol. Sci.* **2023**, *24*, 13259. [CrossRef]
73. Kircheis, R. In Silico Analyses Indicate a Lower Potency for Dimerization of TLR4/MD-2 as the Reason for the Lower Pathogenicity of Omicron Compared to Wild-Type Virus and Earlier SARS-CoV-2 Variants. *Int. J. Mol. Sci.* **2024**, *25*, 5451. [CrossRef] [PubMed]
74. Kircheis, R.; Planz, O. Could a Lower Toll-like Receptor (TLR) and NF-κB Activation Due to a Changed Charge Distribution in the Spike Protein Be the Reason for the Lower Pathogenicity of Omicron? *Int. J. Mol. Sci.* **2022**, *23*, 5966. [CrossRef]
75. Colleselli, K.; Stierschneider, A.; Wiesner, C. An Update on Toll-like Receptor 2, Its Function and Dimerization in Pro- and Anti-Inflammatory Processes. *Int. J. Mol. Sci.* **2023**, *24*, 12464. [CrossRef]
76. Gorecki, A.M.; Anyaegbu, C.C.; Anderton, R.S. TLR2 and TLR4 in Parkinson's disease pathogenesis: The environment takes a toll on the gut. *Transl. Neurodegener.* **2021**, *10*, 47. [CrossRef]

77. Dutta, D.; Jana, M.; Majumder, M.; Mondal, S.; Roy, A.; Pahan, K. Selective targeting of the TLR2/MyD88/NF-κB pathway reduces α-synuclein spreading in vitro and in vivo. *Nat. Commun.* **2021**, *12*, 5382. [CrossRef]
78. Sobrinho, H.M.d.R.; da Silva, D.J.; Gomides, L.F.; Dorta, M.L.; de Oliveira, M.A.P.; Ribeiro-Dias, F. TLR4 and TLR2 activation is differentially associated with age during Parkinson's disease. *Immunol. Investig.* **2018**, *47*, 71–88. [CrossRef]
79. Dutta, D.; Jana, M.; Paidi, R.K.; Majumder, M.; Raha, S.; Dasarathy, S.; Pahan, K. Tau fibrils induce glial inflammation and neuropathology via TLR2 in Alzheimer's disease–related mouse models. *J. Clin. Investig.* **2023**, *133*, e161987. [CrossRef]
80. Dallas, M.L.; Widera, D. TLR2 and TLR4-mediated inflammation in Alzheimer's disease: Self-defense or sabotage? *Neural Regen. Res.* **2021**, *16*, 1552–1553. [CrossRef] [PubMed] [PubMed Central]
81. Lax, N.; Fainstein, N.; Nishri, Y.; Ben-Zvi, A.; Ben-Hur, T. Systemic microbial TLR2 agonists induce neurodegeneration in Alzheimer's disease mice. *J. Neuroinflamm.* **2020**, *17*, 55. [CrossRef]
82. Abarca-Merlin, D.M.; Martínez-Durán, J.A.; Medina-Pérez, J.D.; Rodríguez-Santos, G.; Alvarez-Arellano, L. From Immunity to Neurogenesis: Toll-like Receptors as Versatile Regulators in the Nervous System. *Int. J. Mol. Sci.* **2024**, *25*, 5711. [CrossRef] [PubMed]
83. Squillace, S.; Salvemini, D. Toll-like receptor-mediated neuroinflammation: Relevance for cognitive dysfunctions. *Trends Pharmacol. Sci.* **2022**, *43*, 726–739. [CrossRef]
84. Zhao, J.; Bi, W.; Xiao, S.; Lan, X.; Cheng, X.; Zhang, J.; Lu, D.; Wei, W.; Wang, Y.; Li, H.; et al. Neuroinflammation induced by lipopolysaccharide causes cognitive impairment in mice. *Sci. Rep.* **2019**, *9*, 5790. [CrossRef]
85. Dubik, M.; Marczynska-Grzelak, J.; Sørensen, M.Z.; Dieu, R.S.; Rusin, D.; Schiöth, E.S.; Ramazani, B.; Belal, R.; Ojha, B.; Krieger, J.; et al. Synergistic Targeting of Innate Receptors TLR7 and NOD2 for Therapeutic Intervention in Multiple Sclerosis. *Int. J. Mol. Sci.* **2024**, *25*, 7462. [CrossRef]
86. Hernandez, J.; Schäffer, J.; Herden, C.; Pflieger, F.J.; Reiche, S.; Körber, S.; Kitagawa, H.; Welter, J.; Michels, S.; Culmsee, C.; et al. n-3 Polyunsaturated Fatty Acids Modulate LPS-Induced ARDS and the Lung–Brain Axis of Communication in Wild-Type versus Fat-1 Mice Genetically Modified for Leukotriene B4 Receptor 1 or Chemerin Receptor 23 Knockout. *Int. J. Mol. Sci.* **2023**, *24*, 13524. [CrossRef]
87. Höpfinger, A.; Schmid, A.; Schweitzer, L.; Patz, M.; Weber, A.; Schäffler, A.; Karrasch, T. Regulation of Cathelicidin Antimicrobial Peptide (CAMP) Gene Expression by TNFα and cfDNA in Adipocytes. *Int. J. Mol. Sci.* **2023**, *24*, 15820. [CrossRef]
88. Mylonas, K.S.; Peroulis, M.; Schizas, D.; Kapelouzou, A. MYD88 and Proinflammatory Chemokines in Aortic Atheromatosis: Exploring Novel Statin Effects. *Int. J. Mol. Sci.* **2023**, *24*, 9248. [CrossRef]
89. Loaiza, R.; Fattahi, F.; Kalbitz, M.; Grailer, J.J.; Russell, M.W.; Jalife, J.; Valdivia, H.H.; Zetoune, F.S.; Ward, P.A. The Impact of Extracellular Histones and Absence of Toll-like Receptors on Cardiac Functional and Electrical Disturbances in Mouse Hearts. *Int. J. Mol. Sci.* **2024**, *25*, 8653. [CrossRef]
90. Bosshart, H.; Heinzelmann, M. Targeting bacterial endotoxin: Two sides of a coin. *Ann. N. Y. Acad. Sci.* **2007**, *1096*, 1–17. [CrossRef] [PubMed]
91. Salaun, B.; Romero, P.; Lebecque, S. Toll-like receptors' two-edged sword: When immunity meets apoptosis. *Eur. J. Immunol.* **2007**, *37*, 3311–3318. [CrossRef] [PubMed]
92. Cario, E. Therapeutic impact of toll-like receptors on inflammatory bowel diseases: A multiple-edged sword. *Inflamm. Bowel Dis.* **2008**, *14*, 411–421. [CrossRef]
93. Killeen, S.D.; Wang, J.H.; Andrews, E.J.; Redmond, H.P. Exploitation of the Toll-like receptor system in cancer: A doubled-edged sword? *Br. J. Cancer* **2006**, *95*, 247–252. [CrossRef] [PubMed]
94. Lien, E.; Zipris, D. The Role of Toll-like receptor pathways in the mechanism of type 1 diabetes. *Curr. Mol. Med.* **2009**, *9*, 52–68. [CrossRef] [PubMed]
95. Basith, S.; Manavalan, B.; Yoo, T.H.; Kim, S.G.; Choi, S. Roles of toll-like receptors in cancer: A double-edged sword for defense and offense. *Arch. Pharmacal Res.* **2012**, *35*, 1297–1316. [CrossRef]
96. Marsh, B.J.; Stenzelpoore, M. Toll-like receptors: Novel pharmacological targets for the treatment of neurological diseases. *Curr. Opin. Pharmacol.* **2008**, *8*, 8–13. [CrossRef]
97. Greene, C.; Connolly, R.; Brennan, D.; Laffan, A.; O'keeffe, E.; Zaporojan, L.; O'callaghan, J.; Thomson, B.; Connolly, E.; Argue, R.; et al. Blood–brain barrier disruption and sustained systemic inflammation in individuals with long COVID-associated cognitive impairment. *Nat. Neurosci.* **2024**, *27*, 421–432. [CrossRef]
98. Su, W.; Ju, J.; Gu, M.; Wang, X.; Liu, S.; Yu, J.; Mu, D. SARS-CoV-2 envelope protein triggers depression-like behaviors and dysosmia via TLR2-mediated neuroinflammation in mice. *J. Neuroinflamm.* **2023**, *20*, 110. [CrossRef]
99. Fontes-Dantas, F.L.; Fernandes, G.G.; Gutman, E.G.; De Lima, E.V.; Antonio, L.S.; Hammerle, M.B.; Mota-Araujo, H.P.; Colodeti, L.C.; Araújo, S.M.; Froz, G.M.; et al. SARS-CoV-2 Spike protein induces TLR4-mediated long-term cognitive dysfunction recapitulating post-COVID-19 syndrome in mice. *Cell Rep.* **2023**, *42*, 112189. [CrossRef]
100. Kaushik, D.; Bhandari, R.; Kuhad, A. TLR4 as a therapeutic target for respiratory and neurological complications of SARS-CoV-2. *Expert Opin. Ther. Targets* **2021**, *25*, 491–508. [CrossRef]
101. Frank, M.G.; Fleshner, M.; Maier, S.F. Exploring the immunogenic properties of SARS-CoV-2 structural proteins: PAMP:TLR signaling in the mediation of the neuroinflammatory and neurologic sequelae of COVID-19. *Brain Behav. Immun.* **2023**, *111*, 259–269. [CrossRef] [PubMed]

102. Burnett, F.N.; Coucha, M.; Bolduc, D.R.; Hermanns, V.C.; Heath, S.P.; Abdelghani, M.; Macias-Moriarity, L.Z.; Abdelsaid, M. SARS-CoV-2 Spike Protein Intensifies Cerebrovascular Complications in Diabetic hACE2 Mice through RAAS and TLR Signaling Activation. *Int. J. Mol. Sci.* **2023**, *24*, 16394. [CrossRef]
103. Theoharides, T.C. Could SARS-CoV-2 Spike Protein Be Responsible for Long-COVID Syndrome? *Mol. Neurobiol.* **2022**, *59*, 1850–1861. [CrossRef] [PubMed]

International Journal of
Molecular Sciences

MDPI

Review

Unraveling the Complexities of Toll-like Receptors: From Molecular Mechanisms to Clinical Applications

Yi-Hsin Chen [1,2,3], Kang-Hsi Wu [4,5,*] and Han-Ping Wu [6,7,*]

1 Department of Nephrology, Taichung Tzu Chi Hospital, Taichung 427, Taiwan; yishin0819@gmail.com
2 School of Medicine, Tzu Chi University, Hualien 97004, Taiwan
3 Department of Artificial Intelligence and Data Science, National Chung Hsing University, Taichung 40227, Taiwan
4 Department of Pediatrics, Chung Shan Medical University Hospital, Taichung 402, Taiwan
5 School of Medicine, Chung Shan Medical University, Taichung 402, Taiwan
6 College of Medicine, Chang Gung University, Taoyuan 33302, Taiwan
7 Department of Pediatrics, Chiayi Chang Gung Memorial Hospital, Chiayi 613016, Taiwan
* Correspondence: cshy1903@gmail.com (K.-H.W.); arthur1226@gmail.com (H.-P.W.); Tel.: +886-9-18250636 (H.-P.W.)

Abstract: Toll-like receptors (TLRs) are vital components of the innate immune system, serving as the first line of defense against pathogens by recognizing a wide array of molecular patterns. This review summarizes the critical roles of TLRs in immune surveillance and disease pathogenesis, focusing on their structure, signaling pathways, and implications in various disorders. We discuss the molecular intricacies of TLRs, including their ligand specificity, signaling cascades, and the functional consequences of their activation. The involvement of TLRs in infectious diseases, autoimmunity, chronic inflammation, and cancer is explored, highlighting their potential as therapeutic targets. We also examine recent advancements in TLR research, such as the development of specific agonists and antagonists, and their application in immunotherapy and vaccine development. Furthermore, we address the challenges and controversies surrounding TLR research and outline future directions, including the integration of computational modeling and personalized medicine approaches. In conclusion, TLRs represent a promising frontier in medical research, with the potential to significantly impact the development of novel therapeutic strategies for a wide range of diseases.

Keywords: Toll-like receptors; immunity; adaptive immunity; innate immunity; ligands

Citation: Chen, Y.-H.; Wu, K.-H.; Wu, H.-P. Unraveling the Complexities of Toll-like Receptors: From Molecular Mechanisms to Clinical Applications. *Int. J. Mol. Sci.* **2024**, *25*, 5037. https://doi.org/10.3390/ijms25095037

Academic Editors: Ralf Kircheis and Oliver Planz

Received: 28 March 2024
Revised: 2 May 2024
Accepted: 3 May 2024
Published: 5 May 2024

1. Introduction

1.1. Overview of Innate Immunity and Pattern Recognition Receptors (PRRs)

Toll-like receptors (TLRs) are integral to the innate immune system as they recognize diverse molecular patterns that trigger immune responses. Humans possess ten functional TLRs (TLR1–10), each specializing in detecting different signals and pathogen-associated molecular patterns, including damage-associated molecular patterns (DAMPs), microbial-associated molecular patterns (MAMPs), pathogen-associated molecular patterns (PAMPs), and xenobiotic-associated molecular patterns (XAMPs). These receptors share a common structural organization, comprising an N-terminal domain with leucine-rich repeats (LRRs), a single transmembrane helix (TM), and a C-terminal cytoplasmic Toll-interleukin-1 receptor (TIR) domain [1]. The aim of this review is to delineate the multifaceted roles of TLRs within the scope of the innate immune system, elucidating their structural characteristics, signaling pathways, and implications in various diseases. This review also aims to discuss recent advancements in TLR research that have clinical relevance, particularly focusing on their therapeutic potential.

In addition to TLRs, the innate immune system is equipped with a robust array of pattern recognition receptors that ensure a comprehensive and adaptable response to pathogens. These include RIG-I-like receptors, which are cytoplasmic detectors of viral RNA, initiating crucial antiviral defenses by producing type I interferons [2]. C-type lectin receptors play a pivotal role in recognizing fungal carbohydrates and bacterial elements, enhancing phagocytosis and mediating direct antimicrobial actions [3]. NOD-like receptors within the cell cytoplasm activate upon sensing microbial motifs or cellular stress signals, triggering inflammasomes that promote cytokine production and drive inflammatory responses [3]. Furthermore, the cGAS-STING pathway marks a critical defense mechanism against cytosolic DNA from viruses or damaged host cells, orchestrating a potent immune activation that includes interferon responses [4].

Studies have elucidated the critical and multifaceted roles of TLRs in the innate immune system. Notably, TLR9 and the cyclic GMP-AMP synthase-stimulator of interferon genes (cGAS-STING) pathway possess DNA-sensing capabilities and play roles in initiating innate immune responses, underscoring the complexity and adaptability of TLR-mediated signaling in response to diverse stimuli [5]. The significance of TLRs in viral infections is further highlighted by reports that TLR3, TLR7, and TLR8, along with cytoplasmic retinoic acid-inducible gene I (RIG-I)-like receptors, are instrumental in initiating innate immune responses against viral nucleic acids [6]. Additionally, research on TLR signaling pathways has emphasized their importance in disease progression such as in herpes simplex virus 1 (HSV-1) infection wherein TLRs and their signaling pathways in the trigeminal ganglia play crucial roles in controlling the progression to encephalitis [7]. This underscores the vital function of TLRs in antiviral defense and their potential as therapeutic targets.

TLRs have been targeted in numerous clinical trials due to their key role in immune system activation and disease pathogenesis, particularly in infectious diseases, cancer, and inflammatory conditions. Clinical trials have systematically explored the use of TLR agonists and antagonists, grouping them based on the targeted pathology and therapy combinations. As illustrated in Table 1, the range of TLRs' applications is extensive, encompassing not only agonists and antagonists but also their use as vaccine adjuvants, which are critical in infectious disease contexts. For instance, TLR7 and TLR9 are often activated in therapies against viral infections and as adjuvants in cancer vaccines, while TLR2 and TLR4 are targeted by antagonists to mitigate inflammatory responses in sepsis and other inflammatory diseases [8]. Additionally, advancements in understanding TLR-mediated signaling, such as its regulation by ubiquitination and microRNAs, suggest further therapeutic possibilities for targeted manipulation in clinical settings [9].

Table 1. Therapeutic potential of TLRs.

Approach	Examples	Diseases	Reference
TLR Agonists	Poly-ICLC, MEDI9197, QbG10, Resiquimod, Imiquimod	Cancer, infectious diseases, allergies, HPV	[10–12]
TLR Antagonists	TAC5, MHV370	Autoimmune diseases (psoriasis, SLE), inflammatory conditions	[13,14]
Vaccine Adjuvants	Monophosphoryl Lipid A (MPL), CpG-1018, Resiquimod, Imiquimod	Infectious diseases (HBV, HCV, HIV, SARS-CoV-2, influenza, HPV)	[10–12]

1.2. Brief Introduction to TLRs

TLRs are integral components of the immune system that detect PAMPs and DAMPs. Their activation activates signaling pathways that culminate in inflammatory and immune responses. TLRs are ubiquitously present across various cell types and tissues and play a pivotal role in immune surveillance [6,15]. Upon ligand binding, TLRs initiate a cascade of events that lead to the production of pro-inflammatory cytokines and type I interferons, which are essential for shaping the immune response. Regulation of TLR signaling is crucial for maintaining immune homeostasis; dysregulation is linked to several diseases, including autoimmunity, chronic inflammation, and cancer, positioning TLRs as prime therapeutic targets [6,16]. The significance of TLRs extends to the pathogenesis of lifestyle-related diseases wherein TLR9 recognizes DNA fragments, highlighting the intricate cross-regulation of type 1 interferon signaling pathways [17]. This is corroborated by studies illustrating the association of endosomal transmembrane TLRs (such as TLR3) with various diseases that can activate the interferon signaling pathway [6]. Furthermore, the role of TLRs in diabetic encephalopathy underlines their impact on signaling pathways by interacting with receptors such as RAGE/Toll-like receptors, which are implicated in several conditions, including cancer, cardiovascular disease, and sepsis [18].

Therapeutic modulation of TLR signaling holds promise for the treatment of a range of diseases. For instance, LAMP3-induced ectopic TLR7 expression in salivary gland epithelial cells and amplification of type I IFN production highlight potential therapeutic strategies for Sjögren's disease [19]. Moreover, compounds such as piperine, which attenuates hepatic ischemia/reperfusion injury by suppressing the TLR4 signaling cascade, point towards safer options for the clinical treatment and prevention of ischemia-related diseases [20]. In conclusion, the critical role of TLRs in immune surveillance and their involvement in various diseases underscores the importance of proper regulation of TLR signaling. Their potential as therapeutic targets offers avenues for the treatment of autoimmune, inflammatory, and chronic diseases, emphasizing the need for further research in this domain.

1.3. Structure and Characterization of TLRs

1.3.1. Structural Features of TLRs

TLRs are type I transmembrane proteins with distinctive structures that play critical roles in the innate immune system. They are known for their extracellular leucine-rich repeats (LRRs), which range from 20 to 27 per TLR, enabling the detection of PAMPs and DAMPs [21]. Recent findings highlight the significant impact of gene polymorphisms within TLRs on their function. For example, polymorphisms in TLR4 have been shown to influence milk production traits in livestock by altering immune responsiveness [22]. Additionally, specific TLR gene polymorphisms are associated with differential susceptibility to autoimmune diseases, further underscoring the importance of considering genetic variability in TLR function [23]. These LRRs form horseshoe-like structures that recognize a variety of molecular signatures of microbial intruders or indicators of cell damage. A pivotal component of TLRs is the Toll/interleukin-1 receptor (TIR) domain, which is located within the intracellular region and is crucial for signal transduction that leads to the activation of genes that drive inflammatory responses [24,25]. Furthermore, the extracellular domains of TLRs are modified by glycan moieties, which are presumed to play a role in ligand binding; however, the exact functions of these modifications remain to be elucidated [21,26].

1.3.2. Differences between TLR Family Members

TLRs are a class of proteins instrumental in the immune system's first line of defense. Each TLR recognizes specific PAMPs or DAMPs because of their LRR structures, which confer ligand specificity. This specificity enables the immune system to differentiate between various pathogens and damaged host cells, thereby initiating the appropriate immune responses [21]. The TLR family is functionally diverse, with members located on the cell membrane or within intracellular compartments such as the endoplasmic reticulum, endosomes, and lysosomes. Cell membrane TLRs, such as TLR2 in heterodimeric association with TLR1 or TLR6, TLR4, TLR5, and TLR10, primarily respond to microbial membrane components such as lipoproteins, flagellin, and lipopolysaccharides. In contrast, intracellular TLRs, including TLR3, TLR7, TLR8, and TLR9, are specialized in sensing nucleic acids, highlighting the system's adaptability to various types of molecular signatures indicative of infection or cellular damage [21]. Humans possess ten functional TLRs (TLR1–TLR10), each tailored to detect specific pathogen-associated and damage-associated molecular patterns, reflecting their role in immune surveillance and response. These receptors share a common structural organization comprising an N-terminal domain with leucine-rich repeats (LRRs), a single transmembrane helix (TM), and a C-terminal cytoplasmic Toll-interleukin-1 receptor (TIR) domain [27]. Expression mapping shows widespread but distinct patterns, with variations depending on cell type and tissue. For example, TLR4 is broadly expressed across many cell types and tissues, whereas TLR9 is more restricted to immune cell populations within endosomal compartments, illustrating the functional specialization of TLR locations. Notably, TLR11, TLR12, and TLR13, though not present in humans, are expressed in mice and other species, playing crucial roles in immune responses to various pathogens such as protozoans and bacteria [28]. These TLRs are primarily involved in recognizing specific bacterial components and are essential in initiating appropriate immune responses in these species.

TLR signaling pathways diverge to initiate tailored immune responses. For instance, TLR4 recognition of lipopolysaccharides (LPSs) from Gram-negative bacteria triggers signaling cascades involving adapters such as MyD88 and TRIF, leading to the production of pro-inflammatory cytokines and type I interferons [21]. These mechanisms underscore the role of TLRs in bridging innate and adaptive immunity and ensuring coherent and effective immune defense. The subcellular localization of TLRs plays a crucial role in their function. Cell surface TLRs detect extracellular threats, whereas intracellular TLRs monitor the interior of cells for viral and bacterial nucleic acids, demonstrating a sophisticated surveillance system that protects against a wide array of pathogens [29]. Table 2 demonstrates the primary locations, principal ligands, signaling pathways, and functions of each TLR, offering a detailed comparison of their roles in immune surveillance and disease pathogenesis.

Table 2. Overview of Toll-like receptors (TLRs).

TLR	Primary Location	Principal Ligand	Signaling Pathway	Functions and Application Examples
TLR1/2	Cell Surface	Triacylated lipopeptides (Pam3CSK4)	MyD88-dependent	Involved in immune response activation, potential applications in cancer and infectious disease treatments [1]
TLR2/6	Cell Surface	Lipopeptides	MyD88-dependent	Regulates inflammation and immune responses, potential treatments for allergic and autoimmune diseases [30]
TLR3	Endosomes	Double-stranded RNA (dsRNA)	TRIF-dependent	Antiviral immune responses, applications as vaccine adjuvants and in tumor treatments [31]
TLR4	Cell Surface	Lipopolysaccharide (LPS)	MyD88-dependent and TRIF-dependent	Regulates inflammatory responses, potential treatments for infectious diseases and cancer [32]
TLR7	Endosomes	Single-stranded RNA (ssRNA)	MyD88-dependent	Antiviral responses and regulation of autoimmune diseases such as SLE [13]
TLR8	Endosomes	Single-stranded RNA (ssRNA)	MyD88-dependent	Antiviral and anti-tumor therapeutic applications [13]
TLR9	Endosomes	CpG DNA	MyD88-dependent	Applications in antiviral vaccine adjuvants, treatments for infectious diseases and tumors [30]
TLR10	Cell Surface	Double-stranded RNA (dsRNA) (presumed)	MyD88-dependent	Potential roles in antiviral responses and immune regulation, further research needed to clarify specific functions [33]

1.4. Ligands and Signaling of TLRs

1.4.1. Exogenous Ligands (PAMPs)

PAMPs include a variety of microbial molecular structures that trigger immune responses, such as flagellin, lipoteichoic acid (LTA), and peptidoglycan (PGN) from Gram-positive bacteria and LPS from Gram-negative bacteria. These components play significant roles in inducing various immune and inflammatory responses, as explored in a study that investigated the synergistic effects of PGN, LTA, and LPS on bovine mammary epithelial cells, revealing insights into transcriptome changes, inflammatory responses, and associated epigenetic mechanisms [34].

1.4.2. Endogenous Ligands (DAMPs)

Endogenous ligands, also known as DAMPs, play a crucial role in immune system activation in response to non-physiological cell death and injury. These molecules, including extracellular matrix components such as hyaluronan and fibrinogen, plasma membrane constituents, nuclear and cytosolic proteins such as high-mobility group box protein 1 (HMGB1) and heat shock proteins, and elements from damaged or fragmented organelles such as mitochondrial DNA. Importantly, galectins, particularly Galectin-3, have been recognized as critical mediators in inflammation by binding to TLRs, notably TLR-4, to modulate immune cell activation and tissue damage responses [35]. Their recognition is primarily mediated by pattern recognition receptors (PRRs) such as TLRs [36,37].

Upon recognizing DAMPs, TLRs initiate signaling cascades that result in the production of inflammatory cytokines, type I interferons, and other mediators, facilitating an effective immune response against damage [36]. These processes underscore the sophisticated mechanisms by which the immune system distinguishes between self and non-self as well as between different types of cell death to maintain homeostasis and defend against various pathological conditions [37].

1.4.3. Adaptor Proteins and Signaling Pathways and Functional Consequences of TLR Activation

TLR signaling pathways are central to initiating both innate and adaptive immune responses. These pathways involve the recruitment of specific adaptor proteins, leading to the activation of downstream transcription factors that regulate immune and inflammatory responses. Each TLR can initiate distinct signaling cascades through one or both of the main pathways: the MyD88-dependent pathway and the TRIF-dependent pathway. All TLRs, except for TLR3, primarily utilize the MyD88-dependent pathway. This pathway involves the adaptor protein MyD88 (myeloid differentiation primary response 88) and leads to the activation of NF-κB, a key transcription factor that drives the expression of genes involved in inflammation and immunity, such as TNF-α, IL-1, and IL-6. The rapid activation of NF-κB facilitates immediate inflammatory responses, which are crucial for the initial phase of pathogen defense [38].

In contrast, the TRIF-dependent pathway, which is exclusive to TLR3 and is also utilized by TLR4, is initiated from TLR4-containing endosomes and results in the activation of interferon regulatory factors (IRFs), particularly IRF3. This pathway leads to the production of type I interferons, like IFN-α and IFN-β, which are vital for antiviral defense. The activation of IRFs contributes to a sustained immune response, suitable for combating viral infections and for modulating the immune system over a longer period [39]. The distinct and non-redundant roles of these pathways are critical for tailored immune responses, enabling the immune system to effectively handle a variety of pathogens. Understanding these pathways' unique and overlapping functions is crucial for developing therapeutic interventions for immune-related diseases.

Upon recognition of PAMPs or DAMPs, TLRs trigger these signaling pathways through the recruitment of specific adaptor proteins, namely, MyD88, TIRAP (MAL), TRIF, TRAM, and SARM, which are crucial for downstream signaling. This cascade results in the activation of various transcription factors, including NF-κB, interferon regulatory factors (IRFs), and activator protein 1 (AP-1), leading to the expression of genes that drive the production of immune response molecules such as pro-inflammatory cytokines (e.g., TNF-α, IL-1β, IL-6), type 1 interferons (IFN-α and IFN-β), chemokines (CXCL8, CXCL10), and antimicrobial peptides [21,40]. The intricate network of TLR signaling pathways is illustrated in Figure 1. The consequences of these signaling events are significant, not only in combating infections but also in influencing host behavior and physiology. Cytokines produced in response to TLR activation can induce systemic effects such as weakness, lethargy, fatigue, and anorexia. These cytokines also orchestrate metabolic shifts towards catabolism, affecting muscle protein synthesis, inducing muscle wasting, and altering lipid metabolism by promoting lipolysis, upregulating fatty acid synthesis, increasing hepatic triglyceride production, and elevating serum triglyceride levels. Such extensive physiological changes underscore the pivotal role of TLRs in managing not just the immune response but also the metabolic adaptations during infections [21].

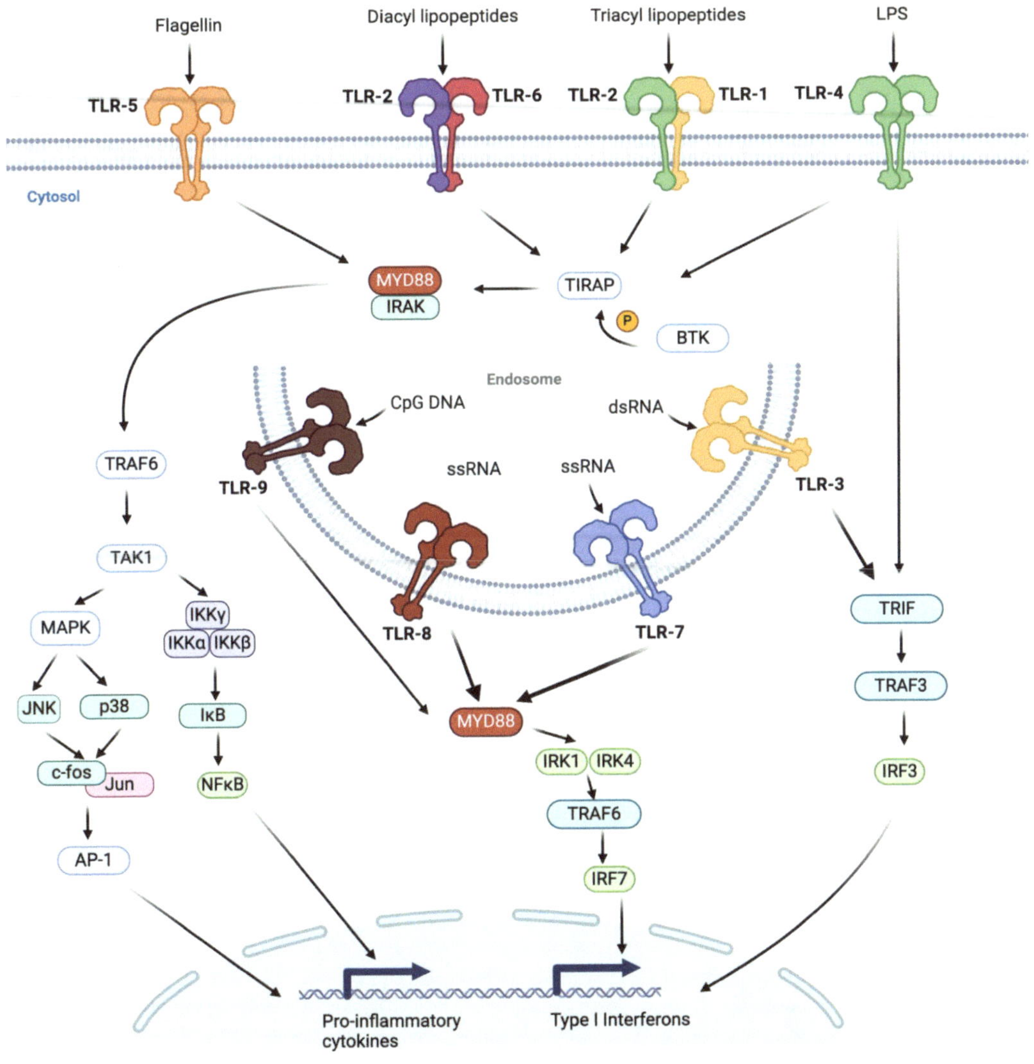

Figure 1. Overview of Toll-like receptor (TLR) signaling pathways. This schematic depicts the cell-surface TLRs (TLR-1, TLR-2, TLR-4, TLR-5, and TLR-6) and their ligands (flagellin, diacyl lipopeptides, triacyl lipopeptides, and LPS). It also shows endosomal TLRs (TLR-3, TLR-7, TLR-8, and TLR-9) recognizing CpG DNA and RNA molecules. The diagram illustrates the subsequent signaling cascades involving adaptor proteins such as MYD88, TRIF, and TRAF6, leading to the activation of NF-κB and IRFs and resulting in the expression of pro inflammatory cytokines and type I interferons.

Furthermore, TLRs bridge the gap between innate and adaptive immunity, highlighting their importance in a comprehensive immune defense strategy. The complexity and regulation of these pathways ensure that the immune responses are appropriate and proportional to the threat, preventing overactivation and potential autoimmunity [41].

1.5. Role of TLRs in Infection

1.5.1. Recognition of Microbial Pathogens

TLRs play a crucial role in the immune system by recognizing PAMPs derived from microbes, thus initiating key signaling pathways that lead to immune responses. This recognition involves the specific binding of PAMPs, such as lipopolysaccharides from bacteria and viral RNA, to corresponding TLRs, which then activate transcription factors like NF-κB and IRFs. These factors are vital for the production of pro-inflammatory and antiviral cytokines and chemokines, essential for orchestrating the first line of defense in the immune system. The maturation of dendritic cells (DCs), triggered by these signaling events, is critical for T cell activation and the effective clearance of pathogens. These interactions highlight the indispensable role of TLRs in linking innate and adaptive immunity, underscoring their importance in immune surveillance. Enhanced understanding of TLR pathways, particularly their role in detecting microbial components, can significantly influence therapeutic strategies aimed at modulating immune responses for better disease management. The interaction between mature DCs and T cells is a fundamental aspect of the immune response that facilitates the activation and differentiation of T cells, which are critical for clearing pathogens [41–44].

TLR signaling pathways directly regulate effector and regulatory T cells (Tregs), highlighting the diverse roles of TLRs in both the innate and adaptive arms of the immune system. These pathways are involved in various diseases, including infectious diseases, autoimmune diseases, and cancer, highlighting the importance of TLRs in maintaining immune homeostasis and pathogenesis [45]. The interaction between TLRs and T cells is not unidirectional; T cell responses can be modulated by TLR signaling in a complex interplay that involves not only direct effects on T cells but also indirect effects mediated through antigen-presenting cells such as DCs. TLR-induced DC activation requires intrinsic complement pathways, further underscoring the complexity of TLR signaling and its impact on the immune response [41].

This intricate network of interactions among TLRs, DCs, and T cells underscores the essential role of TLRs in orchestrating a coherent immune response, linking innate immunity with the adaptive immune system to effectively respond to infections and other challenges.

1.5.2. Initiation of Innate Immune Responses

TLRs play a pivotal role in the first line of immune system defense, distinguishing between self and non-self through the recognition of PAMPs derived from microbial entities and damaged cell components. Among these, TLRs located on the cell surface, such as TLR1, TLR2, TLR4, and TLR5, and those found in intracellular compartments, such as TLR3, TLR7, TLR8, and TLR9, provide a comprehensive system that allows for nuanced and highly specific responses to a wide array of microbial entities.

TLR2 and TLR4 have been recognized for their roles in identifying the components of the bacterial cell wall and distinguishing between Gram-positive and Gram-negative bacteria. TLR4 is known for its response to LPS, a major component of the outer membrane of Gram-negative bacteria, which initiates a signaling cascade that leads to the production of pro-inflammatory cytokines. Conversely, TLR2 is implicated in the recognition of peptidoglycans, lipoteichoic acids, and other components, predominantly in Gram-positive bacteria [46,47].

The interaction of TLR2 with its ligands plays a crucial role in the immune response to various pathogens, including Helicobacter pylori, a Gram-negative bacterium associated with gastric ulcers and cancer. This demonstrates the versatility and specificity of TLR2 in recognizing and responding to different microbial patterns [48]. Research has expanded our understanding of TLR2's role in immune responses across different species, highlighting its evolutionary importance in host defense mechanisms [49].

The signaling pathways activated by TLR engagement are complex and highly regulated and involve a range of adaptor proteins and downstream signaling molecules. These pathways ensure that the immune response is tailored appropriately to the nature of the microbial threat. TLR signaling specificity is often determined by TIR domain-containing adaptors, which play a crucial role in translating the recognition of microbial components into appropriate cellular responses [50].

In conclusion, TLRs, particularly TLR2 and TLR4, are critical for the immune system to distinguish between different types of pathogens and initiate specific responses. Their ability to recognize a wide range of PAMPs underlies the effectiveness of the innate immune response as the first line of defense against microbial infections.

1.5.3. Connection between Innate and Adaptive Immunity

Intracellular TLRs, specifically TLR3, TLR7, TLR8, and TLR9, are located in the endosomal compartments of cells and play pivotal roles in the recognition of foreign nucleic acids by the innate immune system. This process is essential for initiating a robust antiviral response via the production of type I interferons and inflammatory cytokines, as illustrated in Figure 1. These TLRs serve as crucial sentinels in detecting pathogen invasions, enabling the body to mount an immediate defense against infections and to effectively bridge innate and adaptive immune responses [51–53].

TLR3 is unique in its ability to recognize double-stranded RNA, a molecular pattern associated with viral infection. TLR3 activation leads to the recruitment of the adaptor molecule TRIF, initiating a signaling cascade that ultimately results in the production of type I interferons and pro-inflammatory cytokines, thereby establishing an antiviral state within the host [53]. The crosstalk between TLRs and other receptor families, such as RIG-I-like receptors, further emphasizes the integrated nature of the innate immune response, allowing for the coordinated activation of antiviral defenses [52].

Moreover, conventional dendritic cells (cDCs) expressing TLR2, TLR4, TLR5, and TLR6 play significant roles in connecting innate and adaptive immunity. These TLRs, located on the cell surface, are crucial for the recognition of microbial components such as lipoproteins and lipopolysaccharides, leading to the activation of signaling pathways that promote the maturation of dendritic cells. This maturation is vital for effective antigen presentation to T cells, facilitating the activation and differentiation of T cells into effector cells that are central to adaptive immunity. Specifically, TLR4-mediated signaling through the NF-kB pathway is essential for the production of key cytokines and the upregulation of co-stimulatory molecules, which are critical for initiating a robust immune response [54]. Additionally, aqueous extracts of Artemisia rupestris L. have been shown to activate TLR4/TLR2, enhancing the phenotypic maturation of dendritic cells and promoting their ability to stimulate T-cell proliferation, further underscoring the pivotal role of these TLRs in the immune-modulating activities of dendritic cells [55].

Engagement of these intracellular TLRs not only triggers immediate innate immune responses but also influences the adaptive immune system. Activation of dendritic cells by TLRs results in the maturation of these antigen-presenting cells, enhancing their ability to present antigens to T cells and facilitating the development of a tailored adaptive immune response. The interplay between innate and adaptive immunity underscores the critical role of TLRs in maintaining immune surveillance and defense mechanisms against pathogens [10].

TLRs are instrumental in the development of immunotherapeutic strategies, including the formulation of vaccine adjuvants that target these receptors to enhance vaccine efficacy. By activating TLRs, these adjuvants can augment the immunogenicity of vaccines and promote stronger and more durable immune responses against cancers and infectious diseases [10].

In conclusion, the integration of innate and adaptive immune responses mediated by intracellular TLRs highlights the sophistication of immune system defense mechanisms. Through the recognition of pathogen-derived nucleic acids and the subsequent activation of immune responses, TLRs exemplify the intricate balance and communication between the body's first line of defense and its more specific adaptive immune responses [56].

1.6. TLRs in Inflammation

1.6.1. TLR-Mediated Inflammatory Pathways

TLRs are crucial for orchestrating the inflammatory pathways that are vital for immune responses to pathogens. Recent studies have underscored the crucial role of TLR4 in mediating severe inflammatory responses in conditions such as COVID-19, where inflammatory signaling pathways are markedly upregulated, contributing to disease severity [57]. Furthermore, innovative therapeutic strategies involving peptides derived from TIRAP, which act as decoys for TLR4, have shown promise in reducing inflammatory and autoimmune symptoms in murine models, offering potential for clinical applications [58]. They recognize distinct PAMPs, which initiate signaling cascades that activate transcription factors such as NF-κB and AP-1. These transcription factors drive the production of pro-inflammatory cytokines, chemokines, and other mediators involved in the inflammatory response. TIRAP (MAL), a key adaptor in TLR signaling, plays a critical role in recruiting and activating MyD88, which is essential for signaling through TLR2 and TLR4, thus promoting inflammatory cytokine production via NF-κB and AP-1 activation [59,60].

Intracellular signaling in the context of TLR activation involves a complex network of adaptors and kinases. Among these, the interaction between TIRAP and TRAF6 is critical for TLR2- and TLR4-mediated pro-inflammatory responses, illustrating the importance of protein–protein interactions in the specificity and amplification of TLR signaling pathways [61]. MyD88, a central adaptor molecule in innate immune signaling, plays a pivotal role in the downstream signaling of most TLRs as well as the IL-1 receptor family, connecting receptor engagement with the activation of IRAK family kinases. The organization of MyD88 domains, including the death domain, intermediate domain, and Toll-interleukin-1 receptor domain (TIR), is essential for its function in signal transduction [62].

Furthermore, TIRAP's preferential localization to the cytoplasmic membrane through a PIP2-binding domain highlights the spatial regulation of TLR signaling, emphasizing the importance of membrane localization in the initiation of specific signal transduction pathways. Myristoylated TRAM, another adaptor molecule, localizes to endosomes and triggers type I interferon production via TRIF, delineating a clear distinction in the signaling roles and pathways mediated by different TLR adaptors [63].

This sophisticated regulatory mechanism ensures that the immune response is appropriately modulated, thereby preventing excessive inflammation that can lead to tissue damage or autoimmunity. The intricate interplay between these molecules exemplifies the complexity of the innate immune response and provides multiple targets for therapeutic intervention in diseases characterized by dysregulated inflammation or immune responses.

1.6.2. Impact on Autoimmune Diseases

TLRs play complex roles in the development of autoimmune diseases, demonstrating distinct and varying implications across different conditions. TLR2, TLR4, TLR7, and TLR9 recognize PAMPs from infectious agents and DAMPs from tissue damage, contributing uniquely to the pathogenesis of specific autoimmune diseases. For example, TLR7 is prominently involved in the immune response mechanisms of systemic lupus erythematosus (SLE) by promoting the production of type I interferons, which are crucial in the disease's progression. Similarly, TLR9 has been linked to the stimulation of autoreactive B cells in SLE, illustrating a direct influence on autoimmune pathology [64].

Alterations in TLR signaling, such as those resulting from gene polymorphisms, can significantly affect the immune balance, leading to diverse clinical outcomes in autoimmune conditions. Dysfunctions in TLR2, for instance, often result in decreased autoimmune inflammation, while overexpression might exacerbate it, as seen in diseases like rheumatoid arthritis and multiple sclerosis [23]. TLR signaling extends beyond innate cells to adaptive immune cells such as T and B lymphocytes, which utilize TLRs to modulate responses and affect autoimmune disease progression.

Recent research has further demonstrated that inhibitors like TAC5 can selectively target endosomal TLRs (TLR3, TLR7, TLR8, and TLR9) to modulate immune responses and potentially treat autoimmune diseases like psoriasis and systemic lupus erythematosus. For instance, a novel selective Toll-like receptor 7/8 inhibitor, known as M5049, has shown potential in treating autoimmunity by blocking multiple TLR7/8 RNA ligands with significant efficacy in murine lupus models [65]. Additionally, modulation of TLR signaling by small-molecular inhibitors such as those targeting TLR7, TLR8, and TLR9 has shown promise in ameliorating symptoms of autoimmune disorders, as illustrated by the discovery of potent and orally bioavailable small-molecular antagonists of TLR7/8/9, which have demonstrated efficacy from oral dosing in the preclinical models of autoimmune diseases [66]. The complex interaction between various TLRs, including their non-canonical roles in immune system modulation, continues to be a crucial area of study with significant therapeutic potential.

1.7. Advancements in TLR Research

Recent Discoveries and Technological Advancements

The recognition of PAMPs and DAMPs by TLRs initiates signaling cascades that significantly affect the immune balance, leading to either the mitigation or exacerbation of autoimmune inflammation. TLRs, such as TLR2, TLR4, TLR7, and TLR9, are involved in this process, and their dysfunction or overexpression in immune cells plays a pivotal role in the pathogenesis of autoimmune diseases.

TLR signaling pathways are complex and extend beyond innate immune cells to adaptive immune cells, such as T and B lymphocytes, affecting autoimmune disease progression. For instance, alterations in TLR signaling can lead to the inappropriate triggering of TLR pathways by exogenous or endogenous ligands, causing the initiation and/or perpetuation of autoimmune reactions and tissue damage. This underlines the importance of TLRs in both regulatory and inflammatory pathways within the context of autoimmune disease pathogenesis [63,64,67]. Recent studies have further highlighted novel technological advancements in TLR research that significantly enhance our understanding and potential therapeutic applications. Notably, the development of TLR therapeutics has shown promise in the treatment of chronic viral infections, with TLR7 agonists currently in clinical trials for chronic HBV infection and a TLR9 ligand used as an adjuvant in a superior HBV vaccine [68]. Additionally, the creation of a highly sensitive TLR reporter platform represents a significant technological advancement. This platform employs a sensitive NF-κB::eGFP reporter cell line for the detection of TLR ligands with high specificity, which is crucial for ensuring the purity of therapeutic compounds and understanding TLR ligand interactions [69]. These technological innovations not only deepen our understanding of TLR roles in immune modulation but also open new avenues for the development of targeted therapies for autoimmune and infectious diseases. Given their crucial roles in infectious and non-infectious disease processes, TLRs and their signaling pathways have emerged as attractive targets for therapeutic interventions. Strategies aimed at modulating TLR signaling could offer new avenues for treating autoimmune diseases by restoring immune balance and preventing tissue damage [70,71].

Recent advancements include the discovery of small-molecular antagonists, such as 7f, which exhibits strong on-target potency and selectivity against TLR7 and TLR9, demonstrating efficacy in the preclinical models of autoimmune diseases by oral administration [66]. Furthermore, the identification of the novel small-molecular TLR-inhibitor, TAC5, has shown to significantly reduce inflammation and alleviate symptoms in autoimmune disease models, indicating a potential therapeutic application for diseases such as psoriasis and SLE [13]. These insights into TLR signaling pathways in the context of autoimmune diseases highlight the need for further research to explore therapeutic options that target these pathways. Such approaches may provide novel treatments for autoimmune conditions by leveraging our understanding of TLR-mediated immune responses.

1.8. TLRs as Therapeutic Targets

1.8.1. Potential in Drug Development

TLRs have significant potential in drug development, particularly for their roles in immunotherapy, cancer treatment, and inflammation control. Recent advances in immunotherapy have highlighted the efficacy of TLR agonists, such as Poly-ICLC, in enhancing anticancer immune responses. These agents function by stimulating TLR3, leading to enhanced expression of inflammatory genes and facilitating the infiltration of T cells into tumors, thus demonstrating significant potential in cancer treatment [72]. Their capacity to recognize PAMPs and DAMPs makes them attractive targets for modulating immune responses as illustrated in Figure 1.

Poly-ICLC, a well-studied TLR3 agonist, has been highlighted for its immunostimulatory effects, which can be leveraged to induce anticancer immune responses. Poly ICLC functions by stimulating TLR3 and cytosolic MDA5, leading to the activation of various pattern recognition receptors. This activation results in the enhanced expression of inflammatory genes and infiltration of T cells into tumors, illustrating its multifaceted role in cancer immunotherapy. The uniqueness of poly-ICLC lies in its ability to engage different receptors, thereby eliciting a broad immunomodulatory response that can overcome barriers to effective cancer treatment [72].

Further exploration of TLR3 agonists, such as RGC100, ARNAX, and poly-IC, revealed distinct mechanisms and shared characteristics. These agonists mimic double-stranded RNA and trigger pathways that contribute to cancer regression. ARNAX activates TLR3 but not the cytoplasmic receptor MDA5/RIG-I, inducing the Toll-like receptor 3–Toll-interleukin-1 receptor domain-containing adaptor molecule 1–interferon regulatory factor 3–interferon-α/β receptor (TLR3-TICAM-1-IRF3-IFNAR) signaling axis, which is crucial in dendritic cells for cytotoxic T lymphocyte induction. Such specificity and ability to enhance anti-tumor responses, especially in conjunction with PD-1/PD-L1 blockade, underscore the therapeutic promise of TLR3 agonists in cancer treatment [73].

Additionally, the development of chemically defined TLR3 agonists, such as TL-532, further demonstrates ongoing innovation in this field. TL-532, which is composed of poly(I:C) and poly(A:U) blocks, has shown potential because of its favorable pharmacokinetic, pharmacodynamic, and toxicological properties. In preclinical models, TL-532 reduces tumor growth and restores the efficacy of immunogenic chemotherapy, making it a promising candidate for anticancer immunotherapy [74].

Moreover, several TLR9 ligands have been effectively utilized as vaccine adjuvants and in combination with other anti-tumor drugs for cancer therapy. The dual-adjuvant effect of liposomes loaded with TLR9 and other immune stimulants has shown significant potential in regressing tumor development and enhancing Th1 immune responses in clinical settings [75]. The strategic use of these TLR9 ligands underscores their adaptability and efficacy in a broad range of therapeutic applications, providing a significant enhancement to both innate and adaptive immunity [10].

Furthermore, recent advancements in the development of dual TLR7/9 antagonists have opened new avenues for therapeutic applications in autoimmune diseases due to their selective inhibition properties [76]. Additionally, sugar-conjugated TLR7 ligands have shown increased immunostimulatory activity and have potential applicability in vaccine adjuvant and cancer therapy contexts [77]. As summarized in Table 3, TLRs play crucial roles in infection and inflammation. These examples highlight the evolving landscape of TLR-targeted therapies and their broadening applicability across different medical fields.

Table 3. TLRs in infection and inflammation.

Role	Mechanisms	Implications
Innate Immunity	Recognize PAMPs and DAMPs, initiate signaling cascades	First line of defense against pathogens
Adaptive Immunity	Activate dendritic cells, modulate T cell responses	Bridge innate and adaptive immunity
Inflammatory Diseases	Aberrant TLR signaling, persistent activation	Autoimmunity, chronic inflammation
Cancer	Dual role in tumor progression and anti-tumor immunity	Potential therapeutic targets

Overall, advancements in TLR-targeting strategies, including the development and clinical trials of specific ligands and biologics, illustrate the dynamic and promising fields of immunotherapy. Through modulation of TLR activity, these approaches offer a versatile toolkit for enhancing immune responses against cancers, further enriching the landscape of therapeutic options available for cancer treatment. For a detailed list of these clinical trials, please refer to Table 4.

Table 4. Future directions in TLR research.

TLR	Molecule	Type	Clinical Trial Status	Indications	Reference
TLR3	Poly-ICLC	Agonist	In Use	Cancer Immunotherapy	[72]
TLR3	RGC100	Agonist	Research Phase	Cancer Treatment	[73]
TLR3	ARNAX	Agonist	Research Phase	Cancer Treatment	[73]
TLR3	poly-IC	Agonist	Research Phase	Cancer Treatment	[73]
TLR3	TL-532	Agonist	Preclinical	Cancer Immunotherapy	[74]
TLR9	Various Ligands	Agonist	Clinical Settings	Cancer Therapy, Vaccine Adjuvant	[75]
TLR7	GSK2245035	Agonist	Clinical Trial	Respiratory Allergies	[78]
TLR7/8	MEDI9197	Agonist	Clinical Trial	Solid Tumors	[79]
TLR9	CYT003-QbG10	Agonist	Clinical Trial	Allergic Bronchial Asthma	[80]
TLR7	Imidazo derivatives	Antagonist	Phase I	Autoimmune, Infectious Diseases	[81]
TLR7	Cholesterolized Liposomes	Agonist	Preclinical	Cancer Treatment	[82]

1.8.2. Current Clinical Trials and Future Prospects

Recent clinical trials have shown the potential of TLR agonists and antagonists in treating a broad spectrum of diseases by modulating the immune system. These developments have led to significant progress in leveraging the immunomodulatory capabilities of TLRs for therapeutic interventions.

One study investigated the safety and pharmacodynamics of the TLR7 agonist GSK2245035 in patients with respiratory allergies. This study aimed to explore the potential of TLR7 agonists in altering immune responses in the upper airways, which may contribute to the reduction in allergic reactivity [78]. Similarly, another study evaluated MEDI9197, a TLR7/8 agonist, when administered alone or in combination with durvalumab and/or palliative radiation in subjects with solid tumors, highlighting the exploration of TLR agonists in cancer therapy [73,79]. Moreover, the study on CYT003-QbG10 for treating allergic bronchial asthma is another example of TLR-targeted therapy aiming to improve asthma symptoms by potentially modulating the immune system [80]. Additionally, TLR agonists are also being utilized as vaccine adjuvants targeting cancer and infectious diseases, demonstrating their significant potential in enhancing the immune response to purified antigens, thus broadening protection against highly variable pathogens [10].

Moreover, recent trials have explored the use of various TLR antagonists and agonists. Among the promising TLR7 antagonists in development are compounds from the imidazo[1,2-a]pyrazine, imidazo[1,5-a]quinoxaline, and pyrazolo[1,5-a]quinoxaline series, which have been shown to be potent and selective without any TLR7/8 agonistic activity, offering potential therapeutic applications in autoimmune and infectious diseases [81].

Additional ongoing studies include the examination of cholesterolized liposome formulations of TLR7 agonists, which have shown potential in reducing tumor growth through effective delivery and activation of immune responses within the lymph nodes [82]. These clinical trials represent only a fraction of the ongoing research efforts to harness TLRs' immunomodulatory potential across various diseases, including allergies, asthma, cancer, and chronic infections. The therapeutic applicability of TLR agonists and antagonists across a wide range of conditions underscores their promise of TLRs as targets for future medical interventions, potentially offering new treatments for chronic inflammation, autoimmunity, and other conditions. The results of these trials could pave the way for novel therapeutic strategies that leverage the intricate mechanisms of the immune system to fight diseases.

These clinical trials represent only a fraction of the ongoing research efforts to harness TLRs' immunomodulatory potential across various diseases, including allergies, asthma, cancer, and chronic infections. The therapeutic applicability of TLR agonists and antagonists across a wide range of conditions underscores their promise of TLRs as targets for future medical interventions, potentially offering new treatments for chronic inflammation, autoimmunity, and other conditions. The results of these trials could pave the way for novel therapeutic strategies that leverage the intricate mechanisms of the immune system to fight diseases.

1.9. Challenges and Controversies

1.9.1. Limitations in Current Research

TLRs are critical components of the innate immune system that recognize a broad spectrum of molecular patterns to initiate an immune response. Each of the ten functional TLRs (TLR1–10) in humans plays a pivotal role in detecting various dangers and pathogen-associated molecular patterns, including DAMPs, MAMPs, PAMPs, and XAMPs. These receptors share a common domain organization that includes an N-terminal domain with LRRs, a single TM, and a C-terminal cytoplasmic TIR domain [1,64].

Recent studies have elucidated the significance and multi-functionality of TLRs in innate immunity. TLR9 and cyclic GMP-AMP synthase-stimulator of interferon genes (cGAS-STING) have been recognized for their roles in DNA sensing and the initiation of innate immune responses, highlighting the complexity and adaptability of TLR-mediated signaling in response to various stimuli [83]. Furthermore, the importance of TLRs in viral infections has been underscored by the finding that TLR3, TLR7, and TLR8, along with cytoplasmic RIG-I-like receptors, facilitate the initiation of innate immune responses by recognizing viral nucleic acids [84].

Moreover, research on TLR signaling has revealed its impact on the progression of diseases, such as herpes simplex virus 1 (HSV-1) infection, where TLRs and their signaling pathways in the trigeminal ganglia are critical for controlling the progression of the virus to encephalitis [85]. These findings demonstrate the essential role of TLRs in antiviral defense and their potential as therapeutic targets.

Activation of TLR pathways, along with other innate immune pathways, has been exploited for therapeutic purposes, such as in the development of adjuvant-free subunit vaccines. One study demonstrated the use of a polymer that activates both TLR and cGAS pathways, inducing a strong humoral immune response that is pivotal for effective vaccination strategies [86].

In summary, TLRs are indispensable for recognizing a wide array of molecular patterns that initiate critical defense responses in the innate immune system. Ongoing research and discoveries in the field of TLR signaling pathways continue to reveal the intricate mechanisms through which the innate immune system senses and responds to microbial and non-microbial threats, thereby reinforcing the potential for TLR-targeted therapeutic interventions.

1.9.2. Debated Roles of TLRs in Various Diseases

The complex roles of Toll-like receptors (TLRs) in various diseases have been a subject of ongoing debate, as evidenced by their contradictory effects on disease progression and therapeutic outcomes. In autoimmune diseases, particularly in systemic lupus erythematosus (SLE), TLRs may worsen the condition by promoting inflammatory responses. Conversely, in cases of infection-induced autoimmunity, TLRs might serve a protective role by helping to eliminate pathogens. This ambiguity is highlighted in neuroimmune diseases, where TLRs can both exacerbate and mitigate disease symptoms depending on the context [87]. In the context of cancer, TLRs exhibit a dual functionality; their activation might suppress tumor growth by enhancing immune surveillance, yet it could also facilitate tumor progression by creating an immune-tolerant environment [88,89]. The role of TLRs in neurodegenerative diseases remains particularly complex and is still under investigation. Some studies suggest that TLR activation could be neuroprotective by aiding in the clearance of pathogenic proteins, while others argue that it may lead to neuroinflammation and neuronal damage [90]. These examples underscore the need for a more refined understanding of TLRs' roles, which is crucial for advancing both research and clinical interventions in these areas.

1.10. Future Directions

1.10.1. Emerging Research Areas

Future research is focused on their therapeutic potential, particularly in immunotherapies. Studies have focused on TLR agonists and inhibitors as dual-purpose agents in cancer therapy, infectious diseases, and autoimmune conditions, aiming to refine immune responses in various diseases with a keen interest in personalized medicine [91]. Research is progressing, with clinical trials and computational models to enhance the efficacy and specificity of TLR-based therapies.

TLR agonists have shown promise for enhancing the immune response to purified antigens in vaccines against life-threatening and complex diseases such as cancer, AIDS, and malaria [10]. In head and neck squamous cell carcinoma (HNSCC), a combination of intratumoral injections of TLR7 and TLR9 agonists with PD-1 blockade suppresses tumor growth and promotes systemic adaptive immunity. This approach activates tumor-associated macrophages, induces tumor-specific immune responses, and inhibits metastasis in HNSCC models [91].

In oncological treatments, TLR agonists elevate PD-L1 levels within the tumor microenvironment, potentially augmenting the efficacy of checkpoint inhibitor therapies [92]. Employing TLR agonists in tandem with checkpoint inhibitors can effectively mitigate head and neck cancer, demonstrating the efficacy of localized TLR agonist application alongside anti-PD-1 therapy in fostering systemic adaptive immune responses [91].

For infectious diseases, TLR agonists have been formulated as licensed vaccines for their adjuvant activity, and other TLR agonists have been developed for this purpose [93]. Monophosphoryl lipid A and CpG-1018 have been used as adjuvants in vaccines against viral infections, including hepatitis B virus, hepatitis C virus, human immunodeficiency virus, SARS-CoV-2, influenza virus, and flaviviruses [94].

In autoimmune conditions, TLR agonists have been linked to the perpetuation of inflammation in chronic inflammatory diseases due to the permanent exposure of the immune system to TLR-specific stimuli [95]. However, recent studies have demonstrated that MSCs are activated by TLR ligands, leading to the modulation of differentiation, migration, proliferation, survival, and immunosuppression. This activation has not been reported to modulate the "immunoprivileged" phenotype of MSCs, which is of particular relevance for the use of allogeneic MSC-based therapies [95].

In summary, future TLR research should focus on their therapeutic potential in immunotherapies, particularly in cancer therapy, infectious diseases, and autoimmune conditions. Studies are progressing with clinical trials and computational models to enhance the efficacy and specificity of TLR-based therapies, with a keen interest in personalized medicine.

1.10.2. Potential Breakthroughs

TLRs are increasingly being recognized as critical players in the development of innovative therapeutic strategies, particularly in oncology and autoimmune disease management. As key components of the immune system, TLRs are involved in the recognition of pathogen-associated molecular patterns (PAMPs) and damage-associated molecular patterns (DAMPs), which are crucial in initiating immune responses [96,97]. Their ability to modulate both innate and adaptive immunity makes them valuable targets for new cancer therapies aimed at harnessing the body's defense mechanisms to combat malignancies [98].

In the context of cancer, TLRs can influence tumor progression and the ability of the immune system to recognize and destroy cancer cells. The dual nature of TLRs in cancer, in which they can either promote or inhibit tumor growth, underscores the complexity of their roles in oncogenesis and the need for precise therapeutic targeting [96,97]. For example, TLR agonists have shown promise in enhancing the efficacy of cancer vaccines in combination with checkpoint inhibitors to stimulate anti-tumor immunity [96,98].

Dysregulation of TLR signaling can lead to persistent inflammation and tissue damage in autoimmune diseases. TLRs are being explored as potential therapeutic targets to dampen excessive immune responses without compromising the body's ability to fight infections [23]. Understanding the genetic polymorphisms within TLRs that contribute to autoimmune susceptibility emphasizes the importance of personalized medical approaches for treating these conditions [23].

Computational modeling has revolutionized TLR research by providing sophisticated tools for predicting the outcomes of TLR signaling pathways and the impact of potential therapeutics [99]. These models can simulate the dynamic interactions within the immune system, allowing researchers to explore the effects of TLR modulation in silico before proceeding with in vivo studies. This approach can streamline drug development, reduce the risk of adverse effects, and accelerate the translation of research findings into clinical applications [99].

In summary, the integration of TLR biology with computational innovations is expanding therapeutic possibilities for a range of medical conditions, from infectious diseases to cancer and autoimmune disorders. Researchers are paving the way for more targeted and effective treatments that can transform patient care by leveraging the ability of TLRs to regulate immune responses.

2. Conclusions

We have summarized the critical roles that TLRs play within the innate immune system. These receptors are instrumental in detecting microbial patterns and serve as the first line of defense against pathogens. Our discussion extends the molecular intricacies of TLRs, including their structure and signaling pathways, which are fundamental to their function in immune surveillance and disease pathogenesis.

The implications of TLRs in a spectrum of diseases are the focus of our review. In autoimmune disorders, aberrant TLR signaling can lead to inappropriate immune responses against tissues, whereas in chronic inflammatory diseases, persistent TLR activation contributes to ongoing tissue damage. In oncology, TLRs can have a dual role, either supporting tumor progression through inflammatory pathways or enhancing anti-tumor immunity, depending on the context and specific TLRs involved.

The therapeutic potential of TLRs is vast, particularly highlighted by the development of controlled TLR agonists and antagonists that modulate immune responses. These agents are being actively explored for their capacity to either boost the immune response in cases such as cancer and infectious diseases or dampen it in autoimmune and inflammatory conditions. Specifically, TLR agonists such as those targeting TLR7/8 have shown promise in enhancing the immunogenicity of vaccines against severe diseases like cancer, AIDS, and malaria by linking innate with adaptive immune responses [10]. Additionally, novel therapeutic applications are being developed for TLR10 due to its unique anti-inflammatory properties, suggesting its potential use in therapeutics targeting inflammatory and autoimmune diseases [100].

As we look to the future, the ongoing research into TLRs is set to revolutionize the field of medicine. The integration of TLR-based therapies with other modalities, such as checkpoint inhibitors in cancer or biologics in autoimmune diseases, is likely to enhance treatment efficacy and patient outcomes. Moreover, continuous advancements in computational biology and genomics will enable a deeper understanding of TLR signaling and its implications in health and disease, paving the way for more personalized and targeted therapeutic interventions.

In conclusion, TLRs represent a promising frontier in medical research with the potential to significantly impact the development of novel therapeutic strategies. Their roles in the immune system and disease pathogenesis offer a rich landscape for scientific exploration, with the promise of yielding new insights that could transform the practice of medicine.

Author Contributions: Conceptualization, Y.-H.C. and H.-P.W.; writing—original draft, Y.-H.C.; writing—review and editing, H.-P.W. and K.-H.W.; supervision, H.-P.W. and K.-H.W. All authors have read and agreed to the published version of the manuscript.

Funding: This research received no external funding.

Institutional Review Board Statement: Not applicable.

Informed Consent Statement: Not applicable.

Data Availability Statement: Not applicable.

Acknowledgments: The research was supported in part by the Chang Gung Memorial hospital at Chiayi, Taiwan (CMRPG6M0271-3).

Conflicts of Interest: The authors declare no conflicts of interest.

References

1. Bzówka, M.; Bagrowska, W.; Góra, A. Recent advances in studying toll-like receptors with the use of computational methods. *J. Chem. Inf. Model.* **2023**, *63*, 3669–3687. [CrossRef] [PubMed]
2. Lee, J.-H.; Chiang, C.; Gack, M.U. Endogenous nucleic acid recognition by RIG-I-like receptors and cGAS. *J. Interferon Cytokine Res.* **2019**, *39*, 450–458. [CrossRef] [PubMed]
3. McKernan, D.P. Pattern recognition receptors as potential drug targets in inflammatory disorders. *Adv. Protein Chem. Struct. Biol.* **2020**, *119*, 65–109. [PubMed]
4. Ge, Z.; Ding, S. Regulation of cGAS/STING signaling and corresponding immune escape strategies of viruses. *Front. Cell Infect. Microbiol.* **2022**, *12*, 954581. [CrossRef] [PubMed]
5. Valentin, R.; Wong, C.; Alharbi, A.S.; Pradeloux, S.; Morros, M.P.; Lennox, K.A.; Ellyard, J.I.; Garcin, A.J.; Ullah, T.R.; Kusuma, G.D.; et al. Sequence-Dependent Inhibition of CGAS and TLR9 DNA Sensing by 2′-O-Methyl Gapmer Oligonucleotides. *Nucleic Acids Res.* **2021**, *49*, 6082–6099. [CrossRef] [PubMed]
6. Sengupta, P.; Chattopadhyay, S. Interferons in viral infections. *Viruses* **2024**, *16*, 451. [CrossRef]
7. Campos, M.A.; de Pádua Zolini, G.; Kroon, E.G. Impact of toll-like receptors (TLRs) and TLR signaling proteins in trigeminal ganglia impairing herpes simplex virus 1 (HSV-1) progression to encephalitis: Insights from mouse models. *Front. Biosci.-Landmark* **2024**, *29*, 102. [CrossRef] [PubMed]
8. Hennessy, E.J.; Parker, A.E.; O'Neill, L.A.J. Targeting Toll-like receptors: Emerging therapeutics? *Nat. Rev. Drug Discov.* **2010**, *9*, 293–307. [CrossRef]
9. Savva, A.; Roger, T. Targeting toll-like receptors: Promising therapeutic strategies for the management of sepsis-associated pathology and infectious diseases. *Front. Immunol.* **2013**, *4*, 387. [CrossRef]
10. Luchner, M.; Reinke, S.; Milicic, A. TLR agonists as vaccine adjuvants targeting cancer and infectious diseases. *Pharmaceutics* **2021**, *13*, 142. [CrossRef]

11. Gupta, N.; Regar, H.; Verma, V.K.; Prusty, D.; Mishra, A.; Prajapati, V.K. Receptor-ligand based molecular interaction to discover adjuvant for immune cell TLRs to develop next-generation vaccine. *Int. J. Biol. Macromol.* **2020**, *152*, 535–545. [CrossRef] [PubMed]
12. Gential, G.P.P.; Hogervorst, T.P.; Tondini, E.; van de Graaff, M.J.; Overkleeft, H.S.; Codée, J.D.C.; van der Marel, G.A.; Ossendorp, F.; Filippov, D.V. Peptides conjugated to 2-alkoxy-8-oxo-adenine as potential synthetic vaccines triggering TLR7. *Bioorg Med. Chem. Lett.* **2019**, *29*, 1340–1344. [CrossRef] [PubMed]
13. Patra, M.C.; Achek, A.; Kim, G.-Y.; Panneerselvam, S.; Shin, H.-J.; Baek, W.-Y.; Lee, W.H.; Sung, J.; Jeong, U.; Cho, E.Y.; et al. A novel small-molecule inhibitor of endosomal TLRs reduces inflammation and alleviates autoimmune disease symptoms in Murine models. *Cells* **2020**, *9*, 1648. [CrossRef] [PubMed]
14. Alper, P.; Betschart, C.; André, C.; Boulay, T.; Cheng, D.; Deane, J.; Faller, M.; Feifel, R.; Glatthar, R.; Han, D.; et al. Discovery of the TLR7/8 antagonist MHV370 for treatment of systemic autoimmune diseases. *ACS Med. Chem. Lett.* **2023**, *14*, 1054–1062. [CrossRef] [PubMed]
15. Federico, S.; Pozzetti, L.; Papa, A.; Carullo, G.; Gemma, S.; Butini, S.; Campiani, G.; Relitti, N. Modulation of the Innate Immune Response by Targeting Toll-like Receptors: A Perspective on Their Agonists and Antagonists. *J. Med. Chem.* **2020**, *63*, 13466–13513. [CrossRef] [PubMed]
16. Cui Sun, M.; Otálora-Alcaraz, A.; Prenderville, J.A.; Downer, E.J. Toll-like receptor signalling as a cannabinoid target. *Biochem. Pharmacol.* **2024**, *222*, 116082. [CrossRef] [PubMed]
17. Dias de Oliveira, N.F.; Santi, C.G.; Maruta, C.W.; Miyamoto, D.; Pereira, N.V.; Sotto, M.N.; da Silva, L.F.F.; Aoki, V. Increased expression of in situ CD123 and reduced Toll-like receptor 7/9 signalling pathway suggest impaired activation of plasmacytoid dendritic cells in recurrent erythema multiforme. *J. Eur. Acad. Dermatol. Venereol.* **2024**. [CrossRef]
18. Dash, U.K.; Mazumdar, D.; Singh, S. High mobility group box protein (HMGB1): A potential therapeutic target for diabetic encephalopathy. *Mol. Neurobiol.* **2024**. [CrossRef] [PubMed]
19. Nakamura, H.; Tanaka, T.; Zheng, C.; Afione, S.A.; Atsumi, T.; Noguchi, M.; Oliveira, F.R.; Motta, A.C.F.; Chahud, F.; Rocha, E.M.; et al. LAMP3 induces ectopic TLR7 expression in salivary gland epithelial cells and amplifies type I IFN production in Sjögren's disease. *Arthritis Rheumatol.* **2024**. [CrossRef]
20. Zhang, L.; Kuang, G.; Gong, X.; Huang, R.; Zhao, Z.; Li, Y.; Wan, J.; Wang, B. Piperine attenuates hepatic ischemia/reperfusion injury via suppressing the TLR4 signaling cascade in mice. *Transpl. Immunol.* **2024**, *84*, 102033. [CrossRef]
21. El-Zayat, S.R.; Sibaii, H.; Mannaa, F.A. Toll-like receptors activation, signaling, and targeting: An overview. *Bull. Natl. Res. Cent.* **2019**, *43*, 187. [CrossRef]
22. Wang, M.; Song, H.; Zhu, X.; Xing, S.; Zhang, M.; Zhang, H.; Wang, X.; Yang, Z.; Ding, X.; Karrow, N.A.; et al. Toll-like receptor 4 gene polymorphisms influence milk production traits in Chinese Holstein cows. *J. Dairy Res.* **2018**, *85*, 407–411. [CrossRef] [PubMed]
23. Zhang, Y.; Liu, J.; Wang, C.; Liu, J.; Lu, W. Toll-like receptors gene polymorphisms in autoimmune disease. *Front. Immunol.* **2021**, *12*, 672346. [CrossRef] [PubMed]
24. Areal, H.; Abrantes, J.; Esteves, P.J. Signatures of positive selection in Toll-like receptor (TLR) genes in mammals. *BMC Evol. Biol.* **2011**, *11*, 368. [CrossRef]
25. Sartorius, R.; Trovato, M.; Manco, R.; D'Apice, L.; De Berardinis, P. Exploiting viral sensing mediated by Toll-like receptors to design innovative vaccines. *NPJ Vaccines* **2021**, *6*, 127. [CrossRef] [PubMed]
26. Schjoldager, K.T.; Narimatsu, Y.; Joshi, H.J.; Clausen, H. Global view of human protein glycosylation pathways and functions. *Nat. Rev. Mol. Cell Biol.* **2020**, *21*, 729–749. [CrossRef]
27. Fore, F.; Budipranama, M.; Destiawan, R.A. TLR10 and its role in immunity. *Handb. Exp. Pharmacol.* **2022**, *276*, 161–174. [PubMed]
28. Shi, S.; Xu, C.; Fang, X.; Zhang, Y.; Li, H.; Wen, W.; Yang, G. Expression profile of Toll-like receptors in human breast cancer. *Mol. Med. Rep.* **2020**, *21*, 786–794. [CrossRef] [PubMed]
29. Barton, G.M.; Kagan, J.C. A cell biological view of Toll-like receptor function: Regulation through compartmentalization. *Nat. Rev. Immunol.* **2009**, *9*, 535–542. [CrossRef]
30. Fonte, E.; Agathangelidis, A.; Reverberi, D.; Ntoufa, S.; Scarfò, L.; Ranghetti, P.; Cutrona, G.; Tedeschi, A.; Xochelli, A.; Caligaris-Cappio, F.; et al. Toll-like receptor stimulation in splenic marginal zone lymphoma can modulate cell signaling, activation and proliferation. *Haematologica* **2015**, *100*, 1460–1468. [CrossRef]
31. Månsson Kvarnhammar, A.; Tengroth, L.; Adner, M.; Cardell, L.-O. Innate immune receptors in human airway smooth muscle cells: Activation by TLR1/2, TLR3, TLR4, TLR7 and NOD1 agonists. *PLoS ONE* **2013**, *8*, e68701. [CrossRef] [PubMed]
32. Kawamoto, T.; Ii, M.; Kitazaki, T.; Iizawa, Y.; Kimura, H. TAK-242 selectively suppresses Toll-like receptor 4-signaling mediated by the intracellular domain. *Eur. J. Pharmacol.* **2008**, *584*, 40–48. [CrossRef] [PubMed]
33. Lee, S.M.-Y.; Yip, T.-F.; Yan, S.; Jin, D.-Y.; Wei, H.-L.; Guo, R.-T.; Peiris, J.S.M. Recognition of double-stranded RNA and regulation of interferon pathway by toll-like receptor 10. *Front. Immunol.* **2018**, *9*, 516. [CrossRef] [PubMed]
34. Wu, Y.; Sun, Y.; Dong, X.; Chen, J.; Wang, Z.; Chen, J.; Dong, G. The synergism of PGN, LTA and LPS in inducing transcriptome changes, inflammatory responses and a decrease in lactation as well as the associated epigenetic mechanisms in bovine mammary epithelial cells. *Toxins* **2020**, *12*, 387. [CrossRef] [PubMed]
35. Stojanovic, M.D.; Stojanovic, B.; Arsenijevic, N.; Stojanovic, B. The role of TLR-4 and Galectin-3 interaction in acute pancreatitis. *Exp. Appl. Biomed. Res. (EABR)* **2020**. [CrossRef]

36. Di Lorenzo, A.; Bolli, E.; Tarone, L.; Cavallo, F.; Conti, L. Toll-like receptor 2 at the crossroad between cancer cells, the immune system, and the Microbiota. *Int. J. Mol. Sci.* **2020**, *21*, 9418. [CrossRef] [PubMed]
37. Kono, H.; Rock, K.L. How dying cells alert the immune system to danger. *Nat. Rev. Immunol.* **2008**, *8*, 279–289. [CrossRef]
38. Wang, K.L.; Chen, S.N.; Li, L.; Huo, H.J.; Nie, P. Functional characterization of four TIR domain-containing adaptors, MyD88, TRIF, MAL, and SARM in mandarin fish Siniperca chuatsi. *Dev. Comp. Immunol.* **2021**, *122*, 104110. [CrossRef]
39. Richard, K.; Perkins, D.J.; Harberts, E.M.; Song, Y.; Gopalakrishnan, A.; Shirey, K.A.; Lai, W.; Vlk, A.; Mahurkar, A.; Nallar, S.; et al. Dissociation of TRIF bias and adjuvanticity. *Vaccine* **2020**, *38*, 4298–4308. [CrossRef]
40. O'Neill, L.A.J.; Bowie, A.G. The family of five: TIR-domain-containing adaptors in Toll-like receptor signalling. *Nat. Rev. Immunol.* **2007**, *7*, 353–364. [CrossRef]
41. Sheen, J.-H.; Strainic, M.G.; Liu, J.; Zhang, W.; Yi, Z.; Medof, M.E.; Heeger, P.S. TLR-induced Murine dendritic cell (DC) activation requires DC-intrinsic complement. *J. Immunol.* **2017**, *199*, 278–291. [CrossRef] [PubMed]
42. Kawai, T.; Akira, S. TLR signaling. *Cell Death Differ.* **2006**, *13*, 816–825. [CrossRef] [PubMed]
43. Xia, P.; Wu, Y.; Lian, S.; Yan, L.; Meng, X.; Duan, Q.; Zhu, G. Research progress on Toll-like receptor signal transduction and its roles in antimicrobial immune responses. *Appl. Microbiol. Biotechnol.* **2021**, *105*, 5341–5355. [CrossRef]
44. Daly, K.; Burdyga, G.; Al-Rammahi, M.; Moran, A.W.; Eastwood, C.; Shirazi-Beechey, S.P. Toll-like receptor 9 expressed in proximal intestinal enteroendocrine cells detects bacteria resulting in secretion of cholecystokinin. *Biochem. Biophys. Res. Commun.* **2020**, *525*, 936–940. [CrossRef] [PubMed]
45. Duan, T.; Du, Y.; Xing, C.; Wang, H.Y.; Wang, R.-F. Toll-like receptor signaling and its role in cell-mediated immunity. *Front. Immunol.* **2022**, *13*, 812774. [CrossRef]
46. 46 Takeuchi, O.; Hoshino, K.; Kawai, T.; Sanjo, H.; Takada, H.; Ogawa, T.; Takeda, K.; Akira, S. Differential roles of TLR2 and TLR4 in recognition of gram-negative and gram-positive bacterial cell wall components. *Immunity* **1999**, *11*, 443–451. [CrossRef] [PubMed]
47. Kirschning, C.J.; Schumann, R.R. TLR2, Cellular sensor for microbial and endogenous molecular patterns. In *Current Topics in Microbiology and Immunology*; Springer: Berlin/Heidelberg, Germany, 2002; pp. 121–144.
48. Nemati, M.; Larussa, T.; Khorramdelazad, H.; Mahmoodi, M.; Jafarzadeh, A. Toll-like receptor 2, An important immunomodulatory molecule during Helicobacter pylori infection. *Life Sci.* **2017**, *178*, 17–29. [CrossRef] [PubMed]
49. Samanta, M.; Swain, B.; Basu, M.; Panda, P.; Mohapatra, G.B.; Sahoo, B.R.; Maiti, N.K. Molecular characterization of toll-like receptor 2 (TLR2), analysis of its inductive expression and associated down-stream signaling molecules following ligands exposure and bacterial infection in the Indian major carp, rohu (*Labeo rohita*). *Fish. Shellfish. Immunol.* **2012**, *32*, 411–425. [CrossRef] [PubMed]
50. Takeda, K.; Akira, S. Microbial recognition by Toll-like receptors. *J. Dermatol. Sci.* **2004**, *34*, 73–82. [CrossRef]
51. Krieg, A.M.; Vollmer, J. Toll-like receptors 7, 8, and 9, linking innate immunity to autoimmunity. *Immunol. Rev.* **2007**, *220*, 251–269. [CrossRef]
52. Szabo, A.; Rajnavolgyi, E. Collaboration of Toll-like and RIG-I-like receptors in human dendritic cells: tRIGgering antiviral innate immune responses. *Am. J. Clin. Exp. Immunol.* **2013**, *2*, 195–207. [PubMed]
53. Chen, Y.; Lin, J.; Zhao, Y.; Ma, X.; Yi, H. Toll-like receptor 3 (TLR3) regulation mechanisms and roles in antiviral innate immune responses. *J. Zhejiang Univ. Sci. B* **2021**, *22*, 609–632. [CrossRef] [PubMed]
54. Jang, A.-R.; Kim, G.; Hong, J.J.; Kang, S.M.; Shin, S.J.; Park, J.-H. Mycobacterium tuberculosis ESAT6 drives the activation and maturation of bone marrow-derived dendritic cells via TLR4-mediated signaling. *Immune Netw.* **2019**, *19*, e13. [CrossRef] [PubMed]
55. Yang, Y.; Wang, D.; Li, Q.; He, J.; Wang, B.; Li, J.; Zhang, A. Immune-enhancing activity of aqueous extracts from Artemisia rupestris, L. via MAPK and NF-kB pathways of TLR4/TLR2 downstream in dendritic cells. *Vaccines* **2020**, *8*, 525. [CrossRef] [PubMed]
56. Pasare, C.; Medzhitov, R. Toll-like receptors: Linking innate and adaptive immunity. In *Mechanisms of Lymphocyte Activation and Immune Regulation, X*; Springer: Boston, MA, USA, 2005; pp. 11–18.
57. Sohn, K.M.; Lee, S.G.; Kim, H.J.; Cheon, S.; Jeong, H.; Lee, J.; Kim, I.S.; Silwal, P.; Kim, Y.J.; Paik, S.; et al. COVID-19 patients upregulate toll-like receptor 4-mediated inflammatory signaling that mimics bacterial sepsis. *J. Korean Med. Sci.* **2020**, *35*, e343. [CrossRef] [PubMed]
58. Achek, A.; Kwon, H.-K.; Patra, M.C.; Shah, M.; Hong, R.; Lee, W.H.; Baek, W.-Y.; Choi, Y.S.; Kim, G.-Y.; Pham, T.L.; et al. A peptide derived from the core β-sheet region of TIRAP decoys TLR4 and reduces inflammatory and autoimmune symptoms in murine models. *EBioMedicine* **2020**, *52*, 102645. [CrossRef] [PubMed]
59. Takeda, K.; Akira, S. TLR signaling pathways. *Semin. Immunol.* **2004**, *16*, 3–9. [CrossRef] [PubMed]
60. Horng, T.; Barton, G.M.; Flavell, R.A.; Medzhitov, R. The adaptor molecule TIRAP provides signalling specificity for Toll-like receptors. *Nature* **2002**, *420*, 329–333. [CrossRef] [PubMed]
61. Verstak, B.; Nagpal, K.; Bottomley, S.P.; Golenbock, D.T.; Hertzog, P.J.; Mansell, A. MyD88 adapter-like (Mal)/TIRAP interaction with TRAF6 is critical for TLR2- and TLR4-mediated NF-κB proinflammatory responses. *J. Biol. Chem.* **2009**, *284*, 24192–24203. [CrossRef]
62. Deguine, J.; Barton, G.M. MyD88, a central player in innate immune signaling. *F1000Prime Rep.* **2014**, *6*, 97. [CrossRef]

63. Lannoy, V.; Côté-Biron, A.; Asselin, C.; Rivard, N. TIRAP, TRAM, and toll-like receptors: The untold story. *Mediat. Inflamm.* **2023**, *2023*, 2899271. [CrossRef] [PubMed]
64. Chen, J.-Q.; Szodoray, P.; Zeher, M. Toll-like receptor pathways in autoimmune diseases. *Clin. Rev. Allergy Immunol.* **2016**, *50*, 1–17. [CrossRef] [PubMed]
65. Vlach, J.; Bender, A.T.; Przetak, M.; Pereira, A.; Deshpande, A.; Johnson, T.L.; Reissig, S.; Tzvetkov, E.; Musil, D.; Morse, N.T.; et al. Discovery of M5049, A novel selective toll-like receptor 7/8 inhibitor for treatment of autoimmunity. *J. Pharmacol. Exp. Ther.* **2021**, *376*, 397–409. [CrossRef] [PubMed]
66. Mussari, C.P.; Dodd, D.S.; Sreekantha, R.K.; Pasunoori, L.; Wan, H.; Posy, S.L.; Critton, D.; Ruepp, S.; Subramanian, M.; Watson, A.; et al. Discovery of potent and orally bioavailable small molecule antagonists of toll-like receptors 7/8/9 (TLR7/8/9). *ACS Med. Chem. Lett.* **2020**, *11*, 1751–1758. [CrossRef] [PubMed]
67. Mohammad Hosseini, A.; Majidi, J.; Baradaran, B.; Yousefi, M. Toll-like receptors in the pathogenesis of autoimmune diseases. *Adv. Pharm. Bull.* **2015**, *5*, 605–614. [CrossRef] [PubMed]
68. Du, K.; Liu, J.; Broering, R.; Zhang, X.; Yang, D.; Dittmer, U.; Lu, M. Recent advances in the discovery and development of TLR ligands as novel therapeutics for chronic HBV and HIV infections. *Expert. Opin. Drug Discov.* **2018**, *13*, 661–670. [CrossRef] [PubMed]
69. Radakovics, K.; Battin, C.; Leitner, J.; Geiselhart, S.; Paster, W.; Stöckl, J.; Hoffmann-Sommergruber, K.; Steinberger, P. A highly sensitive cell-based TLR reporter platform for the specific detection of bacterial TLR ligands. *Front. Immunol.* **2021**, *12*, 817604. [CrossRef] [PubMed]
70. Loiarro, M.; Ruggiero, V.; Sette, C. Targeting TLR/IL-1R signalling in human diseases. *Mediators Inflamm.* **2010**, *2010*, 674363. [CrossRef] [PubMed]
71. Li, M.; Zhou, Y.; Feng, G.; Su, S. The critical role of toll-like receptor signaling pathways in the induction and progression of autoimmune diseases. *Curr. Mol. Med.* **2009**, *9*, 365–374. [CrossRef]
72. Sultan, H.; Salazar, A.M.; Celis, E. Poly-ICLC, a multi-functional immune modulator for treating cancer. *Semin. Immunol.* **2020**, *49*, 101414. [CrossRef]
73. Komal, A.; Noreen, M.; El-Kott, A.F. TLR3 agonists: RGC100, ARNAX, and poly-IC: A comparative review. *Immunol. Res.* **2021**, *69*, 312–322. [CrossRef] [PubMed]
74. Le Naour, J.; Thierry, S.; Scuderi, S.A.; Boucard-Jourdin, M.; Liu, P.; Bonnin, M.; Pan, Y.; Perret, C.; Zhao, L.; Mao, M.; et al. A chemically defined TLR3 agonist with anticancer activity. *Oncoimmunology* **2023**, *12*, 2227510. [CrossRef] [PubMed]
75. Kocabas, B.B.; Almacioglu, K.; Bulut, E.A.; Gucluler, G.; Tincer, G.; Bayik, D.; Gursel, M.; Gursel, I. Dual-adjuvant effect of pH-sensitive liposomes loaded with STING and TLR9 agonists regress tumor development by enhancing Th1 immune response. *J. Control. Release* **2020**, *328*, 587–595. [CrossRef] [PubMed]
76. Das, N.; Bandopadhyay, P.; Roy, S.; Sinha, B.P.; Dastidar, U.G.; Rahaman, O.; Pal, S.; Ganguly, D.; Talukdar, A. Development, optimization, and in vivo validation of new imidazopyridine chemotypes as dual TLR7/TLR9 antagonists through activity-directed sequential incorporation of relevant structural subunits. *J. Med. Chem.* **2022**, *65*, 11607–11632. [CrossRef] [PubMed]
77. Baba, A.; Wakao, M.; Shinchi, H.; Chan, M.; Hayashi, T.; Yao, S.; Cottam, H.B.; Carson, D.A.; Suda, Y. Synthesis and immunostimulatory activity of sugar-conjugated TLR7 ligands. *Bioorg Med. Chem. Lett.* **2020**, *30*, 126840. [CrossRef] [PubMed]
78. Tsitoura, D.; Ambery, C.; Price, M.; Powley, W.; Garthside, S.; Biggadike, K.; Quint, D. Early clinical evaluation of the intranasal TLR7 agonist GSK2245035, Use of translational biomarkers to guide dosing and confirm target engagement. *Clin. Pharmacol. Ther.* **2015**, *98*, 369–380. [CrossRef] [PubMed]
79. Siu, L.; Brody, J.; Gupta, S.; Marabelle, A.; Jimeno, A.; Munster, P.; Grilley-Olson, J.; Rook, A.H.; Hollebecque, A.; Wong, R.K.S.; et al. Safety and clinical activity of intratumoral MEDI9197 alone and in combination with durvalumab and/or palliative radiation therapy in patients with advanced solid tumors. *J. Immunother. Cancer* **2020**, *8*, e001095. [CrossRef] [PubMed]
80. Beeh, K.-M.; Kanniess, F.; Wagner, F.; Schilder, C.; Naudts, I.; Hammann-Haenni, A.; Willers, J.; Stocker, H.; Mueller, P.; Bachmann, M.F.; et al. The novel TLR-9 agonist QbG10 shows clinical efficacy in persistent allergic asthma. *J. Allergy Clin. Immunol.* **2013**, *131*, 866–874. [CrossRef] [PubMed]
81. Bou Karroum, N.; Moarbess, G.; Guichou, J.-F.; Bonnet, P.-A.; Patinote, C.; Bouharoun-Tayoun, H.; Chamat, S.; Cuq, P.; Diab-Assaf, M.; Kassab, I.; et al. Novel and selective TLR7 antagonists among the imidazo [1,2-a]pyrazines, imidazo [1,5-a]quinoxalines, and pyrazolo [1,5-a]quinoxalines series. *J. Med. Chem.* **2019**, *62*, 7015–7031. [CrossRef]
82. Wan, D.; Que, H.; Chen, L.; Lan, T.; Hong, W.; He, C.; Yang, J.; Wei, Y.; Wei, X. Lymph-node-targeted cholesterolized TLR7 agonist liposomes provoke a safe and durable antitumor response. *Nano Lett.* **2021**, *21*, 7960–7969. [CrossRef]
83. Motwani, M.; Pesiridis, S.; Fitzgerald, K.A. DNA sensing by the cGAS–STING pathway in health and disease. *Nat. Rev. Genet.* **2019**, *20*, 657–674. [CrossRef] [PubMed]
84. Vlaming, K.E.; van Wijnbergen, K.; Kaptein, T.M.; Nijhuis, M.; Kootstra, N.J.; de Bree, G.J.; Geijtenbeek, T.B. Crosstalk between TLR8 and RIG-I-like receptors enhances antiviral immune responses. *Front. Med.* **2023**, *10*, 1146457. [CrossRef] [PubMed]
85. Lima, G.K.; Zolini, G.P.; Mansur, D.S.; Freire Lima, B.H.; Wischhoff, U.; Astigarraga, R.G.; Dias, M.F.; Silva, M.d.G.A.; Béla, S.R.; Antonelli, L.R.D.V.; et al. Toll-like receptor (TLR) 2 and TLR9 expressed in trigeminal ganglia are critical to viral control during herpes simplex virus 1 infection. *Am. J. Pathol.* **2010**, *177*, 2433–2445. [CrossRef] [PubMed]
86. Chen, H.; Wang, L.; Zhao, X.; Jiang, H.; Wu, M.; Ding, Y.; Jia, X.; Zhang, Y.; Li, T.; Zhang, Y.; et al. A polymer-based antigen carrier activates two innate immune pathways for adjuvant-free subunit vaccines. *ACS Nano* **2024**, *18*, 9160–9175. [CrossRef] [PubMed]

87. Li, H.; Liu, S.; Han, J.; Li, S.; Gao, X.; Wang, M.; Zhu, J.; Jin, T. Role of toll-like receptors in neuroimmune diseases: Therapeutic targets and problems. *Front. Immunol.* **2021**, *12*, 777606. [CrossRef] [PubMed]
88. Farooq, M.; Batool, M.; Kim, M.S.; Choi, S. Toll-like receptors as a therapeutic target in the era of immunotherapies. *Front. Cell Dev. Biol.* **2021**, *9*, 756315. [CrossRef] [PubMed]
89. Kashani, B.; Zandi, Z.; Pourbagheri-Sigaroodi, A.; Bashash, D.; Ghaffari, S.H. The role of toll-like receptor 4 (TLR4) in cancer progression: A possible therapeutic target? *J. Cell Physiol.* **2021**, *236*, 4121–4137. [CrossRef]
90. Kouli, A.; Horne, C.B.; Williams-Gray, C.H. Toll-like receptors and their therapeutic potential in Parkinson's disease and α-synucleinopathies. *Brain Behav. Immun.* **2019**, *81*, 41–51. [CrossRef] [PubMed]
91. Sato-Kaneko, F.; Yao, S.; Ahmadi, A.; Zhang, S.S.; Hosoya, T.; Kaneda, M.M.; Varner, J.A.; Pu, M.; Messer, K.S.; Guiducci, C.; et al. Combination immunotherapy with TLR agonists and checkpoint inhibitors suppresses head and neck cancer. *JCI Insight* **2017**, *2*, e93397. [CrossRef]
92. Keshavarz, A.; Pourbagheri-Sigaroodi, A.; Zafari, P.; Bagheri, N.; Ghaffari, S.H.; Bashash, D. Toll-like receptors (TLRs) in cancer; with an extensive focus on TLR agonists and antagonists. *IUBMB Life* **2021**, *73*, 10–25. [CrossRef]
93. Yang, J.-X.; Tseng, J.-C.; Yu, G.-Y.; Luo, Y.; Huang, C.-Y.F.; Hong, Y.-R.; Chuang, T.-H. Recent advances in the development of toll-like receptor agonist-based vaccine adjuvants for infectious diseases. *Pharmaceutic* **2022**, *14*, 423. [CrossRef] [PubMed]
94. Kayesh, M.E.H.; Kohara, M.; Tsukiyama-Kohara, K. TLR agonists as vaccine adjuvants in the prevention of viral infections: An overview. *Front. Microbiol.* **2023**, *14*, 1249718. [CrossRef] [PubMed]
95. DelaRosa, O.; Lombardo, E. Modulation of adult mesenchymal stem cells activity by toll-like receptors: Implications on therapeutic potential. *Mediat. Inflamm.* **2010**, *2010*, 865601. [CrossRef] [PubMed]
96. Giurini, E.F.; Madonna, M.B.; Zloza, A.; Gupta, K.H. Microbial-derived Toll-like receptor agonism in cancer treatment and progression. *Cancers* **2022**, *14*, 2923. [CrossRef] [PubMed]
97. Cui, L.; Wang, X.; Zhang, D. TLRs as a promise target along with immune checkpoint against gastric cancer. *Front. Cell Dev. Biol.* **2020**, *8*, 611444. [CrossRef] [PubMed]
98. Javaid, N.; Choi, S. Toll-like receptors from the perspective of cancer treatment. *Cancers* **2020**, *12*, 297. [CrossRef] [PubMed]
99. Manes, N.P.; Armstrong, A.A.; Fraser, I.D.C.; Germain, R.; Meier-Schellersheim, M.; Nita-Lazar, A. Targeted proteomics-driven computational modeling of the mouse macrophage toll-like receptor signaling pathway. *J. Immunol.* **2023**, *210*, 161.09. [CrossRef]
100. Fore, F.; Indriputri, C.; Mamutse, J.; Nugraha, J. TLR10 and its unique anti-inflammatory properties and potential use as a target in therapeutics. *Immune Netw.* **2020**, *20*, e21. [CrossRef]

International Journal of *Molecular Sciences*

MDPI

Review

Toll-like Receptor Response to Human Immunodeficiency Virus Type 1 or Co-Infection with Hepatitis B or C Virus: An Overview

Mohammad Enamul Hoque Kayesh [1], Michinori Kohara [2] and Kyoko Tsukiyama-Kohara [3,*]

[1] Department of Microbiology and Public Health, Faculty of Animal Science and Veterinary Medicine, Patuakhali Science and Technology University, Barishal 8210, Bangladesh; mehkayesh@pstu.ac.bd

[2] Department of Microbiology and Cell Biology, Tokyo Metropolitan Institute of Medical Science, Tokyo 156-8506, Japan; kohara-mc@igakuken.or.jp

[3] Transboundary Animal Diseases Centre, Joint Faculty of Veterinary Medicine, Kagoshima University, Kagoshima 890-0065, Japan

* Correspondence: kkohara@vet.kagoshima-u.ac.jp; Tel.: +81-99-285-3589

Abstract: Toll-like receptors (TLRs) are evolutionarily conserved pattern recognition receptors that play important roles in the early detection of pathogen-associated molecular patterns and shaping innate and adaptive immune responses, which may influence the consequences of infection. Similarly to other viral infections, human immunodeficiency virus type 1 (HIV-1) also modulates the host TLR response; therefore, a proper understanding of the response induced by human HIV-1 or co-infection with hepatitis B virus (HBV) or hepatitis C virus (HCV), due to the common mode of transmission of these viruses, is essential for understanding HIV-1 pathogenesis during mono- or co-infection with HBV or HCV, as well as for HIV-1 cure strategies. In this review, we discuss the host TLR response during HIV-1 infection and the innate immune evasion mechanisms adopted by HIV-1 for infection establishment. We also examine changes in the host TLR response during HIV-1 co-infection with HBV or HCV; however, this type of study is extremely scarce. Moreover, we discuss studies investigating TLR agonists as latency-reverting agents and immune stimulators towards new strategies for curing HIV. This understanding will help develop a new strategy for curing HIV-1 mono-infection or co-infection with HBV or HCV.

Keywords: toll-like receptor; TLR agonist; HIV-1; co-infection; HBV; HCV

Citation: Kayesh, M.E.H.; Kohara, M.; Tsukiyama-Kohara, K. Toll-like Receptor Response to Human Immunodeficiency Virus Type 1 or Co-Infection with Hepatitis B or C Virus: An Overview. *Int. J. Mol. Sci.* **2023**, *24*, 9624. https://doi.org/10.3390/ijms24119624

Academic Editors: Ralf Kircheis and Oliver Planz

Received: 6 April 2023
Revised: 29 May 2023
Accepted: 31 May 2023
Published: 1 June 2023

1. Introduction

Approximately 10% of human immunodeficiency virus (HIV)-infected individuals have chronic hepatitis B virus (HBV) co-infection in highly endemic areas of HBV infection [1–3]. In addition, approximately 2.3 million individuals are co-infected with HIV and the hepatitis C virus (HCV) worldwide [4]. However, specific molecular techniques are important for the accurate detection of particular viral agents with high sensitivity and high specificity [5,6]. During HIV co-infection, HBV infection shows an increased rate of persistence, higher HBV DNA levels, lower rates of hepatitis B e antigen loss, increased cirrhosis and liver-related mortality, and an increased risk of hepatocellular carcinoma (HCC) [2]. In addition, HIV/HBV co-infection accelerates the progression of liver cirrhosis and HCC compared to HIV or HBV mono-infected individuals [3,7]. Similarly, HIV co-infection also has a detrimental effect on the natural history of HCV, resulting in higher rates of HCV persistence following acute infection, higher viral loads, enhanced liver fibrosis, and development of HCC compared to HCV mono-infection [4,8,9]. Therefore, a proper understanding of the response of toll-like receptors (TLR) in HIV mono-infection or HIV/HBV and HIV/HCV co-infection is critical for developing HIV cure strategies in HIV mono-infection or HIV/HBV or HIV/HCV co-infection using TLR agonists as latency-reversing agents.

2. Toll-like Receptors

The innate immune system, a key component of host immunity, acts as the first line of defense against invading pathogens, including viruses [10]. The innate immune system detects pathogen-associated molecular patterns (PAMPs) or microbe-associated molecular patterns (MAMPs) and damage-associated molecular patterns (DAMPs) via germline-encoded pattern recognition receptors (PRRs) [11,12]. Toll-like receptors (TLRs) are key components of innate immunity and are evolutionarily conserved germline-encoded receptors [10]. TLRs can activate the transcription factor nuclear factor kappa B (NF-κB) and interferon regulatory factors (IRFs), which regulate the outcome of the innate immune response [13]. It is now becoming increasingly evident that TLRs are the most important family of PRRs and are among the key players in determining the outcome of microbial infections, including viruses [14–18]. Although TLRs play a vital role in early interactions with invading pathogens by detecting microbial PAMPs [19,20], a dysregulated TLR response may enhance immune-mediated pathology instead of providing protection [12,21,22]. Therefore, a proper understanding of TLR responses to pathogens, including viruses, is crucial.

TLRs are type I transmembrane proteins that contain a conserved N-terminal ectodomain of leucine-rich repeats, a single transmembrane domain, and a cytosolic Toll/interleukin (IL)-1 receptor (TIR) domain [10]. The TIR domain is common and identical among TLRs and the involvement of ILs binds them together as a unified structure [23]. TLRs are encoded by a large gene family containing 10 TLR members (TLR1–TLR10) in humans and 12 TLR members (TLR1–TLR9 and TLR11–TLR13) in mice [13]. They are localized on the cell surface or in intracellular compartments of the cell. TLRs localized on the cell surface include TLR1, TLR2, TLR4, TLR5, TLR6, and TLR10, and TLRs localized in intracellular compartments—including the endoplasmic reticulum, endosomes, lysosomes, or endolysosomes—comprise TLR3, TLR7, TLR8, and TLR9 [24,25]. Different TLRs are involved in recognizing various viral components. For example, TLR3, TLR7/8, and TLR9 are involved in the recognition of double-stranded viral RNA, single-stranded RNA, and DNA, respectively [26–29]. In contrast, TLR1, TLR2, TLR4, TLR5, and TLR6 are involved in the recognition of viral proteins [30].

3. Toll-like Receptors in Men and Women

Due to differences in sex, there may be a stronger or weaker immune response to viral infections [31,32]. TLRs exist in various immune cells and liver parenchymal cells, playing an important role in the host immune response [33]. The sex-related variability of the host immune response based on TLR response in viral infections has been observed [34]. It has been reported that the sensing of viral RNA by TLR7 is sex-biased, where greater expression of TLR7 is observed in female immune cells than in male immune cells [35,36]. TLR-mediated activation of pDCs in HIV-1 infection may induce higher immune activation in women than in men [37]. The TLR9 rs187084 C allele was found to be involved in spontaneous virus clearance in HCV infection in women, suggesting sex-specific effects of the TLR response in natural immunity against HCV infection [38].

4. Human Immunodeficiency Virus, Hepatitis B Virus, and Hepatitis C Virus

HIV, the etiologic agent of acquired immunodeficiency syndrome (AIDS) [39], remains one of the world's most serious public health challenges. According to WHO estimates, 38.4 million [33.9–43.8 million] people live with HIV, and 650,000 people died of HIV-related illnesses in 2021 globally [40]. HIV infections can be categorized into two subgroups, type 1 and type 2. HIV type 1 (HIV-1) is the most prevalent cause of AIDS globally and has a higher virulence than HIV type 2 [41,42]. Therefore, most studies on pathogenesis and host immune responses have focused on HIV-1.

HIV-1 is an enveloped virus with a 9.6 kb positive-sense RNA genome that codes for three polyproteins (Gag, Pol, and Env) and six accessory proteins (Tat, Rev, Nef, Vpr, Vif, and Vpu) [43]. The HIV-1 ssRNA can be recognized by different PRRs, such as TLR7/TLR8 and retinoic acid inducible gene I (RIG-I)/melanoma differentiation-associated protein 5

(MDA5), and may induce a profound antiviral inflammatory response [44–46] (Figure 1). HIV-1 replication intermediates, such as cDNA, ssDNA, DNA/RNA hybrids, and dsDNA, can be recognized by different host innate immune receptors, including DDX41, cGAS, and IFI16, thereby contributing to the antiviral response through the production of IFNs [47]. However, many reports have described the key role of innate immunity in the modulation of susceptibility to HIV infection, and TLR activation represents the first line of defense against invading pathogens [48–51]. However, TLR signaling may play a critical role in the outcome of HIV-1 infection (Figure 1). The rate of progression of HIV infection may vary among individuals [52]. Retained immune function and low levels of virological parameters indicate non-progressive HIV disease, whereas the loss of immune function and high levels of virus indicate rapid progression of the disease [53], suggesting a key role of the immune response in determining the rates of disease progression.

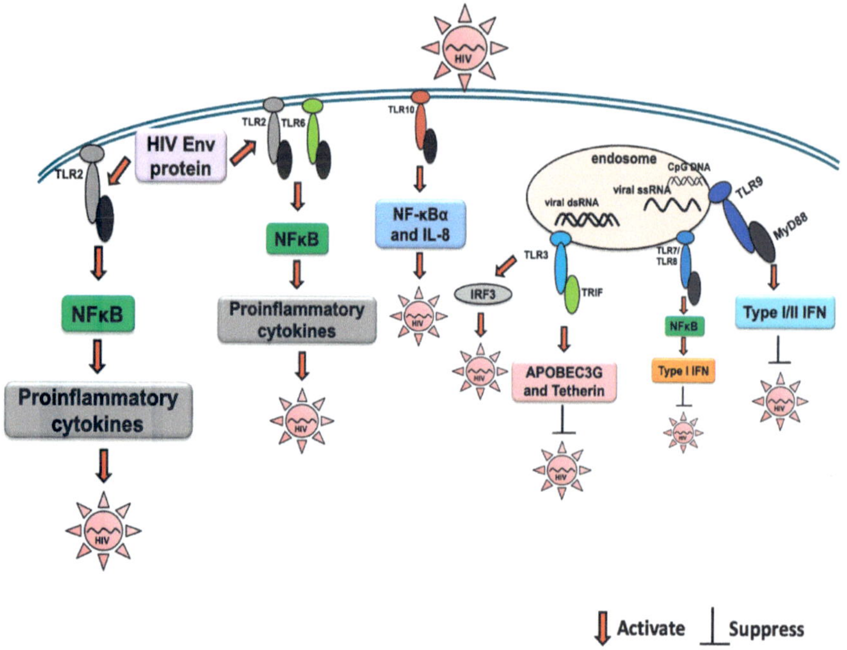

Figure 1. Activation of TLR signaling during HIV infection and the effect on viral replication are indicated. The activation and suppression of signaling is indicated by arrows.

Due to the shared transmission routes of HBV and HCV, HIV/HBV or HIV/HCV co-infections are commonly observed [3,54]. HBV is an enveloped, circular, and partially double-stranded relaxed circular DNA (rcDNA) virus with a genome of approximately 3.2 kb that possesses four overlapping open reading frames that encode seven proteins, including polymerase, core, precore, three envelope/surface proteins (large, middle, and small), and X protein [55,56]. HCV is an enveloped, positive-sense, single-stranded RNA virus with a genome size of ~10 kb that encodes a large polyprotein that is cleaved into three structural (core, E1, and E2) and seven nonstructural proteins (p7, NS2, NS3, NS4A, NS4B, NS5A, and NS5B) by host and viral proteases [57]. Although HBV can behave as a stealth virus, several studies have reported the detection of HBV PAMPs using TLRs [56]. Notably, many reports have suggested a suppressed host TLR response in chronic HBV infection, and activation of the TLR response using TLR agonists as immunomodulators could improve HBV cure strategies [56]. HCV PAMPs are stronger inducers of the innate immune response compared to HBV, and differential expression of TLRs has been reported in HCV-infected patients [17,56,58,59]. Notably, a significant induction of intrahepatic TLR3,

TLR7, and TLR8 mRNA was also observed in the livers of chronically infected tree shrews, squirrel-like non-primate small mammals of the Tupaiidae family that were found to be naturally susceptible to HCV infection [60–62]. Although further studies are warranted, it has been reported that TLR3 and TLR7 play a protective role in HCV infection, and TLR agonists show promise for their use in HCV infection as potential immunomodulators as well as vaccine adjuvants in HCV vaccine development [17]. Activation of TLR signaling by HBV and HCV viral PAMPs are depicted and reviewed in our recently published papers and can be read for further information [17,56].

5. TLR Response to Human Immunodeficiency Virus (HIV) Infection

Early innate viral recognition by host surface TLRs is important for a rapid antiviral response and for preventing infection. Although TLRs can be activated by viruses even before host cells become infected, the antiviral immune response is much more complicated where different cell types, including T cells, B cells, monocytes, NK cells, and neutrophils become involved, producing proinflammatory cytokines and chemokines for viral clearance or possibly promoting infection upon TLR engagement [63–65]. PAMP/MAMP-induced activation of TLR signaling suppresses the replication and spread of invading pathogens/microbes by the production of antimicrobial molecules such as type I interferon (IFN) and tumor necrosis factor (TNF)-α [66–68]. TLRs are important innate immune receptors that mediate immune activation during HIV-1 infection [69–71]. Through their specific ligands, TLRs initially activate the innate immune system, which helps shape the adaptive immune system. Differential regulation of TLR pathways has been observed in HIV-1 infection in monocytes, myeloid dendritic cells, and plasmacytoid dendritic cells, which are dependent on both individual receptors and cell types [69].

In a recent study of HIV infection, TLR2 and TLR4 were found to be differentially expressed [72]. In an earlier study, compared to healthy subjects, increased expression of TLR2 mRNA has been reported in monocytes isolated from HIV-1-infected patients [71]. TLR2 and TLR4 are involved in the regulation of the expression of proinflammatory cytokines in HIV-infected people. In ex vivo analysis, increased expression of TLR2 and TLR4 mRNA was observed in myeloid dendritic cells (mDCs) [71]. It has been shown that the triggering of TLR2 and TLR4 in response to productive HIV-1 infection of DCs and viral spreading to CD4+ T cells exert contrasting effects, where HIV-1 transmission into CD4+ T cells was increased by TLR2 stimulation but transmission to CD4+ T cells was reduced upon TLR4 triggering [73]. Another study reported that soluble TLR2 played a significant role in inhibiting cell-free HIV-1 infection, HIV-induced NF-κB activation, and inflammation in infants breastfeeding from their HIV-infected mother [74]. Increased expression of TLR6, TLR7, and TLR8 mRNA has been observed in chronic untreated HIV-1 infections [70]. TLR2, TLR3, and TLR4 mRNA expression was elevated in patients with advanced disease [70], suggesting sensitization of TLR signaling during HIV infection (Figure 2A). Prior addition of TLR3, TLR7, TLR8, and TLR9 agonists was shown to inhibit HIV-1 infection in peripheral blood mononuclear cells (PBMCs) (Figure 2B); TLR8 and TLR9 agonists were found to be more effective in blocking HIV-1 replication, along with the activation of other antiviral sensors [75].

Henrick et al. showed that the HIV-1 structural proteins p17, p24, and gp41 acted as viral PAMPs that initiated signaling through TLR2 and its heterodimers, resulting in increased immune activation via the NF-κB signaling pathway [76]. In an in vitro study with monocyte-derived macrophages, it has been shown that activation of TLR3 with poly (I:C), a synthetic ligand for TLR3, inhibits HIV replication by the induction of type I interferon-inducible anti-HIV cellular proteins, including APOBEC3G and tetherin [77]. Moreover, TLR3 activation induces CC chemokines and several cellular miRNAs, such as miRNA-28, -125b, -150, -223, and -382, which are known HIV restriction factors [77]. It has been reported that TLR3 agonists, poly (I:C), and bacterial ribosomal RNA could potently reactivate HIV in human microglial cells. Notably, Alvarez-Carbonell et al. reported a new mechanism of poly (I:C), where poly (I:C) did not activate NF-κB; instead virus induction was mediated by IRF3 [78].

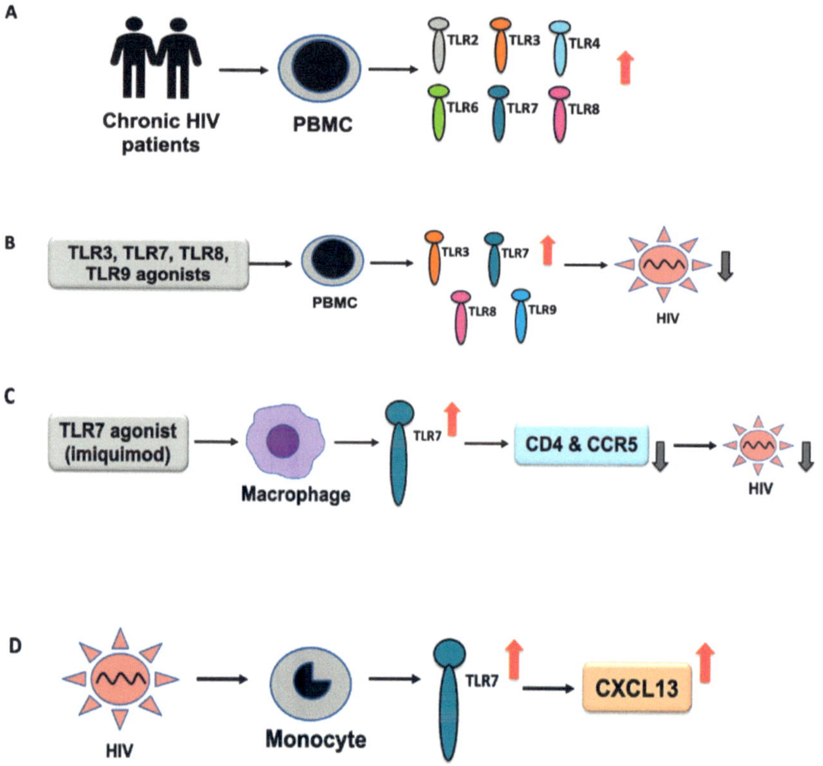

Figure 2. An overview of differential TLR expression in HIV infection. (**A**) Increased expression of TLRs in PBMCs obtained from patients with chronic HIV infection. (**B**) Induction of TLR responses using TLR agonists for TLR3, 7, 8, and 9 inhibited HIV replication in PBMCs. (**C**) Induction of the TLR7 response in macrophages by imiquimod, a TLR7 agonist, inhibited HIV replication by inhibiting CD4 and CCR5. (**D**) HIV-induced production of CXCL13 through TLR7 activation in monocytes.

It has been shown that TLR3 agonist poly (I:C) and CD40-targeting HIV-1 vaccine of a string of five highly conserved CD4+ and CD8+ T cell epitope-rich regions of HIV-1 Gag, Nef, and Pol fused to the C-terminus of a recombinant anti-human CD40 antibody (αCD40.HIV5pep) enhanced anti-HIV-1 immunity and reduced HIV-1 reservoirs in patients with suppressive combined antiretroviral therapy [79]. It has been reported that HIV-1 infection results in dysregulation of the TLR4 response to lipopolysaccharide (LPS) ex vivo [80]. Brenchley et al. demonstrated that microbial translocation was linked to increased plasma levels of LPS, the TLR4 ligand, which has been reported in chronic HIV-1 infection as well as in chronically SIV-infected rhesus macaques [81].

TLR7 recognizes RNA viruses, including HIV [82]. An early study reported the recognition of uridine-rich HIV-1 ssRNA by intracellular TLR7/8 [76]. It has been shown that the TLR7 agonist GS-9620 can activate HIV replication in PBMCs obtained from HIV-infected individuals on suppressive antiretroviral therapy [83]. Additionally, GS-9620 activated HIV-specific T cells and the antibody-mediated clearance of HIV-infected cells [83]. It was observed that imiquimod-based induction of TLR7 by macrophages could inhibit HIV infection by decreasing the expression of viral entry cellular factors, such as CD4 and CCR5 (Figure 2C) [84]. It has been reported that plasma CXCL13 levels increase with

HIV disease progression [85]. HIV-1 induces CXCL13 production upon TLR7 activation (Figure 2D) [86]; however, the exact role of TLR7 in disease progression remains unknown. Although genetic polymorphisms provide a compelling link to the disease [87], the exact mechanism by which TLR polymorphisms affect HIV infection remains to be investigated. However, the association of different TLRs, including TLR1, TLR2, TLR3, TLR4, TLR6, TLR7, TLR8, and TLR9, in HIV infection has been observed [88,89]. TLR7 polymorphisms influence the susceptibility and progression of HIV-1 infection in Chinese men who have sex with men (MSM), suggesting an active role of TLR7 in HIV infection [82].

pDCs play an important role in host defense against viral infection, including HIV through TLR7 and TLR9 signaling [90–92]. HIV-1 encodes multiple uridine-rich TLR7/8 ligands, activates pDCs through TLR7/8, and produces IFN-I [29,44,93,94]. TLR agonists are reported to enhance HIV-specific T cell response, as well as inducing HIV-1 restriction factors and IFN-α production by pDCs [92]. During HIV chronic infection, there is reduced number of circulating pDCs, which is correlated with reduced CD4+ T cell counts in people living with HIV-1 (PLWH) [95,96]. TLR7-driven IFN-I production in pDCs is higher in women than in men, which could be due to the cell-intrinsic actions of estrogen and the X-chromosome complement [97]. Upon TLR7 activation, compared to uninfected controls, an increased expression of IFN-α and TNF-α by pDCs obtained from HIV-infected women under antiretroviral therapy has been reported [98]. In a previous study, the expression of TLR6 and TLR7 was found to be significantly correlated with plasma HIV-RNA load [70].

TLR8 recognizes RNA viruses including HIV [99,100]. The most frequent TLR8 polymorphism, TLR8 A1G (rs3764880), has been reported to confer protection against HIV progression, suggesting the role of TLR8 in infection [101]. Another study assessed TLR7 and eight gene polymorphisms associated with susceptibility to HIV-1 infection and found that TLR8 polymorphisms could affect HIV-1 infection [102]. TLR7 IVS2-151-A and TLR8 Met alleles affect plasma HIV viral loads [102].

Synthetic ligands of TLR8 also boost T cell receptor signaling, resulting in increased cytokine production and upregulation of surface activation markers, suggesting sensing of HIV-1 by TLR8 [100]. In addition, endosomal HIV-induced cytokine secretion from CD4+ T cells in a TLR8-specific manner has also been reported [100].

In a phase I study, MGN1703, a TLR9 agonist administered for 24 weeks, was found to be safe and to enhance innate as well as HIV-1-specific adaptive immunity in HIV-1+ individuals [103]. It has been reported that HIV-1 alone is a poor inducer of IFN-I response [104]; however, if supplemented with suboptimal levels of bacterial LPS, which may be due to microbial translocation, it may trigger increased production of IFN-I and upregulation of interferon-stimulated genes [105]. It has been reported that HIV infection can activate cyclic guanosine monophosphate-adenosine monophosphate (cGAMP) synthase (cGAS) to produce cGAMP, activating STING and induce type I IFNs and other cytokines [106]. LPS, in association with TLR2 or TLR4, may synergistically enhance IFN-I production by cGAMP, a secondary messenger of cGAS [105].

Until 2014, TLR10 remained without a defined ligand or function. Lee et al. first demonstrated a functional role of TLR10 in detecting influenza virus infection [107]. A role of TLR10 in HIV-1 infection has recently been described, whereby HIV-1 gp41 was recognized as a TLR10 ligand, leading to the induction of IL-8 and NF-κBα activation [108]. In vitro studies using TZMbl cells demonstrated that TLR10 overexpression could enhance HIV-1 infection and proviral DNA integration [108]. The interaction between TLRs and HIV-1 infection has been summarized in Table 1.

Table 1. Interaction between TLRs and HIV-1 infection.

HIV-1 Infection System	TLR Ligands	Activation of TLR	Effects on Host Response	Outcome	Reference
Monocyte-derived macrophage (MDM) and PBMCs	HIV proteins	TLR2, TLR4	Increases production of proinflammatory cytokines	Promotes HIV-1 replication	[71]
MDM	Poly (I:C), a TLR3 ligand	TLR3	Induces anti-HIV cellular proteins APOBEC3G and tetherin	Inhibits HIV replication	[77]
PBMCs	TLR3, TLR7, TLR8 and TLR9 agonists	TLR3, TLR7, TLR8, TLR9	Induces type I and type II IFNs and ISGs	Reduces viral replication	[75]
T cells and TZMbl-2 cells (stably expressing TLR2)	HIV-1 structural proteins p17, p24, and gp41	TLR2 or TLR2/1 or TLR2/6 heterodimers	Activates NF-κB signaling pathway	Enhances proviral DNA	[76]
Microglial cells	Poly (I:C)	TLR3	Selective induction of IRF3	Reactivates HIV transcription	[78]
PBMCs	GS-9620, a TLR7 agonist	TLR7	Enhances HIV-specific cellular immunity	Enhances cytokines-mediated HIV replication	[83]
Macrophage	Imiquimod, a TLR7 agonist	TLR7	Reduces viral entry factors such as CD4 and CCR5	Inhibits HIV replication	[84]
DCs	Free HIV-1	TLR8	Activation of IRF1, p38, ERK, PI3K, and NF-κB pathways	Enhances infection	[109]
DCs and macrophages	HIV ssRNA	TLR7, TLR8	Enhances immune activation and secretion of cytokines	May inhibit viral replication	[29,44]
Macrophage	CpG ODN 2216	TLR9	Increases expression of IFN-α, IRF-7, MyD88, and myxovirus resistance gene A	Inhibits HIV replication	[110]
TZMbl cells	HIV-1 gp41	TLR10	Induces IL-8 and NF-κBα activation	Enhances HIV-1 infection	[108]

TLR response in HBV infection may modulate HBV-specific T and B cell responses, resulting in the termination of HBV infection [67]. Single nucleotide polymorphisms (SNPs) of TLRs may play an important role in HBV and HCV infection; however, the mechanisms remain to be fully understood [111]. The role of TLRs in HBV and HCV infection has been reviewed in detail in recently published papers [17,56]. TLR response-related data on HIV/HBV or HIV/HCV co-infection are scarce. Increased expression of TLRs in innate immune cells is observed in patients with a high HIV-1 load who are co-infected with opportunistic pathogens. Increased expression of TLR2 and TLR4 has been reported in the myeloid dendritic cells of HIV-1 patients co-infected with opportunistic infections (without HAART). Increased expression of TLR4 has been reported in plasmacytoid dendritic cells compared to both HIV-1 patients without opportunistic infections and healthy subjects [112].

6. Role of Interleukins and Other Cytokines in HIV-1 Infection

Virus infections can cause a proinflammatory response including expression of cytokines and chemokines [113]. Cytokines, signaling polypeptides, play an important role in immune activation, influencing viral pathogenesis [113,114]. HIV-1 infection may induce differential expression of cytokines and influence clinical outcomes [115]. A decreased secretion of Th1 cytokines such as IL-2 and antiviral IFN-γ, and an increased secretion of Th2 cytokines IL-4 and IL-10, and proinflammatory cytokines IL-1, IL-6, IL-8, and TNF-α have been reported in HIV-1 infection [115]. An increased levels of serum IL-6 has been reported in HIV chronic infection [116], and IL-6 was reported to enhance transcriptional levels of HIV-1 [117]. A previous study showed that HIV-1 Tat protein could activate TLR4 signaling and induce production of proinflammatory cytokines IL-6, IL-8, and TNF-α, and immunosuppressive cytokine IL-10 [118–120]. An increased circulating IL-7 level has been reported in HIV-1 infection, associating CD4+ T lymphopenia, and an increased IL-7 production was due to a homeostatic response to T cell depletion [121].

HIV-1 Nef protein can cause an induction of IL-15 induction and enhance viral replication [122]. A significant increase in IL-15 level has been reported in HIV-1 infected patients with viral loads >100,000 copies/ml compared to healthy subjects [123]. IL-15 has been shown to regulate the susceptibility of CD4+ T cells in HIV infection, and an inverse correlation between IL-15 levels and CD4+ T cell counts has been reported [123,124]. An increased serum level of IL-18 has been reported in the advanced and late stages of the disease, which significantly decreased after highly active antiretroviral therapy; however, in early stage there was no increase in IL-18 in HIV-1 infected patients [125]. Different cytokines such as IFN-alpha, IFN-beta, IL-10, IL-13, and IL-16 were reported to inhibit HIV-1 replication in T cells and/or monocyte-derived macrophages (MDM) [115]. Dysregulation of cytokines in HIV-1 infection is evident, and strategies targeting cytokines may open up new therapeutic intervention [126], requiring further investigation.

7. Inhibition of Innate Immune Response by HIV Infection

To infect susceptible hosts and ensure viral propagation towards establishing an infection, many viruses have developed strategies to evade or inhibit signaling through PRRs, including toll-like receptor pathways. HIV-1 can persistently infect humans and subvert their innate and adaptive immune systems, as HIV-1 primarily infects the CD4+ T lymphocytes and macrophages. The interaction between the trans-membrane domain (TMD) of the HIV-1 envelope protein and the TLR2 TMD may induce a reduced secretion of TNF-α, IL-6, and MCP-1 in macrophage [127]. It has been shown that HIV-1 gp120 can suppress the TLR9-mediated induction of proinflammatory cytokines and the expression of CD83, a marker of DC activation, resulting in a decreased ability of pDCs to secrete antiviral and inflammatory factors [128].

Sterile-motif/histidine-aspartate domain-containing protein 1 (SAMHD1), a host restriction factor possessing dNTP triphosphohydrolase activity, has been reported to restrict HIV-1 infection in monocyte-derived macrophages (MDMs) by reducing the intracellular dNTP pool [129,130]. SAMHD1 was also reported to inhibit an efficient viral DNA synthesis in non-cycling resting CD4+ T cells [131]. However, the HIV-1 accessory protein Vpx counteracts this restriction by directing SAMHD1 toward proteasomal degradation [132,133]. In addition, SAMHD1 suppresses the innate immune response to viral infections, including HIV-1 infection, and inflammatory stimuli by inhibiting NF-κB activation and type I IFN induction [134]. Moreover, HIV-1 capsids may inhibit viral DNA sensing via cGAS [135].

APOBEC3G, a cellular restriction factor, inhibits HIV-1 replication by inhibiting the elongation of HIV-1 reverse transcripts [136]. However, the viral protein Vif helps evade the host antiviral factor APOBEC3G by binding and inducing its degradation in virus-producing cells [137–140]. The inhibition of HIV-1 release mediated by tetherin, a cellular factor, can be antagonized by HIV-1 Vpu [141–144]. It has been shown that HIV-1 Vpr can inhibit IRF3 and NF-κB nuclear transport by interacting with karyopherins [145].

However, different accessory proteins may induce differential IFN response, including a potentiated type 1 IFN response by Vpr, and a suppressed IFN response by Vpu was observed in HIV-1 infected $CD4^+$ T cells [47]. The antiviral mechanisms adopted by HIV-1 largely depend on HIV-1 accessory proteins and their ability to degrade/inhibit/downregulate host antiviral targets [47]. The inhibition of host response by HIV-1 proteins is summarized in Table 2. For information on the inhibition/avoidance of innate immune response by HBV or HCV infection, other recently published papers can be studied [17,56].

Table 2. Inhibition of host innate immune response by HIV-1.

HIV Protein	Targeting Protein/Signaling Pathway	Effects on Host Immunity	Effects on Virus	Reference
HIV-1 Env	Binds TLR2	Inhibits secretion of pro-inflammatory cytokines	-	[127]
HIV-1 Vpr	Interacts with karyopherins	Inhibits IRF3 and NF-κB nuclear transport	Enhances virus replication	[145]

Table 2. *Cont.*

HIV Protein	Targeting Protein/Signaling Pathway	Effects on Host Immunity	Effects on Virus	Reference
HIV-1 Vif	Degrades APOBEC3G	Inhibits host antiviral activity of APOBEC3G	Overcomes inhibition of virus replication	[138,140,146,147]
HIV-1 Vpu	Degrades/downregulates tetherin/BST-2	Inhibits host antiviral activity of tetherin	Facilitates virus release	[141,142,148]
HIV-1 Vpu	Host restriction factors	Inhibits ISGylation of cellular machinery	Enhances late-stage virus replication	[149]
HIV-1 Vpu	Inhibits NF-κB activation	Reduces expression of restriction factors and IFNs release	May promote viral replication	[150–152]
HIV-1 Vpx	SAMHD1	Degrades SAMHD1	Facilitates reverse transcription	[132,133]
HIV-1 Nef	Activates IL-15 synthesis	Modulates cytokine response	Enhances HIV-1 replication	[122]
HIV-1 Nef	SERINC5 and SERINC3 proteins	Inhibits host's retroviral restriction of SERINC5 and SERINC3	Promotes HIV-1 infectivity	[153,154]
HIV-1 Nef	CD36, a scavenger receptor	Downregulates CD36 expression	Enhances opportunistic infection	[155]
HIV-1	Cellular 2′-O-methyltransferase	Avoids innate immune recognition	Acquires 2′-O-methylation at multiple distinct sites of HIV-1 RNA	[156]

8. Enhanced Pathogenesis of HIV/HBV or HIV/HCV Co-Infection

During HIV-1 infection, the immune system fails to completely eliminate the infecting virus, and complete inhibition of viral replication by the immune system is achieved in only a very small fraction of infected individuals [157]. Moreover, the elimination of viral infection or the complete prevention of viral replication cannot be achieved with current antiretroviral (ART) therapies [158]. Both reduction in the amount of virus persisting on antiretroviral therapy and enhanced anti-HIV immune surveillance are critical for curing HIV infection [159]. Different interventions, including the reversion of HIV latency by apoptosis-promoting agents, non-histone deacetylase inhibitor compounds, and immunotherapies, such as immune checkpoint inhibitors and TLR agonists, have been investigated to enhance antiviral immunity in curing HIV infection [159].

HIV infection can disrupt the gut mucosal barrier, resulting in microbial translocation and increased exposure to microbial products that act as TLR agonists [160]. Recent studies suggest that TLR signaling is an underlying mechanism for persistent immune activation, which is linked to progressive HIV-1 disease [161,162]. Further details on the immunopathogenesis of HIV-1 infection can be found in previously published reviews [163,164].

Continuous surveillance is critical for tracking viral diversity, as there is an increase in recombinant viruses, resulting in co-infection and superinfection by divergent HIV strains [41]. Individuals infected with HIV may show a higher tendency to develop chronic HBV infections [165]. Hepatocytes express CCR5 and CXCR4—two co-receptors for HIV entry—and multiple cells in the liver can be infected with HIV, which is associated with liver dysfunction [99,166,167]. HIV infection may induce increased levels of HBV DNA in chronic HBV infection compared with HBV mono-infection [7]. HIV/HBV co-infection enhances HBV-related hepatotoxicity mediated by an immune response [2]. Higher liver-related mortality was observed in HIV/HBV co-infection with lower CD4+ T cell counts than in HIV or HBV mono-infection [168]. An approximately five- to six-fold higher risk of HCC has been reported in HBV/HIV co-infected individuals [169,170].

Increased liver fibrosis is observed in HIV-1/HCV co-infection; however, the underlying mechanism is complex. HIV-1 enteropathy may cause increased microbial translocation and systemic immune activation, which may promote liver fibrosis either by direct interaction with Kupffer cells and hepatic stellate cells or indirectly via the induction of systemic immune activation and activation-induced apoptotic cell death [171]. Moreover, an increased risk of liver fibrosis in HIV-1/HCV co-infection may be associated with reduced CD4 + T cell counts, elevated lipopolysaccharide levels, and/or the depletion of hepatic Kupffer cells [171]. Although oral direct-acting antivirals have become very effective and can cure almost all treated patients infected with HCV, as there is no protective HCV immunity, PLWH with high-risk behaviors may experience HCV reinfections [172].

It has been reported that HIV co-infection may increase profibrogenic cytokine production with enhanced oxidative stress and hepatocyte apoptosis, which may be further augmented by increased microbial translocation of HIV [173]. Another study reported that HIV/HCV co-infection significantly increased epimorphin expression, which may result in increased proliferation and invasion capabilities of hepatic stellate cells [174]. Epimorphin induces the expression of profibrogenic tissue inhibitor of metalloproteinase 1 (TIMP1) in an extracellular signal-regulated kinase (ERK)-dependent manner [174].

HCV infection can induce transforming growth factor-beta1 (TGF-β1) expression in hepatocytes, which has been reported to be further enhanced by HIV co-infection [175]. HIV-1 Tat could activate HCV replication by upregulating interferon gamma-induced protein-10 (IP-10) production [176]. In addition, increased expression of IP-10 mRNA in peripheral blood mononuclear cells has been reported in HIV-1/HCV co-infection. Higher HCV RNA levels were found in the PBMCs of patients with HIV-1/HCV co-infection than in those with HCV mono-infection [176]. Viral infections, including HIV, HBV, and HCV, have been reported to cause the upregulation of programmed cell death protein (PD-1) and its ligands PD-L1 and PD-L2, indicating that PD-1/PD-L1 are key players in the pathogenic process of viruses [177]. Notably, the induction of anti-HBs titers in HBV vaccines is influenced by HIV infection, with a high percentage (90% to 95%) of healthy adults inducing protective anti-HBs titers after standard doses of HBV vaccines; however, during HIV co-infection, only 17.5% to 71% could retain protective anti-HBs [178].

9. TLR Agonists and Antagonists in the Cure of HIV Infection

The presence of latent and persistent viral reservoirs in transcriptionally silent immune cells remains a challenge to a successful HIV cure [179]. The persistent HIV-1 reservoir is predominantly found in long-lived memory CD4+ T cells, where the provirus remains transcriptionally silent and undetected by the host immune system [180,181]. Although ART can suppress HIV-1 replication to undetectable levels, it cannot eliminate latent and persistent viral reservoirs and the use of ART may result in multiple side effects including the development of multidrug-resistant escape mutants [182]. Therefore, it is critical to develop a strategy towards eliminating latently infected cells persisting in people with HIV on ART and identifying an ideal latency-reversing agent (LRA) that can induce viral reactivation, leading to immune cell recognition and the elimination of latently infected cells, which may improve HIV cure strategies [183]. Notably, many studies have indicated that TLR agonists are potential immunomodulators of the host immune response and influence the outcome of chronic HBV and HCV infections, suggesting the merit of using TLR agonists in curing chronic HIV-1 infection [17,56,183].

TLR agonists have the potential to reactivate latent HIV, induce immune activation, and promote an antiviral response by improving HIV-specific cytotoxic CD8+ T cell immunity, making TLR agonists unique LRA [179,183–186]. Pam3CSK4, a TLR1/2 agonist, has been shown to induce viral reactivation in a Tat-dependent manner and in the absence of T cell activation and/or proliferation [185]. TLR7 agonists have been reported to facilitate the reduction of viral reservoirs in a subset of SIV-infected rhesus macaques [187]. In a phase 1b study, vesatolimod, an oral TLR7 agonist, was found to be well tolerated, and immune stimulation was observed, providing a rationale for future combination trials in people living with HIV [188]. Dual TLR2/7 agonists have been shown to have a greater ability to induce the production of soluble factors for the reactivation of latent HIV than the combination of a single TLR2 and TLR7 agonist [189]. Another recent study of the synthesized TLR7/TLR8 agonist 159 pyrido [3,2-d] pyrimidine and pyridine-2-amine-based derivatives reported a high potential for activating HIV-1 latent reservoirs in cell lines and PBMCs of patients with persistent and durable virologic controls [182]. A previous study showed that the TLR9 agonist MGN1703 enhanced HIV-1 transcription and cytotoxic NK cell activation, which could contribute to HIV-1 eradication therapy [190].

The stimulation of TLR3, TLR4, TLR7, and TLR7/8 by the agonists poly (I:C), LPS, imiquimod, and ssRNA40, respectively, in PBMCs isolated from HIV-1-exposed seronegative (ESN) individuals induced a robust release of immunological factors in ESN individuals and showed a protective phenotype against HIV-1 [191]. The combination strategy using TLR agonist-directed innate effector functions and broadly neutralizing antibodies (bNAbs) may help enhance the elimination of HIV-1 infected cells [192,193]. Although there is hope for the success of curing HIV-1 infection using TLR agonists, further research is required to obtain clinical data and select the most suitable candidates for future use. Moreover, studies on the use of TLR agonists in HBV/HIV or HCV/HIV co-infection are scarce. However, a study was initiated to investigate the effect of the oral TLR8 agonist Selgantolimod on HBsAg in participants with both chronic HBV and HIV infections (ClinicalTrials.gov Identifier: NCT05551273), which may provide new data.

TLRs may act as a double-edged sword, and its double-sided effects have been reported in viral infections [21,194,195]. TLR ligand-induced chronic immune activation due to HIV-1 infection or microbial translocation from the gut, resulting in the chronic production of proinflammatory cytokines in HIV-1–infected individuals, may progress the disease. An abnormal activation of TLR7 may stimulate HIV infection [196]; therefore, the reduction or inhibition of chronic immune activation via TLR stimulation could be investigated for a potential target in reducing HIV-related immunopathology [197]. CXCR4 acts as a co-receptor for CXCR4-tropic HIV entry to CD4+ T cell [198]. CXCR4 has recently been shown to be involved in an additional regulatory function by directly inhibiting the induction of TLR signaling and resulting in a direct influence on IFN-I levels [199,200]. An engagement of CXCR4 can be used for inhibiting pDC activation and the control of TLR7-mediated IFN-I signaling [201,202]. Antibody-based treatment strategies, such as the use of anti-CCR5 antibodies targeting coreceptor CCR5, could be considered as alternative approaches for curing HIV-1 infection [203]. Although much knowledge has been gathered, further investigations are warranted for the clinical use of TLR agonists and antagonists in the cure of HIV-1 infection. The role of TLR agonists and antagonists in the cure of HIV-1 infection has been summarized in Table 3.

Table 3. Immunomodulatory effects of TLR agonists and antagonists in the cure of chronic HIV infection.

Compound (TLR Agonist or Antagonist)	Sponsor	Target TLR	Clinical Phase	Effects on Host Immunity	Effects on Virus	Reference/ ClinicalTrials.gov ID
Pam3CSK4	-	TLR1/2	-	Transcription factors NFκB, NFAT, and AP-1 cooperate to induce viral reactivation	Enhances virus replication	[185]
Vesatolimod	-	TLR7	Phase 1b	Safe and well tolerated; enhances immune stimulation	No significant effect on plasma viral load	[188]
GS-9620 (Vesatolimod)	Gilead Sciences	TLR7	Phase 2	No clinical data available	No clinical data available	NCT05281510
MGN1703 (Lefitolimod)	University of Aarhus	TLR9 agonists	Phase 2	No clinical data available	No clinical data available	NCT03837756
MGN1703 (Lefitolimod)		TLR9	Phase 1b/2a	Enhances innate immunity; increases transcription of IFNAR1; no concomitant general inflammatory response in the intestines; reduction of HIV-1 viral reservoir	Decreases integrated HIV-1 DNA	[204]
MGN1703	University of Aarhus	TLR9	Phase 1 Phase 2	Safe and well-tolerated; enhances both innate and adaptive immunity; increases HIV-1-specific T cell responses; enhances cytotoxic NK cell activation	Enhances HIV-1 replication	[103,190]

10. Conclusions

TLRs can produce redundant responses such as inflammatory or antiviral responses; however, there are differences in outcomes, largely influenced by the roles of ligands and tissue-dependent TLR expression [205–207]. The findings of different studies indicate that TLRs are active participants and key regulators of the antiretroviral immune regulation. However, the interaction between HIV-1 and the TLR system is full of complexity, and the dual role of the TLR system affecting viral replication and immune cell functioning requires detailed investigation to understand the specific roles of individual TLRs and their agonists and antagonists in both HIV-1 pathogenesis and protection. Robust immune surveillance is required for eradicating persistent HIV [208]. TLR agonists can be used in reactivating latent HIV-1 reservoirs to make them visible to the immune system. Although further studies are warranted, the use of TLR agonists and antagonists in treating HIV infection may become promising with future research works; however, there are very limited data for HIV/HBV or HIV/HCV co-infection, requiring further studies for their clinical use.

Author Contributions: Conceptualization, M.E.H.K., M.K. and K.T.-K.; writing—original draft preparation, M.E.H.K. and K.T.-K.; writing—review and editing, M.E.H.K., M.K. and K.T.-K.; All authors have read and agreed to the published version of the manuscript.

Funding: This study was supported by grants from the Japan Agency for Medical Research and Development (AMED) and The Tokyo Metropolitan Government.

Institutional Review Board Statement: Not applicable.

Informed Consent Statement: Not applicable.

Acknowledgments: Not applicable.

Conflicts of Interest: The authors declare no conflict of interest.

References

1. Shire, N.J.; Sherman, K.E. Management of HBV/HIV-coinfected Patients. *Semin. Liver Dis.* **2005**, *25* (Suppl. S1), 48–57. [CrossRef] [PubMed]
2. Thio, C.L. Hepatitis B and human immunodeficiency virus coinfection. *Hepatology* **2009**, *49*, S138–S145. [CrossRef] [PubMed]
3. Cheng, Z.; Lin, P.; Cheng, N. HBV/HIV Coinfection: Impact on the Development and Clinical Treatment of Liver Diseases. *Front. Med.* **2021**, *8*, 713981. [CrossRef]
4. Gobran, S.T.; Ancuta, P.; Shoukry, N.H. A Tale of Two Viruses: Immunological Insights into HCV/HIV Coinfection. *Front. Immunol.* **2021**, *12*, 726419. [CrossRef] [PubMed]
5. Behzadi, P.; Ranjbar, R.; Alavian, S.M. Nucleic Acid-based approaches for detection of viral hepatitis. *Jundishapur. J. Microbiol.* **2015**, *8*, e17449. [CrossRef] [PubMed]
6. Behzadi, P.; Behzadi, E.; Alavian, S.M. DNA microarray technology in HBV genotyping. *Minerva Med.* **2017**, *108*, 473–476. [CrossRef] [PubMed]
7. Singh, K.P.; Crane, M.; Audsley, J.; Avihingsanon, A.; Sasadeusz, J.; Lewin, S.R. HIV-hepatitis B virus coinfection: Epidemiology, pathogenesis, and treatment. *AIDS* **2017**, *31*, 2035–2052. [CrossRef]
8. Di Martino, V.; Rufat, P.; Boyer, N.; Renard, P.; Degos, F.; Martinot-Peignoux, M.; Matheron, S.; Le Moing, V.; Vachon, F.; Degott, C.; et al. The influence of human immunodeficiency virus coinfection on chronic hepatitis C in injection drug users: A long-term retrospective cohort study. *Hepatology* **2001**, *34*, 1193–1199. [CrossRef]
9. Benhamou, Y.; Bochet, M.; Di Martino, V.; Charlotte, F.; Azria, F.; Coutellier, A.; Vidaud, M.; Bricaire, F.; Opolon, P.; Katlama, C.; et al. Liver fibrosis progression in human immunodeficiency virus and hepatitis C virus coinfected patients. The Multivirc Group. *Hepatology* **1999**, *30*, 1054–1058. [CrossRef]
10. Medzhitov, R.; Janeway, C., Jr. Innate immunity. *N. Engl. J. Med.* **2000**, *343*, 338–344. [CrossRef]
11. Akira, S.; Uematsu, S.; Takeuchi, O. Pathogen recognition and innate immunity. *Cell* **2006**, *124*, 783–801. [CrossRef]
12. Kumar, V.; Barrett, J.E. Toll-Like Receptors (TLRs) in Health and Disease: An Overview. In *Handbook of Experimental Pharmacology*; Springer: Cham, Switzerland, 2022; Volume 276, pp. 1–21. [CrossRef]
13. Kawasaki, T.; Kawai, T. Toll-like receptor signaling pathways. *Front. Immunol.* **2014**, *5*, 461. [CrossRef]
14. Thompson, J.M.; Iwasaki, A. Toll-like receptors regulation of viral infection and disease. *Adv. Drug Deliv. Rev.* **2008**, *60*, 786–794. [CrossRef]
15. Rahman, M.M.; McFadden, G. Modulation of NF-kappaB signalling by microbial pathogens. *Nat. Rev. Microbiol.* **2011**, *9*, 291–306. [CrossRef]
16. Hayden, M.S.; Ghosh, S. NF-kappaB in immunobiology. *Cell Res.* **2011**, *21*, 223–244. [CrossRef]

17. Kayesh, M.E.H.; Kohara, M.; Tsukiyama-Kohara, K. Toll-like Receptor Response to Hepatitis C Virus Infection: A Recent Overview. *Int. J. Mol. Sci.* **2022**, *23*, 5475. [CrossRef]
18. Deeks, S.G.; Walker, B.D. Human immunodeficiency virus controllers: Mechanisms of durable virus control in the absence of antiretroviral therapy. *Immunity* **2007**, *27*, 406–416. [CrossRef]
19. Janeway, C.A., Jr.; Medzhitov, R. Innate immune recognition. *Annu. Rev. Immunol.* **2002**, *20*, 197–216. [CrossRef]
20. Akira, S.; Takeda, K.; Kaisho, T. Toll-like receptors: Critical proteins linking innate and acquired immunity. *Nat. Immunol.* **2001**, *2*, 675–680. [CrossRef]
21. Kayesh, M.E.H.; Kohara, M.; Tsukiyama-Kohara, K. An Overview of Recent Insights into the Response of TLR to SARS-CoV-2 Infection and the Potential of TLR Agonists as SARS-CoV-2 Vaccine Adjuvants. *Viruses* **2021**, *13*, 2302. [CrossRef]
22. Modhiran, N.; Watterson, D.; Muller, D.A.; Panetta, A.K.; Sester, D.P.; Liu, L.; Hume, D.A.; Stacey, K.J.; Young, P.R. Dengue virus NS1 protein activates cells via Toll-like receptor 4 and disrupts endothelial cell monolayer integrity. *Sci. Transl. Med.* **2015**, *7*, 304ra142. [CrossRef]
23. Behzadi, P.; Sameer, A.S.; Nissar, S.; Banday, M.Z.; Gajdacs, M.; Garcia-Perdomo, H.A.; Akhtar, K.; Pinheiro, M.; Magnusson, P.; Sarshar, M.; et al. The Interleukin-1 (IL-1) Superfamily Cytokines and Their Single Nucleotide Polymorphisms (SNPs). *J. Immunol. Res.* **2022**, *2022*, 2054431. [CrossRef]
24. O'Neill, L.A.; Golenbock, D.; Bowie, A.G. The history of Toll-like receptors—Redefining innate immunity. *Nat. Rev. Immunol.* **2013**, *13*, 453–460. [CrossRef]
25. Chaturvedi, A.; Pierce, S.K. How location governs toll-like receptor signaling. *Traffic* **2009**, *10*, 621–628. [CrossRef]
26. Kawai, T.; Akira, S. Toll-like receptor and RIG-I-like receptor signaling. *Ann. N. Y. Acad. Sci.* **2008**, *1143*, 1–20. [CrossRef]
27. Alexopoulou, L.; Holt, A.C.; Medzhitov, R.; Flavell, R.A. Recognition of double-stranded RNA and activation of NF-kappaB by Toll-like receptor 3. *Nature* **2001**, *413*, 732–738. [CrossRef]
28. Diebold, S.S.; Kaisho, T.; Hemmi, H.; Akira, S.; Reis e Sousa, C. Innate antiviral responses by means of TLR7-mediated recognition of single-stranded RNA. *Science* **2004**, *303*, 1529–1531. [CrossRef]
29. Heil, F.; Hemmi, H.; Hochrein, H.; Ampenberger, F.; Kirschning, C.; Akira, S.; Lipford, G.; Wagner, H.; Bauer, S. Species-specific recognition of single-stranded RNA via toll-like receptor 7 and 8. *Science* **2004**, *303*, 1526–1529. [CrossRef]
30. Prinz, M.; Heikenwalder, M.; Schwarz, P.; Takeda, K.; Akira, S.; Aguzzi, A. Prion pathogenesis in the absence of Toll-like receptor signalling. *EMBO Rep.* **2003**, *4*, 195–199. [CrossRef]
31. Jacobsen, H.; Klein, S.L. Sex Differences in Immunity to Viral Infections. *Front. Immunol.* **2021**, *12*, 720952. [CrossRef]
32. Klein, S.L.; Flanagan, K.L. Sex differences in immune responses. *Nat. Rev. Immunol.* **2016**, *16*, 626–638. [CrossRef]
33. Redwan, E.M.; Aljadawi, A.A.; Uversky, V.N. Hepatitis C Virus Infection and Intrinsic Disorder in the Signaling Pathways Induced by Toll-Like Receptors. *Biology* **2022**, *11*, 1091. [CrossRef]
34. Torcia, M.G.; Nencioni, L.; Clemente, A.M.; Civitelli, L.; Celestino, I.; Limongi, D.; Fadigati, G.; Perissi, E.; Cozzolino, F.; Garaci, E.; et al. Sex differences in the response to viral infections: TLR8 and TLR9 ligand stimulation induce higher IL10 production in males. *PLoS ONE* **2012**, *7*, e39853. [CrossRef]
35. Souyris, M.; Cenac, C.; Azar, P.; Daviaud, D.; Canivet, A.; Grunenwald, S.; Pienkowski, C.; Chaumeil, J.; Mejia, J.E.; Guery, J.C. TLR7 escapes X chromosome inactivation in immune cells. *Sci. Immunol.* **2018**, *3*, eaap8855. [CrossRef]
36. Hagen, S.H.; Henseling, F.; Hennesen, J.; Savel, H.; Delahaye, S.; Richert, L.; Ziegler, S.M.; Altfeld, M. Heterogeneous Escape from X Chromosome Inactivation Results in Sex Differences in Type I IFN Responses at the Single Human pDC Level. *Cell Rep.* **2020**, *33*, 108485. [CrossRef]
37. Meier, A.; Chang, J.J.; Chan, E.S.; Pollard, R.B.; Sidhu, H.K.; Kulkarni, S.; Wen, T.F.; Lindsay, R.J.; Orellana, L.; Mildvan, D.; et al. Sex differences in the Toll-like receptor-mediated response of plasmacytoid dendritic cells to HIV-1. *Nat. Med.* **2009**, *15*, 955–959. [CrossRef]
38. Fischer, J.; Weber, A.N.R.; Bohm, S.; Dickhofer, S.; El Maadidi, S.; Deichsel, D.; Knop, V.; Klinker, H.; Moller, B.; Rasenack, J.; et al. Sex-specific effects of TLR9 promoter variants on spontaneous clearance of HCV infection. *Gut* **2017**, *66*, 1829–1837. [CrossRef]
39. Clements, J.E.; Zink, M.C. Molecular biology and pathogenesis of animal lentivirus infections. *Clin. Microbiol. Rev.* **1996**, *9*, 100–117. [CrossRef]
40. World Health Organization. HIV. Updated July 2022. Available online: https://www.who.int/data/gho/data/themes/hiv-aids (accessed on 12 October 2022).
41. Bbosa, N.; Kaleebu, P.; Ssemwanga, D. HIV subtype diversity worldwide. *Curr. Opin. HIV AIDS* **2019**, *14*, 153–160. [CrossRef]
42. Azevedo-Pereira, J.M.; Santos-Costa, Q.; Moniz-Pereira, J. HIV-2 infection and chemokine receptors usage—Clues to reduced virulence of HIV-2. *Curr. HIV Res.* **2005**, *3*, 3–16. [CrossRef]
43. German Advisory Committee Blood, Subgroup 'Assessment of Pathogens Transmissible by Blood'. Human Immunodeficiency Virus (HIV). *Transfus. Med. Hemother.* **2016**, *43*, 203–222. [CrossRef] [PubMed]
44. Meier, A.; Alter, G.; Frahm, N.; Sidhu, H.; Li, B.; Bagchi, A.; Teigen, N.; Streeck, H.; Stellbrink, H.J.; Hellman, J.; et al. MyD88-dependent immune activation mediated by human immunodeficiency virus type 1-encoded Toll-like receptor ligands. *J. Virol.* **2007**, *81*, 8180–8191. [CrossRef] [PubMed]
45. Towers, G.J.; Noursadeghi, M. Interactions between HIV-1 and the cell-autonomous innate immune system. *Cell Host Microbe* **2014**, *16*, 10–18. [CrossRef] [PubMed]

46. Wang, M.Q.; Huang, Y.L.; Huang, J.; Zheng, J.L.; Qian, G.X. RIG-I detects HIV-1 infection and mediates type I interferon response in human macrophages from patients with HIV-1-associated neurocognitive disorders. *Genet. Mol. Res.* **2015**, *14*, 13799–13811. [CrossRef] [PubMed]
47. Yin, X.; Langer, S.; Zhang, Z.; Herbert, K.M.; Yoh, S.; Konig, R.; Chanda, S.K. Sensor Sensibility-HIV-1 and the Innate Immune Response. *Cells* **2020**, *9*, 254. [CrossRef]
48. Scott-Algara, D.; Truong, L.X.; Versmisse, P.; David, A.; Luong, T.T.; Nguyen, N.V.; Theodorou, I.; Barre-Sinoussi, F.; Pancino, G. Cutting edge: Increased NK cell activity in HIV-1-exposed but uninfected Vietnamese intravascular drug users. *J. Immunol.* **2003**, *171*, 5663–5667. [CrossRef] [PubMed]
49. Montoya, C.J.; Velilla, P.A.; Chougnet, C.; Landay, A.L.; Rugeles, M.T. Increased IFN-gamma production by NK and CD3+/CD56+ cells in sexually HIV-1-exposed but uninfected individuals. *Clin. Immunol.* **2006**, *120*, 138–146. [CrossRef]
50. Kebba, A.; Stebbing, J.; Rowland, S.; Ingram, R.; Agaba, J.; Patterson, S.; Kaleebu, P.; Imami, N.; Gotch, F. Expression of the common heat-shock protein receptor CD91 is increased on monocytes of exposed yet HIV-1-seronegative subjects. *J. Leukoc. Biol.* **2005**, *78*, 37–42. [CrossRef]
51. Browne, E.P. The Role of Toll-Like Receptors in Retroviral Infection. *Microorganisms* **2020**, *8*, 1787. [CrossRef]
52. Haynes, B.F.; Pantaleo, G.; Fauci, A.S. Toward an understanding of the correlates of protective immunity to HIV infection. *Science* **1996**, *271*, 324–328. [CrossRef]
53. Pantaleo, G.; Demarest, J.F.; Schacker, T.; Vaccarezza, M.; Cohen, O.J.; Daucher, M.; Graziosi, C.; Schnittman, S.S.; Quinn, T.C.; Shaw, G.M.; et al. The qualitative nature of the primary immune response to HIV infection is a prognosticator of disease progression independent of the initial level of plasma viremia. *Proc. Natl. Acad. Sci. USA* **1997**, *94*, 254–258. [CrossRef]
54. Bodsworth, N.; Donovan, B.; Nightingale, B.N. The effect of concurrent human immunodeficiency virus infection on chronic hepatitis B: A study of 150 homosexual men. *J. Infect. Dis.* **1989**, *160*, 577–582. [CrossRef]
55. Lee, W.M. Hepatitis B virus infection. *N. Engl. J. Med.* **1997**, *337*, 1733–1745. [CrossRef] [PubMed]
56. Kayesh, M.E.H.; Kohara, M.; Tsukiyama-Kohara, K. Toll-Like Receptor Response to Hepatitis B Virus Infection and Potential of TLR Agonists as Immunomodulators for Treating Chronic Hepatitis B: An Overview. *Int. J. Mol. Sci.* **2021**, *22*, 10462. [CrossRef]
57. Moradpour, D.; Penin, F.; Rice, C.M. Replication of hepatitis C virus. *Nat. Rev. Microbiol.* **2007**, *5*, 453–463. [CrossRef]
58. Sato, K.; Ishikawa, T.; Okumura, A.; Yamauchi, T.; Sato, S.; Ayada, M.; Matsumoto, E.; Hotta, N.; Oohashi, T.; Fukuzawa, Y.; et al. Expression of Toll-like receptors in chronic hepatitis C virus infection. *J. Gastroenterol. Hepatol.* **2007**, *22*, 1627–1632. [CrossRef]
59. He, Q.; Graham, C.S.; Durante Mangoni, E.; Koziel, M.J. Differential expression of toll-like receptor mRNA in treatment non-responders and sustained virologic responders at baseline in patients with chronic hepatitis C. *Liver Int.* **2006**, *26*, 1100–1110. [CrossRef]
60. Kayesh, M.E.H.; Sanada, T.; Kohara, M.; Tsukiyama-Kohara, K. Tree Shrew as an Emerging Small Animal Model for Human Viral Infection: A Recent Overview. *Viruses* **2021**, *13*, 1641. [CrossRef] [PubMed]
61. Kayesh, M.E.H.; Ezzikouri, S.; Sanada, T.; Chi, H.; Hayashi, Y.; Rebbani, K.; Kitab, B.; Matsuu, A.; Miyoshi, N.; Hishima, T.; et al. Oxidative Stress and Immune Responses During Hepatitis C Virus Infection in Tupaia belangeri. *Sci. Rep.* **2017**, *7*, 9848. [CrossRef] [PubMed]
62. Xie, Z.C.; Riezu-Boj, J.I.; Lasarte, J.J.; Guillen, J.; Su, J.H.; Civeira, M.P.; Prieto, J. Transmission of hepatitis C virus infection to tree shrews. *Virology* **1998**, *244*, 513–520. [CrossRef]
63. Wu, J.; Huang, S.; Zhao, X.; Chen, M.; Lin, Y.; Xia, Y.; Sun, C.; Yang, X.; Wang, J.; Guo, Y.; et al. Poly(I:C) treatment leads to interferon-dependent clearance of hepatitis B virus in a hydrodynamic injection mouse model. *J. Virol.* **2014**, *88*, 10421–10431. [CrossRef] [PubMed]
64. Huang, L.R.; Wohlleber, D.; Reisinger, F.; Jenne, C.N.; Cheng, R.L.; Abdullah, Z.; Schildberg, F.A.; Odenthal, M.; Dienes, H.P.; van Rooijen, N.; et al. Intrahepatic myeloid-cell aggregates enable local proliferation of CD8(+) T cells and successful immunotherapy against chronic viral liver infection. *Nat. Immunol.* **2013**, *14*, 574–583. [CrossRef]
65. Hatton, A.A.; Guerra, F.E. Scratching the Surface Takes a Toll: Immune Recognition of Viral Proteins by Surface Toll-like Receptors. *Viruses* **2022**, *15*, 52. [CrossRef] [PubMed]
66. Li, C.; Kuang, W.D.; Qu, D.; Wang, J.H. Toll-interacting protein inhibits HIV-1 infection and regulates viral latency. *Biochem. Biophys. Res. Commun.* **2016**, *475*, 161–168. [CrossRef]
67. Ma, Z.; Zhang, E.; Yang, D.; Lu, M. Contribution of Toll-like receptors to the control of hepatitis B virus infection by initiating antiviral innate responses and promoting specific adaptive immune responses. *Cell Mol. Immunol.* **2015**, *12*, 273–282. [CrossRef] [PubMed]
68. Zhang, E.; Ma, Z.; Lu, M. Contribution of T- and B-cell intrinsic toll-like receptors to the adaptive immune response in viral infectious diseases. *Cell Mol. Life Sci.* **2022**, *79*, 547. [CrossRef]
69. Chang, J.J.; Lacas, A.; Lindsay, R.J.; Doyle, E.H.; Axten, K.L.; Pereyra, F.; Rosenberg, E.S.; Walker, B.D.; Allen, T.M.; Altfeld, M. Differential regulation of toll-like receptor pathways in acute and chronic HIV-1 infection. *AIDS* **2012**, *26*, 533–541. [CrossRef]
70. Lester, R.T.; Yao, X.D.; Ball, T.B.; McKinnon, L.R.; Kaul, R.; Wachihi, C.; Jaoko, W.; Plummer, F.A.; Rosenthal, K.L. Toll-like receptor expression and responsiveness are increased in viraemic HIV-1 infection. *AIDS* **2008**, *22*, 685–694. [CrossRef]
71. Hernandez, J.C.; Stevenson, M.; Latz, E.; Urcuqui-Inchima, S. HIV type 1 infection up-regulates TLR2 and TLR4 expression and function in vivo and in vitro. *AIDS Res. Hum. Retrovir.* **2012**, *28*, 1313–1328. [CrossRef]

72. Tarasova, O.; Biziukova, N.; Shemshura, A.; Filimonov, D.; Kireev, D.; Pokrovskaya, A.; Poroikov, V.V. Identification of Molecular Mechanisms Involved in Viral Infection Progression Based on Text Mining: Case Study for HIV Infection. *Int. J. Mol. Sci.* **2023**, *24*, 1465. [CrossRef]
73. Thibault, S.; Fromentin, R.; Tardif, M.R.; Tremblay, M.J. TLR2 and TLR4 triggering exerts contrasting effects with regard to HIV-1 infection of human dendritic cells and subsequent virus transfer to CD4+ T cells. *Retrovirology* **2009**, *6*, 42. [CrossRef] [PubMed]
74. Henrick, B.M.; Nag, K.; Yao, X.D.; Drannik, A.G.; Aldrovandi, G.M.; Rosenthal, K.L. Milk matters: Soluble Toll-like receptor 2 (sTLR2) in breast milk significantly inhibits HIV-1 infection and inflammation. *PLoS ONE* **2012**, *7*, e40138. [CrossRef] [PubMed]
75. Buitendijk, M.; Eszterhas, S.K.; Howell, A.L. Toll-like receptor agonists are potent inhibitors of human immunodeficiency virus-type 1 replication in peripheral blood mononuclear cells. *AIDS Res. Hum. Retrovir.* **2014**, *30*, 457–467. [CrossRef] [PubMed]
76. Henrick, B.M.; Yao, X.D.; Rosenthal, K.L.; The INFANT Study Team. HIV-1 Structural Proteins Serve as PAMPs for TLR2 Heterodimers Significantly Increasing Infection and Innate Immune Activation. *Front. Immunol.* **2015**, *6*, 426. [CrossRef]
77. Zhou, Y.; Wang, X.; Liu, M.; Hu, Q.; Song, L.; Ye, L.; Zhou, D.; Ho, W. A critical function of toll-like receptor-3 in the induction of anti-human immunodeficiency virus activities in macrophages. *Immunology* **2010**, *131*, 40–49. [CrossRef]
78. Alvarez-Carbonell, D.; Garcia-Mesa, Y.; Milne, S.; Das, B.; Dobrowolski, C.; Rojas, R.; Karn, J. Toll-like receptor 3 activation selectively reverses HIV latency in microglial cells. *Retrovirology* **2017**, *14*, 9. [CrossRef]
79. Cheng, L.; Wang, Q.; Li, G.; Banga, R.; Ma, J.; Yu, H.; Yasui, F.; Zhang, Z.; Pantaleo, G.; Perreau, M.; et al. TLR3 agonist and CD40-targeting vaccination induces immune responses and reduces HIV-1 reservoirs. *J. Clin. Investig.* **2018**, *128*, 4387–4396. [CrossRef]
80. Lester, R.T.; Yao, X.D.; Ball, T.B.; McKinnon, L.R.; Omange, W.R.; Kaul, R.; Wachihi, C.; Jaoko, W.; Rosenthal, K.L.; Plummer, F.A. HIV-1 RNA dysregulates the natural TLR response to subclinical endotoxemia in Kenyan female sex-workers. *PLoS ONE* **2009**, *4*, e5644. [CrossRef]
81. Brenchley, J.M.; Price, D.A.; Schacker, T.W.; Asher, T.E.; Silvestri, G.; Rao, S.; Kazzaz, Z.; Bornstein, E.; Lambotte, O.; Altmann, D.; et al. Microbial translocation is a cause of systemic immune activation in chronic HIV infection. *Nat. Med.* **2006**, *12*, 1365–1371. [CrossRef]
82. Zhang, T.; Zhu, J.; Su, B.; Cao, L.; Li, Z.; Wei, H.; Huang, X.; Zheng, K.; Li, A.; Chen, N.; et al. Effects of TLR7 Polymorphisms on the Susceptibility and Progression of HIV-1 Infection in Chinese MSM Population. *Front. Immunol.* **2020**, *11*, 589010. [CrossRef]
83. Tsai, A.; Irrinki, A.; Kaur, J.; Cihlar, T.; Kukolj, G.; Sloan, D.D.; Murry, J.P. Toll-Like Receptor 7 Agonist GS-9620 Induces HIV Expression and HIV-Specific Immunity in Cells from HIV-Infected Individuals on Suppressive Antiretroviral Therapy. *J. Virol.* **2017**, *91*, e02166-16. [CrossRef] [PubMed]
84. Meng, F.Z.; Liu, J.B.; Wang, X.; Wang, P.; Hu, W.H.; Hou, W.; Ho, W.Z. TLR7 Activation of Macrophages by Imiquimod Inhibits HIV Infection through Modulation of Viral Entry Cellular Factors. *Biology* **2021**, *10*, 661. [CrossRef] [PubMed]
85. Mehraj, V.; Ramendra, R.; Isnard, S.; Dupuy, F.P.; Lebouche, B.; Costiniuk, C.; Thomas, R.; Szabo, J.; Baril, J.G.; Trottier, B.; et al. CXCL13 as a Biomarker of Immune Activation during Early and Chronic HIV Infection. *Front. Immunol.* **2019**, *10*, 289. [CrossRef] [PubMed]
86. Cohen, K.W.; Dugast, A.S.; Alter, G.; McElrath, M.J.; Stamatatos, L. HIV-1 single-stranded RNA induces CXCL13 secretion in human monocytes via TLR7 activation and plasmacytoid dendritic cell-derived type I IFN. *J. Immunol.* **2015**, *194*, 2769–2775. [CrossRef] [PubMed]
87. Hennessy, E.J.; Parker, A.E.; O'Neill, L.A. Targeting Toll-like receptors: Emerging therapeutics? *Nat. Rev. Drug. Discov.* **2010**, *9*, 293–307. [CrossRef]
88. Rozman, M.; Zidovec-Lepej, S.; Jambrosic, K.; Babic, M.; Drmic Hofman, I. Role of TLRs in HIV-1 Infection and Potential of TLR Agonists in HIV-1 Vaccine Development and Treatment Strategies. *Pathogens* **2023**, *12*, 92. [CrossRef]
89. Yadav, S.; Verma, V.; Singh Dhanda, R.; Yadav, M. Insights into the toll-like receptors in sexually transmitted infections. *Scand. J. Immunol.* **2021**, *93*, e12954. [CrossRef]
90. Ito, T.; Wang, Y.H.; Liu, Y.J. Plasmacytoid dendritic cell precursors/type I interferon-producing cells sense viral infection by Toll-like receptor (TLR) 7 and TLR9. *Springer Semin. Immunopathol.* **2005**, *26*, 221–229. [CrossRef]
91. Mandl, J.N.; Barry, A.P.; Vanderford, T.H.; Kozyr, N.; Chavan, R.; Klucking, S.; Barrat, F.J.; Coffman, R.L.; Staprans, S.I.; Feinberg, M.B. Divergent TLR7 and TLR9 signaling and type I interferon production distinguish pathogenic and nonpathogenic AIDS virus infections. *Nat. Med.* **2008**, *14*, 1077–1087. [CrossRef]
92. Jimenez-Leon, M.R.; Gasca-Capote, C.; Tarancon-Diez, L.; Dominguez-Molina, B.; Lopez-Verdugo, M.; Ritraj, R.; Gallego, I.; Alvarez-Rios, A.I.; Vitalle, J.; Bachiller, S.; et al. Toll-like receptor agonists enhance HIV-specific T cell response mediated by plasmacytoid dendritic cells in diverse HIV-1 disease progression phenotypes. *EBioMedicine* **2023**, *91*, 104549. [CrossRef]
93. Sandgren, K.J.; Smed-Sorensen, A.; Forsell, M.N.; Soldemo, M.; Adams, W.C.; Liang, F.; Perbeck, L.; Koup, R.A.; Wyatt, R.T.; Karlsson Hedestam, G.B.; et al. Human plasmacytoid dendritic cells efficiently capture HIV-1 envelope glycoproteins via CD4 for antigen presentation. *J. Immunol.* **2013**, *191*, 60–69. [CrossRef] [PubMed]
94. Beignon, A.S.; McKenna, K.; Skoberne, M.; Manches, O.; DaSilva, I.; Kavanagh, D.G.; Larsson, M.; Gorelick, R.J.; Lifson, J.D.; Bhardwaj, N. Endocytosis of HIV-1 activates plasmacytoid dendritic cells via Toll-like receptor-viral RNA interactions. *J. Clin. Investig.* **2005**, *115*, 3265–3275. [CrossRef] [PubMed]

95. Barron, M.A.; Blyveis, N.; Palmer, B.E.; MaWhinney, S.; Wilson, C.C. Influence of plasma viremia on defects in number and immunophenotype of blood dendritic cell subsets in human immunodeficiency virus 1-infected individuals. *J. Infect. Dis.* **2003**, *187*, 26–37. [CrossRef]
96. Donaghy, H.; Pozniak, A.; Gazzard, B.; Qazi, N.; Gilmour, J.; Gotch, F.; Patterson, S. Loss of blood CD11c(+) myeloid and CD11c(-) plasmacytoid dendritic cells in patients with HIV-1 infection correlates with HIV-1 RNA virus load. *Blood* **2001**, *98*, 2574–2576. [CrossRef]
97. Guery, J.C. Sex Differences in Primary HIV Infection: Revisiting the Role of TLR7-Driven Type 1 IFN Production by Plasmacytoid Dendritic Cells in Women. *Front. Immunol.* **2021**, *12*, 729233. [CrossRef] [PubMed]
98. Abbas, F.; Cenac, C.; Youness, A.; Azar, P.; Delobel, P.; Guery, J.C. HIV-1 infection enhances innate function and TLR7 expression in female plasmacytoid dendritic cells. *Life Sci. Alliance* **2022**, *5*, e202201452. [CrossRef]
99. Crane, M.; Visvanathan, K.; Lewin, S.R. HIV Infection and TLR Signalling in the Liver. *Gastroenterol. Res. Pract.* **2012**, *2012*, 473925. [CrossRef]
100. Meas, H.Z.; Haug, M.; Beckwith, M.S.; Louet, C.; Ryan, L.; Hu, Z.; Landskron, J.; Nordbo, S.A.; Tasken, K.; Yin, H.; et al. Sensing of HIV-1 by TLR8 activates human T cells and reverses latency. *Nat. Commun.* **2020**, *11*, 147. [CrossRef]
101. Oh, D.Y.; Taube, S.; Hamouda, O.; Kucherer, C.; Poggensee, G.; Jessen, H.; Eckert, J.K.; Neumann, K.; Storek, A.; Pouliot, M.; et al. A functional toll-like receptor 8 variant is associated with HIV disease restriction. *J. Infect. Dis.* **2008**, *198*, 701–709. [CrossRef]
102. Zaidane, I.; Ouladlahsen, A.; Bensghir, R.; Chihab, H.; Jadid, F.Z.; Fihry, R.E.; Baba, H.; Filali, K.M.E.; Oudghiri, M.; Wakrim, L.; et al. An assessment of toll-like receptor 7 and 8 gene polymorphisms with susceptibility to HIV-1 infection, AIDS development and response to antiretroviral therapy. *Immunol. Lett.* **2020**, *227*, 88–95. [CrossRef]
103. Vibholm, L.K.; Konrad, C.V.; Schleimann, M.H.; Frattari, G.; Winckelmann, A.; Klastrup, V.; Jensen, N.M.; Jensen, S.S.; Schmidt, M.; Wittig, B.; et al. Effects of 24-week Toll-like receptor 9 agonist treatment in HIV type 1+ individuals. *AIDS* **2019**, *33*, 1315–1325. [CrossRef] [PubMed]
104. Cingoz, O.; Goff, S.P. HIV-1 Is a Poor Inducer of Innate Immune Responses. *mBio* **2019**, *10*, e02834-18. [CrossRef] [PubMed]
105. Siddiqui, M.A.; Yamashita, M. Toll-Like Receptor (TLR) Signaling Enables Cyclic GMP-AMP Synthase (cGAS) Sensing of HIV-1 Infection in Macrophages. *mBio* **2021**, *12*, e0281721. [CrossRef] [PubMed]
106. Gao, D.; Wu, J.; Wu, Y.T.; Du, F.; Aroh, C.; Yan, N.; Sun, L.; Chen, Z.J. Cyclic GMP-AMP synthase is an innate immune sensor of HIV and other Retrovir. *Science* **2013**, *341*, 903–906. [CrossRef]
107. Lee, S.M.; Kok, K.H.; Jaume, M.; Cheung, T.K.; Yip, T.F.; Lai, J.C.; Guan, Y.; Webster, R.G.; Jin, D.Y.; Peiris, J.S. Toll-like receptor 10 is involved in induction of innate immune responses to influenza virus infection. *Proc. Natl. Acad. Sci. USA* **2014**, *111*, 3793–3798. [CrossRef]
108. Henrick, B.M.; Yao, X.D.; Zahoor, M.A.; Abimiku, A.; Osawe, S.; Rosenthal, K.L. TLR10 Senses HIV-1 Proteins and Significantly Enhances HIV-1 Infection. *Front. Immunol.* **2019**, *10*, 482. [CrossRef]
109. Ellegard, R.; Crisci, E.; Burgener, A.; Sjowall, C.; Birse, K.; Westmacott, G.; Hinkula, J.; Lifson, J.D.; Larsson, M. Complement opsonization of HIV-1 results in decreased antiviral and inflammatory responses in immature dendritic cells via CR3. *J. Immunol.* **2014**, *193*, 4590–4601. [CrossRef]
110. Cen, P.; Ye, L.; Su, Q.J.; Wang, X.; Li, J.L.; Lin, X.Q.; Liang, H.; Ho, W.Z. Methamphetamine inhibits Toll-like receptor 9-mediated anti-HIV activity in macrophages. *AIDS Res. Hum. Retrovir.* **2013**, *29*, 1129–1137. [CrossRef]
111. Xu, Y.; Xue, W.; Gao, H.; Cui, J.; Zhao, L.; You, C. Association of toll-like receptors single nucleotide polymorphisms with HBV and HCV infection: Research status. *PeerJ* **2022**, *10*, e13335. [CrossRef]
112. Hernandez, J.C.; Arteaga, J.; Paul, S.; Kumar, A.; Latz, E.; Urcuqui-Inchima, S. Up-regulation of TLR2 and TLR4 in dendritic cells in response to HIV type 1 and coinfection with opportunistic pathogens. *AIDS Res. Hum. Retrovir.* **2011**, *27*, 1099–1109. [CrossRef]
113. Mogensen, T.H.; Paludan, S.R. Molecular pathways in virus-induced cytokine production. *Microbiol. Mol. Biol. Rev.* **2001**, *65*, 131–150. [CrossRef] [PubMed]
114. Catalfamo, M.; Le Saout, C.; Lane, H.C. The role of cytokines in the pathogenesis and treatment of HIV infection. *Cytokine Growth Factor Rev.* **2012**, *23*, 207–214. [CrossRef]
115. Kedzierska, K.; Crowe, S.M. Cytokines and HIV-1: Interactions and clinical implications. *Antivir Chem. Chemother.* **2001**, *12*, 133–150. [CrossRef]
116. Borges, A.H.; O'Connor, J.L.; Phillips, A.N.; Ronsholt, F.F.; Pett, S.; Vjecha, M.J.; French, M.A.; Lundgren, J.D.; Insight, S.; Groups, E.S.; et al. Factors Associated with Plasma IL-6 Levels during HIV Infection. *J. Infect. Dis.* **2015**, *212*, 585–595. [CrossRef]
117. Rollenhagen, C.; Asin, S.N. Enhanced HIV-1 replication in ex vivo ectocervical tissues from post-menopausal women correlates with increased inflammatory responses. *Mucosal. Immunol.* **2011**, *4*, 671–681. [CrossRef]
118. Ben Haij, N.; Leghmari, K.; Planes, R.; Thieblemont, N.; Bahraoui, E. HIV-1 Tat protein binds to TLR4-MD2 and signals to induce TNF-alpha and IL-10. *Retrovirology* **2013**, *10*, 123. [CrossRef]
119. Ben Haij, N.; Planes, R.; Leghmari, K.; Serrero, M.; Delobel, P.; Izopet, J.; BenMohamed, L.; Bahraoui, E. HIV-1 Tat Protein Induces Production of Proinflammatory Cytokines by Human Dendritic Cells and Monocytes/Macrophages through Engagement of TLR4-MD2-CD14 Complex and Activation of NF-kappaB Pathway. *PLoS ONE* **2015**, *10*, e0129425. [CrossRef]
120. Planes, R.; Ben Haij, N.; Leghmari, K.; Serrero, M.; BenMohamed, L.; Bahraoui, E. HIV-1 Tat Protein Activates both the MyD88 and TRIF Pathways to Induce Tumor Necrosis Factor Alpha and Interleukin-10 in Human Monocytes. *J. Virol.* **2016**, *90*, 5886–5898. [CrossRef]

121. Napolitano, L.A.; Grant, R.M.; Deeks, S.G.; Schmidt, D.; De Rosa, S.C.; Herzenberg, L.A.; Herndier, B.G.; Andersson, J.; McCune, J.M. Increased production of IL-7 accompanies HIV-1-mediated T-cell depletion: Implications for T-cell homeostasis. *Nat. Med.* **2001**, *7*, 73–79. [CrossRef] [PubMed]
122. Quaranta, M.G.; Camponeschi, B.; Straface, E.; Malorni, W.; Viora, M. Induction of interleukin-15 production by HIV-1 nef protein: A role in the proliferation of uninfected cells. *Exp. Cell Res.* **1999**, *250*, 112–121. [CrossRef] [PubMed]
123. Swaminathan, S.; Qiu, J.; Rupert, A.W.; Hu, Z.; Higgins, J.; Dewar, R.L.; Stevens, R.; Rehm, C.A.; Metcalf, J.A.; Sherman, B.T.; et al. Interleukin-15 (IL-15) Strongly Correlates with Increasing HIV-1 Viremia and Markers of Inflammation. *PLoS ONE* **2016**, *11*, e0167091. [CrossRef]
124. Manganaro, L.; Hong, P.; Hernandez, M.M.; Argyle, D.; Mulder, L.C.F.; Potla, U.; Diaz-Griffero, F.; Lee, B.; Fernandez-Sesma, A.; Simon, V. IL-15 regulates susceptibility of CD4(+) T cells to HIV infection. *Proc. Natl. Acad Sci. USA* **2018**, *115*, E9659–E9667. [CrossRef]
125. Torre, D.; Pugliese, A. Interleukin-18: A proinflammatory cytokine in HIV-1 infection. *Curr. HIV Res.* **2006**, *4*, 423–430. [CrossRef] [PubMed]
126. Al-Harthi, L.; Roebuck, K.A.; Landay, A. Induction of HIV-1 replication by type 1-like cytokines, interleukin (IL)-12 and IL-15: Effect on viral transcriptional activation, cellular proliferation, and endogenous cytokine production. *J. Clin. Immunol.* **1998**, *18*, 124–131. [CrossRef] [PubMed]
127. Reuven, E.M.; Ali, M.; Rotem, E.; Schwarzer, R.; Gramatica, A.; Futerman, A.H.; Shai, Y. The HIV-1 envelope transmembrane domain binds TLR2 through a distinct dimerization motif and inhibits TLR2-mediated responses. *PLoS Pathog.* **2014**, *10*, e1004248. [CrossRef]
128. Martinelli, E.; Cicala, C.; Van Ryk, D.; Goode, D.J.; Macleod, K.; Arthos, J.; Fauci, A.S. HIV-1 gp120 inhibits TLR9-mediated activation and IFN-alpha secretion in plasmacytoid dendritic cells. *Proc. Natl. Acad. Sci. USA* **2007**, *104*, 3396–3401. [CrossRef] [PubMed]
129. Hollenbaugh, J.A.; Gee, P.; Baker, J.; Daly, M.B.; Amie, S.M.; Tate, J.; Kasai, N.; Kanemura, Y.; Kim, D.H.; Ward, B.M.; et al. Host factor SAMHD1 restricts DNA viruses in non-dividing myeloid cells. *PLoS Pathog.* **2013**, *9*, e1003481. [CrossRef]
130. Lahouassa, H.; Daddacha, W.; Hofmann, H.; Ayinde, D.; Logue, E.C.; Dragin, L.; Bloch, N.; Maudet, C.; Bertrand, M.; Gramberg, T.; et al. SAMHD1 restricts the replication of human immunodeficiency virus type 1 by depleting the intracellular pool of deoxynucleoside triphosphates. *Nat. Immunol.* **2012**, *13*, 223–228. [CrossRef]
131. Descours, B.; Cribier, A.; Chable-Bessia, C.; Ayinde, D.; Rice, G.; Crow, Y.; Yatim, A.; Schwartz, O.; Laguette, N.; Benkirane, M. SAMHD1 restricts HIV-1 reverse transcription in quiescent CD4(+) T-cells. *Retrovirology* **2012**, *9*, 87. [CrossRef]
132. Hrecka, K.; Hao, C.; Gierszewska, M.; Swanson, S.K.; Kesik-Brodacka, M.; Srivastava, S.; Florens, L.; Washburn, M.P.; Skowronski, J. Vpx relieves inhibition of HIV-1 infection of macrophages mediated by the SAMHD1 protein. *Nature* **2011**, *474*, 658–661. [CrossRef]
133. Laguette, N.; Sobhian, B.; Casartelli, N.; Ringeard, M.; Chable-Bessia, C.; Segeral, E.; Yatim, A.; Emiliani, S.; Schwartz, O.; Benkirane, M. SAMHD1 is the dendritic- and myeloid-cell-specific HIV-1 restriction factor counteracted by Vpx. *Nature* **2011**, *474*, 654–657. [CrossRef]
134. Chen, S.; Bonifati, S.; Qin, Z.; St Gelais, C.; Kodigepalli, K.M.; Barrett, B.S.; Kim, S.H.; Antonucci, J.M.; Ladner, K.J.; Buzovetsky, O.; et al. SAMHD1 suppresses innate immune responses to viral infections and inflammatory stimuli by inhibiting the NF-kappaB and interferon pathways. *Proc. Natl. Acad Sci. USA* **2018**, *115*, E3798–E3807. [CrossRef] [PubMed]
135. Wang, W.; Li, Y.; Zhang, Z.; Wei, W. Human immunodeficiency virus-1 core: The Trojan horse in virus-host interaction. *Front. Microbiol.* **2022**, *13*, 1002476. [CrossRef] [PubMed]
136. Bishop, K.N.; Verma, M.; Kim, E.Y.; Wolinsky, S.M.; Malim, M.H. APOBEC3G inhibits elongation of HIV-1 reverse transcripts. *PLoS Pathog.* **2008**, *4*, e1000231. [CrossRef] [PubMed]
137. Yu, X.; Yu, Y.; Liu, B.; Luo, K.; Kong, W.; Mao, P.; Yu, X.F. Induction of APOBEC3G ubiquitination and degradation by an HIV-1 Vif-Cul5-SCF complex. *Science* **2003**, *302*, 1056–1060. [CrossRef]
138. Sheehy, A.M.; Gaddis, N.C.; Malim, M.H. The antiretroviral enzyme APOBEC3G is degraded by the proteasome in response to HIV-1 Vif. *Nat. Med.* **2003**, *9*, 1404. [CrossRef] [PubMed]
139. Guo, F.; Cen, S.; Niu, M.; Saadatmand, J.; Kleiman, L. Inhibition of tRNA(3)(Lys)-primed reverse transcription by human APOBEC3G during human immunodeficiency virus type 1 replication. *J. Virol.* **2006**, *80*, 11710–11722. [CrossRef]
140. Kobayashi, M.; Takaori-Kondo, A.; Miyauchi, Y.; Iwai, K.; Uchiyama, T. Ubiquitination of APOBEC3G by an HIV-1 Vif-Cullin5-Elongin B-Elongin C complex is essential for Vif function. *J. Biol. Chem.* **2005**, *280*, 18573–18578. [CrossRef]
141. Neil, S.J.; Zang, T.; Bieniasz, P.D. Tetherin inhibits retrovirus release and is antagonized by HIV-1 Vpu. *Nature* **2008**, *451*, 425–430. [CrossRef]
142. Van Damme, N.; Goff, D.; Katsura, C.; Jorgenson, R.L.; Mitchell, R.; Johnson, M.C.; Stephens, E.B.; Guatelli, J. The interferon-induced protein BST-2 restricts HIV-1 release and is downregulated from the cell surface by the viral Vpu protein. *Cell Host Microbe* **2008**, *3*, 245–252. [CrossRef]
143. Mangeat, B.; Gers-Huber, G.; Lehmann, M.; Zufferey, M.; Luban, J.; Piguet, V. HIV-1 Vpu neutralizes the antiviral factor Tetherin/BST-2 by binding it and directing its beta-TrCP2-dependent degradation. *PLoS Pathog.* **2009**, *5*, e1000574. [CrossRef]
144. Dube, M.; Roy, B.B.; Guiot-Guillain, P.; Binette, J.; Mercier, J.; Chiasson, A.; Cohen, E.A. Antagonism of tetherin restriction of HIV-1 release by Vpu involves binding and sequestration of the restriction factor in a perinuclear compartment. *PLoS Pathog.* **2010**, *6*, e1000856. [CrossRef] [PubMed]

145. Khan, H.; Sumner, R.P.; Rasaiyaah, J.; Tan, C.P.; Rodriguez-Plata, M.T.; Van Tulleken, C.; Fink, D.; Zuliani-Alvarez, L.; Thorne, L.; Stirling, D.; et al. HIV-1 Vpr antagonizes innate immune activation by targeting karyopherin-mediated NF-kappaB/IRF3 nuclear transport. *eLife* **2020**, *9*, e60821. [CrossRef] [PubMed]
146. Marin, M.; Rose, K.M.; Kozak, S.L.; Kabat, D. HIV-1 Vif protein binds the editing enzyme APOBEC3G and induces its degradation. *Nat. Med.* **2003**, *9*, 1398–1403. [CrossRef]
147. Zhang, H.; Yang, B.; Pomerantz, R.J.; Zhang, C.; Arunachalam, S.C.; Gao, L. The cytidine deaminase CEM15 induces hypermutation in newly synthesized HIV-1 DNA. *Nature* **2003**, *424*, 94–98. [CrossRef] [PubMed]
148. Arias, J.F.; Heyer, L.N.; von Bredow, B.; Weisgrau, K.L.; Moldt, B.; Burton, D.R.; Rakasz, E.G.; Evans, D.T. Tetherin antagonism by Vpu protects HIV-infected cells from antibody-dependent cell-mediated cytotoxicity. *Proc. Natl. Acad Sci. USA* **2014**, *111*, 6425–6430. [CrossRef]
149. Jain, P.; Boso, G.; Langer, S.; Soonthornvacharin, S.; De Jesus, P.D.; Nguyen, Q.; Olivieri, K.C.; Portillo, A.J.; Yoh, S.M.; Pache, L.; et al. Large-Scale Arrayed Analysis of Protein Degradation Reveals Cellular Targets for HIV-1 Vpu. *Cell Rep.* **2018**, *22*, 2493–2503. [CrossRef]
150. Langer, S.; Hammer, C.; Hopfensperger, K.; Klein, L.; Hotter, D.; De Jesus, P.D.; Herbert, K.M.; Pache, L.; Smith, N.; van der Merwe, J.A.; et al. HIV-1 Vpu is a potent transcriptional suppressor of NF-kappaB-elicited antiviral immune responses. *eLife* **2019**, *8*, e41930. [CrossRef] [PubMed]
151. Manganaro, L.; de Castro, E.; Maestre, A.M.; Olivieri, K.; Garcia-Sastre, A.; Fernandez-Sesma, A.; Simon, V. HIV Vpu Interferes with NF-kappaB Activity but Not with Interferon Regulatory Factor 3. *J. Virol.* **2015**, *89*, 9781–9790. [CrossRef]
152. Bour, S.; Perrin, C.; Akari, H.; Strebel, K. The human immunodeficiency virus type 1 Vpu protein inhibits NF-kappa B activation by interfering with beta TrCP-mediated degradation of Ikappa B. *J. Biol. Chem.* **2001**, *276*, 15920–15928. [CrossRef]
153. Rosa, A.; Chande, A.; Ziglio, S.; De Sanctis, V.; Bertorelli, R.; Goh, S.L.; McCauley, S.M.; Nowosielska, A.; Antonarakis, S.E.; Luban, J.; et al. HIV-1 Nef promotes infection by excluding SERINC5 from virion incorporation. *Nature* **2015**, *526*, 212–217. [CrossRef] [PubMed]
154. Usami, Y.; Wu, Y.; Gottlinger, H.G. SERINC3 and SERINC5 restrict HIV-1 infectivity and are counteracted by Nef. *Nature* **2015**, *526*, 218–223. [CrossRef] [PubMed]
155. Olivetta, E.; Tirelli, V.; Chiozzini, C.; Scazzocchio, B.; Romano, I.; Arenaccio, C.; Sanchez, M. HIV-1 Nef impairs key functional activities in human macrophages through CD36 downregulation. *PLoS ONE* **2014**, *9*, e93699. [CrossRef] [PubMed]
156. Ringeard, M.; Marchand, V.; Decroly, E.; Motorin, Y.; Bennasser, Y. FTSJ3 is an RNA 2′-O-methyltransferase recruited by HIV to avoid innate immune sensing. *Nature* **2019**, *565*, 500–504. [CrossRef]
157. Mellors, J.W.; Rinaldo, C.R., Jr.; Gupta, P.; White, R.M.; Todd, J.A.; Kingsley, L.A. Prognosis in HIV-1 infection predicted by the quantity of virus in plasma. *Science* **1996**, *272*, 1167–1170. [CrossRef]
158. Sadowski, I.; Hashemi, F.B. Strategies to eradicate HIV from infected patients: Elimination of latent provirus reservoirs. *Cell Mol. Life Sci.* **2019**, *76*, 3583–3600. [CrossRef]
159. Rasmussen, T.A.; Anderson, J.L.; Wightman, F.; Lewin, S.R. Cancer therapies in HIV cure research. *Curr. Opin. HIV AIDS* **2017**, *12*, 96–104. [CrossRef]
160. Yu, X.; Li, Z.; Zhou, Z.; Kilby, J.M.; Jiang, W. Microbial TLR Agonists and Humoral Immunopathogenesis in HIV Disease. *Epidemiology* **2013**, *3*, 120. [CrossRef]
161. Meier, A.; Altfeld, M. Toll-like receptor signaling in HIV-1 infection: A potential target for therapy? *Expert Rev. Anti. Infect. Ther.* **2007**, *5*, 323–326. [CrossRef]
162. Baenziger, S.; Heikenwalder, M.; Johansen, P.; Schlaepfer, E.; Hofer, U.; Miller, R.C.; Diemand, S.; Honda, K.; Kundig, T.M.; Aguzzi, A.; et al. Triggering TLR7 in mice induces immune activation and lymphoid system disruption, resembling HIV-mediated pathology. *Blood* **2009**, *113*, 377–388. [CrossRef]
163. Mogensen, T.H.; Melchjorsen, J.; Larsen, C.S.; Paludan, S.R. Innate immune recognition and activation during HIV infection. *Retrovirology* **2010**, *7*, 54. [CrossRef] [PubMed]
164. Douek, D.C.; Roederer, M.; Koup, R.A. Emerging concepts in the immunopathogenesis of AIDS. *Annu. Rev. Med.* **2009**, *60*, 471–484. [CrossRef] [PubMed]
165. Koblin, B.A.; Taylor, P.E.; Rubinstein, P.; Stevens, C.E. Effect of duration of hepatitis B virus infection on the association between human immunodeficiency virus type-1 and hepatitis B viral replication. *Hepatology* **1992**, *15*, 590–592. [CrossRef] [PubMed]
166. Vlahakis, S.R.; Villasis-Keever, A.; Gomez, T.S.; Bren, G.D.; Paya, C.V. Human immunodeficiency virus-induced apoptosis of human hepatocytes via CXCR4. *J. Infect. Dis.* **2003**, *188*, 1455–1460. [CrossRef]
167. Yoong, K.F.; Afford, S.C.; Jones, R.; Aujla, P.; Qin, S.; Price, K.; Hubscher, S.G.; Adams, D.H. Expression and function of CXC and CC chemokines in human malignant liver tumors: A role for human monokine induced by gamma-interferon in lymphocyte recruitment to hepatocellular carcinoma. *Hepatology* **1999**, *30*, 100–111. [CrossRef]
168. Thio, C.L.; Seaberg, E.C.; Skolasky, R., Jr.; Phair, J.; Visscher, B.; Munoz, A.; Thomas, D.L.; Multicenter, A.C.S. HIV-1, hepatitis B virus, and risk of liver-related mortality in the Multicenter Cohort Study (MACS). *Lancet* **2002**, *360*, 1921–1926. [CrossRef]
169. Robbins, H.A.; Shiels, M.S.; Pfeiffer, R.M.; Engels, E.A. Epidemiologic contributions to recent cancer trends among HIV-infected people in the United States. *AIDS* **2014**, *28*, 881–890. [CrossRef]
170. Shiels, M.S.; Engels, E.A. Evolving epidemiology of HIV-associated malignancies. *Curr. Opin. HIV AIDS* **2017**, *12*, 6–11. [CrossRef]

171. Page, E.E.; Nelson, M.; Kelleher, P. HIV and hepatitis C coinfection: Pathogenesis and microbial translocation. *Curr. Opin. HIV AIDS* **2011**, *6*, 472–477. [CrossRef]
172. Soriano, V.; Moreno-Torres, V.; Mendoza, C.; Corral, O.; Barreiro, P. Viral hepatitis in persons living with HIV in the post-COVID era. *AIDS Rev.* **2023**, *25*, 1–13. [CrossRef]
173. Lin, W.; Weinberg, E.M.; Chung, R.T. Pathogenesis of accelerated fibrosis in HIV/HCV co-infection. *J. Infect. Dis.* **2013**, *207* (Suppl. S1), S13–S18. [CrossRef] [PubMed]
174. Shi, L.; Qin, E.; Zhou, J.; Zhao, J.; Nie, W.; Jiang, T.; Chen, W.; Wu, D.; Huang, L.; Liu, L.; et al. HIV and HCV Co-Culture Promotes Profibrogenic Gene Expression through an Epimorphin-Mediated ERK Signaling Pathway in Hepatic Stellate Cells. *PLoS ONE* **2016**, *11*, e0158386. [CrossRef] [PubMed]
175. Lin, W.; Weinberg, E.M.; Tai, A.W.; Peng, L.F.; Brockman, M.A.; Kim, K.A.; Kim, S.S.; Borges, C.B.; Shao, R.X.; Chung, R.T. HIV increases HCV replication in a TGF-beta1-dependent manner. *Gastroenterology* **2008**, *134*, 803–811. [CrossRef] [PubMed]
176. Qu, J.; Zhang, Q.; Li, Y.; Liu, W.; Chen, L.; Zhu, Y.; Wu, J. The Tat protein of human immunodeficiency virus-1 enhances hepatitis C virus replication through interferon gamma-inducible protein-10. *BMC Immunol.* **2012**, *13*, 15. [CrossRef]
177. Schonrich, G.; Raftery, M.J. The PD-1/PD-L1 Axis and Virus Infections: A Delicate Balance. *Front. Cell. Infect. Microbiol.* **2019**, *9*, 207. [CrossRef]
178. Sun, H.Y.; Sheng, W.H.; Tsai, M.S.; Lee, K.Y.; Chang, S.Y.; Hung, C.C. Hepatitis B virus coinfection in human immunodeficiency virus-infected patients: A review. *World J. Gastroenterol.* **2014**, *20*, 14598–14614. [CrossRef]
179. Martinsen, J.T.; Gunst, J.D.; Hojen, J.F.; Tolstrup, M.; Sogaard, O.S. The Use of Toll-Like Receptor Agonists in HIV-1 Cure Strategies. *Front. Immunol.* **2020**, *11*, 1112. [CrossRef]
180. Chun, T.W.; Carruth, L.; Finzi, D.; Shen, X.; DiGiuseppe, J.A.; Taylor, H.; Hermankova, M.; Chadwick, K.; Margolick, J.; Quinn, T.C.; et al. Quantification of latent tissue reservoirs and total body viral load in HIV-1 infection. *Nature* **1997**, *387*, 183–188. [CrossRef]
181. Rodari, A.; Darcis, G.; Van Lint, C.M. The Current Status of Latency Reversing Agents for HIV-1 Remission. *Annu. Rev. Virol.* **2021**, *8*, 491–514. [CrossRef]
182. Li, Y.; Wang, Z.; Hou, Y.; Liu, X.; Hong, J.; Shi, X.; Huang, X.; Zhang, T.; Liao, X.; Zhang, L. Novel TLR7/8 agonists promote activation of HIV-1 latent reservoirs and human T and NK cells. *Front. Microbiol.* **2023**, *14*, 1033448. [CrossRef]
183. Tanaka, K.; Kim, Y.; Roche, M.; Lewin, S.R. The role of latency reversal in HIV cure strategies. *J. Med. Primatol.* **2022**, *51*, 278–283. [CrossRef]
184. Macedo, A.B.; Novis, C.L.; Bosque, A. Targeting Cellular and Tissue HIV Reservoirs with Toll-Like Receptor Agonists. *Front. Immunol.* **2019**, *10*, 2450. [CrossRef] [PubMed]
185. Novis, C.L.; Archin, N.M.; Buzon, M.J.; Verdin, E.; Round, J.L.; Lichterfeld, M.; Margolis, D.M.; Planelles, V.; Bosque, A. Reactivation of latent HIV-1 in central memory CD4(+) T cells through TLR-1/2 stimulation. *Retrovirology* **2013**, *10*, 119. [CrossRef]
186. Winckelmann, A.A.; Munk-Petersen, L.V.; Rasmussen, T.A.; Melchjorsen, J.; Hjelholt, T.J.; Montefiori, D.; Ostergaard, L.; Sogaard, O.S.; Tolstrup, M. Administration of a Toll-like receptor 9 agonist decreases the proviral reservoir in virologically suppressed HIV-infected patients. *PLoS ONE* **2013**, *8*, e62074. [CrossRef] [PubMed]
187. Lim, S.Y.; Osuna, C.E.; Hraber, P.T.; Hesselgesser, J.; Gerold, J.M.; Barnes, T.L.; Sanisetty, S.; Seaman, M.S.; Lewis, M.G.; Geleziunas, R.; et al. TLR7 agonists induce transient viremia and reduce the viral reservoir in SIV-infected rhesus macaques on antiretroviral therapy. *Sci. Transl. Med.* **2018**, *10*, eaao4521. [CrossRef] [PubMed]
188. Riddler, S.A.; Para, M.; Benson, C.A.; Mills, A.; Ramgopal, M.; DeJesus, E.; Brinson, C.; Cyktor, J.; Jacobs, J.; Koontz, D.; et al. Vesatolimod, a Toll-like Receptor 7 Agonist, Induces Immune Activation in Virally Suppressed Adults Living with Human Immunodeficiency Virus-1. *Clin. Infect. Dis.* **2021**, *72*, e815–e824. [CrossRef] [PubMed]
189. Macedo, A.B.; Novis, C.L.; De Assis, C.M.; Sorensen, E.S.; Moszczynski, P.; Huang, S.H.; Ren, Y.; Spivak, A.M.; Jones, R.B.; Planelles, V.; et al. Dual TLR2 and TLR7 agonists as HIV latency-reversing agents. *JCI Insight* **2018**, *3*, e122673. [CrossRef]
190. Vibholm, L.; Schleimann, M.H.; Hojen, J.F.; Benfield, T.; Offersen, R.; Rasmussen, K.; Olesen, R.; Dige, A.; Agnholt, J.; Grau, J.; et al. Short-Course Toll-Like Receptor 9 Agonist Treatment Impacts Innate Immunity and Plasma Viremia in Individuals with Human Immunodeficiency Virus Infection. *Clin. Infect. Dis.* **2017**, *64*, 1686–1695. [CrossRef]
191. Biasin, M.; Piacentini, L.; Lo Caputo, S.; Naddeo, V.; Pierotti, P.; Borelli, M.; Trabattoni, D.; Mazzotta, F.; Shearer, G.M.; Clerici, M. TLR activation pathways in HIV-1-exposed seronegative individuals. *J. Immunol.* **2010**, *184*, 2710–2717. [CrossRef]
192. Hvilsom, C.T.; Sogaard, O.S. TLR-Agonist Mediated Enhancement of Antibody-Dependent Effector Functions as Strategy for an HIV-1 Cure. *Front. Immunol.* **2021**, *12*, 704617. [CrossRef]
193. Board, N.L.; Moskovljevic, M.; Wu, F.; Siliciano, R.F.; Siliciano, J.D. Engaging innate immunity in HIV-1 cure strategies. *Nat. Rev. Immunol.* **2022**, *22*, 499–512. [CrossRef]
194. Liu, Z.M.; Yang, M.H.; Yu, K.; Lian, Z.X.; Deng, S.L. Toll-like receptor (TLRs) agonists and antagonists for COVID-19 treatments. *Front. Pharmacol.* **2022**, *13*, 989664. [CrossRef] [PubMed]
195. Perales-Linares, R.; Navas-Martin, S. Toll-like receptor 3 in viral pathogenesis: Friend or foe? *Immunology* **2013**, *140*, 153–167. [CrossRef] [PubMed]
196. Zheng, H.; Wu, P.; Bonnet, P.A. Recent Advances on Small-Molecule Antagonists Targeting TLR7. *Molecules* **2023**, *28*, 634. [CrossRef]
197. Chang, J.J.; Altfeld, M. TLR-mediated immune activation in HIV. *Blood* **2009**, *113*, 269–270. [CrossRef]

198. Feng, Y.; Broder, C.C.; Kennedy, P.E.; Berger, E.A. HIV-1 entry cofactor: Functional cDNA cloning of a seven-transmembrane, G protein-coupled receptor. *Science* **1996**, *272*, 872–877. [CrossRef]
199. McHugh, J. Hitting the right spot on CXCR4. *Nat. Rev. Rheumatol.* **2019**, *15*, 514. [CrossRef] [PubMed]
200. Caspar, B.; Cocchiara, P.; Melet, A.; Van Emelen, K.; Van der Aa, A.; Milligan, G.; Herbeuval, J.P. CXCR4 as a novel target in immunology: Moving away from typical antagonists. *Future Drug Discov.* **2022**, *4*, FDD77. [CrossRef]
201. Smith, N.; Pietrancosta, N.; Davidson, S.; Dutrieux, J.; Chauveau, L.; Cutolo, P.; Dy, M.; Scott-Algara, D.; Manoury, B.; Zirafi, O.; et al. Natural amines inhibit activation of human plasmacytoid dendritic cells through CXCR4 engagement. *Nat. Commun.* **2017**, *8*, 14253. [CrossRef]
202. Smith, N.; Rodero, M.P.; Bekaddour, N.; Bondet, V.; Ruiz-Blanco, Y.B.; Harms, M.; Mayer, B.; Bader-Meunier, B.; Quartier, P.; Bodemer, C.; et al. Control of TLR7-mediated type I IFN signaling in pDCs through CXCR4 engagement-A new target for lupus treatment. *Sci. Adv.* **2019**, *5*, eaav9019. [CrossRef]
203. Siracusano, G.; Lopalco, L. Immunotherapy with Cell-Based Biological Drugs to Cure HIV-1 Infection. *Cells* **2021**, *11*, 77. [CrossRef] [PubMed]
204. Krarup, A.R.; Abdel-Mohsen, M.; Schleimann, M.H.; Vibholm, L.; Engen, P.A.; Dige, A.; Wittig, B.; Schmidt, M.; Green, S.J.; Naqib, A.; et al. The TLR9 agonist MGN1703 triggers a potent type I interferon response in the sigmoid colon. *Mucosal. Immunol.* **2018**, *11*, 449–461. [CrossRef] [PubMed]
205. Dowling, D.; Hamilton, C.M.; O'Neill, S.M. A comparative analysis of cytokine responses, cell surface marker expression and MAPKs in DCs matured with LPS compared with a panel of TLR ligands. *Cytokine* **2008**, *41*, 254–262. [CrossRef] [PubMed]
206. Deifl, S.; Kitzmuller, C.; Steinberger, P.; Himly, M.; Jahn-Schmid, B.; Fischer, G.F.; Zlabinger, G.J.; Bohle, B. Differential activation of dendritic cells by toll-like receptors causes diverse differentiation of naive CD4+ T cells from allergic patients. *Allergy* **2014**, *69*, 1602–1609. [CrossRef] [PubMed]
207. Ioannidis, I.; Ye, F.; McNally, B.; Willette, M.; Flano, E. Toll-like receptor expression and induction of type I and type III interferons in primary airway epithelial cells. *J. Virol.* **2013**, *87*, 3261–3270. [CrossRef] [PubMed]
208. Shan, L.; Deng, K.; Shroff, N.S.; Durand, C.M.; Rabi, S.A.; Yang, H.C.; Zhang, H.; Margolick, J.B.; Blankson, J.N.; Siliciano, R.F. Stimulation of HIV-1-specific cytolytic T lymphocytes facilitates elimination of latent viral reservoir after virus reactivation. *Immunity* **2012**, *36*, 491–501. [CrossRef] [PubMed]

International Journal of
Molecular Sciences

MDPI

Review

TLRs: Innate Immune Sentries against SARS-CoV-2 Infection

Stefania Mantovani [1,*], Barbara Oliviero [1], Stefania Varchetta [1], Alessandra Renieri [2,3,4] and Mario U. Mondelli [1,5,*]

1 Department of Research, Division of Clinical Immunology—Infectious Diseases, Fondazione IRCCS Policlinico San Matteo, 27100 Pavia, Italy; b.oliviero@smatteo.pv.it (B.O.); s.varchetta@smatteo.pv.it (S.V.)
2 Medical Genetics, University of Siena, 53100 Siena, Italy; alessandra.renieri@unisi.it
3 Med Biotech Hub and Competence Center, Department of Medical Biotechnologies, University of Siena, 53100 Siena, Italy
4 Genetica Medica, Azienda Ospedaliero-Universitaria Senese, 53100 Siena, Italy
5 Department of Internal Medicine and Therapeutics, University of Pavia, 27100 Pavia, Italy
* Correspondence: s.mantovani@smatteo.pv.it (S.M.); mario.mondelli@unipv.it (M.U.M.); Tel.: +39-0382-501808 (S.M.); +39-0382-502636 (M.U.M.); Fax: +39-0382-526450 (M.U.M.)

Abstract: Coronavirus disease 2019 (COVID-19), caused by severe acute respiratory syndrome coronavirus 2 (SARS-CoV-2), has been responsible for a devastating pandemic since March 2020. Toll-like receptors (TLRs), crucial components in the initiation of innate immune responses to different pathogens, trigger the downstream production of pro-inflammatory cytokines, interferons, and other mediators. It has been demonstrated that they contribute to the dysregulated immune response observed in patients with severe COVID-19. TLR2, TLR3, TLR4 and TLR7 have been associated with COVID-19 severity. Here, we review the role of TLRs in the etiology and pathogenesis of COVID-19, including TLR7 and TLR3 rare variants, the L412F polymorphism in TLR3 that negatively regulates anti-SARS-CoV-2 immune responses, the TLR3-related cellular senescence, the interaction of TLR2 and TLR4 with SARS-CoV-2 proteins and implication of TLR2 in NET formation by SARS-CoV-2. The activation of TLRs contributes to viral clearance and disease resolution. However, TLRs may represent a double-edged sword which may elicit dysregulated immune signaling, leading to the production of proinflammatory mediators, resulting in severe disease. TLR-dependent excessive inflammation and TLR-dependent antiviral response may tip the balance towards the former or the latter, altering the equilibrium that drives the severity of disease.

Keywords: Toll-like receptor; COVID-19; SARS-CoV-2; innate immunity; host genetics

Citation: Mantovani, S.; Oliviero, B.; Varchetta, S.; Renieri, A.; Mondelli, M.U. TLRs: Innate Immune Sentries against SARS-CoV-2 Infection. *Int. J. Mol. Sci.* **2023**, *24*, 8065. https://doi.org/10.3390/ijms24098065

Academic Editors: Ralf Kircheis and Oliver Planz

Received: 23 March 2023
Revised: 21 April 2023
Accepted: 27 April 2023
Published: 29 April 2023

1. Introduction

The damaging effects of pathogens have led to the evolution of a variety of host-defense mechanisms, among which innate immune responses are of prime importance in the early recognition and elimination of invading microorganisms, including severe acute respiratory syndrome coronavirus-2 (SARS-CoV-2) [1].

SARS-CoV-2, a positive-sense single-stranded (ss)RNA Betacoronavirus member of the family *Coronaviridae*, emerged at the end 2019 and is the causative viral pathogen of the coronavirus disease 2019 (COVID-19), which became pandemic in March 2020. Infection with SARS-CoV-2 causes mild to moderate respiratory illness in the majority of patients; however, it may occasionally be responsible for severe interstitial pneumonia, myocarditis, acute kidney injury, acute respiratory distress syndrome (ARDS), multiorgan failure and death, predominantly in elderly patients and in those with several comorbidities such as cardiovascular disease, diabetes, chronic respiratory disease, or cancer or genetic predisposition [2–5] Clearly, beyond SARS-CoV-2 variability that contributes to incomplete immune protection, inflammatory conditions, as well as the immune status of patients, are critical in determining the course of the disease [6,7].

Innate immune recognition is based primarily on complement activation, phagocytosis, autophagy and immune activation by a limited number of germline-encoded pattern recognition receptors (PRRs) that recognize conserved motifs on pathogens named pathogen associated molecular patterns (PAMPs), whose individual variability is genetically determined. Importantly, PRRs are expressed constitutively in the host, and sense the pathogens regardless of their life-cycle stage [8,9]. PRRs are present in both immune and non-immune cells, and induce the release of a selected repertoire of host defense factors, inducing a specific inflammatory response driven by the pathogen's molecular signatures. In mammalian cells, PRRs can be classified into two main classes: membrane-bound receptors, such as Toll-like receptors (TLRs) and C-type lectin receptors (CLRs), and cytoplasmic sensors, including NOD-like receptors (NLRs), absent in melanoma-2-like receptors (ALRs), RIG-I-like receptors (RLRs) and an increasing range of cytosolic nucleic acid sensors [8,9] (Figure 1).

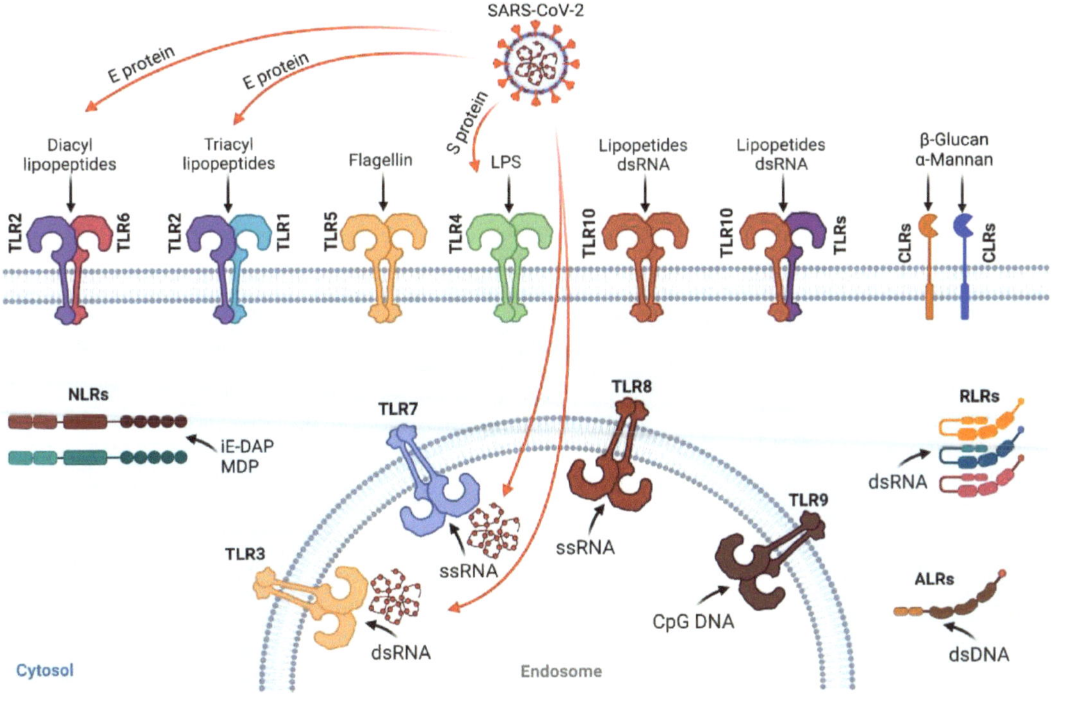

Figure 1. Pattern recognition receptors (PRRs) and their ligands. PRRs can be classified into two main classes: membrane-bound receptors, such as TLRs and CLRs, and cytoplasmic sensors, including NLRs, ALRs and RLRs. In humans, 10 TLRs respond to a variety of Pathogen Associated Molecular Patterns, including lipopolysaccharide (TLR4), lipopeptides (TLR1, 2, 6 and 10), bacterial flagellin (TLR5), viral dsRNA (TLR3 and 10), viral or bacterial ssRNA (TLR7 and 8) and CpG-rich unmethylated DNA (TLR9). TLR1, 2, 4, 5, 6 and 10 are plasma membrane TLRs, whereas TLR3, 7, 8 and 9 are located on endosomes. TLRs can be homodimers, in the cases of TLRs 3, 4, 5, 7, 8 and 9, or heterodimers as in TLRs 1 and 2 or TLR2 and 6. TLR2 and 4 sense the SARS-CoV-2 Envelope protein (E) and the Spike protein (S), respectively, whereas TLR3 and 7 sense SARS-CoV-2 nucleic acid. CLRs are a family of receptors that recognize carbohydrates, such as β-Glucan and α-Mannan, on the surface of pathogenic microorganisms. NLRs are intracellular PRRs that mainly recognize the diaminopimelic acid iE-DAP of the cell wall of Gram-negative bacteria and MDP in all bacterial cell walls. ALRs recognize the dsDNA of bacteria. RLRs recognize short dsRNA (<1000 bp) and long-chain dsRNA (>1000 bp) of different

viruses through ligand-recognition domains. TLRs, Toll-like receptors; CLRs, C-type lectin receptors; NLRs, NOD-like receptors; ALRs, absent in melanoma-2-like receptors; RLRs, RIG-I-like receptors; dsRNA, double-stranded RNA; ssRNA, single-stranded RNA; dsDNA, double-stranded DNA; LPS, lipopolysaccharides; CpG DNA, cytosine-phosphate-guanine DNA; iE-DAP, γ-D-glu-meso-diaminopimelic acid; and MDP, muramyl dipeptide. Created with BioRender.comTLRs are crucial components in the initiation of innate immune responses to a variety of pathogens, triggering the downstream production of pro-inflammatory cytokines, interferons (IFNs), and other mediators [10]. Several findings have highlighted the potential contribution of TLRs to the dysregulated immune response observed in patients with severe COVID-19.

2. Structure and Downstream Signaling Pathway of TLRs

The TLR family is one of the best-characterized PRR families, and is responsible for sensing invading pathogens outside the cell and in intracellular endosomes and lysosomes, leading to a potent immunostimulatory response. In humans, 10 TLRs respond to a variety of PAMPs, including lipopolysaccharide (TLR4), lipopeptides (TLR1, TLR2, TLR6 and TLR10), bacterial flagellin (TLR5), viral dsRNA (TLR3), viral or bacterial ssRNA (TLR7 and TLR8) and CpG-rich unmethylated DNA (TLR9). TLR1, TLR2, TLR4, TLR5, TLR6 and TLR10 are plasma membrane TLRs, whereas TLR3, TLR7, TLR8 and TLR9 are located on endosomes [11–16] (Figure 1). TLRs are type I integral membrane glycoproteins, characterized by extracellular (or extraendosomal) domains containing varying numbers of leucine-rich-repeat (LRR) motifs and a cytoplasmic signaling domain homologous to that of the interleukin 1 receptor (IL-1R), termed the Toll/IL-1R homology (TIR) domain. TLRs share a common structural framework in their extracellular, ligand-binding domains. These domains all adopt horseshoe-shaped structures formed by LRR motifs. Classically, in ligand binding, two extracellular domains form an "m"-shaped dimer, sandwiching the ligand molecule and bringing the transmembrane and cytoplasmic domains into close proximity and triggering a downstream signaling cascade. Although the ligand-induced dimerization of these receptors has many common features, the nature and number of the interactions of the TLR extracellular domains with their ligands varies markedly between TLRs [17]. For instance, crystallographic studies of TLR7 and TLR8 have revealed that these receptors contain two distinct ligand-binding sites that cooperate to maximize dimerization. Site 1, highly conserved in both TLRs, recognizes nucleosides (guanosine for TLR7 and uridine for TLR8) and is essential for receptor dimerization and signaling with ad hoc ligand concentration. The other binding site (Site 2) recognizes nucleic acid sequences and plays an auxiliary role in receptor dimerization by enhancing the binding affinities of site 1 ligands. This process, by which dual ligands (nucleic acid sequences and free nucleosides) concur to maximally dimerize TLRs, takes place only in the endosome. Of note, ssRNA-binding to site 2 is not sufficient for the formation of a signaling competent TLR dimer, highlighting the relevance of these TLRs to sense distinct RNA-degradation products rather than full-length ssRNAs [18–22]. Extracellular domains of TLRs can be homodimers, in the cases of TLRs 3, 4, 5, 7, 8 and 9, or heterodimers as in TLRs 1 and 2 or TLR2 and 6 [14,23,24] (Figure 1).

After ligand-induced dimerization, the cytoplasmic TIR domain associates with TIR domain-containing adaptor molecules to transmit signaling. To date, six adaptor proteins have been identified: MyD88 (myeloid differentiation primary-response gene 88), MAL (MyD88-adaptor-like protein), TRIF (TIR-domain-containing adaptor protein inducing interferon-β (IFNβ)) and TRAM (TRIF-related adaptor molecule) are recruited to TIR domains to initiate signaling, whereas SARM (sterile α- and armadillo-motif-containing protein) and BCAP (B-cell adaptor for PI3K) inhibit TLR responses. All TLRs, except for TLR3, associate with MyD88 and MAL proteins. Otherwise, TLR3 and TLR4 use a TRIF-dependent pathway. Both the MyD88-dependent and TRIF-dependent pathways lead to the activation of downstream molecules: nuclear factor-κB (NF-κB), activating protein-1 (AP-1), and members of the IFN-regulatory factor (IRF) family [17,25–28].

The transcription factor NF-κB is the master regulator of all TLR-induced responses. There are seven NF-κB proteins (p65, c-Rel, RelB, p105, p50, p100 and p52), all of which share the Rel homology domain (RHD) that binds to discrete DNA sequences known as κB sites present in promoter and enhancer regions of various genes. The most frequently activated form of NF-kB in TLR signaling is a heterodimer composed of RelA and p50. The transcriptional activity of NF-κB is regulated by the inhibitory kappaB (IκB) proteins that prevent translocation to the nucleus where NF-κB binds their cognate sites in DNA and activates gene transcription. This pathway, called the 'canonical pathway', is responsible for the TLR-mediated induction of inflammatory cytokines such as tumor necrosis factor-α (TNF-α) and IL-6 [29] (Figure 2).

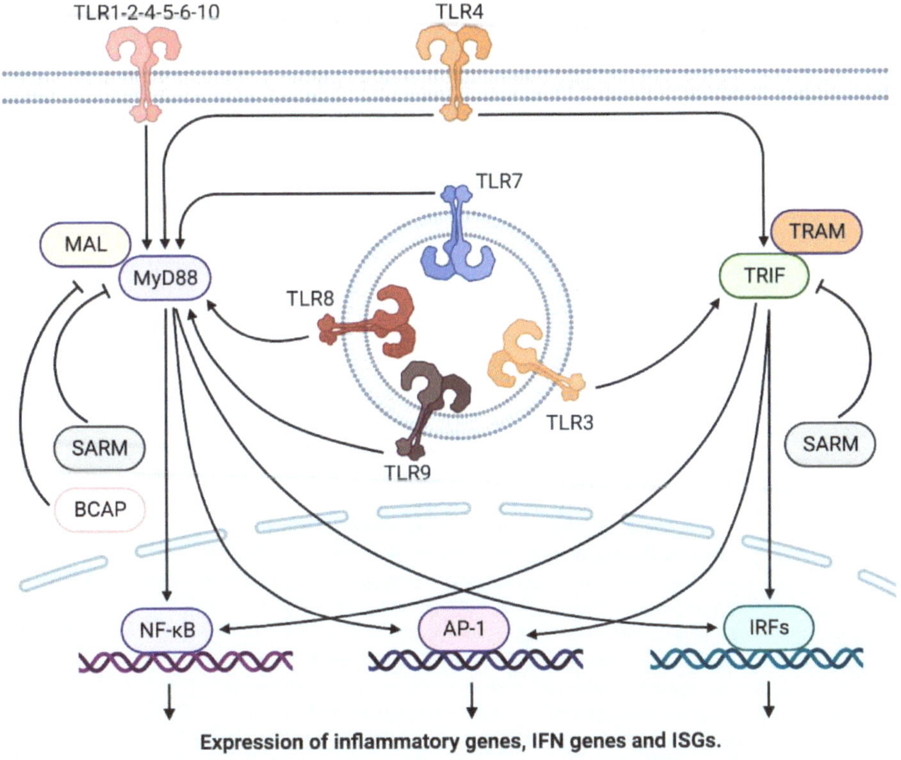

Figure 2. Downstream signaling pathway of TLRs. TLRs are type I integral membrane glycoproteins characterized by the extracellular (or the extraendosomal) domains and a cytoplasmic signaling domain. TLRs share a common structural framework in their extracellular ligand-binding domains. These domains, formed by varying numbers of leucine-rich-repeat motifs, all adopt horseshoe-shaped structures. After ligand binding, two extracellular domains form an "m"-shaped dimer sandwiching the ligand molecule, bringing the transmembrane and cytoplasmic domains into close proximity and triggering a downstream signaling cascade. After ligand-induced dimerization, the cytoplasmic signaling domain associates with adaptor molecules to transmit signaling. Six adaptor proteins have been identified: MyD88, MAL, TRIF and TRAM are recruited to TLR cytoplasmic domains to initiate signaling, whereas SARM and BCAP inhibit TLR responses. All TLRs, except for TLR3, associate with MyD88 and MAL proteins. Otherwise, TLR3 and TLR4 use a TRIF-dependent pathway. Both the MyD88-dependent and TRIF-dependent pathways lead to the activation of downstream molecules: NF-κB, AP-1, and members of the IRF family. These pathways are responsible for the TLR-mediated

expression of inflammatory genes, type I, type II and type III IFN genes and interferon stimulated genes (ISGs). MyD-88, myeloid differentiation primary-response gene 88; MAL, MyD88-adaptor-like protein; TRIF, TIR-domain-containing adaptor protein inducing interferon-β (IFNβ); TRAM, TRIF-related adaptor molecule; SARM, sterile α- and armadillo-motif-containing protein; BCAP, B-cell adaptor for PI3K; NF-κB, nuclear factor-κB; AP-1, activating protein-1; and IRF, IFN-regulatory factor. Created with BioRender.com.All nine members of the IRF family have a conserved amino-terminal DNA-binding domain (DBD) [30–35] that recognizes the consensus DNA sequence element ISRE [36] in the gene promoters of IFNs and interferon-stimulated gene (ISG) genes [37]. These cytokines, in turn, activate antimicrobial and proinflammatory activities, as well as the maturation of antigen-specific adaptive immune responses. IRF3, IRF5 and IRF7 are primarily responsible for the activation of type I IFN genes downstream of TLR activation, although IRF1 and IRF8 can also contribute [37]. IRF1, IRF3, IRF5 and IRF8 also induce the expression of proinflammatory cytokines such as IL-6, TNF-α, CCL5/RANTES, CXCL10 or CCL2, and other genes in response to TLR activation [38].

3. TLRs and SARS-CoV-2

Several TLRs, such as TLR2, TLR3, TLR4 and TLR7, have been associated with COVID-19 severity. Here, we review the roles of TLRs in the pathogenesis of COVID-19, including TLR7 and TLR3 rare variants, as well as the L412F polymorphism in TLR3 that negatively regulates anti-SARS-CoV-2 immune responses, the interaction of TLR2 and TLR4 with SARS-CoV-2 proteins, TLR3-related cellular senescence and the implication of TLR2 in NET formation by SARS-CoV-2 (Figure 3).

3.1. TLR7

The constitutive expression of TLR7 is predominant in plasmacytoid and myeloid dendritic cells (pDCs and mDCs), B cells and monocytes, as compared with Natural Killer (NK) cells and T cells, which showed only marginal levels of TLR7 [39–41]. Low levels of TLR7 have also been observed in non-immune cells such as hepatocytes and epithelial cells, as well as nasal epithelial cells and keratinocytes [42–46].

A bioinformatic analysis showed that the related viruses SARS-CoV-1 and Middle East Respiratory Syndrome Coronavirus (MERS-CoV) possess a high number of binding motifs for TLR7, and that the SARS-CoV-2 genome has even more ssRNA motifs that can potentially interact with TLR7 than its family member SARS-CoV-1 [47].

Van der Made et al. [48] were the first to describe deleterious variants in the X-chromosomal TLR7 gene associated with the pathogenesis of COVID-19. They reported two families with affected males at a mean age of 26 years, with no history of major chronic disease, characterized by rare loss-of-function (LOF) TLR7 variants who suffered from severe COVID-19. Whole-exome sequencing identified both a maternally inherited 4-nucleotide deletion p.(Gln710Argfs*18) and a missense mutation p.(Val795Phe). Stimulation with the TLR7 agonist imiquimod (IMQ) of primary peripheral blood mononuclear cells (PBMC) from the patients resulted in a transcriptionally impaired host type I IFN response downstream of the TLR7 pathway, as evidenced by the impaired upregulation of IRF7, IFNβ1 and interferon-stimulated gene ISG15, and by an abrogated production of IFNγ, supporting the importance of intact TLR7 signaling in COVID-19 pathogenesis. After this sporadic observation, the first paper [49] statistically demonstrating the relevance of TLR7 came from the Italian consortium GEN-COVID (accessed on 16 March 2020, https://sites.google.com/dbm.unisi.it/gen-covid). The relevance of TLR7 was then confirmed by several other studies [39,50,51]. In the Italian cohort study, eight male cases, six aged less than 60 years and two in their mid-60s, with severe COVID-19 carried three rare missense LOF [p.(Ser301Pro), p.(Arg920Lys) and p.(Ala1032Thr)], one hypomorphic p.(Val219Ile) and two less functionally impacting variants [p.(Ala288Val) and p.(Ala448Val)] [47]. Therefore, a new entity of "X-linked Mendelian disorder conditioned by viral infection" was definitively established. The disease segregates in families though females and mutated males are affected.

In a further follow-up study with an enlarged GEN-COVID cohort, Mantovani et al. reported another variant, p.(Asp41Glu), predicted to be deleterious by in silico analysis in a 79-year-old severely affected and deceased male patient, whose not-yet-infected mutated brothers were highly at risk of severe disease if not permanently protected by vaccine. These studies demonstrated an impaired upregulation of IRF7, ISG15, CXCL10, RSAD2, ACOD1, IFIT2, IFNα and IFNγ genes, highlighting a profound impairment of the TLR7 signaling pathway in response to the TLR7 agonist [51]. Solanich et al. [50] identified two additional, previously unreported LOF variants in a cohort of young males affected with severe COVID-19. One, p.(Asn215Ser), was detected in a 30-year-old male patient and the other p.(Trp933Arg) in a 28-year-old male patient. Functional testing of the p.(Trp933Arg) variant revealed decreased type I and II IFN production, similar to previously published findings. Fifteen very rare TLR7 variants were reported in a larger cohort of 1202 male patients aged 0.5 to 99 years (mean: 52.9 years) with unexplained critical COVID-19 pneumonia. In a total of sixteen unrelated male individuals aged 7 to 71 years (mean, 36.7 years), twelve had LOF and three had hypomorphic variants [39]. One variant, p.(Leu988Ser), was recurrently found in three patients, including one patient carrying two very rare variants: p.(Met854Ile) and p.(Leu988Ser). Importantly, this study showed that the patients' peripheral blood pDCs responded poorly to SARS-CoV-2 infection. Indeed, IFNα production by TLR7-deficient pDCs was impaired but not entirely abolished upon SARS-CoV-2 infection. Thus, SARS-CoV-2 triggered type I IFN induction in pDCs in a manner that was at least partially dependent on TLR7.

A de novo hemizygous deleterious mutation in the TLR7 gene, p.(Leu372Met), was reported in a 7-year-old male patient who developed critical COVID-19 pneumonia with preexisting hyper IgM syndrome and Ataxia-Telangiectasia (A-T) [52]. It is unknown if the antibody and T cell defects associated with the ATM deficiency may contribute to severe COVID-19. Although severe acute infections are uncommon [53], viral infections have been reported in 25–30% of A-T patients, mainly in the respiratory tract, predominantly caused by usually asymptomatic Rhinovirus infection [54,55]. Of 247 A-T patients evaluated, 36 had SARS-CoV-2 infection, all with mild or no symptoms, except for the patient carrying the TLR7 variant who had had a critical clinical presentation and was transferred to the intensive care unit (ICU), suggesting that ATM mutation is probably not critical for COVID-19 development [52]. This combined TLR7-ATM deficient case illustrates a potential synergistic impact of defective type I IFN and humoral immune responses which deserves further investigation and should be considered in future studies.

Single-nucleotide polymorphisms (SNPs) of the TLR7 gene have been associated with susceptibility to different infectious diseases, such as HIV-1 and HCV, and with respiratory diseases [56,57]. Among these, the frequent SNP of the TLR7 coding sequence, rs179008 (NM_016562.3:c.32A>T), was associated with higher viral loads and accelerated progression to advanced immune suppression in HIV positive male patients, and with a diminished cytokine response to a TLR7 agonist in leukocytes of male allele T carriers [58]. Azar P et al. [59] found that carriage of the rs179008-T allele in HIV-positive females was associated with both a decrease in TLR7 protein relative to A/A homozygous pDCs and lower per-cell IFN-I production. The TLR7 rs179008A/T SNP also modulates the clearance and progression of HCV infection with different magnitudes between sexes, with the T allele increasing the risk of disease progression in both sexes [60]. TLR7 rs179008 polymorphism showed a strong association with the development of bronchial asthma, with the most significant association in boys, suggesting a predominantly recessive effect of these variants [57]. The minor allele rs179008-T is common worldwide, except in East Asia, and is especially frequent among European populations, where 30–50% of women are homozygous or heterozygous carriers. The rs179008 is a missense SNP substituting a leucine for a glutamine (p.Gln11Leu) at protein level, associated with decreased TLR7 protein synthesis and homodimer formation in vitro [59]. To date, few studies have investigated the association of rs179008 SNP with SARS-CoV-2 infection and COVID-19 development. One study found statistically non-significant differences between patient groups for the TLR7 rs179008 A/T allele in mild, moderate and severe COVID-19 [61]. In contrast, in another study the mutant T allele was associated with an increased risk of COVID-19 pneumonia but not with disease outcome [62]. Moreover, SARS-CoV-2 was associated with hepatitis in a male child carrying the rs179008-T allele, suggesting that hepatitis purportedly caused by SARS-CoV-2 infection could be associated with inefficient initial innate immune responses against the virus caused by the polymorphism [63]. Another TLR7 SNP (rs3853839G/C) had a relationship with Dengue virus infection [64], with HIV-1 infection and prognosis [65] and with Chikungunya virus infection in Indians [66], and was linked in Chinese patients to HCV persistence and predisposition to enterovirus-71-mediated hand, foot and mouth infection [67,68]. The function of rs3853839 and the expression of the TLR7 mRNA transcript in the development, severity and progression of COVID-19 was investigated in a study which suggested that the G/G genotype and the G allele could be a genetic risk factor for COVID-19 development, severe illness and poor clinical outcome in middle-aged individuals without comorbidities. Furthermore, patients with the GG genotype had the highest levels of TLR7 mRNA expression, whereas those with the CC genotype had the lowest [69]. In conclusion, despite several findings indicating that these SNPs are involved in susceptibility to a broad range of infectious diseases and inflammatory phenotypes, few, not-gender-related data were reported in relation to SARS-CoV-2 infection and COVID-19 development, highlighting the need for in-depth studies.

A male sex bias has emerged in the COVID-19 pandemic, fitting to the sex-biased pattern in other viral infections. A meta-analysis of 3,111,714 globally reported cases demonstrated that, whilst there was no difference in the proportion of males and females with confirmed COVID-19, male patients had an almost three times higher probability of requiring ICU admission and death compared to females. With a few exceptions, the sex bias observed in COVID-19 was a worldwide phenomenon [70–73]. The X-linked TLR7 gene escapes X chromosome inactivation in 30% of female immune cells, including pDCs, and was associated with higher TLR7 protein expression in leukocytes from rs179008 A/A women compared with A/0 men [74]. As a consequence, a sex bias was reported in the TLR7-driven production of IFNα by human pDCs, with higher frequencies of IFNα–producing cells in adult females than in adult males [59,75,76]. The sex bias in the TLR7-mediated response of pDCs arises from independent mechanisms implicating estrogen signaling [77] and the X-chromosome complement [78]. The X chromosome inactivation escape of TLR7 and the subsequent increased expression of type I IFN may explain in part the observed sex differences concerning severe COVID-19 susceptibility. Severe COVID-19 has been associated with a reduction in circulating pDCs, as well as a minimal influx of pDCs into the lungs compared with patients with moderate disease and healthy controls, and with an impaired type I IFN response in patients with severe SARS-CoV-2 infection [79–85]. An increase in TLR7-mediated IFN production in females may prevent females from progressing to severe disease. In the case of SARS-CoV-2, in which initially there was no or only mild cross-reactive pre-existing adaptive immune memory, the success of an early antiviral response mediated by IFNα may be a major determinant of disease outcome.

Overall, X-linked recessive TLR7 deficiency was pathogenic in patients by impairing the production of large amounts of type I IFNs by pDCs. PDCs, while not supporting SARS-CoV-2 replication, are recruited to the infection sites where they can phagocytose infected cells and, in response, produce multiple antiviral and inflammatory cytokines that protect epithelial cells from de novo SARS-CoV-2 infection. TLR7-MyD88 signaling was identified as crucial for the production of IFNs in SARS-CoV-2 infection by the targeted deletion of virus-recognition innate immune pathways [79]. Moreover, the evidence that numerous SARS-CoV-2 proteins interfere with the RIG-I pathway, the other major pathway for the activation of antiviral type I IFN response to infection with RNA viruses, suggests dependence on strong signaling via the TLR7-pathway in the early phase of viral infection. Indeed, by acting on various molecules in the RIG-I-MAVS signaling pathway, SARS-CoV-2 proteins can evade host antiviral responses and promote viral replication [86–88]. In this way, dependence on functional TLR7-signaling for initial innate immune responses highlights the crucial role of the X-linked recessive TLR7 deficiency in the pathogenesis of COVID-19. The clinical observations of deleterious TLR7 variants and a poor type I IFN response found in 1–2% across male cohorts of severe cases identify TLR7 as an important PRR in the immune response against COVID-19, especially in younger patients. Therefore, genetic screening for TLR7 primary immunodeficiency may be recommended in young males with severe COVID-19 in the absence of other relevant risk factors. On the other hand, a study from a cohort comprising 5085 participants with severe COVID-19 and 571,737 controls that combined whole-exome and whole-genome sequencing and a rare variant burden test meta-analysis observed that carrying a rare deleterious variant in TLR7 was associated with a 5.3-fold increase in severe disease. Importantly, this observation was consistent across both sexes, suggesting that TLR7-mediated genetic predisposition to severe COVID-19 may be a dominant or co-dominant trait, an observation that cannot be made in cohorts limited to male participants [89].

3.2. TLR3

Toll-like receptor 3 is broadly expressed in humans; indeed, its expression and functionality have been confirmed in immune and non-immune cells such as mDCs [90], effector CD8 T cells [91], NK cells [92], intestinal epithelial cells [93], lung and dermal fibroblasts [94,95] and different CNS-resident cells [95–99]. It was shown that the transcript levels of TLR3 were overexpressed in the nasopharyngeal epithelial cells of COVID-19 patients with clinical symptoms compared with controls [100], and that a lower peripheral blood TLR3 expression was associated with an unfavorable outcome in severe COVID-19 patients [101]. Several lines of evidence support the idea that TLR-mediated signaling pathways, including the TLR3 one, originate in the cell membrane in various cells (reviewed in [102]), and that cell membrane TLR3 contributes to endosomal TLR-mediated inflammatory signaling pathways. Surface expression was demonstrated on human lung epithelial cells [103] and on several cell lines [102]. To our knowledge, there was no evidence of cell membrane TLR3-mediated responses contributing to responses to SARS-CoV-2.

A role for the TLR3 receptor has been described in host defense against numerous viruses such as herpes simplex virus, hepatitis C and B virus, A/H1N1/2009 influenza virus, rotavirus, tick-borne encephalitis virus and HIV-1 (reviewed in [104]). An in silico study on the binding of all SARS-CoV-2 mRNAs with the intracellular TLRs depicts the NSP10 protein of SARS-CoV-2 being a putative PAMP for TLR3, triggering downstream cascade reactions [105].

Bortolotti et al. [106] showed in an in vitro system in which SARS-CoV-2 infection of Calu-3/MRC-5 multicellular spheroids induced the activation of the TLR3 pathway via IRF3, leading to pro-inflammatory cytokine secretion, including IL-1α, IL-1β, IL-4, IL-6, IFNα and IFNβ, during the first 24 h post-infection.

The first evidence of the involvement of TLR3 and type I IFN immunity in the control of SARS-CoV-2 infection was in the early phases of the pandemic, when four LOF variants [p.(Ser339fs/WT), p.(Pro554Ser/WT), p.(Trp769Ter/WT) and p.(Met870Val/WT)] were reported in four patients with life-threatening COVID-19 pneumonia aged 44 to 77 years [107]. Of note, the same p.(Pro554Ser) variant has been previously reported in patients with life-threatening influenza pneumonia [99,108]. IFNL1 mRNA levels were impaired in TLR3-deficient P2.1 fibrosarcoma cells stably transfected with plasmids expressing WT or mutant forms of TLR3 identified in patients. Importantly, serum levels of IFNα in the patient carrying the p.(Pro554Ser/WT) variant during the acute phase of COVID-19 were considerably lower than those found in the cohort of patients hospitalized with unexplained severe COVID-19.

A single subcutaneous injection of Peg-IFNα2 was administered to a 25-year-old woman with known TLR3 deficiency [p.(Pro455Ser)] and COVID-19 who had two episodes of herpes simplex encephalitis in childhood, but had no further serious viral infections.

Thirty-six hours after administration, the viral load had strongly decreased and the patient reported a partial resolution of anosmia within the next 48 h. Although these findings cannot be considered a formal demonstration of the efficacy of IFNα2a therapy for COVID-19, the case report showed that a single subcutaneous injection of Peg-IFNα2a was safe and effective in the patient with impaired production of type I IFNs associated with a predisposition to severe COVID-19 [109].

In a later study, Povysil et al. [110], looking at a substantially larger cohort of 713 patients with severe COVID symptoms, 1151 with mild disease, and a control population of 15,033, found a rare predicted LOF TLR3 variant [p.(Arg867*)] in a mild case and three other variants in controls. However, they found no evidence of the enrichment of rare, protein-altering variants in the TLR3 gene in severe cases. However, it is important to emphasize that in the two studies there were differences in the ancestry and mean age of cohorts, demonstrating how important it is to adequately control for differences between cases and controls, as the frequency of rare variants strongly depends upon ancestry, with Europeans showing higher rates compared with Middle Eastern individuals. Another study identified in the Chinese population a TLR3 variant, the p.(Arg394Ter), in a patient with severe disease [111]. Although they have not undertaken any functional studies, the variant was predicted to be an LOF.

Among TLR3 variants, the rs3775291 L412F polymorphism p.(Leu412Phe) is known to decrease TLR3 expression on the cell surface [112]. This polymorphism leads to poor recognition of SARS-CoV-2 dsRNA during replication, compared to its WT allele [113], and it has been associated with susceptibility to and mortality from SARS-CoV-2 [114]. The L412F polymorphism was also identified as a marker of severity in COVID-19, further demonstrating an increased association in the sub-cohort of males [115]. Indeed, a functional role for the L412F polymorphism has been demonstrated in males with severe COVID-19. Interestingly, the mutation specifically affected TLR3-dependent autophagy and reduced TNF-α production after stimulation of the HEK293 cells transfected with L412F-encoding plasmid with a TLR3 specific agonist. Moreover, a reduced survival rate at 28 days was shown in L412F COVID-19 patients treated with the autophagy-inhibitor hydroxychloroquine, suggesting that the outcome of clinical trials with hydroxychloroquine should be reinterpreted in the light of the TLR3-L412F polymorphism status [116].

Recently, a case–control study reported an association of the mutant T/T genotype of TLR3 (rs3775290) with an increased risk of developing COVID-19 pneumonia independently of disease outcome [62].

The clinical severity of COVID-19 is largely determined by host factors. Recent advances point to cellular senescence, an aging-related switch in cellular state, as a critical regulator of SARS-CoV-2-induced hyperinflammation. SARS-CoV-2, like other viruses, can induce senescence and exacerbates the senescence-associated secretory phenotype (SASP), which is composed largely of pro-inflammatory, extracellular matrix-degrading, complement-activating and pro-coagulatory factors secreted by senescent cells [116]. It has been reported that the surface protein S1 of SARS-CoV-2 induces senescence in human non-senescent cells. Of note, SARS-CoV-2 exacerbates the tissue-destructive SAPS, increasing IL-1α, IL-1β, IL-6, IL-8 and Granulocyte-Macrophage Colony-Stimulating Factor (GM-CSF) mRNA levels in human senescent cells through TLR3. The genetic or pharmacological inhibition of TLR3 prevented senescence induction and SASP amplification by SARS-CoV-2, suggesting that the induction of cellular senescence and SASP amplification through TLR3 could contribute to SARS-CoV-2 morbidity, suggesting that clinical trials with senolytics and/or SASP/TLR3 inhibitors for alleviating acute long-term SARS-CoV-2 sequelae are eagerly warranted [117].

3.3. TLR2

TLR2 is a cell surface sensor expressed in immune cells, endothelial, and epithelial cells, including respiratory epithelial cells [118].

TLR2 expression was associated with COVID-19 severity, being identified as a sensor of SARS-CoV-2 that drives inflammatory cytokine production, potentially contributing to the dysregulated immune response observed in patients with severe COVID-19 [119]. SARS-CoV-2 E protein, but not S or M proteins, interacted directly with TLR2, inducing inflammatory responses and the secretion of TNF-α, IFNγ, IL-1α, IL-6, CXCL10, MCP-1, G-CSF and CCL3 in human PBMC. Of note, some of these cytokines have emerged early as critical parameters in COVID-19 disease progression and predictors of disease severity and death [120,121]. Moreover, elegant experiments using heat-inactivated viruses to block transcription and entry demonstrated that the TLR2-mediated activation of inflammatory signaling was independent of viral entry and replication [119]. In vivo infection models demonstrated that the E protein induced TLR2-dependent inflammation in the lungs, and that blocking TLR2 signaling provided protection against SARS-CoV-2 induced pathology [119]. Another study has identified the S protein, and not the E protein, as a driver of TLR2 activation and subsequent cytokine production in the SARS-CoV-2 infection of human and mouse macrophages [122].

Multiple studies have shown that elevated levels of IL-6 in COVID-19 patients were associated with disease severity [80,123–125]. Interestingly, it has been demonstrated that pDC, while not directly supporting SARS-CoV-2 replication, senses the virus via TLR2, and secretes the inflammatory cytokine IL-6, further highlighting how TLR2 responses are critically involved in the cytokine storm elicited by SARS-CoV-2 infection [79].

COVID-19 also impacts the cardiovascular system, resulting in myocardial damage, and affects the kidneys, leading to renal dysfunction [126]. In a small cohort of 50 healthy control and 100 COVID-19 patients with severe or moderate disease, the expression levels for TLR2 and TLR4 mRNA were positively correlated with kidney and heart function biomarkers in serum [127], suggesting that the TLR2 and TLR4 sensors are possible targets to prevent the severity of COVID-19.

Excessive neutrophil extracellular traps contributed to immunothrombosis, leading to extensive intravascular coagulopathy and multiple organ dysfunction in COVID-19 patients [128]. Sung and colleagues [129] demonstrated that SARS-CoV-2-activated platelets produced extracellular vesicles that enhanced thrombo-inflammation via the C-type lectin member 5A and TLR2, suggesting that TLR2 may represent a potential therapeutic target to reduce the risk of acute respiratory distress syndrome in COVID-19 patients.

The TLR2 sensor was considered a potential target to support vaccination. Indeed, the TLR2 ligand, Pam_2Cys, a highly effective mucosal adjuvant for peptide or protein-based vaccines [130–133], was tested with recombinant trimeric SARS-CoV-2 Spike protein in a novel subunit vaccine in mice delivered via the mucosal or systemic route [134]. Both routes of vaccination induced substantial neutralizing antibody titers; however, mucosal vaccination uniquely generated anti-Spike IgA, increased the neutralizing antibody in the serum and airways, and increased lung CD4+ T-cell responses, supporting mucosal vaccination as a strategy to improve protection in the respiratory tract against SARS-CoV-2 [134]. Moreover, in a SARS-CoV-2 ferret infection model, prophylactic intranasal administration of the TLR2/6 agonist INNA-051 effectively reduced levels of viral RNA in the nose and throat, supporting the clinical development of a therapy based on prophylactic TLR2/6 innate immune activation to reduce SARS-CoV-2 transmission and providing protection against COVID-19 [135].

Overall, several findings indicate that TLR2 may be exploited to improve the mucosal vaccination strategy via TLR2 agonists. This may provide better protection for the respiratory tract and, in contrast, TLR2 antagonists may protect against the cytokine storm and disease progression.

3.4. TLR4

Innate immune cells, such as circulating monocytes, tissue macrophages and DCs, express TLR4. Analyses of total RNA showed that TLR4 mRNA was present only in myeloid cells and was undetectable in resting or activated lymphoid cell subsets. TLR4 is constitutively expressed in adipocytes, microglia and in the macroglial cells such as astrocytes, dermal microvessel endothelial cells and umbilical vein endothelial cells [136]. Despite TLR4 mainly residing in the plasma membrane, it can also be considered as an intracellular TLR because it can be internalized and stimulate intracellular pathways [137].

Computer-based modeling has predicted the binding of TLR4 to the Spike protein [138,139]. Consistently, it has been reported in murine and human macrophages that Spike protein S1 subunit binds to TLR4, inducing pro-inflammatory responses and the activation of transcription factors such as NF-κB, AP-1 and encoding proinflammatory cytokines and IFNs [140–142]. Blocking Spike-TLR4 interaction could therefore be a potential target for regulating excessive inflammatory responses in COVID-19 patients [143].

Endothelial cells (EC) are sentinels uniquely positioned at the interface between circulating blood and underlying tissues that gatekeep micro- and macro-vascular health by sensing pathogen signals and secreting vasoactive molecules. SARS-CoV-2 infection primarily affects the pulmonary system, but accumulating evidence suggests that it also affects the pan-vasculature in the extrapulmonary systems by directly or indirectly generating a cytokine storm, causing endothelial dysfunction such as endotheliitis, endothelialitis and endotheliopathy [144]. To date, the direct causative mechanism of SARS-CoV-2-induced endotheliitis remains unclear. Ma et al. [145] contributed to define a role for TLR4 in the ACE2-independent inflammatory activation of vascular EC. ACE2-deficient ECs respond to SARS-CoV-2 through TLR4, as demonstrated by treatment with its antagonists, CLI095 or dexamethasone, which inhibited p38 MAPK/NF-kB/IL-1β activation after viral exposure. Genome-wide and single-cell RNA-seq analyses further confirmed the activation of the TLR4/MAPK14/RELA/IL-1β axis in circulating ECs in patients with mild and severe COVID-19, indicating that human ECs were not protected from SARS-CoV-2-mediated activation by being ACE2-deficient as they still recognized and responded to the virus through TLR4 activation. Interestingly, dexamethasone has been demonstrated to lower the 28-day mortality rate among patients hospitalized with COVID-19 receiving respiratory support [146], providing a clinical benefit that could also be explained by the inhibition of TLR4-mediated activation in the endothelium.

Neutrophils, among the first line of innate immune defense in acute infection, extravasate rapidly from the blood vessels into tissues to engulf and kill pathogens. Various stimuli, such as microorganisms, pathogens, activated platelets and pro-inflammatory cytokines, can trigger a rapid and active process of cell death in neutrophils, ending up in the formation of neutrophil extracellular traps (NETs) [147]. Numerous pieces of evidence have implicated neutrophils and the imbalance between NET formation and degradation in the pathophysiology of inflammation, coagulopathy, organ damage and immunothrombosis that characterizes severe cases of COVID-19 [148,149]. Moreover, in patients with severe disease, an increased neutrophil-to-lymphocyte ratio, a high expression of the neutrophil-related cytokines IL-8 and IL-6 in serum and elevated levels of circulating calprotectin, a neutrophil-mediated inflammatory protein, have been described as predictors of poor outcome [150–157]. The mechanism controlling calprotectin production during SARS-CoV-2 infection was elucidated by Loh et al. [158]. It has been demonstrated that calprotectin expression in neutrophils is regulated via the TLR4 pathway in response to Spike protein, and that the adaptor Docking Protein 3 restrained the production of calprotectin during infection when TLR4 was engaged.

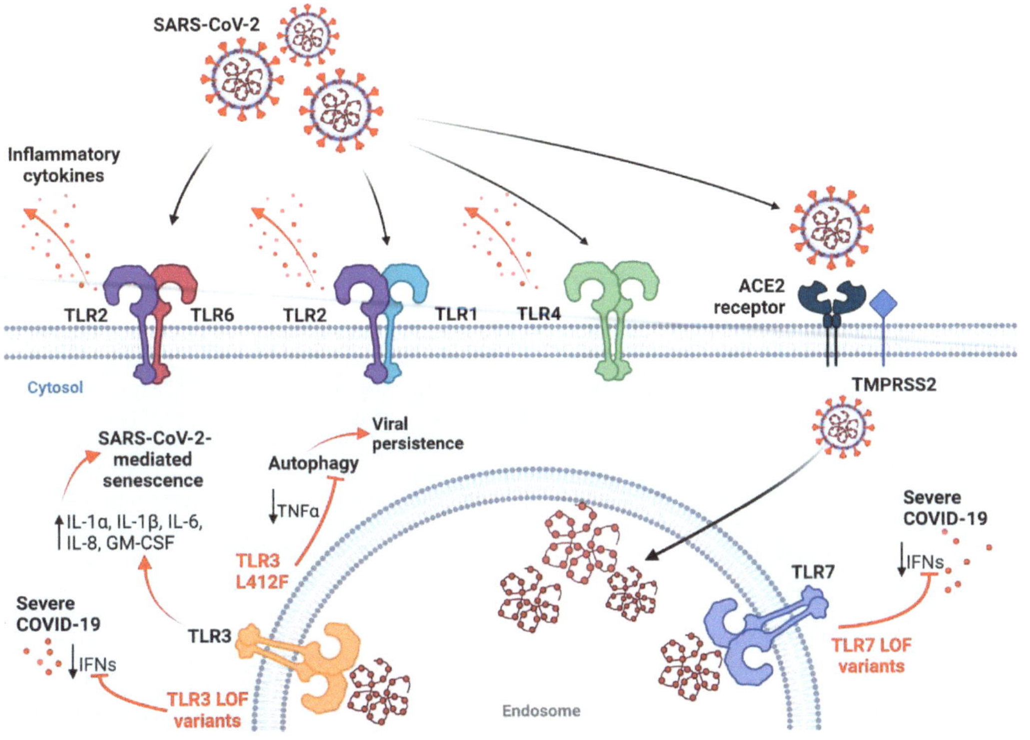

Figure 3. Several TLRs, such as TLR2, TLR3, TLR4 and TLR7, have been associated with the pathogenesis of COVID-19. The interaction of TLR2 and 4 with SARS-CoV-2 proteins triggers downstream pathways leading to the production of pro-inflammatory cytokines, via the activation of multiple adaptor proteins, as shown in Figure 2 [119,122,140–143]. LOF TLR3 and TLR7 variants negatively regulate anti-SARS-CoV-2 immune responses, inducing an impaired IFN response that could delay viral clearance in the initial disease phase and promote a severe COVID-19 clinical course [39,48–51,107]. The L412F polymorphism reduces TNFα production and inhibits TLR3-dependent autophagy, promoting viral persistence [115]. SARS-CoV-2 induces senescence in human non-senescent cells and exacerbates the tissue-destructive SAPS, increasing IL-1α, IL-1β, IL-6, IL-8 and GM-CSF mRNA levels in human senescent cells through TLR3 [117]. SAPS, senescence-associated secretory phenotype. Created with BioRender.com.

4. TLRs, IL-1 and Cytokine Storm Syndrome

During TLR-mediated responses induced by SARS-CoV-2 infection, the release of pro-inflammatory cytokines contributes to cause the cytokine storm syndrome that is associated with heart and lung tissue damage, leading to multiorgan failure and poor outcomes [122,159,160]. Cytokine storm syndrome results from a sudden acute increase in circulating levels of different pro-inflammatory cytokines, including IL-6, IL-1, TNF-α and IFNs [161]. IL-1β, due to its pro-inflammatory and pleiotropic biological activity, plays a central role in orchestrating the innate immune responses following a wide range of conditions, including infections [162,163]. The interleukin-1 (IL-1) family of cytokines and their receptors share a structure and function similar to TLRs. Cytokine members of the IL-1 family trigger innate inflammation via the IL-1 family of receptors, increasing nonspecific resistance to infection and developing immune responses to foreign antigens. Each member of the IL-1 receptor, as for the TLR family, contains a cytoplasmic portion formed by a TIR signaling domain. Once IL-1 binds to its receptor, a structural change occurs that allows binding to the co-receptor, forming a trimeric complex, in which the TIR domains approximate. The binding of MyD88 to the TIR domains triggers a cascade of kinases that produces a strong pro-inflammatory signal, via the activation of NF-κB, that amplifies that mediated by TLRs [163]. Interestingly, longitudinal analyses of immune responses in 113 patients with moderate or severe COVID-19 revealed distinct profiles that influenced the evolution and severity of COVID-19. Indeed, following an early increase in a variety of cytokines, patients with moderate COVID-19 displayed a progressive reduction in inflammatory cytokines; by contrast, patients with severe COVID-19 maintained these elevated responses throughout the course of the disease. Inflammasome-induced cytokines, such as IL-1β and IL-18, were higher in patients with severe disease than in patients with moderate disease [164].

The cytokine storm initiates a variety of inflammatory signaling pathways via their receptors on immune and tissue cells, which lead to fever, capillary leak syndrome, disseminated intravascular coagulation, acute respiratory distress syndrome, and multi-organ failure, ultimately leading to death in the most severe cases [160]. Interestingly, in COVID-19 patients, pain, including headache, widespread myalgia, and back and neck pain, was linked to inflammation and cytokine storm, especially with the rapid release of the pro-inflammatory cytokines IL-6, IL-1RA, IL-10, IL21 and IL-22 [165–167]. Moreover, pro-inflammatory cytokines can act on pain-sensing nociceptors in the neurons of peripheral tissues and in the central nervous system, heightening perceptions of pain and leading to chronic pain [167]. Another critical symptom of COVID-19 is the so-called brain fog, a pathological state represented by cognitive dysfunction and fatigue. Neurologic impairment in COVID-19 might be due to the passage of inflammatory cytokines from the bloodstream through the blood–brain barrier [168,169]. Of note, cerebral hyperinflammation was detected in COVID-19 patients by electroencephalogram [170]. Inflammatory cytokines could be also involved in anosmia, a frequent clinical feature in COVID-19 patients. One study found elevated levels of the proinflammatory cytokine TNF-α in the olfactory epithelium of patients with COVID-19, suggesting that the direct inflammation of the olfactory epithelium could play a role in acute olfactory loss [171]. However, the mechanisms of olfactory dysfunction in COVID-19 are still unclear [172], and there is evidence that isolated anosmia, which is often associated with mildly symptomatic disease, is associated with robust SARS-CoV-2 specific CD4+ and CD8+ T-cell responses [173]. Anosmia is a prominent clinical characteristic of Parkinson's disease (PD); of note, SARS-CoV-2 infection has been predicted as a potential risk factor for developing Parkinsonism-related symptoms in COVID-19 patients [174]. A study reported that two of four patients with long-term olfactory dysfunction, still persisting 4–9 months after SARS-CoV-2 infection, showed damage in the basal ganglia after infection, highlighting the presence of an olfactory-nigral dysfunction, possibly representing a pre-clinical signature of Parkinsonism [175]. Further studies are needed to address the role of inflammatory mediators, activated by SARS-CoV-2 infection, in neurological disorders.

5. Conclusions

Is TLR activation beneficial or harmful in SARS-CoV-2 infection?

An early and strong protective TLR-mediated innate immune response against pathogenic viruses or viral components represents an essential contribution to viral clearance. Indeed, TLR signaling is needed for the secretion of antiviral cytokines, chemokines and IFNs, especially type I IFNs, which help to control infection and to determine the disease outcome. If TLR-mediated responses are deleted or reduced, they can lead to severe disease and fatal outcomes, due to the inability to mount effective innate immune responses. In COVID-19, a peculiar phenotype was observed in critical disease, consisting of a highly impaired IFN type I response (characterized by no IFNβ and low IFNα production and activity) and a low ISG signature in peripheral blood, which was associated with a persistent blood viral load and an exacerbated inflammatory response [80]. The dependence on functional TLR7-signaling for initial innate immune responses was highlighted in male cohorts of severe cases, especially in younger patients. Moreover, SARS-CoV-2 made use of evasion mechanisms to protect against viral dsRNA and ssRNA recognition by TLRs.

On the other hand, TLRs may represent a double-edge sword which may elicit dysregulated immune signaling. Indeed, excessive TLR activation due to overstimulation by viral proteins or by damage-associated molecular patterns released from lysed cells can lead to the untoward production of proinflammatory cytokines and chemokines, resulting in severe disease. Hadjadj J. et al. [80] demonstrated that in COVID-19, besides an impaired type I IFN response, severe disease correlates with an immune signature characterized by an early cytokine- and chemokine-rich inflammatory response, partially driven by NF-κB and characterized by increased TNF-α and IL-6 production and signaling, supporting a role for TLR-mediated responses. Indeed, the SARS-CoV-2 E protein interacts directly with TLR2, inducing inflammatory responses and the secretion of a variety of inflammatory cytokines such as TNF-α and IL-6 [122]. TLR4 recognizes the S protein, inducing pro-inflammatory responses and the activation of NF-κB and encoding proinflammatory cytokines [140–142]. TLR-dependent excessive inflammation and TLR-dependent antiviral responses may tip the balance towards the former or the latter, altering the equilibrium that drives the severity of disease.

SARS-CoV-2 mRNA vaccines have changed the worldwide perspective on the COVID-19 pandemic. BNT162b2 and mRNA-1273 vaccines maintain the potential to be recognized by intracellular RNA sensors, such as TLR3 and TLR7, and may on the one hand boost mRNA vaccine-induced adaptive immune responses and, on the other, elicit exaggerated innate immunity. Notably, to avoid the latter, mRNA COVID-19 vaccines contain purified, in vitro-transcribed single-stranded mRNA with modified nucleotides to reduce binding to TLR and immune sensors, thus limiting the excessive production of type I IFNs and their inhibitory function on cellular translation [176,177]. A recent study longitudinally profiled the B cell response to mRNA vaccination in SARS-CoV-2 naive patients with inherited type I IFN deficiency, such as a 36-yr-old male patient hemizygous for an LOF TLR7 mutation, demonstrating that the RBD-specific memory B cell response in all patients was quantitatively and qualitatively similar to that in healthy donors [178]. Despite the study being carried out on a patient, these findings strengthen the role for mRNA vaccines to protect patients with inherited deficiencies in type I IFN immunity.

Author Contributions: Conceptualization, writing and original draft preparation, S.M. and M.U.M.; review and editing, S.M., A.R., B.O., S.V. and M.U.M.; funding acquisition, M.U.M. All authors have read and agreed to the published version of the manuscript.

Funding: This paper was supported by EU funding within the Next Generation EU-MUR PNRR Extended Partnership initiative on Emerging Infectious Diseases (Project no. PE00000007, INF-ACT).

Conflicts of Interest: The authors declare no conflict of interest.

References

1. Paludan, S.R.; Pradeu, T.; Masters, S.L.; Mogensen, T.H. Constitutive immune mechanisms: Mediators of host defence and immune regulation. *Nat. Rev. Immunol.* **2021**, *21*, 137–150. [CrossRef] [PubMed]
2. Dadras, O.; Seyed-Alinaghi, S.; Karimi, A.; Shamsabadi, A.; Qaderi, K.; Ramezani, M.; Mirghaderi, S.P.; Mahdiabadi, S.; Vahedi, F.; Saeidi, S.; et al. COVID-19 mortality and its predictors in the elderly: A systematic review. *Health Sci. Rep.* **2022**, *5*, e657. [CrossRef] [PubMed]
3. Fallerini, C.; Picchiotti, N.; Baldassarri, M.; Zguro, K.; Daga, S.; Fava, F.; Benetti, E.; Amitrano, S.; Bruttini, M.; Palmieri, M.; et al. Common, low-frequency, rare, and ultra-rare coding variants contribute to COVID-19 severity. *Hum. Genet.* **2022**, *141*, 147–173. [CrossRef] [PubMed]
4. Picchiotti, N.; Benetti, E.; Fallerini, C.; Daga, S.; Baldassarri, M.; Fava, F.; Zguro, K.; Valentino, F.; Doddato, G.; Giliberti, A.; et al. Post-Mendelian Genetic Model in COVID-19. *Cardiol. Cardiovasc. Med.* **2021**, *5*, 673–694. [CrossRef]
5. Fericean, R.M.; Rosca, O.; Citu, C.; Manolescu, D.; Bloanca, V.; Toma, A.O.; Boeriu, E.; Dumitru, C.; Ravulapalli, M.; Barbos, V.; et al. COVID-19 Clinical Features and Outcomes in Elderly Patients during Six Pandemic Waves. *J. Clin. Med.* **2022**, *11*, 6803. [CrossRef]
6. Shi, Y.; Wang, Y.; Shao, C.; Huang, J.; Gan, J.; Huang, X.; Bucci, E.; Piacentini, M.; Ippolito, G.; Melino, G. COVID-19 infection: The perspectives on immune responses. *Cell Death Differ.* **2020**, *27*, 1451–1454. [CrossRef]

7. Li, Q.; Wang, Y.; Sun, Q.; Knopf, J.; Herrmann, M.; Lin, L.; Jiang, J.; Shao, C.; Li, P.; He, X.; et al. Immune response in COVID-19: What is next? *Cell Death Differ.* **2022**, *29*, 1107–1122. [CrossRef]
8. Akira, S.; Uematsu, S.; Takeuchi, O. Pathogen recognition and innate immunity. *Cell* **2006**, *124*, 783–801. [CrossRef]
9. Li, D.; Wu, M. Pattern recognition receptors in health and diseases. *Signal Transduct. Target Ther.* **2021**, *6*, 291. [CrossRef]
10. Janssens, S.; Beyaert, R. Role of Toll-like receptors in pathogen recognition. *Clin. Microbiol. Rev.* **2003**, *16*, 637–646. [CrossRef]
11. Hayashi, F.; Smith, K.D.; Ozinsky, A.; Hawn, T.R.; Yi, E.C.; Goodlett, D.R.; Eng, J.K.; Akira, S.; Underhill, D.M.; Aderem, A. The innate immune response to bacterial flagellin is mediated by Toll-like receptor 5. *Nature* **2001**, *410*, 1099–1103. [CrossRef]
12. Beutler, B.; Rietschel, E.T. Innate immune sensing and its roots: The story of endotoxin. *Nat. Rev. Immunol.* **2003**, *3*, 169–176. [CrossRef]
13. Bell, J.K.; Botos, I.; Hall, P.R.; Askins, J.; Shiloach, J.; Segal, D.M.; Davies, D.R. The molecular structure of the Toll-like receptor 3 ligand-binding domain. *Proc. Natl. Acad. Sci. USA* **2005**, *102*, 10976–10980. [CrossRef]
14. Choe, J.; Kelker, M.S.; Wilson, I.A. Crystal structure of human toll-like receptor 3 (TLR3) ectodomain. *Science* **2005**, *309*, 581–585. [CrossRef]
15. Kang, J.Y.; Nan, X.; Jin, M.S.; Youn, S.J.; Ryu, Y.H.; Mah, S.; Han, S.H.; Lee, H.; Paik, S.G.; Lee, J.O. Recognition of lipopeptide patterns by Toll-like receptor 2-Toll-like receptor 6 heterodimer. *Immunity* **2009**, *31*, 873–884. [CrossRef]
16. Su, S.B.; Tao, L.; Deng, Z.P.; Chen, W.; Qin, S.Y.; Jiang, H.X. TLR10: Insights, controversies and potential utility as a therapeutic target. *Scand. J. Immunol.* **2021**, *93*, e12988. [CrossRef]
17. Botos, I.; Segal, D.M.; Davies, D.R. The structural biology of Toll-like receptors. *Structure* **2011**, *19*, 447–459. [CrossRef]
18. Tanji, H.; Ohto, U.; Shibata, T.; Taoka, M.; Yamauchi, Y.; Isobe, T.; Miyake, K.; Shimizu, T. Toll-like receptor 8 senses degradation products of single-stranded RNA. *Nat. Struct. Mol. Biol.* **2015**, *22*, 109–115. [CrossRef]
19. Zhang, Z.; Ohto, U.; Shibata, T.; Krayukhina, E.; Taoka, M.; Yamauchi, Y.; Tanji, H.; Isobe, T.; Uchiyama, S.; Miyake, K.; et al. Structural Analysis Reveals that Toll-like Receptor 7 Is a Dual Receptor for Guanosine and Single-Stranded RNA. *Immunity* **2016**, *45*, 737–748. [CrossRef]
20. Zhang, Z.; Ohto, U.; Shibata, T.; Taoka, M.; Yamauchi, Y.; Sato, R.; Shukla, N.M.; David, S.A.; Isobe, T.; Miyake, K.; et al. Structural Analyses of Toll-like Receptor 7 Reveal Detailed RNA Sequence Specificity and Recognition Mechanism of Agonistic Ligands. *Cell Rep.* **2018**, *25*, 3371–3381. [CrossRef]
21. Heil, F.; Hemmi, H.; Hochrein, H.; Ampenberger, F.; Kirschning, C.; Akira, S.; Lipford, G.; Wagner, H.; Bauer, S. Species-specific recognition of single-stranded RNA via toll-like receptor 7 and 8. *Science* **2004**, *303*, 1526–1529. [CrossRef] [PubMed]
22. Shibata, T.; Ohto, U.; Nomura, S.; Kibata, K.; Motoi, Y.; Zhang, Y.; Murakami, Y.; Fukui, R.; Ishimoto, T.; Sano, S.; et al. Guanosine and its modified derivatives are endogenous ligands for TLR7. *Int. Immunol.* **2016**, *28*, 211–222. [CrossRef] [PubMed]
23. Jin, M.S.; Kim, S.E.; Heo, J.Y.; Lee, M.E.; Kim, H.M.; Paik, S.G.; Lee, H.; Lee, J.O. Crystal structure of the TLR1-TLR2 heterodimer induced by binding of a tri-acylated lipopeptide. *Cell* **2007**, *130*, 1071–1082. [CrossRef] [PubMed]
24. Park, B.S.; Song, D.H.; Kim, H.M.; Choi, B.S.; Lee, H.; Lee, J.O. The structural basis of lipopolysaccharide recognition by the TLR4-MD-2 complex. *Nature* **2009**, *458*, 1191–1195. [CrossRef]
25. O'Neill, L.A.; Bowie, A.G. The family of five: TIR-domain-containing adaptors in Toll-like receptor signalling. *Nat. Rev. Immunol.* **2007**, *7*, 353–364. [CrossRef]
26. Broz, P.; Monack, D.M. Newly described pattern recognition receptors team up against intracellular pathogens. *Nat. Rev. Immunol.* **2013**, *13*, 551–565. [CrossRef]
27. Guven-Maiorov, E.; Keskin, O.; Gursoy, A.; VanWaes, C.; Chen, Z.; Tsai, C.J.; Nussinov, R. The Architecture of the TIR Domain Signalosome in the Toll-like Receptor-4 Signaling Pathway. *Sci. Rep.* **2015**, *5*, 13128. [CrossRef]
28. Fitzgerald, K.A.; Kagan, J.C. Toll-like Receptors and the Control of Immunity. *Cell* **2020**, *180*, 1044–1066. [CrossRef]
29. Kawai, T.; Akira, S. Signaling to NF-kappaB by Toll-like receptors. *Trends Mol. Med.* **2007**, *13*, 460–469. [CrossRef]
30. Veals, S.A.; Schindler, C.; Leonard, D.; Fu, X.Y.; Aebersold, R.; Darnell, J.E., Jr.; Levy, D.E. Subunit of an alpha-interferon-responsive transcription factor is related to interferon regulatory factor and Myb families of DNA-binding proteins. *Mol. Cell Biol.* **1992**, *12*, 3315–3324. [CrossRef]
31. Honda, K.; Taniguchi, T. IRFs: Master regulators of signalling by Toll-like receptors and cytosolic pattern-recognition receptors. *Nat. Rev. Immunol.* **2006**, *6*, 644–658. [CrossRef]
32. Paun, A.; Pitha, P.M. The IRF family, revisited. *Biochimie* **2007**, *89*, 744–753. [CrossRef]
33. Tamura, T.; Yanai, H.; Savitsky, D.; Taniguchi, T. The IRF family transcription factors in immunity and oncogenesis. *Annu. Rev. Immunol.* **2008**, *26*, 535–584. [CrossRef]
34. Yanai, H.; Negishi, H.; Taniguchi, T. The IRF family of transcription factors: Inception, impact and implications in oncogenesis. *Oncoimmunology* **2012**, *1*, 1376–1386. [CrossRef]
35. Mancino, A.; Natoli, G. Specificity and Function of IRF Family Transcription Factors: Insights from Genomics. *J. Interferon Cytokine Res.* **2016**, *36*, 462–469. [CrossRef]
36. Tanaka, N.; Kawakami, T.; Taniguchi, T. Recognition DNA sequences of interferon regulatory factor 1 (IRF-1) and IRF-2, regulators of cell growth and the interferon system. *Mol. Cell Biol.* **1993**, *13*, 4531–4538. [CrossRef]
37. Jefferies, C.A. Regulating IRFs in IFN Driven Disease. *Front. Immunol.* **2019**, *10*, 325. [CrossRef]
38. Taniguchi, T.; Ogasawara, K.; Takaoka, A.; Tanaka, N. IRF family of transcription factors as regulators of host defense. *Annu. Rev. Immunol.* **2001**, *19*, 623–655. [CrossRef]

39. Asano, T.; Boisson, B.; Onodi, F.; Matuozzo, D.; Moncada-Velez, M.; Maglorius Renkilaraj, M.R.L.; Zhang, P.; Meertens, L.; Bolze, A.; Materna, M.; et al. X-linked recessive TLR7 deficiency in ~1% of men under 60 years old with life-threatening COVID-19. *Sci. Immunol.* **2021**, *6*, eabl4348. [CrossRef]
40. Mele, D.; Mantovani, S.; Oliviero, B.; Grossi, G.; Ludovisi, S.; Mondelli, M.U.; Varchetta, S. Hepatitis C virus inhibits CD4 T cell function via binding to Toll-like receptor 7. *Antivir. Res.* **2017**, *137*, 108–111. [CrossRef]
41. Hornung, V.; Rothenfusser, S.; Britsch, S.; Krug, A.; Jahrsdörfer, B.; Giese, T.; Endres, S.; Hartmann, G. Quantitative expression of toll-like receptor 1-10 mRNA in cellular subsets of human peripheral blood mononuclear cells and sensitivity to CpG oligodeoxynucleotides. *J. Immunol.* **2002**, *168*, 4531–4537. [CrossRef] [PubMed]
42. Lee, J.; Wu, C.C.; Lee, K.J.; Chuang, T.H.; Katakura, K.; Liu, Y.T.; Chan, M.; Tawatao, R.; Chung, M.; Shen, C.; et al. Activation of anti-hepatitis C virus responses via Toll-like receptor 7. *Proc. Natl. Acad. Sci. USA* **2006**, *103*, 1828–1833. [CrossRef] [PubMed]
43. Tengroth, L.; Millrud, C.R.; Kvarnhammar, A.M.; Kumlien, G.S.; Latif, L.; Cardell, L.O. Functional effects of Toll-like receptor (TLR)3, 7, 9, RIG-I and MDA-5 stimulation in nasal epithelial cells. *PLoS ONE* **2014**, *9*, e98239. [CrossRef] [PubMed]
44. Schaefer, T.M.; Desouza, K.; Fahey, J.V.; Beagley, K.W.; Wira, C.R. Toll-like receptor (TLR) expression and TLR-mediated cytokine/chemokine production by human uterine epithelial cells. *Immunology* **2004**, *112*, 428–436. [CrossRef] [PubMed]
45. Li, Z.J.; Sohn, K.C.; Choi, D.K.; Shi, G.; Hong, D.; Lee, H.E.; Whang, K.U.; Lee, Y.H.; Im, M.; Lee, Y. Roles of TLR7 in activation of NF-κB signaling of keratinocytes by imiquimod. *PLoS ONE* **2013**, *8*, e77159. [CrossRef]
46. Cherfils-Vicini, J.; Platonova, S.; Gillard, M.; Laurans, L.; Validire, P.; Caliandro, R.; Magdeleinat, P.; Mami-Chouaib, F.; Dieu-Nosjean, M.C.; Fridman, W.H. Triggering of TLR7 and TLR8 expressed by human lung cancer cells induces cell survival and chemoresistance. *J. Clin. Invest.* **2010**, *120*, 1285–1297. [CrossRef]
47. Moreno-Eutimio, M.A.; López-Macías, C.; Pastelin-Palacios, R. Bioinformatic analysis and identification of single-stranded RNA sequences recognized by TLR7/8 in the SARS-CoV-2, SARS-CoV, and MERS-CoV genomes. *Microbes Infect.* **2020**, *22*, 226–229. [CrossRef]
48. van der Made, C.I.; Simons, A.; Schuurs-Hoeijmakers, J.; van den Heuvel, G.; Mantere, T.; Kersten, S.; van Deuren, R.C.; Steehouwer, M.; van Reijmersdal, S.V.; Jaeger, M.; et al. Presence of Genetic Variants Among Young Men With Severe COVID-19. *JAMA* **2020**, *324*, 663–673. [CrossRef]
49. Fallerini, C.; Daga, S.; Mantovani, S.; Benetti, E.; Picchiotti, N.; Francisci, D.; Paciosi, F.; Schiaroli, E.; Baldassarri, M.; Fava, F.; et al. Association of Toll-like receptor 7 variants with life-threatening COVID-19 disease in males: Findings from a nested case-control study. *Elife* **2021**, *10*, e67569. [CrossRef]
50. Solanich, X.; Vargas-Parra, G.; van der Made, C.I.; Simons, A.; Schuurs-Hoeijmakers, J.; Antolí, A.; Del Valle, J.; Rocamora-Blanch, G.; Setién, F.; Esteller, M.; et al. Genetic Screening for TLR7 Variants in Young and Previously Healthy Men With Severe COVID-19. *Front. Immunol.* **2021**, *12*, 719115. [CrossRef]
51. Mantovani, S.; Daga, S.; Fallerini, C.; Baldassarri, M.; Benetti, E.; Picchiotti, N.; Fava, F.; Gallì, A.; Zibellini, S.; Bruttini, M.; et al. Rare variants in Toll-like receptor 7 results in functional impairment and downregulation of cytokine-mediated signaling in COVID-19 patients. *Genes Immun.* **2022**, *23*, 51–56. [CrossRef]
52. Abolhassani, H.; Vosughimotlagh, A.; Asano, T.; Landegren, N.; Boisson, B.; Delavari, S.; Bastard, P.; Aranda-Guillén, M.; Wang, Y.; Zuo, F.; et al. X-Linked TLR7 Deficiency Underlies Critical COVID-19 Pneumonia in a Male Patient with Ataxia-Telangiectasia. *J. Clin. Immunol.* **2022**, *42*, 1–9. [CrossRef]
53. Bhatt, J.M.; Bush, A.; van Gerven, M.; Nissenkorn, A.; Renke, M.; Yarlett, L.; Taylor, M.; Tonia, T.; Warris, A.; Zielen, S.; et al. ERS statement on the multidisciplinary respiratory management of ataxia telangiectasia. *Eur. Respir. Rev.* **2015**, *24*, 565–581. [CrossRef]
54. Méndez-Echevarría, A.; Caminoa, M.B.; Del Rosal, T.; Casas, I.; Pozo, F.; Pascual-Pascual, S.I.; García-Romero, M.; Cámara, C.; Calvo, C. The Role of Respiratory Viruses in Children with Ataxia-Telangiectasia. *Viruses* **2021**, *13*, 867. [CrossRef]
55. Bhatt, J.M.; Bush, A. Microbiological surveillance in lung disease in ataxia telangiectasia. *Eur. Respir. J.* **2014**, *43*, 1797–1801. [CrossRef]
56. Mukherjee, S.; Huda, S.; Sinha Babu, S.P. Toll-like receptor polymorphism in host immune response to infectious diseases: A review. *Scand. J. Immunol.* **2019**, *90*, e12771. [CrossRef]
57. Møller-Larsen, S.; Nyegaard, M.; Haagerup, A.; Vestbo, J.; Kruse, T.A.; Børglum, A.D. Association analysis identifies TLR7 and TLR8 as novel risk genes in asthma and related disorders. *Thorax* **2008**, *63*, 1064–1069. [CrossRef]
58. Oh, D.Y.; Baumann, K.; Hamouda, O.; Eckert, J.K.; Neumann, K.; Kücherer, C.; Bartmeyer, B.; Poggensee, G.; Oh, N.; Pruss, A. A frequent functional toll-like receptor 7 polymorphism is associated with accelerated HIV-1 disease progression. *AIDS* **2009**, *23*, 297–307. [CrossRef]
59. Azar, P.; Mejía, J.E.; Cenac, C.; Shaiykova, A.; Youness, A.; Laffont, S.; Essat, A.; Izopet, J.; Passaes, C.; Müller-Trutwin, M.; et al. TLR7 dosage polymorphism shapes interferogenesis and HIV-1 acute viremia in women. *JCI Insight* **2020**, *5*, e136047. [CrossRef]
60. Fakhir, F.Z.; Lkhider, M.; Badre, W.; Alaoui, R.; Meurs, E.F.; Pineau, P.; Ezzikouri, S.; Benjelloun, S. Genetic variations in toll-like receptors 7 and 8 modulate natural hepatitis C outcomes and liver disease progression. *Liver Int.* **2018**, *38*, 432–442. [CrossRef]
61. Al-Tamimi, Z.H.D.; Alta'ee, A.H.; Jasim, A.H. Effect of Toll-Like Receptor 7 Gene Polymorphism and ABO Blood Groups on the Severity of COVID-19 Patients. *Acta Inform. Med.* **2022**, *30*, 191–195. [CrossRef] [PubMed]
62. Alseoudy, M.M.; Elgamal, M.; Abdelghany, D.A.; Borg, A.M.; El-Mesery, A.; Elzeiny, D.; Hammad, M.O. Prognostic impact of toll-like receptors gene polymorphism on outcome of COVID-19 pneumonia: A case-control study. *Clin. Immunol.* **2022**, *235*, 108929. [CrossRef] [PubMed]

63. Pessoa, N.L.; Bentes, A.A.; de Carvalho, A.L.; de Souza Silva, T.B.; Alves, P.A.; de Sousa Reis, E.V.; Rodrigues, T.A.; Kroon, E.G.; Campos, M.A. Case report: Hepatitis in a child infected with SARS-CoV-2 presenting toll-like receptor 7 Gln11Leu single nucleotide polymorphism. *Virol. J.* **2021**, *18*, 180. [CrossRef] [PubMed]
64. Mukherjee, S.; Tripathi, A. Contribution of Toll like receptor polymorphisms to dengue susceptibility and clinical outcome among eastern Indian patients. *Immunobiology* **2019**, *224*, 774–785. [CrossRef]
65. Zhang, T.; Zhu, J.; Su, B.; Cao, L.; Li, Z.; Wei, H.; Huang, X.; Zheng, K.; Li, A.; Chen, N.; et al. Effects of TLR7 Polymorphisms on the Susceptibility and Progression of HIV-1 Infection in Chinese MSM Population. *Front. Immunol.* **2020**, *11*, 589010. [CrossRef]
66. Dutta, S.K.; Tripathi, A. Association of toll-like receptor polymorphisms with susceptibility to chikungunya virus infection. *Virology* **2017**, *511*, 207–213. [CrossRef]
67. Yue, M.; Feng, L.; Tang, S.D.; Wang, J.J.; Xue, X.X.; Ding, W.L.; Zhang, Y.; Deng, X.Z. Sex-specific association between X-linked Toll-like receptor 7 with the outcomes of hepatitis C virus infection. *Gene* **2014**, *548*, 244–250. [CrossRef]
68. Tian, H.; Xu, W.; Wen, L.; Tang, L.; Zhang, X.; Song, T.; Yang, C. Association of TLR3 gene 1377C/T (rs3775290) and TLR7 gene C/G (rs3853839) polymorphism with hand, foot, and mouth disease caused by human enterovirus 71 infection susceptibility and severity in the Chinese Han population: A meta-analysis of case-control studies. *Medicine* **2022**, *101*, e29758. [CrossRef]
69. El-Hefnawy, S.M.; Eid, H.A.; Mostafa, R.G.; Soliman, S.S.; Omar, T.A.; Azmy, R.M. COVID-19 susceptibility, severity, clinical outcome and Toll-like receptor (7) mRNA expression driven by TLR7 gene polymorphism (rs3853839) in middle-aged individuals without previous comorbidities. *Gene Rep.* **2022**, *27*, 101612. [CrossRef]
70. Peckham, H.; de Gruijter, N.M.; Raine, C.; Radziszewska, A.; Ciurtin, C.; Wedderburn, L.R.; Rosser, E.C.; Webb, K.; Deakin, C.T. Male sex identified by global COVID-19 meta-analysis as a risk factor for death and ITU admission. *Nat. Commun.* **2020**, *11*, 6317. [CrossRef]
71. Kragholm, K.; Andersen, M.P.; Gerds, T.A.; Butt, J.H.; Østergaard, L.; Polcwiartek, C.; Phelps, M.; Andersson, C.; Gislason, G.H.; Torp-Pedersen, C.; et al. Association Between Male Sex and Outcomes of Coronavirus Disease 2019 (COVID-19)-A Danish Nationwide, Register-based Study. *Clin. Infect. Dis.* **2021**, *73*, e4025–e4030. [CrossRef]
72. Spiering, A.E.; de Vries, T.J. Why Females Do Better: The X Chromosomal TLR7 Gene-Dose Effect in COVID-19. *Front. Immunol.* **2021**, *12*, 756262. [CrossRef]
73. Arnold, C.G.; Libby, A.; Vest, A.; Hopkinson, A.; Monte, A.A. Immune mechanisms associated with sex-based differences in severe COVID-19 clinical outcomes. *Biol. Sex Differ.* **2022**, *13*, 7. [CrossRef]
74. Souyris, M.; Cenac, C.; Azar, P.; Daviaud, D.; Canivet, A.; Grunenwald, S.; Pienkowski, C.; Chaumeil, J.; Mejía, J.E.; Guéry, J.C. TLR7 escapes X chromosome inactivation in immune cells. *Sci. Immunol.* **2018**, *3*, eaap8855. [CrossRef]
75. Berghöfer, B.; Frommer, T.; Haley, G.; Fink, L.; Bein, G.; Hackstein, H. TLR7 ligands induce higher IFN-alpha production in females. *J. Immunol.* **2006**, *177*, 2088–2096. [CrossRef]
76. Meier, A.; Chang, J.J.; Chan, E.S.; Pollard, R.B.; Sidhu, H.K.; Kulkarni, S.; Wen, T.F.; Lindsay, R.J.; Orellana, L.; Mildvan, D.; et al. Sex differences in the Toll-like receptor-mediated response of plasmacytoid dendritic cells to HIV-1. *Nat. Med.* **2009**, *15*, 955–959. [CrossRef]
77. Seillet, C.; Laffont, S.; Trémollières, F.; Rouquié, N.; Ribot, C.; Arnal, J.F.; Douin-Echinard, V.; Gourdy, P.; Guéry, J.C. The TLR-mediated response of plasmacytoid dendritic cells is positively regulated by estradiol in vivo through cell-intrinsic estrogen receptor α signaling. *Blood* **2012**, *119*, 454–464. [CrossRef]
78. Laffont, S.; Rouquié, N.; Azar, P.; Seillet, C.; Plumas, J.; Aspord, C.; Guéry, J.C. X-Chromosome complement and estrogen receptor signaling independently contribute to the enhanced TLR7-mediated IFN-α production of plasmacytoid dendritic cells from women. *J. Immunol.* **2014**, *193*, 5444–5452. [CrossRef]
79. van der Sluis, R.M.; Cham, L.B.; Gris-Oliver, A.; Gammelgaard, K.R.; Pedersen, J.G.; Idorn, M.; Ahmadov, U.; Hernandez, S.S.; Cémalovic, E.; Godsk, S.H.; et al. TLR2 and TLR7 mediate distinct immunopathological and antiviral plasmacytoid dendritic cell responses to SARS-CoV-2 infection. *EMBO J.* **2022**, *41*, e109622. [CrossRef]
80. Hadjadj, J.; Yatim, N.; Barnabei, L.; Corneau, A.; Boussier, J.; Smith, N.; Péré, H.; Charbit, B.; Bondet, V.; Chenevier-Gobeaux, C.; et al. Impaired type I interferon activity and inflammatory responses in severe COVID-19 patients. *Science* **2020**, *369*, 718–724. [CrossRef]
81. Liao, M.; Liu, Y.; Yuan, J.; Wen, Y.; Xu, G.; Zhao, J.; Cheng, L.; Li, J.; Wang, X.; Wang, F.; et al. Single-cell landscape of bronchoalveolar immune cells in patients with COVID-19. *Nat. Med.* **2020**, *26*, 842–844. [CrossRef] [PubMed]
82. Zhou, R.; To, K.K.; Wong, Y.C.; Liu, L.; Zhou, B.; Li, X.; Huang, H.; Mo, Y.; Luk, T.Y.; Lau, T.T.; et al. Acute SARS-CoV-2 Infection Impairs Dendritic Cell and T Cell Responses. *Immunity* **2020**, *53*, 864–877.e5. [CrossRef] [PubMed]
83. Pérez-Gómez, A.; Vitallé, J.; Gasca-Capote, C.; Gutierrez-Valencia, A.; Trujillo-Rodriguez, M.; Serna-Gallego, A.; Muñoz-Muela, E.; Jiménez-Leon, M.L.R.; Rafii-El-Idrissi Benhnia, M.; Rivas-Jeremias, I.; et al. Dendritic cell deficiencies persist seven months after SARS-CoV-2 infection. *Cell Mol. Immunol.* **2021**, *18*, 2128–2139. [CrossRef] [PubMed]
84. Severa, M.; Diotti, R.A.; Etna, M.P.; Rizzo, F.; Fiore, S.; Ricci, D.; Iannetta, M.; Sinigaglia, A.; Lodi, A.; Mancini, N.; et al. Differential plasmacytoid dendritic cell phenotype and type I Interferon response in asymptomatic and severe COVID-19 infection. *PLoS Pathog.* **2021**, *17*, e1009878. [CrossRef] [PubMed]
85. Winheim, E.; Rinke, L.; Lutz, K.; Reischer, A.; Leutbecher, A.; Wolfram, L.; Rausch, L.; Kranich, J.; Wratil, P.R.; Huber, J.E.; et al. Impaired function and delayed regeneration of dendritic cells in COVID-19. *PLoS Pathog.* **2021**, *17*, e1009742. [CrossRef]

86. Wu, J.; Shi, Y.; Pan, X.; Wu, S.; Hou, R.; Zhang, Y.; Zhong, T.; Tang, H.; Du, W.; Wang, L.; et al. SARS-CoV-2 ORF9b inhibits RIG-I-MAVS antiviral signaling by interrupting K63-linked ubiquitination of NEMO. *Cell Rep.* **2021**, *34*, 108761. [CrossRef]
87. Zheng, Y.; Deng, J.; Han, L.; Zhuang, M.W.; Xu, Y.; Zhang, J.; Nan, M.L.; Xiao, Y.; Zhan, P.; Liu, X.; et al. SARS-CoV-2 NSP5 and N protein counteract the RIG-I signaling pathway by suppressing the formation of stress granules. *Signal Transduct. Target Ther.* **2022**, *7*, 22. [CrossRef]
88. Wang, M.; Zhao, Y.; Liu, J.; Li, T. SARS-CoV-2 modulation of RIG-I-MAVS signaling: Potential mechanisms of impairment on host antiviral immunity and therapeutic approaches. *MedComm—Future Med.* **2022**, *1*, e29. [CrossRef]
89. Butler-Laporte, G.; Povysil, G.; Kosmicki, J.A.; Cirulli, E.T.; Drivas, T.; Furini, S.; Saad, C.; Schmidt, A.; Olszewski, P.; Korotko, U.; et al. Exome-wide association study to identify rare variants influencing COVID-19 outcomes: Results from the Host Genetics Initiative. *PLoS Genet.* **2022**, *18*, e1010367. [CrossRef]
90. Kadowaki, N.; Ho, S.; Antonenko, S.; Malefyt, R.W.; Kastelein, R.A.; Bazan, F.; Liu, Y.J. Subsets of human dendritic cell precursors express different toll-like receptors and respond to different microbial antigens. *J. Exp. Med.* **2001**, *194*, 863–869. [CrossRef]
91. Tabiasco, J.; Devêvre, E.; Rufer, N.; Salaun, B.; Cerottini, J.C.; Speiser, D.; Romero, P. Human effector CD8+ T lymphocytes express TLR3 as a functional coreceptor. *J. Immunol.* **2006**, *177*, 8708–8713. [CrossRef]
92. Schmidt, K.N.; Leung, B.; Kwong, M.; Zarember, K.A.; Satyal, S.; Navas, T.A.; Wang, F.; Godowski, P.J. APC-independent activation of NK cells by the Toll-like receptor 3 agonist double-stranded RNA. *J. Immunol.* **2004**, *172*, 138–143. [CrossRef]
93. Cario, E.; Podolsky, D.K. Differential alteration in intestinal epithelial cell expression of toll-like receptor 3 (TLR3) and TLR4 in inflammatory bowel disease. *Infect. Immun.* **2000**, *68*, 7010–7017. [CrossRef]
94. Fang, F.; Ooka, K.; Sun, X.; Shah, R.; Bhattacharyya, S.; Wei, J.; Varga, J. A synthetic TLR3 ligand mitigates profibrotic fibroblast responses by inducing autocrine IFN signaling. *J. Immunol.* **2013**, *191*, 2956–2966. [CrossRef]
95. Gao, D.; Ciancanelli, M.J.; Zhang, P.; Harschnitz, O.; Bondet, V.; Hasek, M.; Chen, J.; Mu, X.; Itan, Y.; Cobat, A.; et al. TLR3 controls constitutive IFN-β antiviral immunity in human fibroblasts and cortical neurons. *J. Clin. Invest.* **2021**, *131*, e134529. [CrossRef]
96. Bsibsi, M.; Ravid, R.; Gveric, D.; van Noort, J.M. Broad expression of Toll-like receptors in the human central nervous system. *J. Neuropathol. Exp. Neurol.* **2002**, *61*, 1013–1021. [CrossRef]
97. Jack, C.S.; Arbour, N.; Manusow, J.; Montgrain, V.; Blain, M.; McCrea, E.; Shapiro, A.; Antel, J.P. TLR signaling tailors innate immune responses in human microglia and astrocytes. *J. Immunol.* **2005**, *175*, 4320–4330. [CrossRef]
98. Préhaud, C.; Mégret, F.; Lafage, M.; Lafon, M. Virus infection switches TLR-3-positive human neurons to become strong producers of beta interferon. *J. Virol.* **2005**, *79*, 12893–12904. [CrossRef]
99. Zhang, S.Y.; Jouanguy, E.; Ugolini, S.; Smahi, A.; Elain, G.; Romero, P.; Segal, D.; Sancho-Shimizu, V.; Lorenzo, L.; Puel, A.; et al. TLR3 deficiency in patients with herpes simplex encephalitis. *Science* **2007**, *317*, 1522–1527. [CrossRef]
100. Bagheri-Hosseinabadi, Z.; Rezazadeh Zarandi, E.; Mirabzadeh, M.; Amiri, A.; Abbasifard, M. mRNA expression of toll-like receptors 3, 7, 8, and 9 in the nasopharyngeal epithelial cells of coronavirus disease 2019 patients. *BMC Infect. Dis.* **2022**, *22*, 448. [CrossRef]
101. Menezes, M.C.S.; Veiga, A.D.M.; Martins de Lima, T.; Kunimi Kubo Ariga, S.; Vieira Barbeiro, H.; de Lucena Moreira, C.; Pinto, A.A.S.; Brandao, R.A.; Marchini, J.F.; Alencar, J.C.; et al. Lower peripheral blood Toll-like receptor 3 expression is associated with an unfavorable outcome in severe COVID-19 patients. *Sci. Rep.* **2021**, *11*, 15223. [CrossRef] [PubMed]
102. Mielcarska, M.B.; Bossowska-Nowicka, M.; Toka, F.N. Cell Surface Expression of Endosomal Toll-Like Receptors-A Necessity or a Superfluous Duplication. *Front. Immunol.* **2021**, *11*, 620972. [CrossRef] [PubMed]
103. Duffy, K.E.; Lamb, R.J.; San Mateo, L.R.; Jordan, J.L.; Canziani, G.; Brigham-Burke, M.; Korteweg, J.; Cunningham, M.; Beck, H.S.; Carton, J.; et al. Down modulation of human TLR3 function by a monoclonal antibody. *Cell Immunol.* **2007**, *248*, 103–114. [CrossRef] [PubMed]
104. Perales-Linares, R.; Navas-Martin, S. Toll-like receptor 3 in viral pathogenesis: Friend or foe. *Immunology* **2013**, *140*, 153–167. [CrossRef]
105. Choudhury, A.; Das, N.C.; Patra, R.; Mukherjee, S. In silico analyses on the comparative sensing of SARS-CoV-2 mRNA by the intracellular TLRs of humans. *J. Med. Virol.* **2021**, *93*, 2476–2486. [CrossRef]
106. Bortolotti, D.; Gentili, V.; Rizzo, S.; Schiuma, G.; Beltrami, S.; Strazzabosco, G.; Fernandez, M.; Caccuri, F.; Caruso, A.; Rizzo, R. TLR3 and TLR7 RNA Sensor Activation during SARS-CoV-2 Infection. *Microorganisms* **2021**, *9*, 1820. [CrossRef]
107. Zhang, Q.; Bastard, P.; Liu, Z.; Le Pen, J.; Moncada-Velez, M.; Chen, J.; Ogishi, M.; Sabli, I.K.D.; Hodeib, S.; Korol, C.; et al. Inborn errors of type I IFN immunity in patients with life-threatening COVID-19. *Science* **2020**, *370*, eabd4570. [CrossRef]
108. Lim, H.K.; Huang, S.X.L.; Chen, J.; Kerner, G.; Gilliaux, O.; Bastard, P.; Dobbs, K.; Hernandez, N.; Goudin, N.; Hasek, M.L.; et al. Severe influenza pneumonitis in children with inherited TLR3 deficiency. *J. Exp. Med.* **2019**, *216*, 2038–2056. [CrossRef]
109. Lévy, R.; Bastard, P.; Lanternier, F.; Lecuit, M.; Zhang, S.Y.; Casanova, J.L. IFN-α2a Therapy in Two Patients with Inborn Errors of TLR3 and IRF3 Infected with SARS-CoV-2. *J. Clin. Immunol.* **2021**, *41*, 26–27. [CrossRef]
110. Povysil, G.; Butler-Laporte, G.; Shang, N.; Wang, C.; Khan, A.; Alaamery, M.; Nakanishi, T.; Zhou, S.; Forgetta, V.; Eveleigh, R.J.; et al. Rare loss-of function variants in type I IFN immunity genes are not associated with severe COVID-19. *J. Clin. Invest.* **2021**, *131*, e147834. [CrossRef]
111. Liu, P.; Fang, M.; Luo, Y.; Zheng, F.; Jin, Y.; Cheng, F.; Zhu, H.; Jin, X. Rare Variants in Inborn Errors of Immunity Genes Associated With Covid-19 Severity. *Front. Cell Infect. Microbiol.* **2022**, *12*, 888582. [CrossRef]

112. Ranjith-Kumar, C.T.; Miller, W.; Sun, J.; Xiong, J.; Santos, J.; Yarbrough, I.; Lamb, R.J.; Mills, J.; Duffy, K.E.; Hoose, S.; et al. Effects of single nucleotide polymorphisms on Toll-like receptor 3 activity and expression in cultured cells. *J. Biol. Chem.* **2007**, *282*, 17696–17705. [CrossRef]
113. Teimouri, H.; Maali, A. Single-nucleotide polymorphisms in host pattern-recognition receptors show association with antiviral responses against SARS-CoV-2, in-silico trial. *JoMMID* **2020**, *8*, 65–70. [CrossRef]
114. Dhangadamajhi, G.; Rout, R. Association of TLR3 functional variant (rs3775291) with COVID-19 susceptibility and death: A population-scale study. *Hum. Cell.* **2021**, *34*, 1025–1027. [CrossRef]
115. Croci, S.; Venneri, M.A.; Mantovani, S.; Fallerini, C.; Benetti, E.; Picchiotti, N.; Campolo, F.; Imperatore, F.; Palmieri, M.; Daga, S.; et al. The polymorphism L412F in TLR3 inhibits autophagy and is a marker of severe COVID-19 in males. *Autophagy* **2022**, *18*, 1662–1672. [CrossRef]
116. Schmitt, C.A.; Tchkonia, T.; Niedernhofer, L.J.; Robbins, P.D.; Kirkland, J.L.; Lee, S. COVID-19 and cellular senescence. *Nat. Rev. Immunol.* **2022**, *23*, 1–13. [CrossRef]
117. Tripathi, U.; Nchioua, R.; Prata, L.G.P.L.; Zhu, Y.; Gerdes, E.O.W.; Giorgadze, N.; Pirtskhalava, T.; Parker, E.; Xue, A.; Espindola-Netto, J.M.; et al. SARS-CoV-2 causes senescence in human cells and exacerbates the senescence-associated secretory phenotype through TLR-3. *Aging* **2021**, *13*, 21838–21854. [CrossRef]
118. Oliveira-Nascimento, L.; Massari, P.; Wetzler, L.M. The Role of TLR2 in Infection and Immunity. *Front. Immunol.* **2012**, *3*, 79. [CrossRef]
119. Zheng, M.; Karki, R.; Williams, E.P.; Yang, D.; Fitzpatrick, E.; Vogel, P.; Jonsson, C.B.; Kanneganti, T.D. TLR2 senses the SARS-CoV-2 envelope protein to produce inflammatory cytokines. *Nat. Immunol.* **2021**, *22*, 829–838. [CrossRef]
120. Del Valle, D.M.; Kim-Schulze, S.; Huang, H.H.; Beckmann, N.D.; Nirenberg, S.; Wang, B.; Lavin, Y.; Swartz, T.H.; Madduri, D.; Stock, A.; et al. An inflammatory cytokine signature predicts COVID-19 severity and survival. *Nat. Med.* **2020**, *26*, 1636–1643. [CrossRef]
121. Buszko, M.; Nita-Lazar, A.; Park, J.H.; Schwartzberg, P.L.; Verthelyi, D.; Young, H.A.; Rosenberg, A.S. Lessons learned: New insights on the role of cytokines in COVID-19. *Nat. Immunol.* **2021**, *22*, 404–411. [CrossRef] [PubMed]
122. Khan, S.; Shafiei, M.S.; Longoria, C.; Schoggins, J.W.; Savani, R.C.; Zaki, H. SARS-CoV-2 spike protein induces inflammation via TLR2-dependent activation of the NF-κB pathway. *Elife* **2021**, *10*, e68563. [CrossRef] [PubMed]
123. Blanco-Melo, D.; Nilsson-Payant, B.E.; Liu, W.C.; Uhl, S.; Hoagland, D.; Møller, R.; Jordan, T.X.; Oishi, K.; Panis, M.; Sachs, D.; et al. Imbalanced Host Response to SARS-CoV-2 Drives Development of COVID-19. *Cell* **2020**, *181*, 1036–1045. [CrossRef] [PubMed]
124. Liu, T.; Zhang, J.; Yang, Y.; Ma, H.; Li, Z.; Zhang, J.; Cheng, J.; Zhang, X.; Zhao, Y.; Xia, Z.; et al. The role of interleukin-6 in monitoring severe case of coronavirus disease 2019. *EMBO Mol. Med.* **2020**, *12*, e12421. [CrossRef]
125. Zhang, J.; Hao, Y.; Ou, W.; Ming, F.; Liang, G.; Qian, Y.; Cai, Q.; Dong, S.; Hu, S.; Wang, W.; et al. Serum interleukin-6 is an indicator for severity in 901 patients with SARS-CoV-2 infection: A cohort study. *J. Transl. Med.* **2020**, *18*, 406. [CrossRef]
126. Lin, L.; Chen, Y.; Han, D.; Yang, A.; Wang, A.Y.; Qi, W. Cardiorenal Syndrome in COVID-19 Patients: A Systematic Review. *Front. Cardiovasc. Med.* **2022**, *9*, 915533. [CrossRef]
127. Sultan, R.H.; Elesawy, B.H.; Ali, T.M.; Abdallah, M.; Assal, H.H.; Ahmed, A.E.; Ahmed, O.M. Correlations between Kidney and Heart Function Bioindicators and the Expressions of Toll-Like, ACE2, and NRP-1 Receptors in COVID-19. *Vaccines* **2022**, *10*, 1106. [CrossRef]
128. Middleton, E.A.; He, X.Y.; Denorme, F.; Campbell, R.A.; Ng, D.; Salvatore, S.P.; Mostyka, M.; Baxter-Stoltzfus, A.; Borczuk, A.C.; Loda, M.; et al. Neutrophil extracellular traps contribute to immunothrombosis in COVID-19 acute respiratory distress syndrome. *Blood* **2020**, *136*, 1169–1179. [CrossRef]
129. Sung, P.S.; Yang, S.P.; Peng, Y.C.; Sun, C.P.; Tao, M.H.; Hsieh, S.L. CLEC5A and TLR2 are critical in SARS-CoV-2-induced NET formation and lung inflammation. *J. Biomed. Sci.* **2022**, *29*, 52. [CrossRef]
130. Hanna, C.C.; Ashhurst, A.S.; Quan, D.; Maxwell, J.W.C.; Britton, W.J.; Payne, R.J. Synthetic protein conjugate vaccines provide protection against Mycobacterium tuberculosis in mice. *Proc. Natl. Acad. Sci. USA.* **2021**, *118*, e2013730118. [CrossRef]
131. Ashhurst, A.S.; McDonald, D.M.; Hanna, C.C.; Stanojevic, V.A.; Britton, W.J.; Payne, R.J. Mucosal Vaccination with a Self-Adjuvanted Lipopeptide Is Immunogenic and Protective against Mycobacterium tuberculosis. *J. Med. Chem.* **2019**, *62*, 8080–8089. [CrossRef]
132. Tyne, A.S.; Chan, J.G.; Shanahan, E.R.; Atmosukarto, I.; Chan, H.K.; Britton, W.J.; West, N.P. TLR2-targeted secreted proteins from Mycobacterium tuberculosis are protective as powdered pulmonary vaccines. *Vaccine* **2013**, *31*, 4322–4329. [CrossRef]
133. Zeng, W.; Ghosh, S.; Lau, Y.F.; Brown, L.E.; Jackson, D.C. Highly immunogenic and totally synthetic lipopeptides as self-adjuvanting immunocontraceptive vaccines. *J. Immunol.* **2002**, *169*, 4905–4912. [CrossRef]
134. Ashhurst, A.S.; Johansen, M.D.; Maxwell, J.W.C.; Stockdale, S.; Ashley, C.L.; Aggarwal, A.; Siddiquee, R.; Miemczyk, S.; Nguyen, D.H.; Mackay, J.P.; et al. Mucosal TLR2-activating protein-based vaccination induces potent pulmonary immunity and protection against SARS-CoV-2 in mice. *Nat. Commun.* **2022**, *13*, 6972. [CrossRef]
135. Proud, P.C.; Tsitoura, D.; Watson, R.J.; Chua, B.Y.; Aram, M.J.; Bewley, K.R.; Cavell, B.E.; Cobb, R.; Dowall, S.; Fotheringham, S.A.; et al. Prophylactic intranasal administration of a TLR2/6 agonist reduces upper respiratory tract viral shedding in a SARS-CoV-2 challenge ferret model. *EBioMedicine* **2021**, *63*, 103153. [CrossRef]
136. Vaure, C.; Liu, Y. A comparative review of toll-like receptor 4 expression and functionality in different animal species. *Front. Immunol.* **2014**, *5*, 316. [CrossRef]

137. Kagan, J.C.; Su, T.; Horng, T.; Chow, A.; Akira, S.; Medzhitov, R. TRAM couples endocytosis of Toll-like receptor 4 to the induction of interferon-beta. *Nat. Immunol.* **2008**, *9*, 361–368. [CrossRef]
138. Bhattacharya, M.; Sharma, A.R.; Mallick, B.; Sharma, G.; Lee, S.S.; Chakraborty, C. Immunoinformatics approach to understand molecular interaction between multi-epitopic regions of SARS-CoV-2 spike-protein with TLR4/MD-2 complex. *Infect. Genet. Evol.* **2020**, *85*, 104587. [CrossRef]
139. Choudhury, A.; Mukherjee, S. In silico studies on the comparative characterization of the interactions of SARS-CoV-2 spike glycoprotein with ACE-2 receptor homologs and human TLRs. *J. Med. Virol.* **2020**, *92*, 2105–2113. [CrossRef]
140. Shirato, K.; Kizaki, T. SARS-CoV-2 spike protein S1 subunit induces pro-inflammatory responses via toll-like receptor 4 signaling in murine and human macrophages. *Heliyon* **2021**, *7*, e06187. [CrossRef]
141. Aboudounya, M.M.; Heads, R.J. COVID-19 and Toll-Like Receptor 4 (TLR4): SARS-CoV-2 May Bind and Activate TLR4 to Increase ACE2 Expression, Facilitating Entry and Causing Hyperinflammation. *Mediat. Inflamm.* **2021**, *2021*, 8874339. [CrossRef] [PubMed]
142. Zhao, Y.; Kuang, M.; Li, J.; Zhu, L.; Jia, Z.; Guo, X.; Hu, Y.; Kong, J.; Yin, H.; Wang, X.; et al. Publisher Correction: SARS-CoV-2 spike protein interacts with and activates TLR4. *Cell Res.* **2021**, *31*, 825. [CrossRef] [PubMed]
143. Liu, Z.M.; Yang, M.H.; Yu, K.; Lian, Z.X.; Deng, S.L. Toll-like receptor (TLRs) agonists and antagonists for COVID-19 treatments. *Front. Pharmacol.* **2022**, *13*, 989664. [CrossRef] [PubMed]
144. Calabretta, E.; Moraleda, J.M.; Iacobelli, M.; Jara, R.; Vlodavsky, I.; O'Gorman, P.; Pagliuca, A.; Mo, C.; Baron, R.M.; Aghemo, A.; et al. COVID-19-induced endotheliitis: Emerging evidence and possible therapeutic strategies. *Br. J. Haematol.* **2021**, *193*, 43–51. [CrossRef]
145. Ma, Z.; Li, X.; Fan, R.L.Y.; Yang, K.Y.; Ng, C.S.H.; Lau, R.W.H.; Wong, R.H.L.; Ng, K.K.; Wang, C.C.; Ye, P.; et al. A human pluripotent stem cell-based model of SARS-CoV-2 infection reveals an ACE2-independent inflammatory activation of vascular endothelial cells through TLR4. *Stem Cell Rep.* **2022**, *17*, 538–555. [CrossRef]
146. RECOVERY Collaborative Group; Horby, P.; Lim, W.S.; Emberson, J.R.; Mafham, M.; Bell, J.L.; Linsell, L.; Staplin, N.; Brightling, C.; Ustianowski, A.; et al. Dexamethasone in Hospitalized Patients with COVID-19. *N. Engl. J. Med.* **2021**, *384*, 693–704. [CrossRef]
147. Amulic, B.; Cazalet, C.; Hayes, G.L.; Metzler, K.D.; Zychlinsky, A. Neutrophil function: From mechanisms to disease. *Annu. Rev. Immunol.* **2012**, *30*, 459–489. [CrossRef]
148. Zuo, Y.; Yalavarthi, S.; Shi, H.; Gockman, K.; Zuo, M.; Madison, J.A.; Blair, C.; Weber, A.; Barnes, B.J.; Egeblad, M.; et al. Neutrophil extracellular traps in COVID-19. *JCI Insight* **2020**, *5*, e138999. [CrossRef]
149. Cesta, M.C.; Zippoli, M.; Marsiglia, C.; Gavioli, E.M.; Cremonesi, G.; Khan, A.; Mantelli, F.; Allegretti, M.; Balk, R. Neutrophil activation and neutrophil extracellular traps (NETs) in COVID-19 ARDS and immunothrombosis. *Eur. J. Immunol.* **2023**, *53*, e2250010. [CrossRef]
150. Huang, C.; Wang, Y.; Li, X.; Ren, L.; Zhao, J.; Hu, Y.; Zhang, L.; Fan, G.; Xu, J.; Gu, X.; et al. Clinical features of patients infected with 2019 novel coronavirus in Wuhan, China. *Lancet* **2020**, *395*, 497–506. [CrossRef]
151. Wang, D.; Hu, B.; Hu, C.; Zhu, F.; Liu, X.; Zhang, J.; Wang, B.; Xiang, H.; Cheng, Z.; Xiong, Y.; et al. Clinical Characteristics of 138 Hospitalized Patients With 2019 Novel Coronavirus-Infected Pneumonia in Wuhan, China. *JAMA* **2020**, *323*, 1061–1069. [CrossRef]
152. Zeng, Z.Y.; Feng, S.D.; Chen, G.P.; Wu, J.N. Predictive value of the neutrophil to lymphocyte ratio for disease deterioration and serious adverse outcomes in patients with COVID-19: A prospective cohort study. *BMC Infect. Dis.* **2021**, *21*, 80. [CrossRef]
153. Seyit, M.; Avci, E.; Nar, R.; Senol, H.; Yilmaz, A.; Ozen, M.; Oskay, A.; Aybek, H. Neutrophil to lymphocyte ratio, lymphocyte to monocyte ratio and platelet to lymphocyte ratio to predict the severity of COVID-19. *Am. J. Emerg. Med.* **2021**, *40*, 110–114. [CrossRef]
154. Li, L.; Li, J.; Gao, M.; Fan, H.; Wang, Y.; Xu, X.; Chen, C.; Liu, J.; Kim, J.; Aliyari, R.; et al. Interleukin-8 as a Biomarker for Disease Prognosis of Coronavirus Disease-2019 Patients. *Front. Immunol.* **2021**, *11*, 602395. [CrossRef]
155. Ma, A.; Zhang, L.; Ye, X.; Chen, J.; Yu, J.; Zhuang, L.; Weng, C.; Petersen, F.; Wang, Z.; Yu, X. High Levels of Circulating IL-8 and Soluble IL-2R Are Associated With Prolonged Illness in Patients With Severe COVID-19. *Front. Immunol.* **2021**, *12*, 626235. [CrossRef]
156. McKenna, E.; Wubben, R.; Isaza-Correa, J.M.; Melo, A.M.; Mhaonaigh, A.U.; Conlon, N.; O'Donnell, J.S.; Ní Cheallaigh, C.; Hurley, T.; Stevenson, N.J.; et al. Neutrophils in COVID-19: Not Innocent Bystanders. *Front. Immunol.* **2022**, *13*, 864387. [CrossRef]
157. Silvin, A.; Chapuis, N.; Dunsmore, G.; Goubet, A.G.; Dubuisson, A.; Derosa, L.; Almire, C.; Hénon, C.; Kosmider, O.; Droin, N.; et al. Elevated Calprotectin and Abnormal Myeloid Cell Subsets Discriminate Severe from Mild COVID-19. *Cell* **2020**, *182*, 1401–1418.e18. [CrossRef]
158. Loh, J.T.; Teo, J.K.H.; Lam, K.P. Dok3 restrains neutrophil production of calprotectin during TLR4 sensing of SARS-CoV-2 spike protein. *Front. Immunol.* **2022**, *13*, 996637. [CrossRef]
159. Wang, C.; Xie, J.; Zhao, L.; Fei, X.; Zhang, H.; Tan, Y.; Nie, X.; Zhou, L.; Liu, Z.; Ren, Y.; et al. Alveolar macrophage dysfunction and cytokine storm in the pathogenesis of two severe COVID-19 patients. *EBioMedicine* **2020**, *57*, 102833. [CrossRef]
160. Fajgenbaum, D.C.; June, C.H. Cytokine Storm. *N. Engl. J. Med.* **2020**, *383*, 2255–2273. [CrossRef]
161. Hu, B.; Huang, S.; Yin, L. The cytokine storm and COVID-19. *J. Med. Virol.* **2021**, *93*, 250–256. [CrossRef] [PubMed]
162. Mantovani, A.; Dinarello, C.A.; Molgora, M.; Garlanda, C. Interleukin-1 and Related Cytokines in the Regulation of Inflammation and Immunity. *Immunity* **2019**, *50*, 778–795. [CrossRef] [PubMed]

163. Dinarello, C.A. Overview of the IL-1 family in innate inflammation and acquired immunity. *Immunol. Rev.* **2018**, *281*, 8–27. [CrossRef] [PubMed]
164. Lucas, C.; Wong, P.; Klein, J.; Castro, T.B.R.; Silva, J.; Sundaram, M.; Ellingson, M.K.; Mao, T.; Oh, J.E.; Israelow, B.; et al. Longitudinal analyses reveal immunological misfiring in severe COVID-19. *Nature* **2020**, *584*, 463–469. [CrossRef]
165. Karadaş, Ö.; Öztürk, B.; Sonkaya, A.R.; Taşdelen, B.; Özge, A.; Bolay, H. Latent class cluster analysis identified hidden headache phenotypes in COVID-19: Impact of pulmonary infiltration and IL-6. *Neurol. Sci.* **2021**, *42*, 1665–1673. [CrossRef]
166. Trigo, J.; García-Azorín, D.; Sierra-Mencía, Á.; Tamayo-Velasco, Á.; Martínez-Paz, P.; Tamayo, E.; Guerrero, A.L.; Gonzalo-Benito, H. Cytokine and interleukin profile in patients with headache and COVID-19: A pilot, CASE-control, study on 104 patients. *J. Headache Pain* **2021**, *22*, 51. [CrossRef]
167. Kritas, S.K. COVID-19 and pain. *Eur. J. Neurodegener. Dis.* **2021**, *10*, 32–35.
168. Yarlagadda, A.; Preston, S.L.; Jeyadhas, R.P.; Lang, A.E.; Hammamieh, R.; Clayton, A.H. Blood-Brain Barrier: COVID-19, Pandemics, and Cytokine Norms. *Innov. Clin. Neurosci.* **2021**, *18*, 21–23.
169. Antoniades, E.; Melissaris, S.; Panagopoulos, D.; Kalloniati, E.; Sfakianos, G. Pathophysiology and neuroinflammation in COVID-19. *Eur. J. Neurodegener. Dis.* **2022**, *11*, 7–9.
170. Monti, G.; Giovannini, G.; Marudi, A.; Bedin, R.; Melegari, A.; Simone, A.M.; Santangelo, M.; Pignatti, A.; Bertellini, E.; Trenti, T.; et al. Anti-NMDA receptor encephalitis presenting as new onset refractory status epilepticus in COVID-19. *Seizure* **2020**, *81*, 18–20. [CrossRef]
171. Torabi, A.; Mohammadbagheri, E.; Akbari Dilmaghani, N.; Bayat, A.H.; Fathi, M.; Vakili, K.; Alizadeh, R.; Rezaeimirghaed, O.; Hajiesmaeili, M.; Ramezani, M.; et al. Proinflammatory Cytokines in the Olfactory Mucosa Result in COVID-19 Induced Anosmia. *ACS Chem. Neurosci.* **2020**, *11*, 1909–1913. [CrossRef]
172. Butowt, R.; Bilinska, K.; von Bartheld, C.S. Olfactory dysfunction in COVID-19: New insights into the underlying mechanisms. *Trends Neurosci.* **2023**, *46*, 75–90. [CrossRef]
173. Mele, D.; Calastri, A.; Maiorano, E.; Cerino, A.; Sachs, M.; Oliviero, B.; Mantovani, S.; Baldanti, F.; Bruno, R.; Benazzo, M.; et al. High Frequencies of Functional Virus-Specific CD4+ T Cells in SARS-CoV-2 Subjects With Olfactory and Taste Disorders. *Front. Immunol.* **2021**, *12*, 748881. [CrossRef]
174. Rethinavel, H.S.; Ravichandran, S.; Radhakrishnan, R.K.; Kandasamy, M. COVID-19 and Parkinson's disease: Defects in neurogenesis as the potential cause of olfactory system impairments and anosmia. *J. Chem. Neuroanat.* **2021**, *115*, 101965. [CrossRef]
175. De Rosa, M.A.; Calisi, D.; Carrarini, C.; Mazzatenta, A.; Mattioli, M.V.; Neri, C.; D'Ardes, C.; Giansante, R.; Onofrj, M.; Stuppia, L.; et al. Olfactory dysfunction as a predictor of the future development of parkinsonism in COVID-19 patients: A 18f-fdopa pet study. *Eur. J. Neurodegener. Dis.* **2023**, *12*.
176. Pardi, N.; Hogan, M.J.; Porter, F.W.; Weissman, D. mRNA vaccines—A new era in vaccinology. *Nat. Rev. Drug Discov.* **2018**, *17*, 261–279. [CrossRef]
177. Nelson, J.; Sorensen, E.W.; Mintri, S.; Rabideau, A.E.; Zheng, W.; Besin, G.; Khatwani, N.; Su, S.V.; Miracco, E.J.; Issa, W.J.; et al. Impact of mRNA chemistry and manufacturing process on innate immune activation. *Sci. Adv.* **2020**, *6*, eaaz6893. [CrossRef]
178. Sokal, A.; Bastard, P.; Chappert, P.; Barba-Spaeth, G.; Fourati, S.; Vanderberghe, A.; Lagouge-Roussey, P.; Meyts, I.; Gervais, A.; Bouvier-Alias, M.; et al. Human type I IFN deficiency does not impair B cell response to SARS-CoV-2 mRNA vaccination. *J. Exp. Med.* **2023**, *220*, e20220258. [CrossRef]

International Journal of
Molecular Sciences

MDPI

Article

The Role of Toll-like Receptor-4 in Macrophage Imbalance in Lethal COVID-19 Lung Disease, and Its Correlation with Galectin-3

Maria Carmela Pedicillo [1,†], Ilenia Sara De Stefano [1,†], Rosanna Zamparese [2], Raffaele Barile [3], Mario Meccariello [3], Alessio Agostinone [1], Giuliana Villani [4], Tommaso Colangelo [3,5], Gaetano Serviddio [3], Tommaso Cassano [3], Andrea Ronchi [6], Renato Franco [6], Paola Pannone [7], Federica Zito Marino [6], Francesco Miele [8], Maurizio Municinò [9] and Giuseppe Pannone [1,*]

1 Department of Clinical and Experimental Medicine, University of Foggia, Viale L.Pinto 1, 71122 Foggia, Italy; mariacamela.pedicillo@unifg.it (M.C.P.); ileniadestefano@hotmail.it (I.S.D.S.); alessio.agostinone@gmail.com (A.A.)
2 Legal Medicine Unit, Ascoli Piceno Hospital C-G. Mazzoni, Viale Degli Iris 13, 63100 Ascoli Piceno, Italy; rosanna.zamparese@sanita.marche.it
3 Department of Medical and Surgical Sciences, University of Foggia, Viale L.Pinto 1, 71122 Foggia, Italy; raffaele_barile.552733@unifg.it (R.B.); mario_meccariello.552823@unifg.it (M.M.); tommaso.colangelo@unifg.it (T.C.); gaetano.serviddio@unifg.it (G.S.); tommaso.cassano@unifg.it (T.C.)
4 Policlinico Riuniti, University-Hospital, Viale L.Pinto 1, 71122 Foggia, Italy; giuliana-91@hotmail.it
5 Cancer Cell Signalling Unit, Institute for Stem-Cell Biology, Regenerative Medicine and Innovative Therapies (ISBReMIT), IRCCS Fondazione Casa Sollievo della Sofferenza, Viale Cappuccini sc.c., San Giovanni Rotondo, 71013 Foggia, Italy
6 Pathology Unit, Department of Mental and Physical Health and Preventive Medicine, University of Campania "L Vanvitelli", via Luciano Armanni, 80138 Naples, Italy; andrea.ronchi@unicampania.it (A.R.); renato.franco@unicampania.it (R.F.); federicazito.marino@libero.it (F.Z.M.)
7 Federico II, Department of Clinical Medicine and Surgery, School of medicine and Surgery, University of Naples, via Sergio Pasini, 80131 Naples, Italy; pao.pannone@studenti.unina.it
8 Department of Surgery, University of Campania "L Vanvitelli", 80138 Naples, Italy; dott.miele@gmail.co
9 Forensic Medicine Unit, "S. Giuliano" Hospital, via Giambattista Basile, 80014 Giugliano in Campania, Italy; maurizio.municino@aslnapoli2nord.it
* Correspondence: giuseppe.pannone@unifg.it; Tel.: +39-081-5608495
† These authors contributed equally to this work.

Citation: Pedicillo, M.C.; De Stefano, I.S.; Zamparese, R.; Barile, R.; Meccariello, M.; Agostinone, A.; Villani, G.; Colangelo, T.; Serviddio, G.; Cassano, T.; et al. The Role of Toll-like Receptor-4 in Macrophage Imbalance in Lethal COVID-19 Lung Disease, and Its Correlation with Galectin-3. *Int. J. Mol. Sci.* **2023**, *24*, 13259. https://doi.org/10.3390/ijms241713259

Academic Editors: Oliver Planz and Ralf Kircheis

Received: 19 July 2023
Revised: 22 August 2023
Accepted: 24 August 2023
Published: 26 August 2023

Abstract: To the current data, there have been 6,955,141 COVID-19-related deaths worldwide, reported to WHO. Toll-like receptors (TLRs) implicated in bacterial and virus sensing could be a crosstalk between activation of persistent innate-immune inflammation, and macrophage's sub-population alterations, implicated in cytokine storm, macrophage over-activation syndrome, unresolved Acute Respiratory Disease Syndrome (ARDS), and death. The aim of this study is to demonstrate the association between Toll-like-receptor-4 (TLR-4)-induced inflammation and macrophage imbalance in the lung inflammatory infiltrate of lethal COVID-19 disease. Twenty-five cases of autopsy lung tissues were studied by digital pathology-based immunohistochemistry to evaluate expression levels of TLR-4 (CD 284), pan-macrophage marker CD68 (clone KP1), sub-population marker related to alveolar macrophage Galectin-3 (GAL-3) (clone 9C4), and myeloid derived CD163 (clone MRQ-26), respectively. SARS-CoV-2 viral persistence has been evaluated by in situ hybridation (ISH) method. This study showed TLR-4 up-regulation in a subgroup of patients, increased macrophage infiltration in both Spike-1$^{(+)}$ and Spike-1$^{(-)}$ lungs ($p < 0.0001$), and a macrophage shift with important down-regulation of GAL-3$^{(+)}$ alveolar macrophages associated with Spike-1 persistence ($p < 0.05$), in favor of CD163$^{(+)}$ myeloid derived monocyte-macrophages. Data show that TLR-4 expression induces a persistent activation of the inflammation, with inefficient resolution, and pathological macrophage shift, thus explaining one of the mechanisms of lethal COVID-19.

Keywords: TLRs; TLR-4; SARS-CoV-2; lethal COVID-19; lung disease; macrophage shift; ARDS; CD68; galectin-3; CD163; M1 macrophages; M2 macrophages

1. Introduction

The innate immune response is a key issue to act as the first line of defense against many viral infections [1]. The host immune response to SARS-CoV-2 infection is critical in determining the severity of COVID-19 [2].

It is assumed that cytokine storms, which are the consequence of hyperinflammation driven by innate immunity, play an important role in the pathogenesis of severe SARS-CoV-2 [3], however, the underlying mechanisms of the altered pathological inflammation in COVID-19 remain largely unknown.

The key innate immune sensing receptors are germ line-encoded pattern-recognition receptors (PRRs), which mediate the initial sensing of infection by recognition of PAMPs (pathogen-associated molecular patterns), upon microbial invasion of the host [4].

PRRs belong to different families, including Toll-like-receptors (TLRs).

TLRs are the most studied in the families of PRRs, and Toll-like receptor-4 (TLR-4) is one of the most studied. TLRs specifically recognizes bacterial lipopolysaccharide (LPS), and its activation mainly leads to the synthesis of pro-inflammatory cytokines and chemokines [5,6].

TLRs are the important mediators of inflammatory pathways in the gut, playing a major role in mediating the immune responses towards a wide variety of pathogen-derived ligands and link adaptive immunity with the innate immunity [7]. TLRs are directly involved in the regulation of inflammatory reactions and activation of the innate or adaptive immune responses for the elimination of infectious pathogens [8,9].

TLR-4 is a protein that in humans is encoded by the TLR-4 gene. TLR-4 is a transmembrane protein, member of the TLRs family, which belongs to the PRR family. TLR-4 activation leads to an intracellular signaling pathway NF-KB and inflammatory cytokine production, which is responsible for activating the innate immune system [10].

TLR-4 expressing cells are myeloid (erythrocytes, granulocytes, macrophages) rather than lymphoid (T-cells, B-cells, NK cells). Most myeloid cells also express high levels of CD14, which facilitates activation of TLR-4 by LPS.

It is well known for recognizing LPS, a component present in many Gram-negative bacteria (e.g., *Neisseria* spp.), and selected Gram-positive bacteria. TLR-4 ligands also include several viral proteins, polysaccharide, and a variety of endogenous proteins such as low-density lipoprotein, beta-defensins, and heat shock protein [11].

A recent study reported that alveolar macrophages activated by COVID-19 through TLR signaling produced interleukin-1 (IL-1), which further stimulates mast cells to produce interleukin-6 (IL-6) [12]. In addition, an association among mast cell, eosinophil activation, and COVID-19 inflammation was reported [13]. A significantly elevated expression of IL-6 and tumor necrosis factor-α (TNF-α) was found to be associated with TLR expression in obese individuals but not in the controls [14], a well-known cofactor associated with increased risk of mortality in COVID-19 disease.

In silico studies indicated that cell surface TLRs, mainly TLR-4, are most likely to be involved in sensing molecular patterns, including SARS-CoV-2 S protein, to induce inflammatory responses [15,16]. It was been reported that the SARS-CoV-2 S protein S1 subunit can induce pro-inflammatory cytokines through TLR-4 signalling in murine and human macrophages, and inhibition of TLR-4 by using its antagonist attenuates pro-inflammatory cytokine induction [17].

It was considered that pro-inflammatory cytokines greatly contribute to the pathogenesis of SARS-CoV-2 and its severity. Zheng et al., reported the sensing of SARS-CoV-2 E protein by Toll-like receptor-2 (TLR-2), which results in the hyper-expression of pro-inflammatory cytokines that may contribute to disease severity [18], suggesting that TLR-2 mediated inflammation plays a pathogenic role in SARS-CoV-2 infection.

Therefore, TLR signaling might be involved in the induction of pro-inflammatory mediators in COVID-19.

Recent evidences from current literature showed a central role of macrophages in SARS-CoV-2 infection [19,20]. Among inflammatory cells, macrophages play a crucial role in COVID-19 phlogistic mechanisms, in the induction and non-resolution, which subsequently lead to the persistence and amplification of the phlogistic state, culminating in death. Although monocyte blood cell count results in COVID-19 patients is often within the reference range, a high proportion of activated cells was been shown [21]. Patients with severe and critical COVID-19 disease, admitted to intensive care units (ICU), have elevated blood levels of IL-1β, IL-2, IL-7, IL-9, IL-10, IL-17, Granulocyte Colony-stimulating factor (G-CSF), Granulocyte-Macrophage Colony-Stimulating Factor (GM-CSF), Interferon-γ (IFN-γ), Tumor Necrosis Factor-α (TNF-α), C-X-C motif chemokine ligand 8 (CXCL8), C-X-C motif chemokine ligand 10 (CXCL10). Furthermore, fatal cases of SARS-CoV-1, MERS-CoV and SARS-CoV-2 lung autopsies have revealed extensive cellular infiltration of phlogistic cells with macrophages predominance [19].

Macrophages can actually sub-divided into three functional classes: M1, M2, and M2-like [22,23], surpassing the dichotomous classification in two classes of functionally distinct macrophages [24]. M1 macrophages are activated by intracellular pathogens, bacterial wall, lipoptoteins and many cytokines such as IFN-γ and TNF-α, while activation of M2 is induced by fungi, parasites, immune-complex, interleukins, and many other signals. M2 macrophages have phagocytosis capacity, but they can also mitigate the inflammatory response and promote the injury repair [25]. M2-like macrophage phenotype is under the control of IL-10, which in turn activates Signal Transducer and Activator of Transcription 3 (STAT3)-mediated gene expression of IL-10, Transforming Growth Factor beta 1 (TGF beta 1), Mannose Receptor C-type 1 (MRC 1) [26].

Regarding the anatomical sites, lung macrophages were been divided into three broad categories, since they are located in interstitial, alveolar, and in the vascular compartments respectively [27].

In the context of the alveolar macrophage population, the maintenance of the activation of pathway regulated by PPARγ (peroxisome proliferator-activated receptor gamma) has a fundamental role in the containment of lung inflammatory process, and in the resolution of acute phase, while on the contrary an important PPARγ repression was been demonstrated during activation of lung monocyte-macrophages in severe COVID-19 [28].

Interstitial macrophages also consist of heterogeneous populations, as reported in recent scientific articles [29,30].

An important role in the activation of M2 macrophage is played by Galectin-3 (GAL-3). Indeed, GAL-3 was been formerly called MAC-2 antigen because of its role in activating M2 macrophages [31].

In this study, we have analysed the immunohistochemical expression of TLR-4, GAL -3, CD68 and CD163 in the specimens derived from autopsy procedures of patients died from critical SARS-CoV-2 infection, in order to potentially associate innate immune hyperinflammation with monocyte-macrophage up-regulation and macrophage's sub-population changes.

In details, aim of this work was to evaluate, in the complex context of macrophage activation, the pro-inflammatory role of TLR-4, and association among alveolar GAL-3 expression, innate immune phlogistic pathway activation, and lung macrophage repletion by CD163$^{(+)}$ myeloid derived monocyte-macrophages.

Further aim is to evaluate possible association between a macrophage shift between loss of alveolar GAL -3 macrophages and persistence of Spike-1 viral sequence as a crucial event in lethal SARS-CoV-2 lung disease.

In summary, our study will focus on the innate-immune pathway, the inflammatory profile of macrophages, and composition of alveolar and interstitial infiltrate in lethal SARS-CoV-2 lung disease.

2. Results

2.1. Main Histopathological Findings

SARS-CoV-2 infection determined alveolar congestion, with pneumocytic damage, depicting a particular Diffuse Alveolar Damage (DAD) (Figure 1), representing the histological hallmark for the acute phase of Acute Respiratory Disease Syndrome (ARDS), together with acute phlogistic cell inflammation, interstitial vessel congestion, hemorrhage, and hyaline membrane deposition. Massive alveolar and interstitial phlogistic cell infiltration, with hyaline membrane, and thin fibril deposition was been detected, depicting a network of early-intermediate fibrosis. Furthermore, polymorphonuclear cells infiltrate the lungs, especially in cases associated with bacterial superinfection. Most deceased patients showed lungs already affected by chronic inflammation with lymphocytic infiltration and anthracotic deposits. SARS-CoV-2 infection determined in these lungs alveolar congestion with pneumocyte type I and type II damage. Proteinaceous fibrillar edema induced oxygen exchange impairment, leading to Acute Lung Injury, (ALI), and, in turn, ARDS.

2.2. In Situ Hybridization (ISH)

All patients who died from a lethal form of SARS-CoV-2 infection in our cohort of COVID-19 lethal disease had ARDS due to lung tissue viral persistence demonstrated by PCR as previously reported [32]. Therefore in this study we presented the percentage of Spike-$1^{(+)}$ cases as evaluated by ISH compared to mean values ± Standard Error Means (SEM) of selected studied markers (TLR-4, CD68, CD163, GAL-3) (Table 1 and Figure 2).

Table 1. Summary of Mean ± SEM of TLR4, CD68, CD163, and GAL-3 immunoreactivity values according to COVID-19 status, and Spike-1 ISH detection.

COVID-19 Status	Immunohistochemistry				Spike-1 (ISH) Positive/Total, (%)
	TLR-4 Mean ± SEM	CD68 Mean ± SEM	CD163 Mean ± SEM	GAL-3 Mean ± SEM	
Positive	183.04 ± 54.90	314.92 ± 36.24	111.00 ± 36.65	19.58 ± 9.81	Spike-1 positive 12/25(48)
		316.15 ± 34.63	39.15 ± 4.34	47.00 ± 15.87	Spike-1 negative 13/24(52)
Negative	42.67 ± 10.20	59.43 ± 20.42	34.33 ± 7.82	26.67 ± 6.49	

2.3. TLR-4 Expression Was Up-Regulated in a Subgroup of Patients with Lethal COVID-19 Lung Disease

TLR-4 was strongly up-regulated in damaged pneumocytes in a subgroup of deceased COVID-19 patients (Figure 3A,B). Furthermore, TLR-4 was up-regulated in lung macrophages in a subgroup of deceased COVID-19 patients (Table 1 and Figure 3C,D). The average of TLR-4 stained macrophage per square area is up-regulated as compared to control cases, however, statistically significant values are not reached, in this cohort, due to heterogeneous variability of TLR-4 mediated hyper-inflammation in different deceased subjects. In detail, our digital pathology based immunohistochemistry showed a mean count of TLR-4 positive macrophages of 183.04 ± SEM 54.9 in COVID-19 patients, versus mean count of 42.67 ± SEM 10.2 in control cases. Nevertheless, our study showed a trend to TLR-4 up-regulation in lungs of a subgroup patients with lethal COVID-19, when compared to the control group of SARS-CoV-2 negative lungs. In particular, we showed a sub-group with strong TLR-4 over-expression (Figure 4).

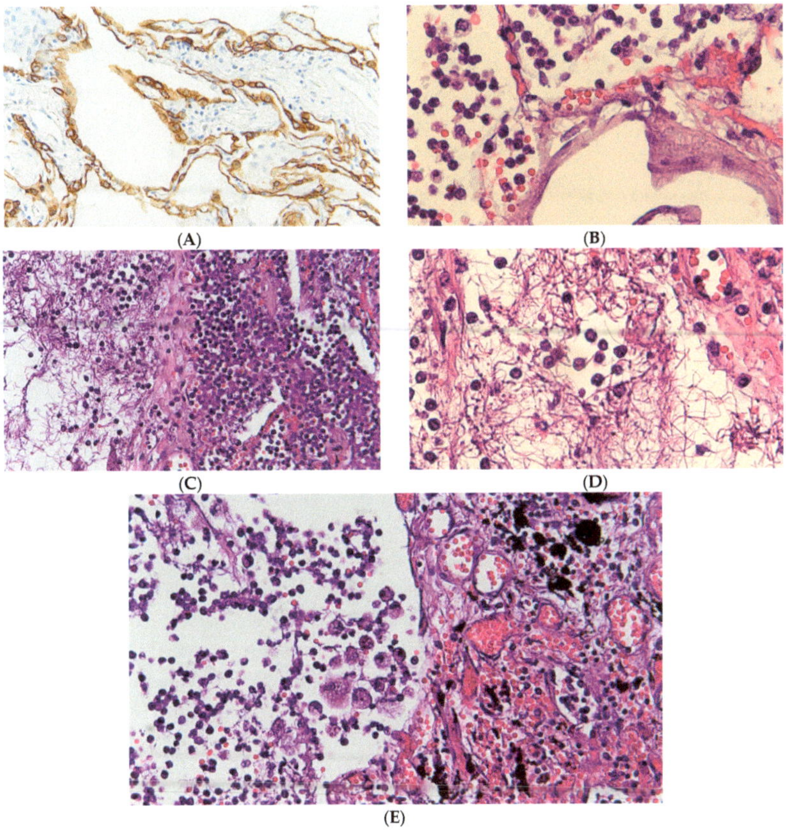

Figure 1. Histopathological picture of lungs in SARS-CoV-2$^{(+)}$ lethal infection (**B–E**) compared to SARS-CoV-2$^{(-)}$ control (**A**). (**A**) Type I and type II pneumocytes are major target of SARS-CoV-2 infection. This picture overview shows human intact pneumocytes stained with pan-cytokeratin marker (clone CK-AE1/AE3/PCK26) by standard immunohistochemistry, and slight interstitial lymphocytic inflammation (unstained cells) in a representative SARS-CoV-2$^{(-)}$ control patient (H&E original magnification, ×20 HPF). (**B**) SARS-CoV-2$^{(+)}$ infection determined alveolar congestion with pneumocyte damage, acute phlogistic cells inflammation, haemorrage, as expression of early stage of Diffuse Alveolar Damage (H&E staining, original magnification, ×40 HPF). (**C**) Massive alveolar and interstitial phlogistic cell infiltration with membrane hyaline and thin fibril deposition, depicting a network of early fibrosis (H&E staining, Original magnification, ×20). (**D**) Massive alveolar and interstitial phlogistic cell infiltration with hyaline membrane and thin fibril deposition. Note interstitial vessel congestion and slight haemorrage. In this case, polymorphonucleates infiltrate lungs, this was been especially observed in cases with Gram-negative bacterial superinfection (H&E, original magnification, ×40 HPF). (**E**) SARS-CoV-2 lethal damage in previous damaged lungs. The lung was been already affected by chronic inflammation with anthracotic deposits (H&E, original magnification, ×40 HPF).

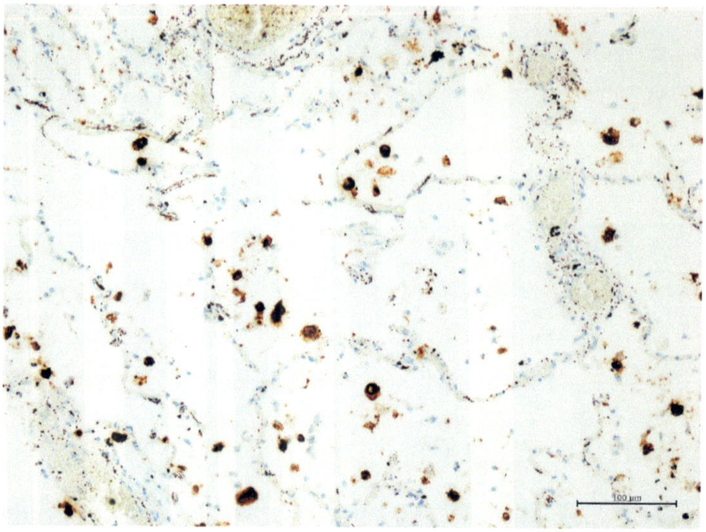

Figure 2. In Situ Hybridization (ISH) for Spike-1 of SARS-CoV-2 in infected human lungs. Representative case of ISH for Spike-1 viral strand in lung alveolar macrophages of COVID-19 deceased case [ISH detection of SARS-CoV-2 has been performed using V-nCoV2019-S probe detect viral sequences of Spike (S) of SARS-CoV-2 in lung tissue using RNAscope 2.5 high-definition detection kit (Advanced Cell Diagnostics), Original magnification, ×20].

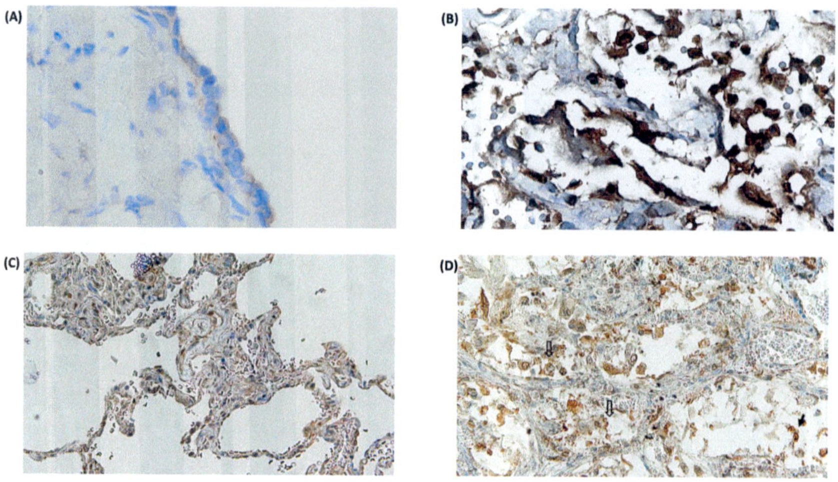

Figure 3. Over-expression of TLR-4 in SARS-CoV-2 injured lungs. Note up-regulation in SARS-CoV-2 damaged lung pneumocytes (**B**), as compared to human normal pneumocytes (**A**), and in pneumocytes and macrophages (**C**) as compared to SARS-CoV-2 negative lung with emphysema and slight interstitial chronic infiltration (**D**) [(↓) Solid black arrows indicate damaged pneumocytes, and (⇓) empty arrows indicate macrophages; standard LSAB-technique, nuclear counterstaining with haematoxylin, original magnification ×20].

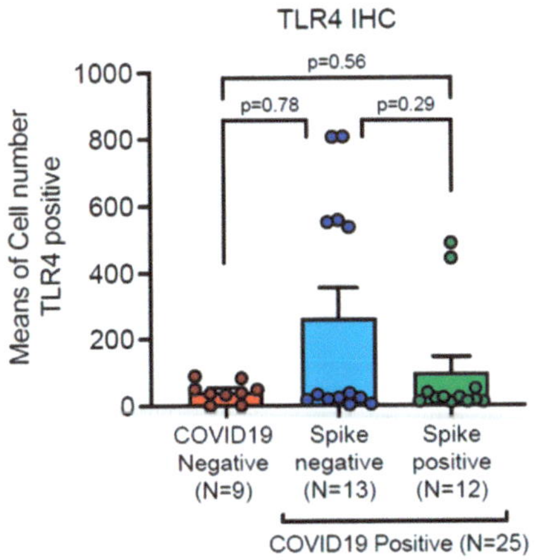

Figure 4. Up-regulation of TLR-4 in a subgroup of lethal COVID-19 lung disease—statistical comparison. Expression of TLR-4 showed a trend to up-regulation in lungs of a subgroup patients with lethal COVID-19, when compared to the control group of SARS-CoV-2 negative lungs. Note that a subgroup of patients demonstrates significant up-regulation of TLR-4 innate-immunity marker, irrespective of Spike-1 persistence. Mean ± SEM were obtained by immunohistochemical expression of TLR-4 (CD 284), as evaluated by digital pathology analysis (Mann-Whitney U-test).

2.4. GAL-3 Macrophages Depletion Is Associated with Persistence of Viral Spike-1 Sequence in COVID-19 Lethal Lungs

GAL-3 positive macrophages were up-regulated in Spike-1$^{(-)}$ and significantly down-regulated in Spike-1$^{(+)}$ SARS-CoV-2 infected lethal lungs respectively. Means and SEM were obtained by immunohistochemical expression of GAL-3 (clone 9C4) (Table 1 and Figure 5), as evaluated by digital pathology analysis; Spike-1$^{(+)}$ status was detected by PCR-based methods and ISH (Figure 2). Immunohistochemistry showed a mean count in SARS-CoV-2$^{(-)}$ controls of 29.67 ± 6.49 SEM, versus mean count of 47.00 ± 15.87 SEM, and 19.58 ± 9.81 SEM in Spike-1$^{(-)}$ and Spike-1$^{(+)}$ COVID-19 deceased patients respectively. Comparisons were statistically significant as evaluated by Mann-U-Whitney test (Table 1 and Figure 6).

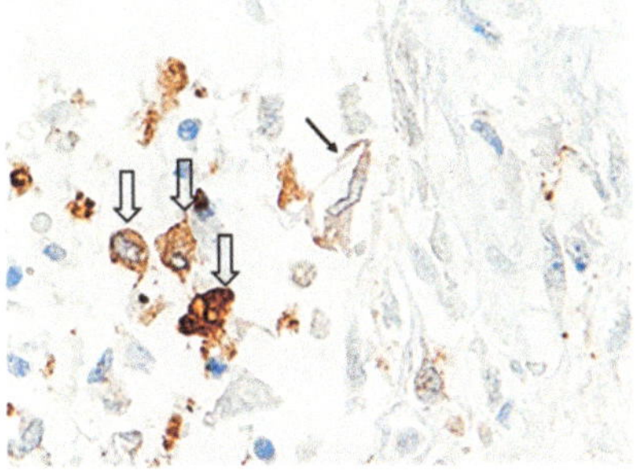

Figure 5. GAL-3 immunostaining with standard LSAB-HRP in representative case of SARS-CoV-2 infected lung in a patient deceased in the early stage of ARDS. The picture was captured in hot spot for GAL-3 in section serially stained with Spike-1 sequence by ISH. Note the persistence of GAL-3 expression in macrophages and damaged pneumocytic cells [(↓) Solid black arrows indicate damaged pneumocytes, and (⇓) empty arrows indicate macrophages; LSAB performed with mononuclear Ab against GAL-3, nuclear counter stainig with Gill's type-II Haematoxylin; original magnification ×20].

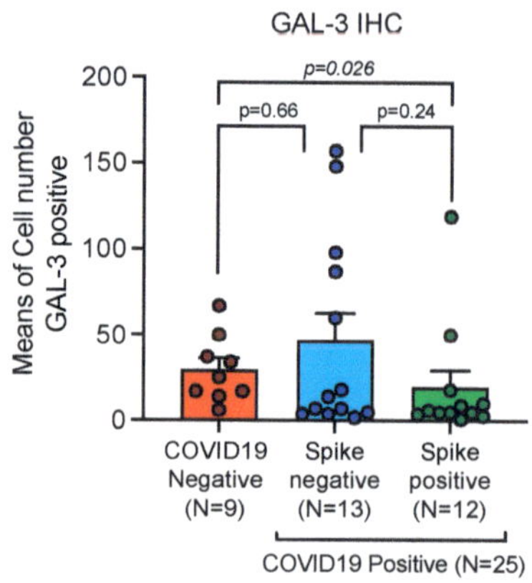

Figure 6. GAL-3 positive macrophage are up-regulated in Spike-1$^{(-)}$ and significantly down-regulated in Spike-1-$^{(+)}$ SARS-CoV-2 infected lethal lungs. GAL-3 positive macrophage infiltration in SARS-CoV-2 shows a trend of increased levels in Spike$^{(-)}$ COVID-19 cases, while a significant down-regulation associated with the presence of viral Spike-1 sequence was demonstrated; means and SEM were obtained by immunohistochemical expression of GAL-3 (clone 9C4), as evaluated by digital pathology analysis; Spike$^{(+)}$ status was detected by PCR-based methods and ISH; Mann-Whitney U-test ($p < 0.05$).

2.5. GAL-3/CD68, and CD163/CD68 Ratios, and Their Role in Altering Balance between M2 and M1 Macrophages in the Lungs

Our data showed an increased pan-macrophage marker CD68 in COVID-19$^{(+)}$ patients associated with imbalance of M2-like sub-populations. In detail, GAL-3/CD68 ratio, was down-regulated in SARS-CoV-2 infected lethal COVID-19 lung disease (Mean 14.92 ± 5 SEM), compared to controls (Mean 84.85 ± 24.9 SEM), as evaluated by Mann-U-Whitney test ($p < 0.001$) (Figure 7A). Furthermore, CD163/CD68 ratio was downregulated in SARS-CoV-2 infected lethal COVID-19 lung disease (Mean 19.9 ± 3.7 SEM), compared to controls (Mean 69.6 ± 14.47 SEM), as evaluated by Mann-U-Whitney test ($p < 0.001$) (Figure 7B).

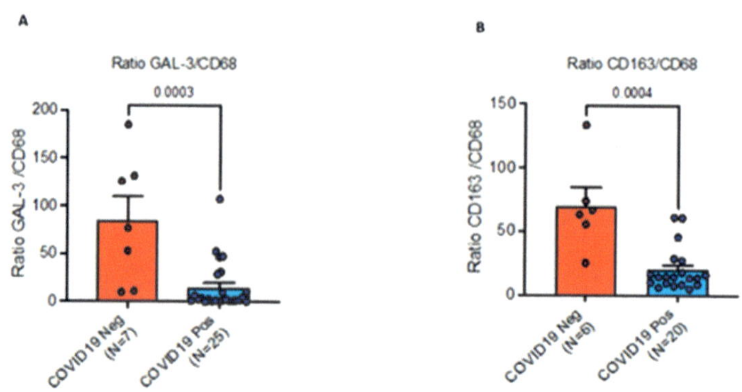

Figure 7. (**A**) GAL-3/CD68 ratio in COVID-19$^{(+)}$, compared to COVID-19$^{(-)}$, as evaluated by Mann-Whitney U-test ($p < 0.001$). These data show increase of pan-macrophage marker CD68 associated with imbalance of M2-like macrophage sub-population infiltrating alveolar and interstitial compartments of SARS-CoV-2$^{(+)}$ lethal lung disease. (**B**) CD163/CD68 ratio in COVID-19$^{(+)}$, compared to COVID-19$^{(-)}$, as evaluated by Mann-Whitney U-test ($p < 0.001$). CD163/CD68 ratio was downregulated in SARS-CoV-2 infected lethal COVID-19 lung disease, compared to controls as evaluated by Mann-U-Whitney test ($p < 0.001$).

2.6. TLR-4 Innate Immunity Marker Is Associated with GAL-3 Expression in SARS-CoV-2 Related Lethal Lung Disease

Expression of TLR-4 and GAL-3 positive macrophages, assessed by digital-pathology-immunohistochemical, showed statistically significant association by Spearman's statistical correlation test ($p < 0.001$) (Figure 8).

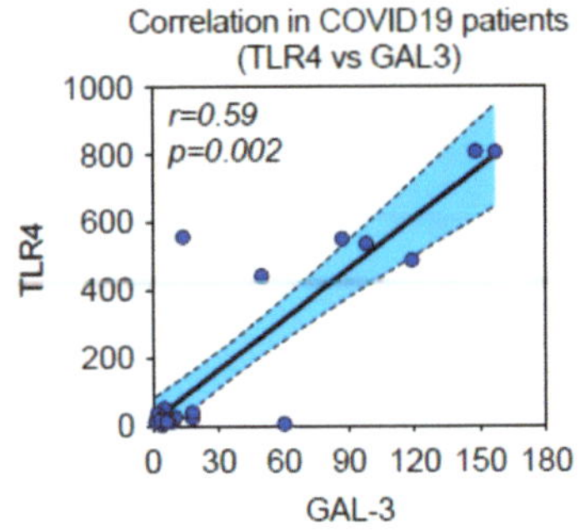

Figure 8. TLR-4 innate immunity marker is associated with GAL-3 expression in SARS-CoV-2 related lethal lung disease. Expression of TLR-4 and GAL-3 positive macrophages, assessed by digital-pathology-IHC, showed statistically significant association by Spearman's statistical correlation test ($p < 0.01$).

2.7. CD68 Positive Macrophages Were Significantly Increased in COVID-19 Lethal Lungs

Our study showed a significant overexpression of pan-macrophage marker CD68 in lungs of deceased COVID-19 cases as compared to SARS-CoV-2 negative control lungs (Table 1). Up-regulation was observed irrespective of Spike-1 persistence. Means and SEM were obtained by immunohistochemical expression of CD68 (clone KP1) as evaluated by digital pathology analysis; Spike-1-$^{(+)}$ status was detected by in ISH. In detail, digital pathology based immunohistochemistry demonstrated a mean count of CD68(+) macrophages in SARS-CoV-2 negative controls of 59.43 $\pm$ 20.42 SEM, versus mean count of 316.15 $\pm$ 34.63 SEM, and 314.92 $\pm$ 36.24 SEM in Spike-1 negative and Spike-1 positive COVID-19 deceased patients respectively (Table 1), comparison were significant (Figure 9).

2.8. Variation of the Macrophage Populations Found in the Lung of COVID-19 Deceased Patients Was Conditioned by the Infiltration of Myeloid Derived Monocytes-Macrophages as Evaluated by CD163

Mean and standard deviation was been obtained by immunohistochemical expression of CD-163 stained cells, as evaluated by digital pathology and Mann-Whitney U-test statistical analysis. At the time of the fatal event, a subgroup showing markedly high values was part of Spike-1 persistent lungs. Comparisons were not statistically significant, due to heterogeneous variability of CD163$^{(+)}$ myeloid derived macrophages in different deceased subjects, but showed an interesting trend, as compared to the limited number of samples examined. In details, our study showed CD163$^{(+)}$ mean count in SARS-CoV-2 negative controls of 34.33 $\pm$ 7.82 SEM, versus mean count of 39.15 $\pm$ 4.34 SEM, and 111.00 $\pm$ 36.65 SEM in Spike-1$^{(-)}$ and Spike-1$^{(+)}$ COVID-19 deceased patients respectively (Table 1 and Figure 10). Furthermore, CD163 and GAL-3 showed an inverse correlation, although not significant, when associated with Spearman's test (Figure 11). This interesting trend requires confirmatory analysis on more representative samples.

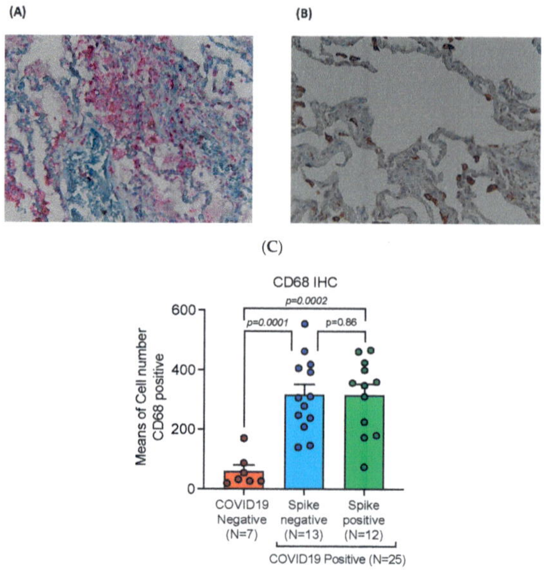

Figure 9. Endoalveolar macrophages exudation as evaluated by CD-68 immunostaining in COVID-19 related lungs (**A**) and SARS-CoV-2 unrelated lung (**B**) (original magnifications, ×20); statistical comparisons (**C**): the increase in the total number of macrophages was associated with SARS-CoV-2 infection with statistically significant values in both groups of Spike-S1$^{(+)}$ and Spike-S1$^{(-)}$ cases, when compared to control group. Means and SEM were obtained by immunohistochemical expression of CD68 (clone KP1), as evaluated by digital pathology analysis; Spike-1-positive status was detected by in ISH; Mann-U-Whitney test ($p < 0.001$).

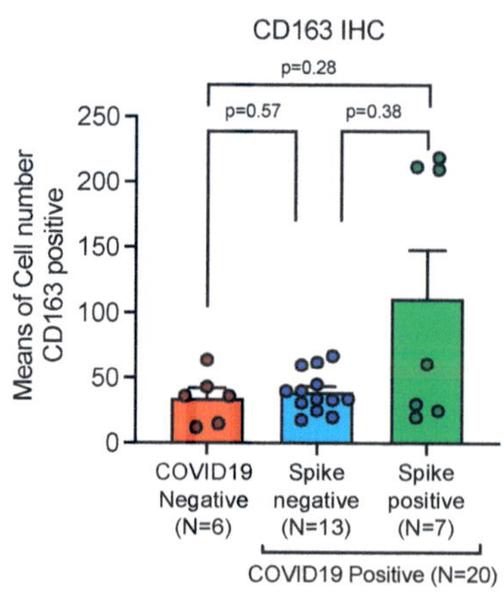

Figure 10. Association between up-regulation of CD-163 macrophage marker and Spike-1 detection in lethal COVID-19 lung disease. Expression of CD163 showed up-regulation in lungs of patients with lethal COVID-19, when compared to the control group of SARS-CoV-2 negative lungs. At the time of the fatal event, a subgroup showing markedly high values was part of Spike-1 persistent lungs. Mean and SEM were obtained by immunohistochemical expression of CD163, as evaluated by digital pathology and statistical unpaired *t*-test/Mann-Whitney U-test analysis. Comparisons were not statistically significant, but showed an interesting trend, in relation to the limited number of samples examined.

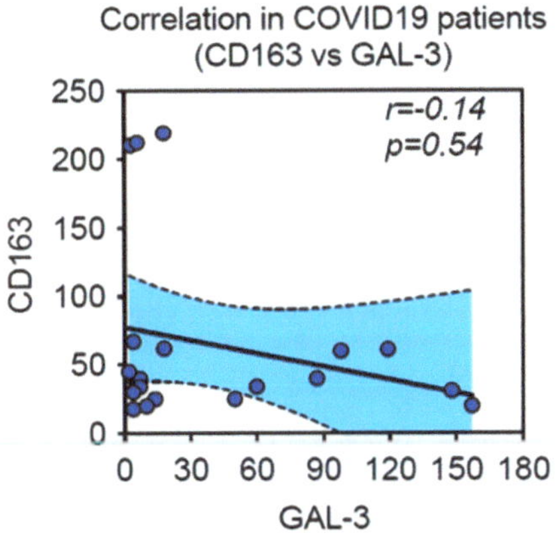

Figure 11. Association between CD163 and GAL-3 in lethal COVID-19 lung disease. CD163 and GAL-3 showed an inverse relationship (r = −0.14), even though not significant, when compared with Spearman's correlation test.

3. Discussion

The involvement of TLRs in inflammation and bacterial infection have been recognized for a long time. There is an increasing number of reports demonstrating the involvement of TLR activation in a variety of viral infections, associated with protective immunity, but also immune hyper-activation and even viral replication.

Recent data show the involvement of TLR activation in various acute respiratory viral infections, including SARS-CoV-2 and indicate an essential role in COVID-19 pathology. A number of studies and meta-analyses conducted in the last 3 years have shown pulmonary involvement in the most of COVID-19 lethal cases [33,34]. However, the role played by monocyte-macrophage subpopulations requires further investigation, taking into account that macrophages were been recognized as a heterogeneous and dynamic population of cells that have the capacity to perform a wide range of critical functions [35].

During severe and critical SARS-CoV-2 infections, as well as in many severe infections leading to ARDS, macrophage destruction and depletion occur during alveolar infection with subsequent repopulation attempts by monocyte-marrow-derived macrophage populations. In response to viral infections, also without co-infections and even more in those with bacterial co-infections, the marrow production of cellular elements of the monocytic-macrophage series may shift towards a pro-inflammatory or immunoregulatory phenotype according to phase progression and interaction with host immune system. On this point, recent evidences suggest that the alterations of the immune responses under the control of monocytes-macrophages are certainly one of the elements affecting the severity of SARS-CoV-2 infections [36–39].

Patients with lethal COVID-19 associated with macrophage counts of >130/High Power Field (HPF) showed shorter survival time in an autopsy cohort studied by Cao W et al. [40].

Our study shows that in lethal cases of COVID-19 associated lung disease, the inflammatory cells that predominate in the histological picture are macrophages, together with lymphocytes and polymorphonuclear neutrophils. Another important result is the finding of a drastic depletion of GAL-3$^{(+)}$ macrophages, associated with down-regulation of GAL-3$^{(+)}$/CD-68$^{(+)}$ ratio, and persistence of Spike-1 sequence.

GAL-3 has several effects in the immune response, especially in lung injuries. GAL-3 is highly expressed in endothelial and alveolar macrophages, playing an important role in mechanisms of lung repair and fibrosis [41]. GAL-3 has been associated with M2-like macrophage activation [42]. Indeed, GAL-3 previous name was MAC-2 antigen because of its role in the activating of M2 macrophages [31].

However, a recent study showed an increased blood concentration of GAL-3 in patients affected by COVID-19, associated with mortality and severity of patient affected by COVID-19 [43]. These apparently conflicting results may be possibly explained hypothesizing a reduction of tissue GAL-3 mediated by apoptosis, necroptosis and necrosis of tissue resident lung cells, together with up-regulation of serum levels of cleaved GAL-3 protein.

A series of evidence from current literature showed that GAL-3 antagonizes generation of a pro-inflammatory macrophage phenotype. In this context Di Gregoli K, et al., reported that the accumulation of GAL-3$^{(-)}$ macrophage plays a central role to the heightened invasive capacity observed in the GAL-3$^{(-)}$/CD-68$^{(+)}$ macrophage sub-population in atherosclerosis. Additionally, they detected a clear shift toward a pro-inflammatory macrophage phenotype in response to GAL-3 silencing as mRNA expression of the pro-inflammatory molecules TNF-alpha, PTGS2 (prostaglandin-endoperoxide synthase 2 cyclooxygenase-2), and IL-6 was increased in response to GAL-3 depletion, implicating GAL-3 as a possible negative regulator of inflammation and proinflammatory macrophage polarization [44].

Our study demonstrates that in the lungs of subjects who died of critical forms of SARS-CoV-2 infection there is a drastic reduction of the GAL-3$^{(+)}$ immune-regulatory macrophages associated with Spike-1 persistence, which instead should be responsible for promoting healing of acute inflammation, favoring the phases of chronic inflammation, and tissue repair-regeneration. These macrophages in subjects with co-morbidities consistently associated with the increased risk of mortality of the disease COVID-19 (neoplasms, COBP) are PD-L1+ and despite their role in the pathophysiology of these pathologies they still have a finalistic significance in reducing hyper-inflammation potentially lethal in the lungs. Therefore we demonstrate that the inability to generate an immunoregulatory phenotype in the lungs (PD-L1+, GAL-3+) is one of the main causes of lethality in the analyzed subjects.

The key innate immune sensing receptors are germ line-encoded PRRs, which mediate the initial sensing of infection by recognition of PAMPs upon microbial invasion of the host [4].

PRRs on the cell surface and endosomal membranes and in the cytosol may respond to SARS-CoV-2 PAMPs to activate innate immune signaling pathways. TLR-11, TLR-2, TLR-4 and TLR-6 can signal through MyD88 to activate Nucelar Factor kappa-light-chain-enhancer of activated b cells (NF-κB) and Mitogen-activated Protein Kinase (MAPK) signaling pathways to induce transcription of genes encoding pro-inflammatory cytokines and other sensors [45]. TLR-3 and TLR-4 can signal through TIR domain containing adapter-inducing interferon (TRIF) to activate Interferon 3 (IRF3), and induce expression of type I and type III IFNs. Strong experimental evidence supports SARS-CoV-2-mediated activation of TLR-22, while activation of TLR-1, TLR-3, TLR-4 and TLR-6 has been suggested bioinformatically and through associative studies [46].

Interestingly, Quero, L et al., showed that TLR-22 stimulation by Pam3 impairs anti-inflammatory activity of M2-like macrophages, generating a chimeric M1/M2 phenotype [47]. In summary, these results indicate that TLR-2, and to a higher degree TLR-4, ligands are able to change the anti-inflammatory M2 gene markers toward an M1-specific expression phenotype. CD14 and CD163 marker expression on M2 macrophages did not change upon TLR-2 and TLR-4 engagement. By contrast, M2 gene markers *HMOX1*, *FOLR2*, and *SLC40A1* were decreased phenotype [47]. NF-κB was, similarly, activated in M1 and M2 by TLR2 and TLR-4 ligands. Thus, following TLR-2 stimulation by its ligand Pam3, the M2 population secreted the pro-inflammatory cytokines IL-6 and IL-8 at levels comparable to Pam3-stimulated M1-polarized macrophages. Therefore, Quero, L et al. concluded that despite this shift toward a proinflammatory M1 function, M2 macrophages continued to express the typical M2 cell surface markers CD14 and CD163 phenotype [47].

Furthermore, we hypothesize that according to Quero L, et al. [47], in deceased patients from this cohort, despite a myeloid response directed to M2 macrophage recovery function, with expression the typical M2 phenotype and cell surface markers CD163, there is a functional pathological shift toward a persistent pro-inflammatory M1 state.

On this topic, recent results support that timely acquisition of a myeloid cell immune-regulatory phenotype might contribute to recovery in severe systemic SARS-CoV-2 infection and suggest that therapeutic agents favoring an innate immune system regulatory shift may represent the best strategy to implement at this stage [48]. Furthermore, on this topic Grassi G. et al. reported a paradoxical effect of Myeloid-Derived Suppressor Cells in COVID-19 [49].

On the basis of experimental data, in our cohort of lethal cases, we demonstrate that the inflammatory microenvironment induced by SARS-CoV-2 infection through innate immunity molecules determines an hyper-inflammatory state capable of influencing the presence of functionalized macrophages in the M1- direction with overexpression of TLR-4, and that, at the same time, macrophages with the M2-like CD-163$^{(+)}$ phenotype persist in the alveolar and interstitial compartments. Furthermore, M2-like activity could be further impaired by TLRs overexpression in local alveolar and interstitial compartments, due to cytokine storm, leading to unresolved inflammation and death.

Expression of TLR-4$^{(+)}$ and GAL-3$^{(+)}$ stained macrophages, assessed by digital-pathology-immunohistochemistry, showed statistically significant association by Spearman's statistical correlation test ($p < 0.001$).

This study demonstrated, in the complex context of macrophage activation, association between GAL-3 expression and innate immune phlogistic pathway activation in patients deceased from SARS-CoV-2 infection.

Further results deriving from our research are the demonstration of a macrophage sub-population shift, and association between loss of alveolar GAL-3 macrophages and persistence of viral sequences Spike-1 as a crucial event in lethal COVID-19 lung disease.

In addition, our study focuses on the variation of pulmonary inflammatory infiltration by bone myeloid-derived monocyte-macrophages, as demonstrated by immunolabeling with CD163.

This study also represents the histological basis for hypothesizing that early, prolonged and dose appropriate therapy with drugs such as high power steroids capable of modulating the innate inflammatory response can potentially reduce the mortality associated with SARS-CoV-2 infection. On this topic, in a recent systematic review and meta-analysis, Chaudhuri D. et al. showed that the use of corticosteroids reduces mortality in patients with ARDS. This effect was consistent between patients with COVID-19, and non-COVID-19 ARDS [50]. Furthermore, translational research in ARDS patients randomized to methylprednisolone has demonstrated the ability of corticosteroid therapy to rescue the cellular concentrations and functions of activated Glucocorticoid Receptors GRα (GC-GRα) leading to downregulation of systemic and pulmonary NF-κB-activated markers of inflammation, coagulation, and fibroproliferation [51–53]

4. Materials and Methods

4.1. Ethics Approval

The study was conducted according to the guide-lines of the Declaration of Helsinki and approved by the Attorney's Office of Naples, protocol number p.p. 509568/020/44, dated 15 May 2020.

4.2. Clinicopathological Data of Selected Patients

A series of 25 patients, the most (24/25, 96%) with co-morbidities (NIDDM, Type II-Non-Insulin-Dependent Diabetes Mellitus; OCBP, Obstructive Chronic Broncho-Pneumopathy; CN, Chronic Nephropathy; O, Obesity; H, Hypertension; CH, Cardiac Hypertrophy), who died from a lethal form of SARS-CoV-2$^{(+)}$ infection (COVID-19$^{(+)}$), of which 13 males and 12 females, aged between 45 and 80 years were studied. The study was conducted exclusively on formalin fixed paraffin embedded (FFPE) lung tissues. Lung tissues of the cohort of COVID-19$^{(+)}$ patients were compared to SARS-CoV-2$^{(-)}$ lung tissues selected from a pool of controls constituting the COVID-19-negative cohort, deriving from lung surgery specimens for primary or metastatic lung tumors, on tumor-free margins and at least 3 cm distant from the latter, or from surgery specimens for non-tumor-related lung surgery. All the surgical procedures were conducted with curative intent, according to standard approved therapeutic protocols. Clinicopathological data, and details regarding co-morbidity/morbidity of both COVID-19$^{(+)}$ and COVID-19$^{(-)}$ cohorts has been collected (Table 2). The control study collection was conducted over a three-year period (2020–2023), and due to the limited amount of lung tissue available, comparisons between COVID-19$^{(+)}$ and COVID-19$^{(-)}$ cases were gradually conducted not on all controls, but on a minimum number of at least 6 of the overall total specimens available (total 11 cases).

Table 2. Summary of clinic-pathological parameters of patients groups according to COVID-19 status.

Variables	COVID-19 Status	
	Positive (Total n = 5)	**Negative (Total n = 11)**
Age, Mean ± SEM	65.76 ± 2.11	68.5 ± 5.40
Gender, Male/Female (%)	13/12 (1.08)	8/3 (2.66)
Pathological co-morbidities, Yes/Not (%)	24/1 (96)	11/11 (100)
Single co-morbidity/morbidity, (%)	12/25 (48)	11/11 (100)
Type of single co-morbidities/morbidity, (%)	Hypertension 18/25 (72)	Lung metastatic carcinoma 2/11 (18.18)
	Cardiac Hypertrophy 8/25 (32)	Primary lung neoplasm 6/11 (54.54)
	OCBP, 3/25 (12)	Lung emphysema 2/11 (18.18)
	Chronic nephropathy 2/25 (8)	Lung hamartoma 1/11 (9.09)
	Obesity 5/25 (20)	0/11 (0)
Multiple co-morbidities/morbidity, (%)	12/25 (48)	0/11(0)
Type of multiple co-morbidities/morbidity, (%)	Hypertension; CH, Cardiac Hypertrophy 6/25 (24)	
	Obesity, and NIDDM 2/25 (8)	
	Obesity, hypertension, and NIDDM 1/25 (4)	
	Hypertension; CH, Cardiac Hypertrophy, and Chronic nephropathy 7/25 (28)	
	Obesity, Hypertension, and Cardiac Hypertrophy 1/25 (4)	

Legend. NIDDM, Type II-Non-Insulin-Dependent Diabetes Mellitus; OCBP, Obstructive Chronic Broncho-Pneumopathy; CH, Cardiac Hypertrophy.

Autopsy cases were been collected between June 2020 and July 2021 at time of wild-type SARS-CoV-2 with early onset of Epsilon and Alpha variants [54].

Patients deceased in a mean interval between diagnosis and death of 62.52 days (minimum of 20 days, maximum 120 days).

4.3. *Autopsy Protocol*

Deceased patients underwent autopsy, through help of a ventilation system with six complete air changes/h (ACH) in a pressure-negative environment, with air exhausted through HEPA filters [Biosafety Level 3 (BSL3)], according to the University of Padova autopsy protocol [55,56].

Specimens of lung tissue from 25 patients deceased for SARS-CoV2 were been used as representative cases of COVID-19 lethal ARDS.

4.4. Methods

Immunohistochemical analysis has been performed on formalin-fixed paraffin embedded specimens derived from autopsy procedures (COVID-19$^{(+)}$ cohort), and from routine lung surgical operations (COVID-19$^{(-)}$ cohort). Single immunostaining has been performed using specific monoclonal antibodies against TLR-4 (clone 76B357.1), GAL-3 (clone 9C4), CD68 (clone KP1), and CD163 (Clone MRQ-26), revealed by standard linked-streptavidin-biotin immunoperoxidase technique (LSAB), developed by 5′diaminobenzidine and/or alkaline phosphatase methods (Table 3).

Table 3. Antibodies used, and experimental conditions.

Antibody	Clone	Method
TLR-4	76B357	LSAB-HRP/AP, Ventana Benchmark® XT autostainer
CD68	KP1	LSAB-HRP/AP, Ventana Benchmark® XT autostainer
CD163	MRQ-26	LSAB-HRP/AP, Ventana Benchmark® XT autostainer
GAL-3	9C4	LSAB-HRP/AP, Ventana Benchmark® XT autostainer

Evaluation was been performed by two blinded observers analyzing standard High Power Field (HPF) area, with 20× magnification, measuring 45,508.61 μm^2. At least eight HPFs were been evaluated for a mean total area of 1,160,000 μm^2. Antibodies used and experimental conditions have been reported in Table 3.

The percentage of positive immunostained cells was been evaluated by digital pathology analyses using cellSens Dimension and Image-Pro Premier Offline (MediaCybernetics, Version 9.1.4 Build 5638) software.

All the controls have pathological morbidity leading to surgical intervention for therapeutic purpose. In details, two out eleven cases had metastasis to the lung from distant carcinomas (one from breast cancer and one from large bowel cancer), six out 11 cases had primary lung adenocarcinomas, one out 11 had lung hamartoma, and two out 11 had complicated lung emphysema. This COVID-19$^{(-)}$ cohort, carrying slight chronic lung pneumonia, have been used to match samples of SARS-CoV-2$^{(+)}$ infected hyper-inflamed lethal cases of pneumonia, leading to not resolved COVID-19 related ARDS and death, representing the COVID-19$^{(+)}$ cohort.

4.5. In Situ Hybridization (ISH) Method for SARS-CoV-2 Detection

Briefly, ISH detection of SARS-CoV-2 was been performed using V-nCoV2019-S probe to detect viral sequences Spike (S) of SARS-CoV-2 in lung tissue using RNAscope 2.5 high-definition detection kit (Advanced Cell Diagnostics). SARS-CoV-2 was been detected in lung tissues of lethal COVID-19 disease by in ISH according to methods reported in previous work of our laboratories in which we compared ISH with PCR methods [32,56,57], and in current literature [58,59].

4.6. Statistical Analyses

Data have been analyzed by SOFA Statistics 1.4.6, and SPSS 28.01.0 Data Analysis and Statistical Software and Windows Operating Systems; Spearman's test and ANOVA statistical analysis were been used to correlate to clinicopathological parameters. Mann-U-Whitney test was been used to compare non-parametrical variables (Table 1).

5. Conclusions

In summary, our study focused on the innate-immune pathway, the inflammatory profile of macrophages and composition of alveolar and interstitial infiltrate in lethal SARS-CoV-2 lung disease.

Our study shows that in lethal cases of COVID-19 associated lung disease, the inflammatory cells that predominate in the histological picture are TLR-4$^{(+)}$ damaged pneumocytes, with phlogistic infiltration of macrophages, together with lymphocytes and polymorphonuclear neutrophils. Another important result is the finding of a drastic depletion of GAL-3$^{(+)}$ macrophages, associated with down-regulation of GAL-3$^{(+)}$/CD-68$^{(+)}$ ratio, and persistence of Spike-1 sequence.

The association between loss of alveolar GAL-3 macrophages and persistence of viral sequences is a crucial event in lethal COVID-19 lung disease in the cohort of analyzed cases.

In this study we showed that inflammatory factors of the pulmonary microenvironment can condition an adverse and lethal response despite the triggering of a bone marrow response potentially favorable to healing. The inflammatory microenvironment induced by SARS-CoV-2 infection through innate immunity molecules determines an hyper-inflammatory state capable of influencing the presence of functionalized macrophages in the M1 direction with overexpression of TLR-4, and that, at the same time, macrophages with the M2-like CD-163$^{(+)}$ phenotype persist in the alveolar and interstitial compartments. Furthermore, M2-like activity could be further impaired by TLRs overexpression in local alveolar and interstitial compartments, due to cytokine storm, necroptosis, and microthrombosis, in turn leading to unresolved inflammation and death.

This study also represents the histological basis for hypothesizing that early, prolonged and dose appropriate therapy with drugs such as high power steroids capable of modulating the innate inflammatory response can potentially reduce the mortality associated with SARS-CoV-2 infection.

Author Contributions: Conceptualization, G.P., T.C. (Tommaso Cassano) and M.C.P.; writing—original draft preparation, R.Z.; software, R.B.; validation, A.A. and I.S.D.S.; formal analysis, G.V., M.M. (Mario Meccariello) and P.P.; investigation, T.C. (Tommaso Colangelo); resources, F.Z.M.; data curation, G.S.; writing—original draft preparation, A.R.; writing—review and editing, R.F.; visualization, F.Z.M.; methodology, F.M.; supervision, M.M. (Maurizio Municinò); project administration, G.P. All authors have read and agreed to the published version of the manuscript.

Funding: This research received no external funding.

Institutional Review Board Statement: The study was conducted according to the guide-lines of the Declaration of Helsinki and approved by Attorney's Office of Naples.

Informed Consent Statement: Not applicable.

Data Availability Statement: Not applicable.

Conflicts of Interest: The authors declare no conflict of interest.

References

1. Zuniga, E.I.; Macal, M.; Lewis, G.M.; Harker, J.A. Innate and Adaptive Immune Regulation during Chronic Viral Infections. *Annu. Rev. Virol.* **2015**, *2*, 573–597. [CrossRef] [PubMed]
2. Liu, B.M.; Hill, H.R. Role of Host Immune and Inflammatory Responses in COVID-19 Cases with Underlying Primary Immunodeficiency: A Review. *J. Interf. Cytokine Res.* **2020**, *40*, 549–554. [CrossRef]
3. Ye, Q.; Wang, B.; Mao, J. The pathogenesis and treatment of the 'Cytokine Storm' in COVID-19. *J. Infect.* **2020**, *80*, 607–613. [CrossRef] [PubMed]
4. Takeuchi, O.; Akira, S. Pattern Recognition Receptors and Inflammation. *Cell* **2010**, *140*, 805–820. [CrossRef] [PubMed]
5. Janssens, S.; Beyaert, R. Role of toll-like receptors in pathogen recognition. *Clin. Microbiol. Rev.* **2003**, *16*, 637–646. [CrossRef] [PubMed]
6. Kanzler, H.; Barrat, F.J.; Hessel, E.M.; Coffan, R.L. Therapeutic targeting of innate immunity with toll-like receptor agonists and antagonists. *Nat. Med.* **2007**, *13*, 552–559. [CrossRef]
7. Medzhitov, R.; Janeway, C. Innate Immunity. *N. Engl. J. Med.* **2000**, *343*, 338–344. [CrossRef]

8. Akira, S.; Takeda, K.; Kaisho, T. Toll-like receptors: Critical proteins linking innate and acquired immunity. *Nat. Immunol.* **2001**, *2*, 675–680. [CrossRef]
9. Delneste, Y.; Beauvillain, C.; Jeannin, P. Innate immunity: Structure and function of TLRs. *Med. Sci.* **2007**, *23*, 67–73.
10. Vaure, C.; Liu, Y. A comparative review of toll-like receptor 4 expression and functionality in different animal species. *Front. Immunol.* **2014**, *5*, 316. [CrossRef]
11. Brubaker, S.W.; Bonham, K.S.; Zanoni, I.; Kagan, J.C. Innate immune pattern recognition: A cell biological perspective. *Annu. Rev. Immunol.* **2015**, *33*, 257–290. [CrossRef] [PubMed]
12. Conti, P.; Caraffa, A.; Tete, G.; Gallenga, C.E.; Ross, R.; Kritas, S.K.; Frydas, I.; Younes, A.; Di Emidio, P.; Ronconi, G. Mast cells activated by SARS-CoV-2 release histamine which increases IL-1 levels causing cytokine storm and inflammatory reaction in COVID-19. *J. Biol. Regul. Homeost. Agents* **2020**, *34*, 1629–1632. [PubMed]
13. Gebremeskel, S.; Schanin, J.; Coyle, K.M.; Butuci, M.; Luu, T.; Brock, E.C.; Xu, A.; Wong, A.; Leung, J.; Korver, W.; et al. Mast Cell and Eosinophil Activation Are Associated with COVID-19 and TLR-Mediated Viral Inflammation: Implications for an Anti-Siglec-8 Antibody. *Front. Immunol.* **2021**, *12*, 641. [CrossRef] [PubMed]
14. Ahmad, R.; Al-Mass, A.; Atizado, V.; Al-Hubail, A.; Al-Ghimlas, F.; Al-Arouj, M.; Bennakhi, A.; Dermime, S.; Behbehani, K. Elevated expression of the toll like receptors 2 and 4 in obese individuals: Its significance for obesity-induced inflammation. *J. Inflamm.* **2012**, *9*, 1–11. [CrossRef] [PubMed]
15. Bhattacharya, M.; Sharma, A.R.; Mallick, B.; Sharma, G.; Lee, S.-S.; Chakraborty, C. Immunoinformatics approach to understand molecular interaction between multi-epitopic regions of SARS-CoV-2 spike-protein with TLR4/MD-2 complex. *Infect. Genet. Evol.* **2020**, *85*, 104587. [CrossRef]
16. Choudhury, A.; Mukherjee, S. In silico studies on the comparative characterization of the interactions of SARS-CoV-2 spike glycoprotein with ACE-2 receptor homologs and human TLRs. *J. Med. Virol.* **2020**, *92*, 2105–2113. [CrossRef]
17. Shirato, K.; Kizaki, T. SARS-CoV-2 spike protein S1 subunit induces pro-inflammatory responses via toll-like receptor 4 signaling in murine and human macrophages. *Heliyon* **2021**, *7*, e06187. [CrossRef]
18. Zheng, M.; Karki, R.; Williams, E.P.; Yang, D.; Fitzpatrick, E.; Vogel, P.; Jonsson, C.B.; Kanneganti, T.-D. TLR2 senses the SARS-CoV-2 envelope protein to produce inflammatory cytokines. *Nat. Immunol.* **2021**, *22*, 829–838. [CrossRef]
19. Kosyreva, A.; Dzhalilova, D.; Lokhonina, A.; Vishnyakova, P.; Fatkhudinov, T. The Role of Macrophages in the Pathogenesis of SARS-CoV-2-Associated Acute Respiratory Distress Syndrome. *Front. Immunol.* **2021**, *12*, 682871. [CrossRef]
20. Wang, C.; Xie, J.; Zhao, L.; Fei, X.; Zhang, H.; Tan, Y.; Nie, X.; Zhou, L.; Liu, Z.; Ren, Y.; et al. Alveolar macrophage dysfunction and cytokine storm in the pathogenesis of two severe COVID-19 patients. *eBioMedicine* **2020**, *57*, 102833. [CrossRef]
21. Zhang, D.; Guo, R.; Lei, L.; Liu, H.; Wang, Y.; Wang, Y.; Qian, H.; Dai, T.; Zhang, T.; Lai, Y.; et al. Frontline Science: COVID-19 infection induces readily detectable morphologic and inflammation-related phenotypic changes in peripheral blood monocytes. *J. Leucoc. Biol.* **2021**, *109*, 13–22. [CrossRef]
22. Sica, A.; Mantovani, A. Macrophage plasticity and polarization: In vivo veritas. *J. Clin. Investig.* **2012**, *122*, 787–795. [CrossRef]
23. Locati, M.; Curtale, G.; Mantovani, A. Diversity, Mechanisms, and Significance of Macrophage Plasticity. *Annu. Rev. Pathol. Mech. Dis.* **2020**, *15*, 123–147. [CrossRef]
24. Mantovani, A.; Sozzani, S.; Locati, M.; Allavena, P.; Sica, A. Macrophage polarization: Tumor-associated macrophages as a paradigm for polarized M2 mononuclear phagocytes. *Trends Immunol.* **2002**, *23*, 549–555. [CrossRef] [PubMed]
25. Tamàs, R. Understanding the mysterious M2 macrophage through activation markers and effector mechanisms. *Mediat. Inflamm.* **2015**, *2015*, 816466. [CrossRef]
26. Chuang, Y.; Hung, M.E.; Cangelose, B.K.; Leonard, J.N. Regulation of the IL-10-driven macrophage phenotype under incoherent stimuli. *Innate Immun.* **2016**, *22*, 647–657. [CrossRef] [PubMed]
27. Evren, E.; Ringqvist, E.; Willinger, T. Origin and ontogeny of lung macrophages: From mice to humans. *Immunology* **2020**, *160*, 126–138. [CrossRef] [PubMed]
28. Smith, M.R.; Standiford, T.J.; Reddy, R.C. PPARs in alveolar macrophage biology. *PPAR Res.* **2007**, *2007*, 23812. [CrossRef] [PubMed]
29. Chakarov, S.; Lim, H.Y.; Tan, L.; Lim, S.Y.; See, P.; Lum, J.; Zhang, X.M.; Foo, S.; Nakamizo, S.; Duan, K.; et al. Two distinct interstitial macrophage populations coexist across tissues in specific subtissular niches. *Science* **2019**, *363*, eaau0964. [CrossRef] [PubMed]
30. Gibbings, S.L.; Thomas, S.M.; Atif, S.; McCubbrey, A.L.; Desch, A.N.; Danhorn, T.; Leach, S.M.; Bratton, D.L.; Henson, P.M.; Janssen, W.J.; et al. Three Unique Interstitial Macrophages in the Murine Lung at Steady State. *Am. J. Respir. Cell Mol. Biol.* **2017**, *57*, 66–76. [CrossRef] [PubMed]
31. Reichert, F.; Rotshenker, S. Galectin-3/MAC-2 in experimental allergic encephalomyelitis. *Exp. Neurol.* **1999**, *160*, 508–514. [CrossRef] [PubMed]
32. Zito Marino, F.; De Cristofaro, T.; Varriale, M.; Zannini, G.; Ronchi, A.; La Mantia, E.; Campobasso, C.P.; De Micco, F.; Mascolo, P.; Municinò, M.; et al. Variable levels of spike and ORF1ab RNA in post-mortem lung samples of SARS-CoV-2-positive subjects: Comparison between ISH and RT-PCR. *Virchows Arch.* **2022**, *480*, 597–607. [CrossRef] [PubMed]
33. Hooper, J.E.; Padera, R.F.; Dolhnikoff, M.; da Silva, L.; Duarte-Neto, A.N.; Kapp, M.E.; Lacy, J.M.; Mauad, T.; Saldiva, P.; Rapkiewicz, A.V.; et al. A Postmortem Portrait of the Coronavirus Disease 2019 (COVID-19) Pandemic: A Large Multi-institutional Autopsy Survey Study. *Arch. Pathol. Lab. Med.* **2021**, *145*, 529–535. [CrossRef] [PubMed]

34. Pannone, G.; Caponio, V.C.A.; De Stefano, I.S.; Ramunno, M.A.; Meccariello, M.; Agostinone, A.; Pedicillo, M.C.; Troiano, G.; Zhurakivska, K.; Cassano, T.; et al. Lung histopathological findings in COVID-19 disease—A systematic review. *Infect. Agents Cancer* **2021**, *16*, 34. [CrossRef]
35. Kwiecień, I.; Polubiec-Kownacka, M.; Dziedzic, D.; Wołosz, D.; Rzepecki, P.; Domagała-Kulawik, J. CD163 and CCR7 as markers for macrophage polarization in lung cancer microenvironment. *Cent. Eur. J. Immunol.* **2019**, *44*, 395–402. [CrossRef] [PubMed]
36. Zingaropoli, M.A.; Nijhawan, P.; Carraro, A.; Pasculli, P.; Zuccalà, P.; Perri, V.; Marocco, R.; Kertusha, B.; Siccardi, G.; Del Borgo, C.; et al. Increased sCD163 and sCD14 Plasmatic Levels and Depletion of Peripheral Blood Pro-Inflammatory Monocytes, Myeloid and Plasmacytoid Dendritic Cells in Patients with Severe COVID-19 Pneumonia. *Front. Immunol.* **2021**, *12*, 627548. [CrossRef]
37. Gómez-Rial, J.; Currás-Tuala, M.J.; Rivero-Calle, I.; Gómez-Carballa, A.; Cebey-López, M.; Rodríguez-Tenreiro, C.; Dacosta-Urbieta, A.; Rivero-Velasco, C.; Rodríguez-Núñez, N.; Trastoy-Pena, R.; et al. Increased Serum Levels of sCD14 and sCD163 Indicate a Preponderant Role for Monocytes in COVID-19 Immunopathology. *Front. Immunol.* **2020**, *11*, 560381. [CrossRef]
38. Chang, Y.; Liu, L.; Wang, H.; Liu, J.; Liu, Y.; Du, C.; Hua, M.; Liu, X.; Liu, J.; Li, A. Presepsin Predicts Severity and Secondary Bacterial Infection in COVID-19 by Bioinformatics Analysis. *Comput. Math. Methods Med.* **2022**, *2022*, 9914927. [CrossRef]
39. Patterson, B.K.; Francisco, E.B.; Yogendra, R.; Long, E.; Pise, A.; Rodrigues, H.; Hall, E.; Herrera, M.; Parikh, P.; Guevara-Coto, J.; et al. Persistence of SARS CoV-2 S1 Protein in CD16+ Monocytes in Post-Acute Sequelae of COVID-19 (PASC) up to 15 Months Post-Infection. *Front. Immunol.* **2022**, *12*, 746021. [CrossRef]
40. Cao, W.; Birkenbach, M.; Chen, S. Patterns of Inflammatory Cell Infiltration and Expression of STAT6 in the Lungs of Patients with COVID-19: An Autopsy Study. *Appl. Immunohistochem. Mol. Morphol.* **2022**, *30*, 350–357. [CrossRef]
41. Kasper, M.; Hughes, C. Immunocytochemical evidence for a modulation of galectin 3 (Mac-2), a carbohydrate binding protein, in pulmonary fibrosis. *J. Pathol.* **1996**, *179*, 309–316. [CrossRef]
42. Bai, L.; Lu, W.; Tang, S.; Tang, H.; Xu, M.; Liang, C.; Zheng, S.; Liu, S.; Kong, M.; Duan, Z.; et al. Galectin-3 critically mediates the hepatoprotection conferred by M2-like macrophages in ACLF by inhibiting pyroptosis but not necroptosis signalling. *Cell Death Dis.* **2022**, *13*, 775. [CrossRef] [PubMed]
43. Portacci, A.; Diaferia, F.; Santomasi, C.; Dragonieri, S.; Boniello, E.; Di Serio, F.; Carpagnano, G.E. Galectin-3 as prognostic biomarker in patients with COVID-19 acute respiratory failure. *Respir. Med.* **2021**, *187*, 106556. [CrossRef] [PubMed]
44. Di Gregoli, K.; Somerville, M.; Bianco, R.; Thomas, A.C.; Frankow, A.; Newby, A.C.; George, S.J.; Jackson, C.L.; Johnson, J.L. Galectin-3 Identifies a Subset of Macrophages With a Potential Beneficial Role in Atherosclerosis. *Arterioscler. Thromb. Vasc. Biol.* **2020**, *40*, 1491–1509. [CrossRef] [PubMed]
45. Duan, T.; Du, Y.; Xing, C.; Wang, H.Y.; Wang, R.F. Toll-like Receptor Signaling and Its Role in Cell-Mediated Immunity. *Front. Immunol.* **2022**, *13*, 812774. [CrossRef] [PubMed]
46. Diamond, M.S.; Kanneganti, T.D. Innate immunity: The first line of defense against SARS-CoV-2. *Nat. Immunol.* **2022**, *23*, 165–176. [CrossRef]
47. Quero, L.; Hanser, E.; Manigold, T.; Tiaden, A.N.; Kyburz, D. TLR2 stimulation impairs anti-inflammatory activity of M2-like macrophages, generating a chimeric M1/M2 phenotype. *Arthritis. Res. Ther.* **2017**, *19*, 245. [CrossRef]
48. Trombetta, A.C.; Farias, G.B.; Gomes, A.M.C.; Godinho-Santos, A.; Rosmaninho, P.; Conceição, C.M.; Laia, J.; Santos, D.F.; Almeida, A.R.M.; Mota, C.; et al. Severe COVID-19 Recovery Is Associated with Timely Acquisition of a Myeloid Cell Immune-Regulatory Phenotype. *Front. Immunol.* **2021**, *12*, 691725. [CrossRef]
49. Grassi, G.; Notari, S.; Gili, S.; Bordoni, V.; Casetti, R.; Cimini, E.; Tartaglia, E.; Mariotti, D.; Agrati, C.; Sacchi, A. Myeloid-Derived Suppressor Cells in COVID-19: The Paradox of Good. *Front. Immunol.* **2022**, *13*, 842949. [CrossRef]
50. Chaudhuri, D.; Sasaki, K.; Karkar, A.; Sharif, S.; Lewis, K.; Mammen, M.J.; Alexander, P.; Ye, Z.; Lozano, L.E.C.; Munch, M.W.; et al. Corticosteroids in COVID-19 and non-COVID-19 ARDS: A systematic review and meta-analysis. *Intensive Care Med.* **2021**, *47*, 521–537. [CrossRef]
51. Meduri, G.U.; Annane, D.; Chrousos, G.P.; Marik, P.E.; Sinclair, S.E. Activation and regulation of systemic inflammation in ARDS: Rationale for prolonged glucocorticoid therapy. *Chest* **2009**, *136*, 1631–1643. [CrossRef] [PubMed]
52. Annane, D.; Pastores, S.M.; Arlt, W.; Balk, R.A.; Beishuizen, A.; Briegel, J.; Carcillo, J.; Christ-Crain, M.; Cooper, M.S.; Marik, P.E.; et al. Critical illness-related corticosteroid insufficiency (CIRCI): A narrative review from a multispecialty task force of the society of critical care medicine (SCCM) and the European society of intensive care medicine (ESICM). *Intensive Care Med.* **2017**, *43*, 1781–1792. [CrossRef] [PubMed]
53. Meduri, G.U.; Golden, E.; Freire, A.X.; Taylor, E.; Zaman, M.; Carson, S.J.; Gibson, M.; Umberger, R. Methylprednisolone infusion in early severe ARDS: Results of a randomized controlled trial. *Chest* **2007**, *131*, 954–963. [CrossRef] [PubMed]
54. Lixin, L.; Zhao, Y.; Chen, B.; He, D. Multiple COVID-19 Waves and Vaccination Effectiveness in the United States. *Int. J. Environ. Res. Public Health* **2022**, *19*, 2282. [CrossRef] [PubMed]
55. Calabrese, F.; Pezzuto, F.; Fortarezza, F.; Hofman, P.; Ker, I.; Panizo, A.; Thusen, V.V.d.; Timofeev, S.; Gorkiewicz, G.; Lunardi, F. Pulmonary pathology and COVID-19: Lessons from autopsy: The experience of European Pulmonary Pathologists. *Virchows Arch.* **2020**, *477*, 359–372. [CrossRef] [PubMed]
56. Basso, C.; Calabrese, F.; Sbaraglia, M.; Del Vecchio, C.; Carretta, G.; Saieva, A.; Donato, D.; Flor, L.; Crisanti, A.; Tos, A.P.D. Feasibility of postmortem examination in the era of COVID-19 pandemic: The experience of a Northeast Italy University Hospital. *Virchows Arch.* **2020**, *477*, 341–347. [CrossRef] [PubMed]

57. Ronchi, A.; Marino, F.Z.; Carraturo, E.; La Mantia, E.; Campobasso, C.P.; De Micco, F.; Mascolo, P.; Municinò, M.; Mucininò, E.; Vestini, F.; et al. PD-L1 Overexpression in the Lungs of Subjects Who Died from COVID-19: Are We on the Way to Understanding the Immune System Exhaustion Induced by SARS-CoV-2? *Crit. Rev. Eukaryot. Gene Expr.* **2022**, *32*, 9–20. [CrossRef]
58. Massoth, L.R.; Desai, N.; Szabolcs, A.; Harris, C.K.; Neyaz, A.; Crotty, R.; Chebib, I.; Rivera, M.N.; Sholl, L.M.; Stone, J.R.; et al. Comparison of RNA In Situ Hybridization and Immunohistochemistry Techniques for the Detection and Localization of SARS-CoV-2 in Human Tissues. *Am. J. Surg. Pathol.* **2021**, *45*, 14–24. [CrossRef]
59. Roden, A.C.; Vrana, J.A.; Koepplin, J.W.; Hudson, A.E.; Norgan, A.P.; Jenkinson, G.; Yamaoka, S.; Ebihara, H.; Monroe, R.; Szabolcs, M.J.; et al. Comparison of In Situ Hybridization, Immunohistochemistry, and Reverse Transcription-Droplet Digital Polymerase Chain Reaction for Severe Acute Respiratory Syndrome Coronavirus 2 (SARS-CoV-2) Testing in Tissue. *Arch. Pathol. Lab. Med.* **2021**, *145*, 785–796. [CrossRef]

International Journal of *Molecular Sciences*

MDPI

Article

In Silico Analyses Indicate a Lower Potency for Dimerization of TLR4/MD-2 as the Reason for the Lower Pathogenicity of Omicron Compared to Wild-Type Virus and Earlier SARS-CoV-2 Variants

Ralf Kircheis

Syntacoll GmbH, 93342 Saal an der Donau, Germany; rkircheis@syntacoll.de; Tel.: +49-151-167-90606

Abstract: The SARS-CoV-2 Omicron variants have replaced all earlier variants, due to increased infectivity and effective evasion from infection- and vaccination-induced neutralizing antibodies. Compared to earlier variants of concern (VoCs), the Omicron variants show high TMPRSS2-independent replication in the upper airway organs, but lower replication in the lungs and lower mortality rates. The shift in cellular tropism and towards lower pathogenicity of Omicron was hypothesized to correlate with a lower toll-like receptor (TLR) activation, although the underlying molecular mechanisms remained undefined. In silico analyses presented here indicate that the Omicron spike protein has a lower potency to induce dimerization of TLR4/MD-2 compared to wild type virus despite a comparable binding activity to TLR4. A model illustrating the molecular consequences of the different potencies of the Omicron spike protein vs. wild-type spike protein for TLR4 activation is presented. Further analyses indicate a clear tendency for decreasing TLR4 dimerization potential during SARS-CoV-2 evolution via Alpha to Gamma to Delta to Omicron variants.

Keywords: toll-like receptor (TLR); myeloid differentiation factor-2 (MD-2); TLR4/MD-2; Omicron; spike protein; SARS-CoV-2; COVID-19; cytokine storm; NF-kappa B

Citation: Kircheis, R. In Silico Analyses Indicate a Lower Potency for Dimerization of TLR4/MD-2 as the Reason for the Lower Pathogenicity of Omicron Compared to Wild-Type Virus and Earlier SARS-CoV-2 Variants. *Int. J. Mol. Sci.* **2024**, *25*, 5451. https://doi.org/10.3390/ijms25105451

Academic Editor: Maria Pascual

Received: 27 March 2024
Revised: 8 May 2024
Accepted: 9 May 2024
Published: 17 May 2024

1. Introduction

After the emergence of the SARS-CoV-2 Omicron variant of concern (VoC) B.1.1.529 at the End of 2021 [1,2], the Omicron successor variants have successfully replaced all other variants worldwide. The appearance of Omicron has been associated with a dramatically increased infectivity, and broad evasion from therapeutic antibodies and neutralizing antibodies after vaccination or earlier infections with the virus, due to multiple mutations, particularly in the N-terminal domain (NTD) and receptor binding domain (RBD) of the spike protein [3–8].

Besides their higher infectivity and evasion from antibody response, all Omicron variants show a dramatic shift in cellular tropism with a lower virus replication in the lungs but faster replication in the upper respiratory tract [4,8,9]. This correlates with Omicron's higher transmission rate within the population while causing less frequently acute respiratory distress syndrome (ARDS) and the severe clinical symptoms of COVID-19 [10]. The underlying molecular mechanisms for these reciprocal changes in cellular tropism with the appearance of the Omicron variants are not completely defined. A lower activation of Toll-like receptors (TLR) for Omicron variant has been hypothesized to be due to a changed charge distribution due to amino acid substitutions [11]. TLR-2 and TLR4 activation has been shown to be critically involved in SARS-CoV-2 induced pathogenicity [12–18], and binding of SARS-CoV-2 spike protein to TLR4 has been demonstrated for the wild-type virus [19]. However, so far no data regarding the dimerization of the TLR4/MD-2 have been shown. Notably, for the archetypal TLR4 ligand, i.e., LPS, dimerization of the TLR4/MD-2 has been demonstrated as the essential step for further downstream signaling [20]. Here, we show in silico analyses which indicate that (a) the spike protein trimer of SARS-CoV-2 can not only bind but also can result in dimerization of human TLR4/MD-2, dependent on the mode of binding, and (b) while the level of binding to TLR4 seems to be comparable to that of the SARS-CoV-2 wild-type, the Omicron spike protein shows a lower potency in inducing dimerization of TLR4/MD-2.

1.1. SARS-CoV-2 and the Pathophysiological Mechanisms of COVID-19

SARS-CoV-2—an enveloped positive-sense, single-stranded RNA virus—binds by its spike (S) protein to the angiotensin-converting enzyme-related carboxypeptidase-2 (ACE-2) receptor expressed on the surface of the target cells [21]. Besides, ACE2 alternative cellular receptors have been indicated as additional potential binding targets for the SARS-CoV-2, such as integrins, sialic acid, heparane sulfate, CD147, and neutropilin-1 [22,23]. The virus can enter the cell via two different routes, i.e., the highly efficient plasma membrane route or the cathepsin L-dependent endosomal entry route, dependent on whether TMPRSS2 (transmembrane serine protease 2) is co-expressed with ACE2 on the host cell membrane or not. While lung alveolar cells highly express TMPRSS2 [21], innate immune cells show no or only minimal expression of TMPRSS2; instead they use primarily the cathepsin L-dependent endosomal entry route [24–26].

The S protein of SARS-CoV-2—a class I viral membrane fusion protein—is present as a trimer, composed of a large ectodomain, a transmembrane anchor, and a C-terminal intracellular tail. The ectodomain comprises the receptor binding S1 subunit and the S2 subunit, which triggers membrane fusion. The receptor binding domain of S1 binds to ACE-2 on the target cells with high affinity, and S2 is responsible for fusion between the viral and host cell membranes. The S2 subunit consists of two heptad repeat (HR) regions (HR1 and HR2), the proteolytic site for the TMPRSS2 serine protease, a fusion peptide (FP), and the transmembrane (TM) domain. Between the S1 and S2 subunits, there is a SARS-CoV-2-unique poly-basic PRRAR furin-like cleavage site [27,28].

The binding receptor for SARS-CoV-2, ACE-2, is widely expressed among pulmonary and cardiovascular tissues. This explains the broad spectrum of pulmonary and extra-pulmonary symptoms of SARS-CoV-2 infection, including cardiac and gastrointestinal manifestations [21,29,30]. While all Omicron variants generally show less clinical severity compared to the original Wuhan strain or earlier VoCs, severe cases of COVID-19 show a clinical picture that is similar to that seen in SARS-CoV-1- and MERS-CoV-infected patients [31–34]. While younger individuals show in most cases mild-to-moderate clinical symptoms, elderly individuals more often show a severe or critical clinical picture, and pre-existing comorbidities, including diabetes, respiratory and cardiovascular diseases, renal failure, sepsis, and male sex, are with higher probability associated with critical disease and mortality [34–37].

The pathophysiological manifestations of critical or fatal COVID-19 include a diffuse alveolar disease with capillary congestion, interstitial edema, platelet-fibrin thrombi, cell necrosis, and infiltrates of macrophages and lymphocytes [38]. Furthermore, severe signs of endotheliitis in various organs (including lungs, heart, kidney, and intestine) have been found as a direct consequence of viral involvement and of the host inflammatory response [29,30].

The molecular and cellular mechanisms of COVID-19 include (i) virus-induced cytopathic effects, (ii) viral evasion from the host immune response correlating with a dysregulated host IFN type I response [39], and (iii) highly dysregulated exuberant inflammatory and immune responses which correlate with the severity of disease and lethality [32–34]. Upregulated release of cytokines and chemokines, i.e., "cytokine storm", plays a central role in the severity and lethality of SARS-CoV-2 infections [40–45]. The cytokine storm shows predominantly a Th1 pattern correlating with a hyperactivation of the NF-κB pathway as one of the critical signaling pathways for SARS-CoV-2 infection-induced proinflammatory cytokine/chemokine response, interconnected with other pathways, such as IL-6/STAT pathway [46–52]. Excessive activation of exuberant inflammatory responses involving endothelial cells, epithelial cells, and immune cells [44,45] can finally entail disturbances of the complement system, coagulopathies, disturbances of the bradikinine systems, finally triggering positive signaling feedback loops which accelerate inflammatory processes [53–59]. Additionally, vascular occlusion by neutrophil extracellular traps (NETs) [60,61] and coagulopathies, including thromboses and multiple micro-thromboses, are also hallmarks of COVID-19 disease [62–65].

1.2. Molecular Changes in the Spike Protein Omicron and Other VoCs

The Omicron variants have gathered multiple amino acid substitutions and deletions, particularly in the spike protein. The general structure of the spike protein and the mutations found in the Omicron variants and earlier VoCs are shown in Figure 1.

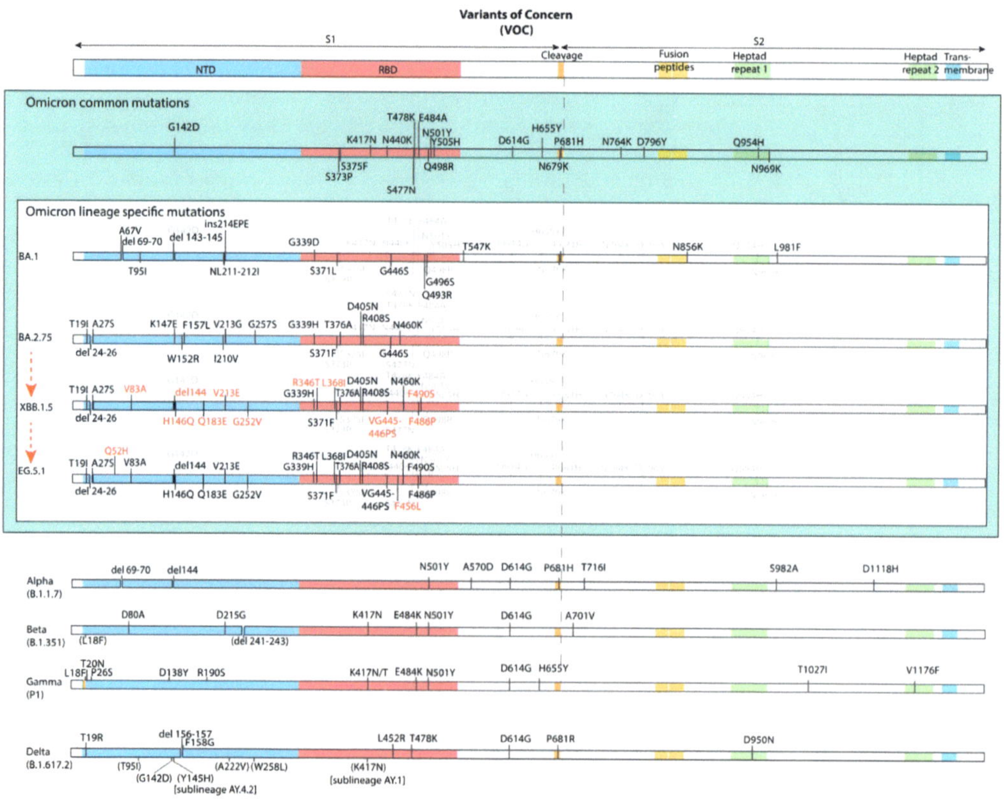

Figure 1. Mutations in the spike protein in Omicron variants compared to wild type SARS-CoV-2 and earlier VoCs. Mutations which are common to all Omicron variants are shown in the upper (green) part. Picture derived from [66].

The highest number of the mutations in the Omicron spike protein are concentrated in the NTD, the RBD, and near to the S1/S2 cleavage site [3–7,67–69]. Two out of the three amino acid substitutions in the S1/S2 cleavage site region, i.e., P681H and H655Y, are found also in earlier VoCs, i.e., alpha (B.1.1.7) and gamma (P.1) variant, respectively (see Figure 1). In contrast, the S2 subunit is rather conserved, with few amino acid substitutions in the FP and HR1 regions and with only four unique mutations, which are common to all Omicron variants, i.e., N764K, D796Y, Q954H, and N969K. Notably, these are exchanges towards positively charged amino acids in three out of these four amino acid substitutions, and one loss of one negative charge (D796Y) [11].

1.3. Lower Pathogenicity of Omicron Compared to Previous VoCs

There is increasing evidence that Omicron, despite a significantly higher transmissibility and infectivity, shows a significantly lower rate of severe clinical courses compared to previous VoCs, including the Delta variant, i.e., the last VoC before Omicron [70]. Early reports on the lower number of severe clinical outcomes in Africa [1,2] are concordant with published studies from other geographical areas on the Omicron pandemic, such as the UK and USA, and have been confirmed by recent reports covering recent Omicron variants, including BA1 (B.1.1.529), BA2, BA4, BA5, XBB1.5 and EG5.1—all showing higher transmissibility but significantly lower pathogenicity [71–80].

The different behaviour of Omicron compared to early viral variants is well illustrated by plotting the ratios of deaths per million against cases per million over time. The resulting continuous line with changing colors (from violet to dark red), visualizing the different waves of infection in the COVID-19 pandemic over time, shows high death rates during earlier phases (steep circles), and a completely different, highly flattening plot for Omicron [81] (Figure 2).

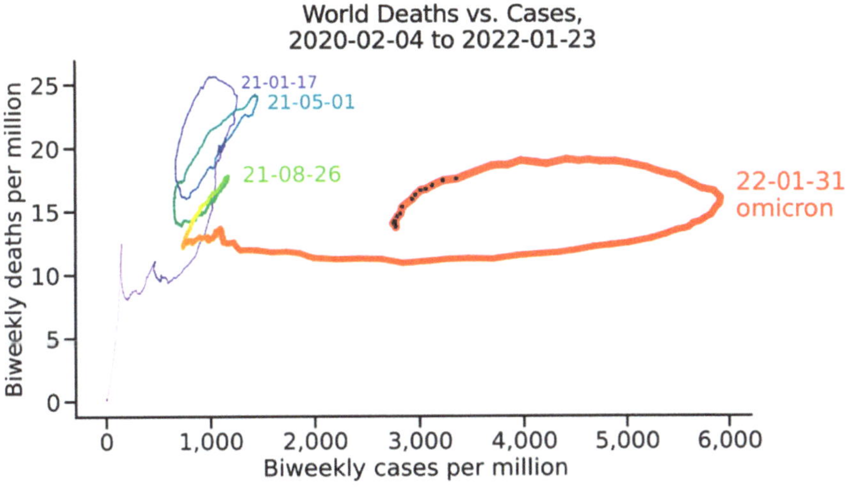

Figure 2. All reported cases and deaths worldwide, reproduced from Ref. [79].

Beside the significantly lower pathogenicity of Omicron, recent studies have indicated a generally narrowing cytokine pattern during evolution of the SARS-CoV-2 from the original Wuhan strain via the various VoCs, e.g., Alpha–Gamma–Delta–Omicron. Comparing the cytokine profiles of patients infected with different VoCs with those infected with the wild-type Wuhan variant, the studies showed that hypercytokinemia and cytokine storm become less threatening with the emergence of new mutations in the viral genome, this tendency being largely visible already with the Delta variant and with the most pronounced change for the Omicron variant [82,83].

The attenuated clinical picture of Omicron is further supported by experimental animal studies, which show equal or higher, mostly TMPRSS2-independent, viral replication in the upper respiratory tract, but lower pathogenicity in animal models. Studies on B.1.1.529 Omicron isolates regarding infection and disease in immunocompetent, human ACE2 expressing mice showed attenuated infection compared to earlier SARS-CoV-2 variants, with lower weight loss and lower viral burden in the lungs [8]. A milder pathological picture was also shown for Omicron infected Syrian golden hamsters compared to infection with earlier VoCs, i.e., Gamma and Delta or ancestral strain 614 G. Omicron infection caused only mild rhinitis and only minimal lesions in the lung. Viral antigens, RNA, and infectious virus titers were lower in the lungs of Omicron infected hamsters [84]. Furthermore, replication of Omicron was shown to be attenuated in human immortalized lung and intestinal cell lines, due to Omicron's inefficient use of TMPRSS2 compared with wild-type SARS-CoV-2 and other VOCs. Omicron showed a significantly lower viral replication in both the upper and lower respiratory tracts of infected K18-hACE2 mice compared to mice infected with the wild-type strain or Delta variant and showed ameliorated lung pathology. In comparison to wild-type SARS-CoV-2 and various VoCs, including Alpha (B.1.1.7), Beta (B. 1.351) and Delta (B.1.617.2) variants, infection by Omicron caused significantly less reduction in body weight and the lowest mortality rate [85]. Furthermore, viral infection course, tissue tropism, and pathogenicity of SARS-CoV-2 D614G (B.1), Delta (B.1.617.2), and Omicron BA.1.1 (B.1.1.529) variants were studied in a feline model. While D614G- and Delta-inoculated cats showed clear signs of lethargy and increase in body temperatures, Omicron-inoculated cats remained subclinical and even showed an increase in weight. Intranasal inoculation of cats with D614G- and Delta variants resulted in high infectious virus shedding in nasal secretions, whereas markedly lower virus shedding was found in Omicron-inoculated animals. The tissue distribution of the Omicron variant was also largely changed in comparison to the D614G and Delta variants, showing lower viral loads in the respiratory tract. Importantly, while D614G- and Delta-inoculated cats showed clear signs of pneumonia, histology of the lungs of Omicron-infected cats showed only mild to modest inflammation [86].

Overall, these studies using different Omicron isolates have demonstrated a shift in cellular tropism with attenuated lung disease in rodents and cats, supporting human clinical data [1,2,71–80]. Still, the molecular mechanisms responsible for this reciprocal change in tropism in the Omicron variant are not completely defined so far.

1.4. SARS-CoV-2 Activates Innate PRRs

Toll-like receptors (TLRs) are essential for recognition and elimination of pathogen-associated molecular patterns (PAMPs) from bacteria, viruses, and of self-derived damage-associated molecular patterns (DAMPs) released from dying or lytic cells. Typical PAMPs are nucleic acids, e.g., viral RNA and DNA, but also various glycoproteins, lipoproteins, and membrane components.

Severity and disease progression in patients with COVID-19 was shown to correlate with the expression of MyD88, i.e., a key adaptor molecule in TLR-signaling, and the expression of TLR1, TLR2, TLR4, TLR5, TLR8 and TLR9 (but not TLR3) [12]. In this context, however, the expression levels of different TLR on the relevant target cells will be of importance. Highest expression levels are found for TLR4/MD-2 and TLR2 on innate immune cells, e.g., macrophages, and endothelial cells, whereas ~10 fold lower expression levels are found for TLR1, 6, and 7. In contrast to macrophages and endothelial cells, alveolar cell (in particular type 1) show generally very low levels of TLRs [11], suggesting that, primarily, the activation of TLR4 and TLR2 on macrophages and endothelial cells may be responsible for the excessive disturbances during severe and critical COVID-19 stages.

The question as to which of the TLRs have a particular high impact to the pathophysiological manifestations in severe and critical COVID-19 has been addressed in several studies. Khan et al. studied the inflammatory potency of different structural proteins of SARS-CoV-2 and showed that the spike protein strongly induces inflammatory cytokines and chemokines, such as IL-6, IL-1β, TNFα, CXCL1, CXCL2, and CCL2, but not IFNs in human and mouse macrophages. The spike protein was shown to activate the NF-κB pathway in a MyD88-dependent manner. Activation of the NF-κB pathway was abrogated in TLR2-deficient macrophages. Consistently, administration of the spike protein triggered the release of IL-6, TNFα, and IL-1β in wild-type mice but not in TLR2-deficient mice. Both, S1 and S2 subunits induced high NF-κB activation, with a higher potency found for S2 when compared on an equimolar basis [87].

Various studies have addressed the role of TLR4 in the pathogenicity of COVID-19. TLR4 is known to recognize multiple PAMPs derived from bacteria, viruses, and other pathogens [20]. Furthermore, TLR4 has been found to recognize certain damage-associated molecular patterns (DAMPs), e.g., high mobility group box 1 (HMGB1) and heat shock proteins (HSPs), which are released from dying or lytic cells during host tissue injury or viral infection [15,88]. TLR4 is mainly expressed on immune cells, in particular on macrophages and dendritic cells, but also on endothelial cells. Activation of TLR4 by PAMPs or DAMPs leads to the production of proinflammatory cytokines and/or the production of type I interferons and anti-inflammatory cytokines [15,88–90]. In contrast to other TLRs, TLR4 is present on both, the cell surface, where it recognizes, e.g., viral proteins before they enter the cell, but also in endosomes [20,91]. As demonstrated in detail with the archetypal TLR4 agonist from gram-negative bacteria, i.e., lipopolysaccharide LPS, dimerization of the TLR4/MD-2 complex represents the central molecular step within the TLR4 activation cascade, with hydrophobic chains and charged groups of the ligand essential for binding and dimerization of TLR4/MD2 [91–95].

The impact of TLR4 in the pathogenies of COVID-19 has been demonstrated in multiple studies [14–19,96]. The release of IL1β triggered after infection by SARS-CoV-2 was shown to be abrogated by the TLR4-specific inhibitor Resatorvid. The SARS-CoV-2 spike trimer was demonstrated by surface plasmon resonance (SPR) to bind to TLR4 with an affinity of ~300 nM, which is comparable to that of many virus-receptor interactions. A monocytic cell line was treated with either the whole spike protein trimer, or the NTD or the RBD of spike protein, respectively. Only the trimeric protein, but not the NTD or RBD alone, induced IL1β and IL6, and this induction was blocked by the TLR4 inhibitor Resatorvid. The spike protein also induced IL1β in a murine macrophage cell line in a TLR4- and MyD88-dependent manner as well as in primary bone marrow-derived macrophages and peritoneal macrophages from wild-type but not in macrophages derived from TLR4-deficient mice. IL1β expression induced by spike protein was blocked by the NF-κB inhibitor JSH-23. Activation of TLR4 by spike protein was shown to be non-related to ACE2, TMPRSS2, or virus entry as shown on macrophages from ACE2-deficient mice or mice overexpressing human ACE2. Treatment with ACE2 inhibitor (MLN-4760) or soluble ACE2 or TMPRSS2-specific inhibitor (Bromhexine hydrochloride) also had no effect on the induction of IL1β by LPS or spike protein. Notably, the induction of IL1β by the spike protein trimer from SARS-CoV-2 or SARS-CoV was comparable to LPS treatment [17].

Exposure of peritoneal exudate-derived murine macrophages to SARS-CoV-2 spike protein S1 subunit induced TNF-α, IL-6, IL-1β, and nitric oxide, and activated NF-κB and c-Jun N-terminal kinase signaling pathways. Treatment with a TLR4 antagonist attenuated pro-inflammatory cytokine release and activation of intracellular signaling by S1 and lipopolysaccharide, suggesting that the S1 subunit activates TLR4 signaling in murine and human macrophages [96]. Furthermore, the SARS-CoV-2 spike S1 domain was shown to be a TLR4 agonist in rat and human cells inducing a pro-inflammatory M1 phenotype in human monocyte-derived macrophages. Adult rat cardiac tissue resident macrophage-derived fibrocytes were treated with either LPS or recombinant SARS-CoV-2 spike S1 glycoprotein and showed TLR4 activation and upregulation of ACE2. The effects of spike S1 and LPS could be blocked by the specific TLR4 inhibitor Resatorvid, confirming the spike S1 subunit as a TLR4 agonist. Confocal immunofluorescence microscopy showed a 1:1 stoichiometric co-localization of spike S1 with TLR4 in rat and human cells, these data confirmed by proximity ligation assays [18].

Moreover, previously published in silico studies have shown the strongest protein–protein interaction of the spike glycoprotein of SARS-CoV-2 with TLR4 compared to other TLRs [19]. Furthermore, a model has been proposed for binding of SARS-CoV-2 spike glycoprotein to TLR4 leading to the activation of TLR4 signaling characterized by increased cell surface expression of ACE2 that facilitates virus entry. Finally, SARS-CoV-2-induced myocarditis and multiple-organ damage was correlated to TLR4 activation and hyperinflammation in COVID-19 patients. All these data indicate that TLR4 may significantly contribute to the pathogenesis of SARS-CoV-2 indicating TLR4 as a promising therapeutic target in COVID-19, which is supported also by the fact that TLR4 antagonists have been previously used in sepsis and in other antiviral contexts [18].

2. Results

2.1. In Silico Analyses of TLR4/MD-2 Binding and Dimerization by Omicron Spike Protein Compared to Wild-Type Virus Spike Protein

Studies with the archetypal TLR4 agonist LPS and its derivatives have demonstrated that agonists share the functional feature of binding both TLR4 and the hydrophobic pocket of the myeloid differentiation factor 2 (MD-2, Lymphocyte antigen 96, Ly96) that is part of the TLR4/MD-2 complex. The high-affinity binding leads to dimerization of the TLR4/MD-2 complex. Dimerization of the TLR4-MD-2 complex is an absolute pre-requisite for TLR4 activation, leading to the recruitment of the intracellular adaptor protein MyD88 coupled to intracellular signaling [20,91–95].

We have analyzed in silico the binding of the spike protein trimer from wild-type SARS-CoV-2 (PDB 6ZGG) (PDB DOI: https://doi.org/10.2210/pdb6ZGG/pdb EM Map EMD-11205: EMDB EMDataResource, accessed on 6 May 2024) [97] to the human TLR4/MD-2 complex (PBD 3FXI) (PDB DOI: https://doi.org/10.2210/pdb3FXI/pdb, accessed on 6 May 2024) [98] and have compared this with binding of the spike protein trimer from Omicron variant (PDB 7TL9) (PDB DOI: https://doi.org/10.2210/pdb7TL9/pdb EM Map EMD-25984: EMDB EMDataResource, accessed on 6 May 2024) [99]. Importantly, for the human TLR4/MD-2, a PDB model based on dimerization of human TLR4/MD-2 by LPS (PDB 3FXI) [98] was chosen in order to analyze for simultaneous binding of the spike protein trimers to two TLR4/MD-2 complexes. Furthermore, for both, wild-type virus, and Omicron, PDB spike protein trimer models with both, monomers in closed ("down") AND in open ("up") state, were chosen to cover the broadest range of configurations of the spike protein trimer.

We have analyzed the potency for dimerization of the human TLR4/MD-2 complex (PDB 3FXI) which will require interaction with two TLR4/MD-2 partners (i.e., TLR4/MD-2 (I) and (II), respectively). Docking studies using HDOCK software (©Lab of Biophysics and Molecular Modeling, huanglab@hust.edu.cn) [100] and PDB driven data showed that both the spike protein trimer from wild-type virus (PDB 6ZGG) as well as from Omicron variant (PDB 7TL9) show a comparable binding to the human TLR4/MD-2 complex with docking scores for the top 10 models ranging between −348 and −290 and −368 and −297, respectively (Tables 1 and 2).

Table 1. Summary of Top 10 Models HDOCK docking for wild type spike protein (PDB 6ZGG) and TLR4/MD2 (PBD 3FXI).

Rank	1	2	3	4	5	6	7	8	9	10
Docking Score	−347.62	−332.77	−313.50	−311.94	−310.11	−305.95	−299.03	−298.52	−291.15	−289.51
Confidence Score	0.9812	0.9748	0.9634	0.9623	0.9609	0.9517	0.9577	0.9512	0.9439	0.9421

Table 2. Summary of Top 10 Models HDOCK docking for Omicron spike protein (PDB 7ZL9) and TLR4/MD2 (PDB 3FXI).

Rank	1	2	3	4	5	6	7	8	9	10
Docking Score	−368.16	−337.33	−326.17	−321.62	−306.41	−305.44	−304.69	−304.24	−300.36	−296.86
Confidence Score	0.9874	0.9769	0.9713	0.9687	0.9580	0.9573	0.9566	0.9563	0.9529	0.9497

However, differences were found when comparing the actual site of binding of the spike protein trimer to the human TLR4/MD-2 complex and regarding the question which of the subunits of the spike protein are involved. For the wild-type SARS-CoV-2 four out of the ten (4/10) top binding models (models 1, 3, 8, 10) (Figure 3A,B) showed no dimerization, whereas two models (2/10) showed dimerization of TLR4/MD-2 induced by the S1 subunit of the spike protein (models 4, 5) (Figure 3C,D). Interestingly, four out of ten (4/10) models showed dimerization of TLR4/MD-2 complex by interaction with both, S1 and S2 subunits (models 2, 6, 7, 9) (Figure 3E–H). In contrast, for the Omicron spike protein, seven out of ten (7/10) top binding models showed no dimerization of TLR4/MD-2 (models 2, 3, 4, 5, 7, 8, 9) (Figure 4A–D), two out of ten (2/10) showed S1 induced dimerization (models 1, 10) (Figure 4E,F), but only one model (1/10) showed TLR4/MD-2 dimerization induced by S1 and S2 subunit (model 6) (Figure 4G).

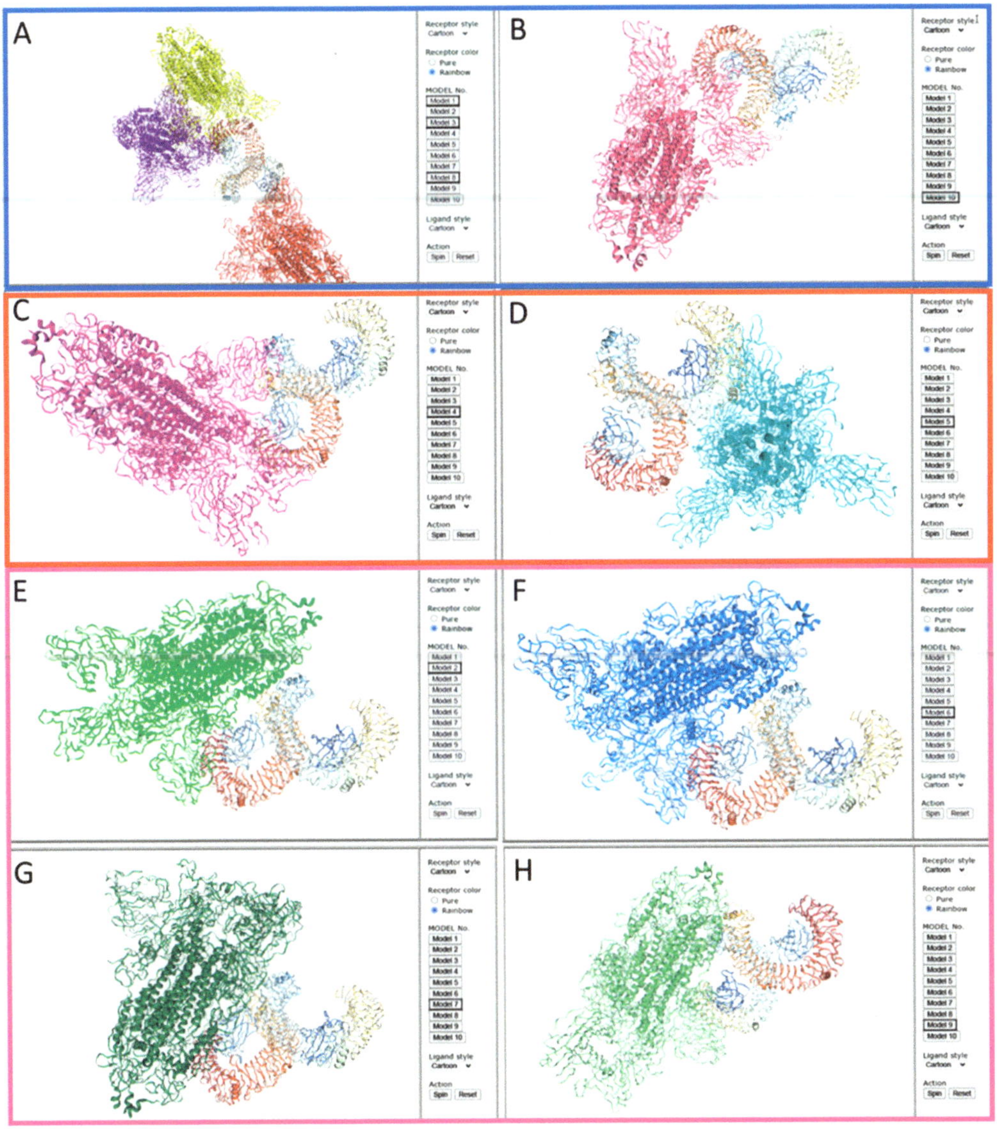

Figure 3. Top ten docking models for spike protein trimer from the wild-type SARS-CoV-2 (PDB 6ZGG) to the human TLR4/MD-2 complex (PBD 3FXI) calculated using HDOCK software (http://hdock.phys.hust.edu.cn/, accessed on 16 December 2023). Four out of the ten top binding models (models 1, 3, 8, 10) show binding to TLR4/MD-2 without dimerization ((**A**,**B**) in blue frame), two models (models 4, 5) show dimerization of TLR4/MD-2 induced by the S1 subunit of the spike protein ((**C**,**D**) in red frame), four out of ten models (models 2, 6, 7, 9) show dimerization induced by both, S1 and S2 subunits ((**E**–**H**) in pink frame).

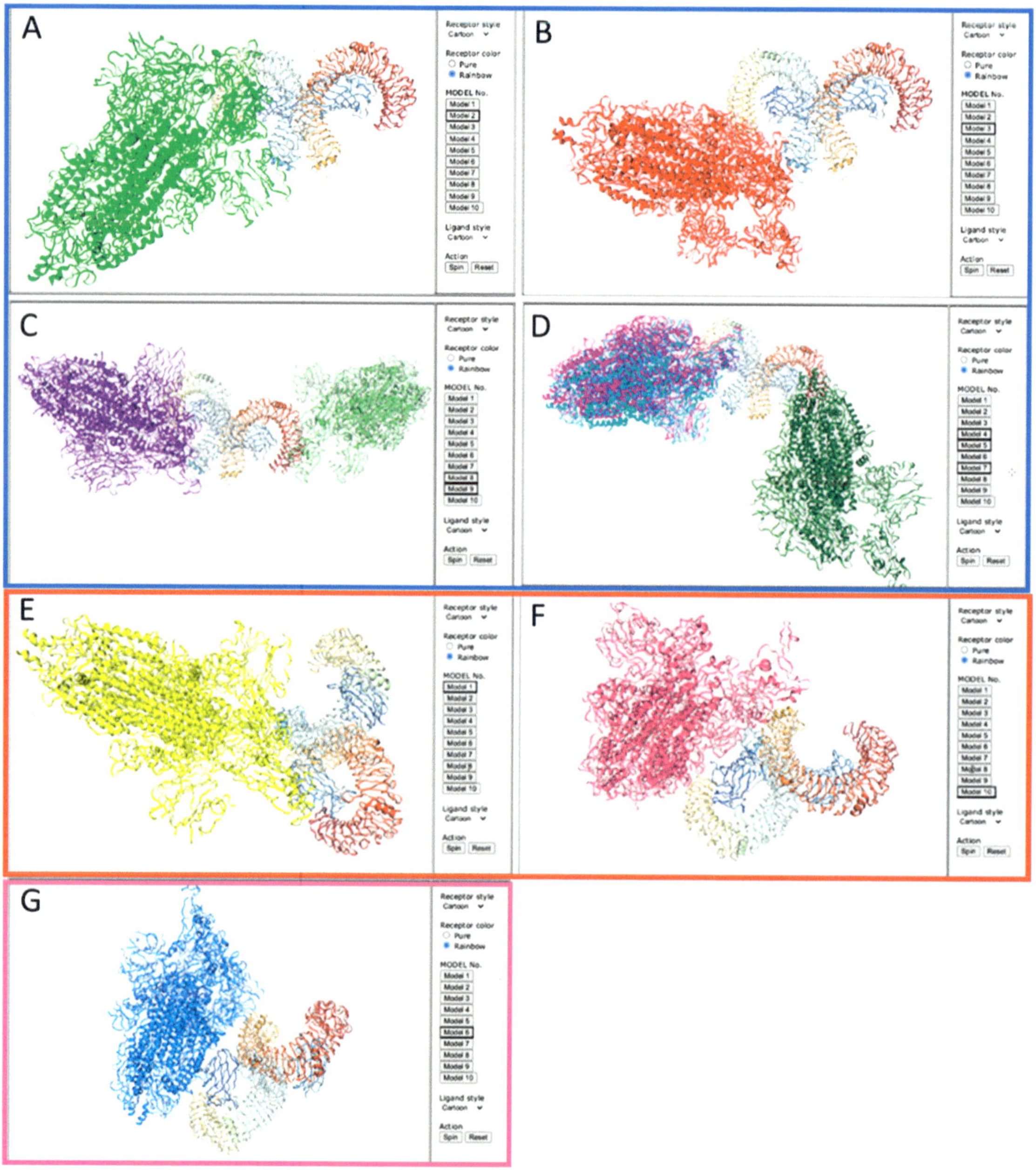

Figure 4. Top ten docking models for spike protein trimer from the Omicron SARS-CoV-2 (PDB 7TL9) to the human TLR4/MD-2 complex (PBD 3FXI) calculated using HDOCK software (http://hdock.phys.hust.edu.cn/, accessed on 16 December 2023). Seven out of ten (7/10) top binding models (models 2, 3, 4, 5, 7, 8, 9) show binding without dimerization of TLR4/MD-2 ((**A**–**D**) in blue frame), two out of ten showed S1 induced dimerization (models 1, 10) ((**E**,**F**) in red frame), one model (model 6) shows TLR4/MD-2 dimerization induced by S1 and S2 subunit ((**G**) in pink frame).

To interpret the differences in dimerization potency vs. binding activity to TLR4/MD-2 between wild-type and Omicron variant calculated for various models and their biological consequences, the following aspects have to be taken into consideration: as the TLR4/MD-2 recognizes a broad variety of danger signals, including large viral proteins [15,20,89–92] that in most cases do not fit spatially into the hydrophobic pocket of MD-2 as used by the relatively small archetypal TLR4 ligand LPS [20,98], alternative binding modalities of recognized danger signals to the TLR4 lycine rich rings and/or the MD-2 have to be assumed. The functional similarity of the various TLR4/MD-2 ligands can therefore not be a structural mimicry of the exact dimensions and configurations of LPS but rather the shared potential to bind both TLR4 and/or MD-2 of the two dimerization partners, i.e., TLR4/MD-2 complexes I and II, resulting in dimerization. It is therefore conceivable that, for the rather large spike protein trimer of wild-type SARS-CoV-2 or Omicron, respectively, there is not only one (i.e., that with the highest binding affinity) binding modus, but rather the sum of all high affinity configurations, which are able to induce dimerization of the TLR4/MD-2 complex. In this context, the higher probability for multiple binding configurations of the wild-type spike protein to the TLR4/MD-2 dimer (I and II) complexes compared to Omicron spike protein can be expected to correlate with the difference in the amplitude of general TLR4 activation in the relevant cells.

Conclusively, these data indicate, that while binding of the spike protein trimers to TLR4/MD-2 seem to be comparable, the mode of binding with respect to dimerization is different for the Omicron variant compared to the wild-type virus. The differences between the virus variants were most striking for those binding configurations which use interaction of both subunits, S1 and S2, with the TLR4/MD-2. Only one configuration with S1 and S2 binding was calculated for the Omicron spike protein trimer compared to four configurations for the wild-type virus. Three configuration found for the wild-type spike protein trimer seem to be lost due to amino acid substitutions in the Omicron variant, including two very similar configurations of binding of the wild-type spike protein with TLR4/MD-2, i.e., models 2 and 6 (see Figure 3E,F).

Performing analogous in silico modelling on other VoCs, including Alpha, Gamma, and Delta variant, showed a decrease in the number of dimerization modes after the Alpha variant, with 6, 7, 5, 4, and 3 modes of TLR4/MD2 dimerization for the wild-type, Alpha, Gamma, Delta, and Omicron VoCs, respectively (Figures for Alpha, Gamma, and Delta variants in Supplementary data, Tables S1–S3 and Figures S1–S3). These data correlate with data from recent studies, which indicate a narrowing cytokine pattern during evolution of the SARS-CoV-2 [82,83].

Analyzing in detail the binding and dimerization regarding the amino acid interactions between the spike trimer of wild-type SARS-CoV-2 with TLR4/MD-2 in model 2, the salt interaction, polar, and non-polar interaction of the spike trimer with TLR/MD-2 dimerization partner I and with the TLR4 part of dimerization partner II, respectively, were calculated. Figure 5. shows the multiple interactions of the S protein trimer with the dimerization partner TLR/MD-2 (I) and significantly fewer interactions with the second dimerization partner, i.e., TLR4 (II), indicating binding to TLR4 (II) as being potentially limiting for dimerization. Notably, ionic interactions with both partners were also identified, e.g., Lys795 of the S protein with Glu563 on TLR4 (II), together with multiple polar interactions S2 region around Lys921, Asn925, Asn928, and Asp936 with TLR4 (II).

Figure 6A,B shows models 2 and 6, respectively, for binding and dimerization of TLR4/MD-2 (I) and TLR4 (II) by the wild-type virus S protein trimer, showing binding of the S1 subunit to TLR4 (I) and MD-2 (I), and S2 binding to TLR4 (II), without involvement of MD-2 (II). Figure 6C shows at a higher zoom the interaction of S protein trimer with TLR4 (I), MD-2 (I) and TLR4 (II), respectively, with the main ionic and polar interacting amino acids indicated (blue—positively charged amino acids, red—negatively charged amino acids, pink—polar amino acids), dotted lines—ionic interactions.

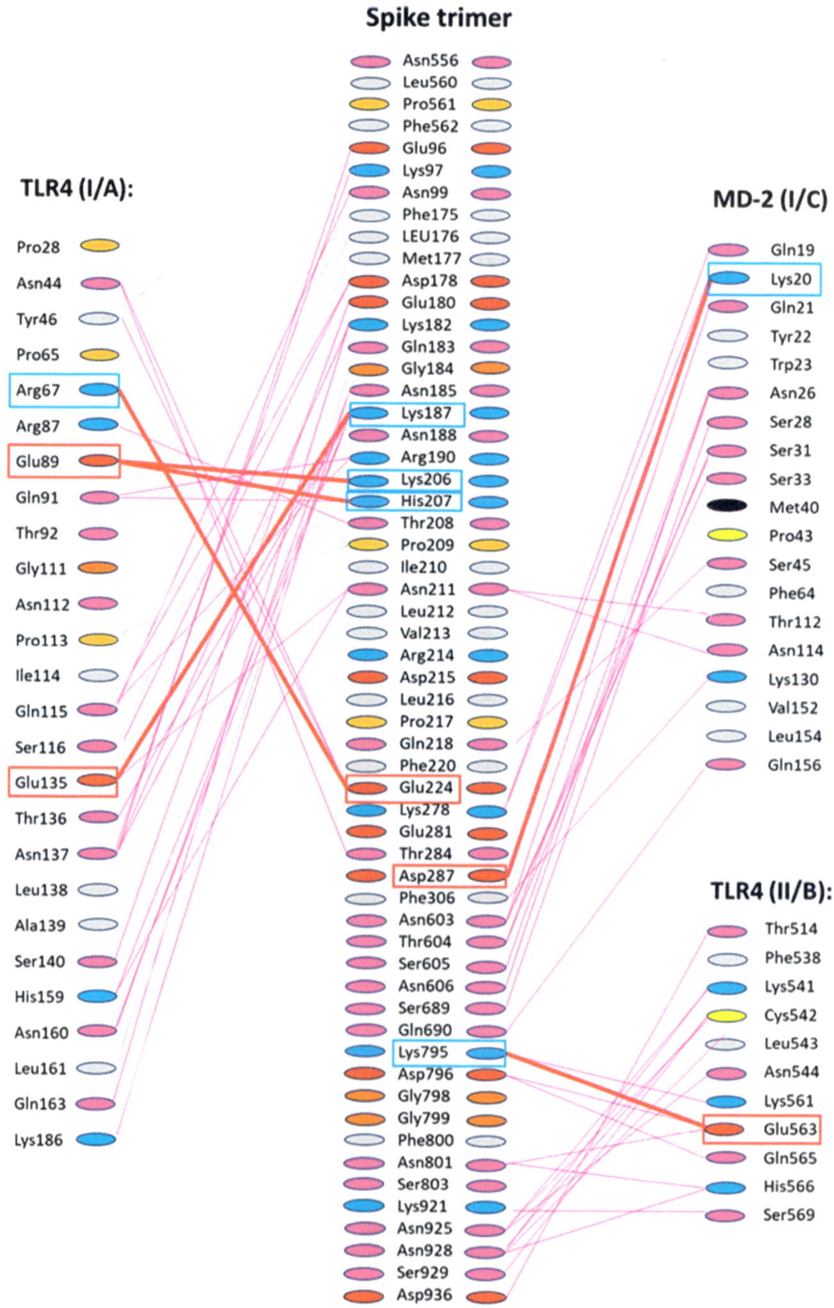

Figure 5. Salt interaction (thick red lines connecting two colored boxes which highlight involved amino acid residues), polar (pink lines), and non-polar (grey lines) interaction of the spike trimer with TLR/MD-2 dimerization partner I and with the TLR4 part of dimerization partner II as calculated by HDOCK software (http://hdock.phys.hust.edu.cn/, accessed on 16 December 2023) are shown.

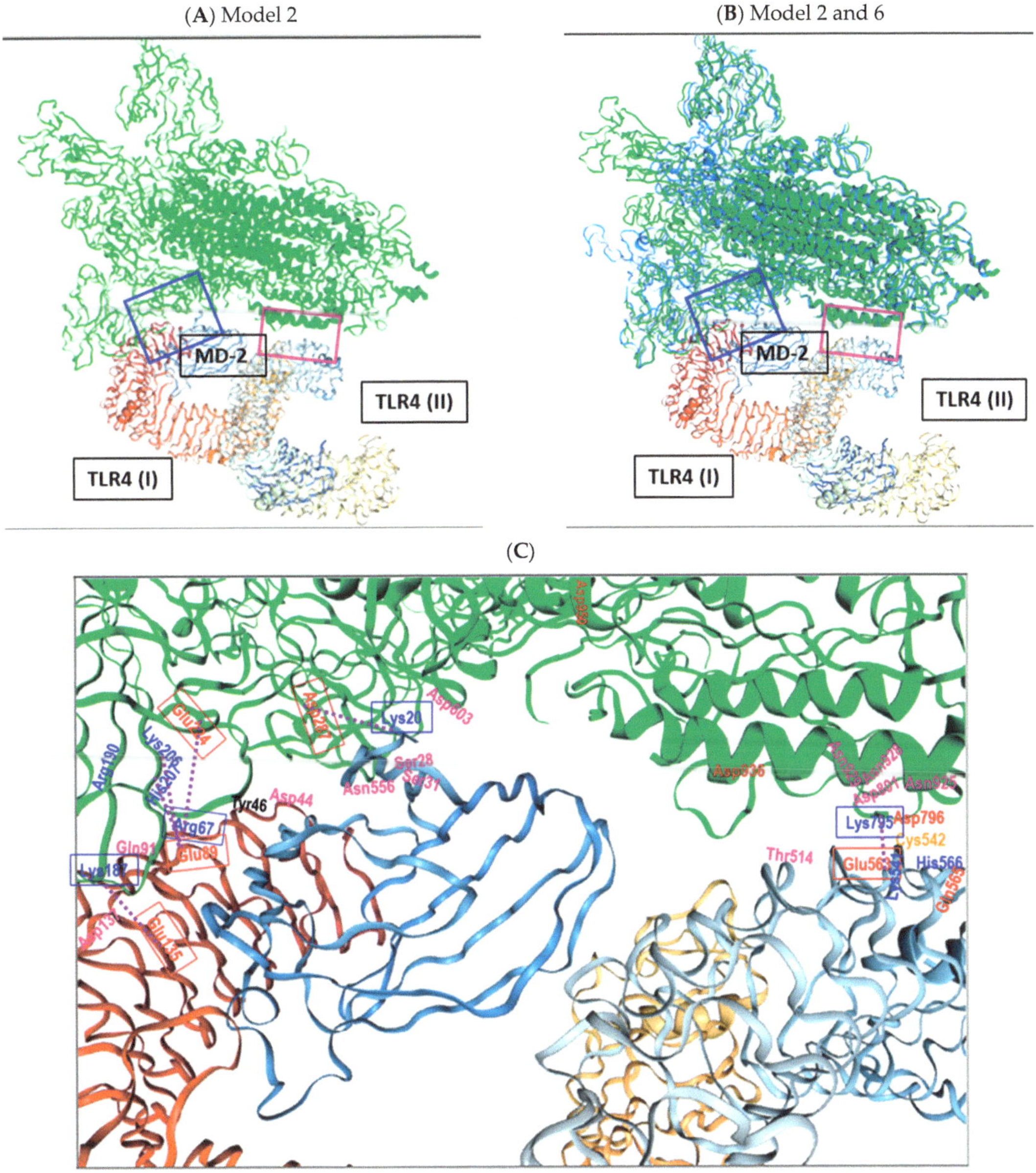

Figure 6. HDOCK generated model 2 (**A**) and models 2 and 6 superimposed (**B**) for binding and dimerization of TLR4/MD-2 (I) and TLR4 (II) by the wild-type virus S protein trimer (model 2—green; model 6—blue). Blue and pink frames show interaction areas with TLR4/MD-2 (I) and TLR4 (II), respectively (**C**) higher zoom of the interaction of S protein trimer (green color, upper part) with TLR4 (I) (spans from left side/red color to right side/yellow color)), MD-2 (I) (left side/center, blue color) and TLR4 (II) (right side, light blue color), colors of amino acids: blue—positively charged amino acids, red—negatively charged amino acids, pink—polar amino acids, orange – cysteine, dotted lines—ionic interactions.

Dimerization of TLR4/MD-2 is supported by binding of the spike protein S1 subunit (mainly NTD) to TLR4 (I) and MD-2 (I) and binding of S2 subunit (FP and HR1 region) to TLR4 (II). Six salt bridges and multiple polar and non-polar interactions are involved in the binding (see Figures 5 and 6).

When the 3D structures of the monomers of the spike proteins of wild type (PDB 6ZGG) and Omicron (PDB 7TL9) are aligned a higher degree of alignment is seen for the S2 subunits compared to the S1 subunit (Figure 7A) except for a region around the Fusion peptide, which seems to be slightly shifted in the Omicron S protein compared to the wild-type S protein (Figure 7B).

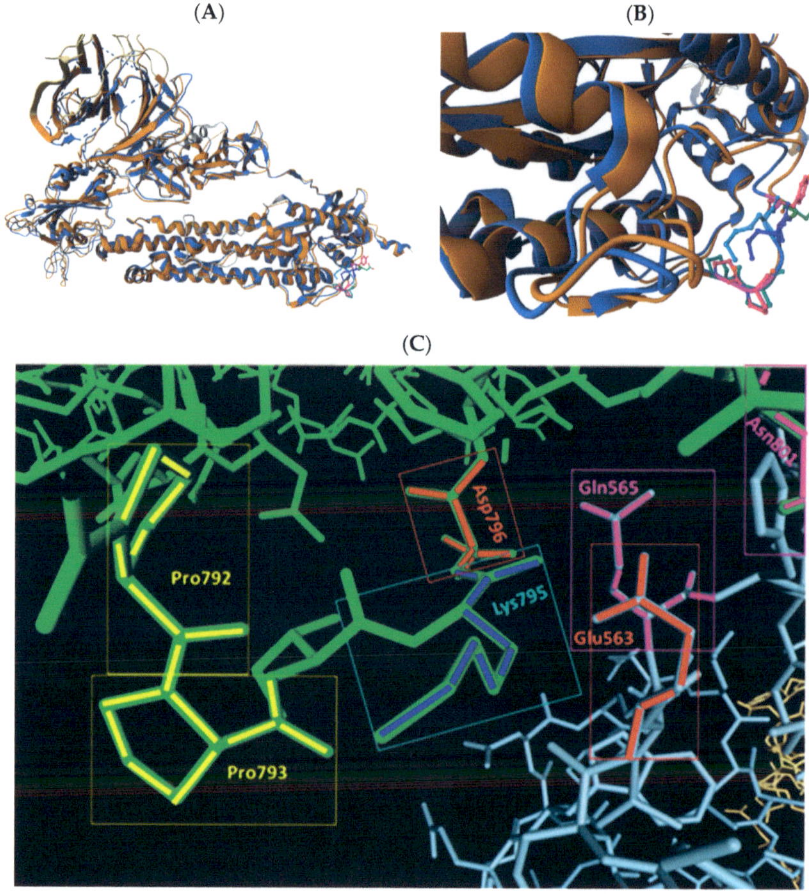

Figure 7. Pairwise alignment of monomers of S proteins of wild type (PDB 6ZGG) and Omicron (PDB 7TL9) (**A**), higher zoom (**B**). The two Lys795 are shown in dark blue vs. light blue and Asp796 and Tyr796 are shown in green and pink in the wild-type (6ZGG) vs. Omicron (7TL9) models, respectively. Pro792 and Pro793 are also shown for both models (**A**,**B**). (**C**) Spatial interaction between Lys795 (dark blue) and Asp796 (red) of wild-type virus spike protein with Glu563 (red) and Gln565 (pink) in TLR4 (II) were calculated and displayed using HDOCK software (http://hdock.phys.hust.edu.cn/, accessed on 9 February 2024). The two proximate prolines (Pro792 and Pro793) are shown in yellow.

And this region seems to be involved in the interaction of the spike protein with TLR4 (II), with salt bridge between Lys795 of the spike protein and Glu563 of TLR4 and multiple polar interactions, including Asp796 (spike protein) and Gln565 (TLR4). This interaction, however, is only found for the wild-type spike protein trimer (PDB 6ZGG) in HDOCK docking studies models 2 and 6 (Figure 7C). In contrast, in the Omicron spike protein, Asp796 is exchanged by Tyr796 (Figure 7B). Importantly, in this region no high affinity interaction between the Omicron spike protein and with TLR4 (II) are found by docking studies using HDOCK among the top binding models, indicating that the spatial shift in FP region (also previously noted by Gobeil et al. [99]) together with the amino acid exchange of Asp796 by Tyr796 are responsible for abrogating an efficient binding to TLR4 in this region.

Conclusively, the amino acid exchanges in the Omicron spike protein in the FP / HR1 region of the S2 subunit are likely to lead to a lower potency in supporting dimerization of the TLR4/MD-2 complex compared to wild-type. Additionally, the docking studies with the other VoCs (Supplementary data, Tables S1–S3 and Figures S1–S3) indicate that also amino acid substitutions present in earlier VoCs, e.g., Delta variant, may contribute to a decreased potential of the spike protein for TLR4 dimerization.

2.2. *Binding and Dimerization of TLR2/TLR1 and TLR2/TLR6 Heterodimers*

Given the high expression of both TLR4 and TLR2 in innate immune cells, the binding and dimerization of TLR2 heterodimers (with TLR1 and TLR6, respectively) have been studied (data shown in Supplementary data, Tables S4–S7 and Figures S4–S7). There was a moderate difference between spike protein of Omicron variant vs. spike protein of wild-type virus regarding the number of dimerization modes, with 8 vs. 7 and 4 vs. 3 dimerization modes, respectively.

In this context, it should be noted that the cellular expression of TLR4/MD-2 complexes is several-fold higher than the expression of TLR1 and TLR6 (see new Figure 8B,C), which serve as dimerization partners for TLR2 heterodimers.

2.3. *Mechanistic Model for Lower Pathogenicity of Omicron Variants Compared to Wild-Type Virus and Early VoCs*

The in silico analyses presented in this study have shown that, despite a comparable binding activity, the Omicron spike protein shows a lower potency for dimerization of human TLR4/MD-2, due to amino acid substitutions and slight conformational shifts in the FP/HR1 region of the S2 subunit [99]. TLR4/MD-2 is preferably expressed on innate immune cells, such as macrophages and dendritic cells, and—to some lower extent—endothelial cells [11]. These cells may be largely responsible for excessive immune stimulation, including cytokine storm found in severe and critical cases of COVID-19 [11,44,101]. Notably, this TLR-mediated immune hyperstimulation is to some extent uncoupled from virus replication, which is largely occurring in ACE2-expressing epithelial cells (such as lung alveolar cells, cells of the intestinal tract etc.) and which depends on proteolytic cleavage of the spike protein by furin. Viral uptake is accelerated by TMPRSS2 co-expression enabling uptake via the highly efficient plasma membrane route. In contrast, in TMPRSS2 negative innate immune cells, such as macrophages, the virus is taken up preferably by clathrin-coated pits and directed into endosomes, where the proteolytic cleavage is taken over by endosomal enzymes, such as cathepsin L [24–26].

Notably, TLR4/MD-2 is the only TLR which is found on the cell surface as well as in the endosomes. Binding and dimerization of TLR4/MD-2 by the SARS-CoV-2 spike protein leads to downstream signaling, resulting in NF-κB pathway activation and cytokine release of TNFα, IL-1, IL-6, IL-8, IL-12, together with HIF-1α, chemokine release, upregulation of adhesion and cell activation molecules, and enhanced expression of furin [11,102,103]. The lower potency for TLR4/MD-2 dimerization of Omicron spike protein compared to wild-type virus or earlier VoCs will—despite comparable binding affinity—results in a lower TLR4-NF-κB—signaling and lower cytokine release (Figure 8).

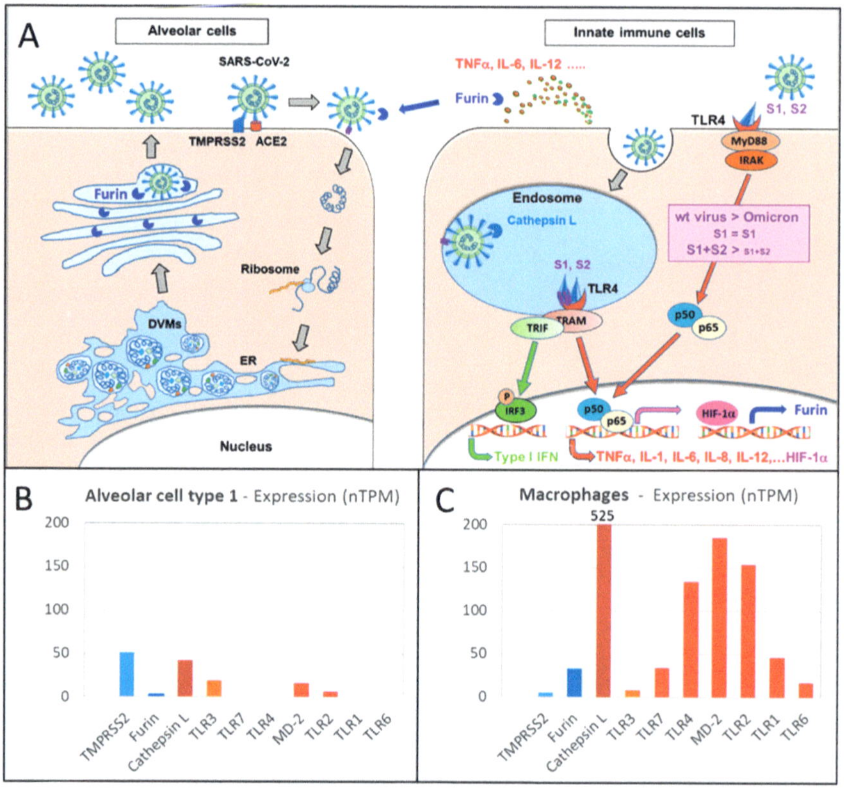

Figure 8. (**A**) The SARS-CoV-2 binds to ACE2 followed by proteolytic cleavage by TMPRRS2 and fusion with the host cell membrane followed by uptake of viral RNA into the host cell. Transcription of the viral RNA and translation of viral non-structural and structural proteins occurs in double-membrane vesicles (DMVs), which are formed after remodeling of the endo-plasmatic reticulum (ER). The newly produced viral components assemble into virus particles which leave the cells via the Golgi apparatus where the spike protein undergoes also proteolytic cleavage by furin. This process preferably occurs in virus producer cells with high TMPRSS2 expression, e.g., alveolar cells (**left side**). Alternatively, the virus can be taken up via clathrin coated pits into endosomes, followed by proteolytic cleavage by cathepsin L. The endosomal uptake is predominant in TMPRSS2-negative, but cathepsin L-rich cells, such as innate immune cells and endothelial cells (**right side**). The spike protein of the SARS-CoV-2 acts as a TLR4 agonist, resulting in dimerization of the TLR4/MD-2 complex triggering downstream signaling, e.g., activation of the NF-κB (p50/p65) pathway. The activation of NF-κB pathway triggers HIF-1α activation and expression of cytokines, such as TNFα, IL-1, IL-6, IL-12. Notably, HIF-1α and IL-12 have been found to enhance furin expression. In contrast to the highly effective TLR4/MD-2 dimerization and activation by the wild-type SARS-CoV-2 spike protein trimer, the amino acid substitutions in the Omicron spike protein decrease the potency of the spike protein for TLR4/MD-2 dimerization, leading to less NF-κB signaling and lower expression of cytokines. (**B**,**C**) The relative expression of TLR4 differs depending on the cell type, with high expression of TLR4/MD-2 on innate immune cells, (**C**) but only low levels expressed on alveolar lung cells (**B**). Figure adapted from [11].

3. Discussion

Omicron variants have successfully replaced all other variants worldwide. The emergence of the Omicron variants has been associated with a dramatically increased infectivity and broad evasion of humoral response induced by vaccination or earlier infections due to multiple mutations in the spike protein [3–8].

In parallel a shift in cellular tropism with a lower virus replication in the lungs but higher replication in the upper respiratory system [8,9], and a lower pathogenicity with less involvement of the lower respiratory system, have been found as general observations in clinical and preclinical studies. The molecular mechanisms responsible for these reciprocal changes in cellular tropism were until now not completely defined, and a lower activation of Toll-like receptors (TLR) for Omicron has been hypothesized based on changes in the charge distribution due to amino acid substitutions [11]. TLR-2 and TLR4 activation have been shown to play an essential role in SARS-CoV-2 induced pathogenicity [12–18], and binding of SARS-CoV-2 spike protein to TLR4 has been demonstrated for the wild-type virus [19] and recently also for the Omicron spike protein [104]. Notably, no dimerization of the TLR4/MD-2 have been demonstrated so far.

In the present in silico study, we show that (a) the spike protein trimer of SARS-CoV-2 can induce not only binding but also dimerization of human TLR4/MD-2, dependent on the mode of binding, and (b) while the binding affinities to TLR4 seems to be comparable to that of the SARS-CoV-2 wild-type, the Omicron spike protein was found to have a lower potency to induce dimerization of TLR4/MD-2. The in silico data from the present study indicate that there is a higher probability for multiple binding configurations, which induce dimerization of the TLR4/MD-2 complex by the wild-type spike protein compared to Omicron spike protein. This difference can be expected to lead to the difference in the amplitude of general TLR4 activation in the relevant cells, preferably innate immune cells, and endothelial cells. The dimerization of TLR4 activating the relevant downstream signaling pathways will lead to excessive gene expression for a broad range of pro-inflammatory cytokines and chemokines, adhesion molecules, and acute phase proteins. Furthermore, the highly activated NF-κB pathway, directly or via HIF-1α activation or via cytokine release (such as IL-12) is expected to stimulate furin expression in the cells. In this line, viral infections, cancer, hypoxia, HIF-1α and cytokines (e.g., IL-12) have been found to significantly enhance furin expression [11,102,103].

In silico modelling of TLR4 dimerization induced by other VoC, including Alpha, Gamma, and Delta variants, showed a decrease of the number of dimerization modes after the Alpha variant, with 6, 7, 5, 4, and 3 modes of TLR4/MD2 dimerization for the wild-type, Alpha, Gamma, Delta, and Omicron VoCs, respectively (see Supplementary data). These data correlate well with reported cytokine release patterns compared in a recent study. Korobova et al. have shown a narrowing cytokine pattern during evolution of the SARS-CoV-2 from the original Wuhan strain via the various VoCs, e.g., Alpha- Delta- Omicron. The study compared the cytokine profiles of patients with different variants of SARS-CoV-2 and showed that hypercytokinemia and cytokine storm become less threatening with the emergence of new mutations in the viral genome, being already visible with the Delta variant and most pronounced for the Omicron variant [82]. A similar tendency of decreasing pathogenicity has been shown by Barh et al. showing decreasing pathogenicity (correlating with decreasing cytokine release) with virus evolution, with Wuhan strain > Gamma > Delta > Omicron [83].

The TLR binding and dimerization pattern under virus evolution from wild-type–Alpha–Gamma–Delta–Omicron may slightly differ for different TLRs, as shown in the present study for TLR2/TLR1 and TLR2/TLR6 heterodimers, which may contribute to a not completely homogenous tendency among different cytokines [82]. However, the multifold higher expression of TLR4/MD-2 compared to the other TLR species (see Figure 8B,C, and Ref. [105]) indicate that activation of TLR4/MD-2 can be expected to largely define the resulting net TLR activation.

It is the opinion of the author that, beside the world-wide vaccination campaign, the emergence of the Omicron variants with their higher infectivity and immune evasion potential outcompeting all earlier VoCs, in combination with the shift in cellular tropism and lower pathogenicity, is one of the two main reasons for successful overcoming of the pandemic. Understanding the underlaying molecular mechanisms will be essential not only for understanding SARS-CoV-2 and COVID-19 but also in preparation for controlling upcoming viral diseases in the future.

The correlation between TLR4 dimerization and activation with pathogenicity of the SARS-CoV-2 variants indicate TLR4 as a potential therapeutic target for treatment of COVID-19. While so far there are no FDA approved TLR4 antagonists available, intensive research [106] on TLR4 as a therapeutic target is ongoing, with emphasis on respiratory and neurological complications of SARS-CoV-2 [107], also with respect to accumulating data for long-term nervous system consequences after COVID-19 [108,109]. Beside various TLR4 antagonists which inhibit binding to or dimerization of TLR4, also inhibitors of the downstream signaling pathways, including inhibitors of the NF-κB pathway, are promising targets for therapeutic intervention [50,51].

Beside chemical TLR4 antagonists, antibody-based therapies targeting TLR4/MD-2 binding and/or dimerization may be highly interesting considering the high binding affinity and specificity, and the good safety profile shown for humanized monoclonal antibodies [110], or recent further developments, such as nanobodies [111]. Still, antibody-based therapies will face similar challenges encountered by previous agents, such as a continued evolution of SARS-CoV-2 and emergence of new escape mutations, in particular for the ACE2 binding RBD of the spike protein, as a major force for generation of new virus variants. Strategies to potentially mitigate this may include alternative target sites of the virus, preferably on the highly exposed spike protein. In this respect epitopes and regions involved in TLR4 binding and/or dimerization may be of particular high interest. While not interfering with the main receptor of SARS-CoV-2 for cellular uptake (thus not exposed to such an extremely high selection pressure as the ACE2 binding sites), these therapies could ameliorate the hyperactivation of TLR4 associated with the cytokine and chemokine release syndrome. In this direction, very encouraging data have been reported from a Phase 2 clinical study with an experimental monoclonal antibody (EB05, Edesa Biotech, Inc. Markham, Ontario L3R 5H6, Canada) in hospitalized COVID-19 patients. A 68% reduction of mortality risk compared to placebo plus standard of care (including dexamethasone, IL-6 inhibitors) at 28 days was reported for this monoclonal antibody which targets TLR4 and inhibits TLR4 signaling [112].

Limitations of the present study regard the focus on a limited panel of TLRs which are relevant for recognizing the spike protein of SARS-CoV-2. Beside spike protein recognition, recognition of the nucleic acid components of the virus by TLR3,7,8,9 and of other structural and non-structural viral proteins can be expected to have an impact on the overall TLR activation by SARS-CoV-2. Furthermore, different TLRs may make different contributions to anti-viral immune activation and to pathophysiological hyperstimulation as two sides of TLR activation [113]. Another limitation of the present study is that all modelling is based on in silico analyses using various data bases (PDB, HDOCK, Human Protein Atlas) in combination with reported experimental data from the literature, without showing direct experimental evidence. Further studies in cell culture and animal models as well as clinical data will be necessary to confirm or further develop the hypotheses raised in the present study.

4. Materials and Methods

The sequences and 3D models of the spike protein trimer from wild-type SARS-CoV-2 (PDB 6ZGG) [97], Omicron variant (PDB 7TL9) [99], Alpha variant (PDB 7LWU) [114], Gamma variant (PDB 7M8K) [115], Delta variant (PDB 2W92) [116], and of the human TLR4/MD-2 complex (PDB 3FXI) [98], TLR2/TLR1 heterodimer (PDB 2Z7X) [117], TLR2/TLR6 heterodimer (PDB 2A79) [118], were derived from the RCSB PDB Protein Data Bank (https://www.rcsb.org/, accessed on December 2023 and January–May 2024). For all spike protein models only 1-RBD-up conformation models, i.e., with monomers in closed ("down") AND in open ("up") state, were chosen to cover the broadest range of configurations of the spike protein trimer.

In silico docking studies were performed using HDOCK docking software (http://hdock.phys.hust.edu.cn/) [100]. Alignments were performed using the RCSB PDB Protein Data Bank (https://www.rcsb.org/)

The expression profiles for relevant receptors on different target cells normalized to nTPM (i.e., Transcripts per million protein coding genes) have been derived from the Human Protein Atlas (https://swww.proteinatlas.org/) [119].

5. Conclusions

The recently appearing Omicron variant shows surprisingly reciprocal changes in cellular tropism of the upper and lower respiratory system, together with lower pathogenicity which cannot be explained simply by the changed binding to the ACE2 receptor or immune escape. Here, we present in silico analyses as part of a mechanistic model, which indicate that amino acids substitutions in the S2 subunit of the spike protein unique to the Omicron variants, as compared to previous VoCs, may disturb the recognition by innate Pattern-recognition receptors (PRRs), such as TLR4/MD-2, resulting in lower activation of downstream signaling pathways, and lower immune hyperactivation.

Supplementary Materials: The following supporting information can be downloaded at: https://www.mdpi.com/article/10.3390/ijms25105451/s1.

Funding: This research received no external funding.

Institutional Review Board Statement: Not applicable.

Informed Consent Statement: Not applicable.

Data Availability Statement: Data is contained within the article and Supplementary Materials.

Conflicts of Interest: Author Ralf Kircheis was employed by the company Syntacoll GmbH. The author declared that the research was conducted in the absence of any commercial or financial relationships that could be construed as a potential conflict of interest.

References

1. Viana, R.; Moyo, S.; Amoako, D.G.; Tegally, H.; Scheepers, C.; Althaus, C.L.; Anyaneji, U.J.; Bester, P.A.; Boni, M.F.; Chand, M.; et al. Rapid epidemic expansion of the SARS-CoV-2 Omicron variant in southern Africa. *Nature* **2022**, *603*, 679–686. [CrossRef]
2. Wolter, N.; Jassat, W.; Walaza, S.; Welch, R.; Moultrie, H.; Groome, M.; Amoako, D.G.; Everatt, J.; Bhiman, J.N.; Scheepers, C.; et al. Early assessment of the clinical severity of the SARS-CoV-2 omicron variant in South Africa: A data linkage study. *Lancet* **2022**, *399*, 437–446. [CrossRef]
3. Dejnirattisai, W.; Huo, J.; Zhou, D.; Zahradnik, J.; Supasa, P.; Liu, C.; Duyvesteyn, H.M.E.; Ginn, H.M.; Mentzer, A.J.; Tuekprakhon, A.; et al. SARS-CoV-2 Omicron-B.1.1.529 leads to widespread escape from neutralizing antibody responses. *Cell* **2022**, *185*, 467–484.e15. [CrossRef]
4. Meng, B.; Abdullahi, A.; Ferreira, I.; Goonawardane, N.; Saito, A.; Kimura, I.; Yamasoba, D.; Gerber, P.P.; Fatihi, S.; Rathore, S.; et al. Altered TMPRSS2 usage by SARS-CoV-2 Omicron impacts infectivity and fusogenicity. *Nature* **2022**, *603*, 706–714. [CrossRef]
5. Kannan, S.R.; Spratt, A.N.; Sharma, K.; Chand, H.S.; Byrareddy, S.N.; Singh, K. Omicron SARS-CoV-2 variant: Unique features and their impact on pre-existing antibodies. *J. Autoimmun.* **2022**, *126*, 102779. [CrossRef] [PubMed]
6. Ye, G.; Liu, B.; Li, F. Cryo-EM structure of a SARS-CoV-2 omicron spike protein ectodomain. *Nat. Commun.* **2022**, *13*, 1214. [CrossRef]
7. Shah, M.; Woo, H.G. Omicron: A Heavily Mutated SARS-CoV-2 Variant Exhibits Stronger Binding to ACE2 and Potently Escapes Approved COVID-19 Therapeutic Antibodies. *Front. Immunol.* **2021**, *12*, 830527. [CrossRef]
8. Halfmann, P.J.; Iida, S.; Iwatsuki-Horimoto, K.; Maemura, T.; Kiso, M.; Scheaffer, S.M.; Darling, T.L.; Joshi, A.; Loeber, S.; Singh, G.; et al. SARS-CoV-2 Omicron virus causes attenuated disease in mice and hamsters. *Nature* **2022**, *603*, 687–692. [CrossRef]
9. Willett, B.J.; Grove, J.; MacLean, O.A.; Wilkie, C.; Logan, N.; Lorenzo, G.D.; Furnon, W.; Scott, S.; Manali, M.; Szemiel, A.; et al. The hyper-transmissible SARS-CoV-2 Omicron variant exhibits significant antigenic change, vaccine escape and a switch in cell entry mechanism. *medRxiv* **2022**. [CrossRef]
10. Aiello, T.F.; Puerta-Alcalde, P.; Chumbita, M.; Monzó, P.; Lopera, C.; Hurtado, J.C.; Meira, F.; Mosquera, M.; Santos, M.; Fernandez-Pittol, M.; et al. Infection with the Omicron variant of SARS-CoV-2 is associated with less severe disease in hospitalized patients with COVID-19. *J. Infect.* **2022**, *85*, e152–e154. [CrossRef] [PubMed]
11. Kircheis, R.; Planz, O. Could a Lower Toll-like Receptor (TLR) and NF-κB Activation Due to a Changed Charge Distribution in the Spike Protein Be the Reason for the Lower Pathogenicity of Omicron? *Int. J. Mol. Sci.* **2022**, *23*, 5966. [CrossRef] [PubMed]

12. Zheng, M.; Karki, R.; Williams, E.P.; Yang, D.; Fitzpatrick, E.; Vogel, P.; Jonsson, C.B.; Kanneganti, T.D. TLR2 senses the SARS-CoV-2 envelope protein to produce inflammatory cytokines. *Nat. Immunol.* **2021**, *22*, 829–838. [CrossRef] [PubMed]
13. Planès, R.; Bert, J.-B.; Tairi, S.; Benmohamed, L.; Bahraoui, E. SARS-CoV-2 Envelope protein (E) binds and activates TLR2: A novel target for COVID-19 interventions. *bioRxiv* **2021**. [CrossRef]
14. Aboudounya, M.M.; Heads, R.J. COVID-19 and Toll-Like Receptor 4 (TLR4): SARS-CoV-2 May Bind and Activate TLR4 to Increase ACE2 Expression, Facilitating Entry and Causing Hyperinflammation. *Mediators Inflamm.* **2021**, *2021*, 8874339. [CrossRef] [PubMed]
15. Mogensen, T.H.; Paludan, S.R. Reading the viral signature by Toll-like receptors and other pattern recognition receptors. *J. Mol. Med.* **2005**, *83*, 180–192. [CrossRef] [PubMed]
16. Sohn, K.M.; Lee, S.G.; Kim, H.J.; Cheon, S.; Jeong, H.; Lee, J.; Kim, I.S.; Silwal, P.; Kim, Y.J.; Paik, S.; et al. COVID-19 Patients Upregulate Toll-like Receptor 4-mediated Inflammatory Signaling That Mimics Bacterial Sepsis. *J. Korean Med. Sci.* **2020**, *35*, e343. [CrossRef] [PubMed] [PubMed Central]
17. Zhao, Y.; Kuang, M.; Li, J.; Zhu, L.; Jia, Z.; Guo, X.; Hu, Y.; Kong, J.; Yin, H.; Wang, X.; et al. SARS-CoV-2 spike protein interacts with and activates TLR41. *Cell Res.* **2021**, *31*, 818–820, Erratum in *Cell Res.* **2021**, *31*, 825. [CrossRef] [PubMed] [PubMed Central]
18. Aboudounya, M.M.; Holt, M.R.; Heads, R.J. SARS-CoV-2 Spike S1 glycoprotein is a TLR4 agonist, upregulates ACE2 expression and induces pro-inflammatory M1 macrophage polarisation. *bioRxiv* **2021**. [CrossRef]
19. Choudhury, A.; Mukherjee, S. In silico studies on the comparative characterization of the interactions of SARS-CoV-2 spike glycoprotein with ACE-2 receptor homologs and human TLRs. *J. Med. Virol.* **2020**, *92*, 2105–2113. [CrossRef]
20. Park, B.S.; Lee, J.O. Recognition of lipopolysaccharide pattern by TLR4 complexes. *Exp. Mol. Med.* **2013**, *45*, e66. [CrossRef]
21. Hoffmann, M.; Kleine-Weber, H.; Schroeder, S.; Kruger, N.; Herrler, T.; Erichsen, S.; Schiergens, T.S.; Herrler, G.; Wu, N.H.; Nitsche, A.; et al. SARS-CoV-2 Cell Entry Depends on ACE2 and TMPRSS2 and Is Blocked by a Clinically Proven Protease Inhibitor. *Cell* **2020**, *181*, 271–280.e8. [CrossRef]
22. Eslami, N.; Aghbash, P.S.; Shamekh, A.; Entezari-Maleki, T.; Nahand, J.S.; Sales, A.J.; Baghi, H.B. SARS-CoV-2: Receptor and Co-receptor Tropism Probability. *Curr. Microbiol.* **2022**, *79*, 133. [CrossRef]
23. Cantuti-Castelvetri, L.; Ojha, R.; Pedro, L.D.; Djannatian, M.; Franz, J.; Kuivanen, S.; van der Meer, F.; Kallio, K.; Kaya, T.; Anastasina, M.; et al. Neuropilin-1 facilitates SARS-CoV-2 cell entry and infectivity. *Science* **2020**, *370*, 856–860. [CrossRef] [PubMed]
24. Matsuyama, S.; Ujike, M.; Morikawa, S.; Tashiro, M.; Taguchi, F. Protease-mediated enhancement of severe acute respiratory syndrome coronavirus infection. *Proc. Natl. Acad. Sci. USA* **2005**, *102*, 12543–12547. [CrossRef] [PubMed]
25. Simmons, G.; Gosalia, D.N.; Rennekamp, A.J.; Reeves, J.D.; Diamond, S.L.; Bates, P. Inhibitors of cathepsin L prevent severe acute respiratory syndrome coronavirus entry. *Proc. Natl. Acad. Sci. USA* **2005**, *102*, 11876–11881. [CrossRef]
26. Belouzard, S.; Millet, J.K.; Licitra, B.N.; Whittaker, G.R. Mechanisms of coronavirus cell entry mediated by the viral spike protein. *Viruses* **2012**, *4*, 1011–1033. [CrossRef]
27. Shah, M.; Woo, H.G. Molecular Perspectives of SARS-CoV-2: Pathology, Immune Evasion, and Therapeutic Interventions. *Mol. Cells* **2021**, *44*, 408–421. [CrossRef] [PubMed]
28. Xia, S.; Zhu, Y.; Liu, M.; Lan, Q.; Xu, W.; Wu, Y.; Ying, T.; Liu, S.; Shi, Z.; Jiang, S.; et al. Fusion mechanism of 2019-nCoV and fusion inhibitors targeting HR1 domain in spike protein. *Cell. Mol. Immunol.* **2020**, *17*, 765–767. [CrossRef]
29. Varga, Z.; Flammer, A.J.; Steiger, P.; Haberecker, M.; Andermatt, R.; Zinkernagel, A.S.; Mehra, M.R.; Schuepbach, R.A.; Ruschitzka, F.; Moch, H. Endothelial cell infection and endotheliitis in COVID-19. *Lancet* **2020**, *395*, 1417–1418. [CrossRef]
30. Ackermann, M.; Verleden, S.E.; Kuehnel, M.; Haverich, A.; Welte, T.; Laenger, F.; Vanstapel, A.; Werlein, C.; Stark, H.; Tzankov, A.; et al. Pulmonary Vascular Endothelialitis, Thrombosis, and Angiogenesis in COVID-19. *N. Engl. J. Med.* **2020**, *383*, 120–128. [CrossRef]
31. Abdelrahman, Z.; Li, M.; Wang, X. Comparative Review of SARS-CoV-2, SARS-CoV, MERS-CoV, and Influenza A Respiratory Viruses. *Front. Immunol.* **2020**, *11*, 552909. [CrossRef] [PubMed]
32. Huang, C.; Wang, Y.; Li, X.; Ren, L.; Zhao, J.; Hu, Y.; Zhang, L.; Fan, G.; Xu, J.; Gu, X.; et al. Clinical features of patients infected with 2019 novel coronavirus in Wuhan, China. *Lancet* **2020**, *395*, 497–506. [CrossRef] [PubMed]
33. Xu, Z.; Shi, L.; Wang, Y.; Zhang, J.; Huang, L.; Zhang, C.; Liu, S.; Zhao, P.; Liu, H.; Zhu, L.; et al. Pathological findings of COVID-19 associated with acute respiratory distress syndrome. *Lancet Respir. Med.* **2020**, *8*, 420–422. [CrossRef] [PubMed]
34. Wang, D.; Hu, B.; Hu, C.; Zhu, F.; Liu, X.; Zhang, J.; Wang, B.; Xiang, H.; Cheng, Z.; Xiong, Y.; et al. Clinical Characteristics of 138 Hospitalized Patients with 2019 Novel Coronavirus-Infected Pneumonia in Wuhan, China. *JAMA* **2020**, *323*, 1061–1069. [CrossRef] [PubMed]
35. Zheng, Z.; Peng, F.; Xu, B.; Zhao, J.; Liu, H.; Peng, J.; Li, Q.; Jiang, C.; Zhou, Y.; Liu, S.; et al. Risk factors of critical & mortal COVID-19 cases: A systematic literature review and meta-analysis. *J. Infect.* **2020**, *81*, e16–e25. [CrossRef] [PubMed]
36. Biswas, M.; Rahaman, S.; Biswas, T.K.; Haque, Z.; Ibrahim, B. Association of Sex, Age, and Comorbidities with Mortality in COVID-19 Patients: A Systematic Review and Meta-Analysis. *Intervirology* **2020**, *64*, 36–47. [CrossRef]
37. O'Driscoll, M.; Ribeiro Dos Santos, G.; Wang, L.; Cummings, D.A.T.; Azman, A.S.; Paireau, J.; Fontanet, A.; Cauchemez, S.; Salje, H. Age-specific mortality and immunity patterns of SARS-CoV-2. *Nature* **2021**, *590*, 140–145. [CrossRef] [PubMed]

38. Carsana, L.; Sonzogni, A.; Nasr, A.; Rossi, R.S.; Pellegrinelli, A.; Zerbi, P.; Rech, R.; Colombo, R.; Antinori, S.; Corbellino, M.; et al. Pulmonary post-mortem findings in a series of COVID-19 cases from northern Italy: A two-centre descriptive study. *Lancet Infect. Dis.* **2020**, *20*, 1135–1140. [CrossRef] [PubMed]
39. Acharya, D.; Liu, G.; Gack, M.U. Dysregulation of type I interferon responses in COVID-19. *Nat. Rev. Immunol.* **2020**, *20*, 397–398. [CrossRef]
40. Tay, M.Z.; Poh, C.M.; Rénia, L.; MacAry, P.A.; Ng, L.F.P. The trinity of COVID-19: Immunity, inflammation and intervention. *Nat. Rev. Immunol.* **2020**, *20*, 363–374. [CrossRef]
41. Schett, G.; Sticherling, M.; Neurath, M.F. COVID-19: Risk for cytokine targeting in chronic inflammatory diseases? *Nat. Rev. Immunol.* **2020**, *20*, 271–272. [CrossRef] [PubMed]
42. Moore, B.J.B.; June, C.H. Cytokine release syndrome in severe COVID-19. *Science* **2020**, *368*, 473–474. [CrossRef] [PubMed]
43. Huang, Y.; Tu, M.; Wang, S.; Chen, S.; Zhou, W.; Chen, D.; Zhou, L.; Wang, M.; Zhao, Y.; Zeng, W.; et al. Clinical characteristics of laboratory confirmed positive cases of SARS-CoV-2 infection in Wuhan, China: A retrospective single center analysis. *Travel Med. Infect. Dis.* **2020**, *36*, 101606. [CrossRef] [PubMed]
44. Chua, R.L.; Lukassen, S.; Trump, S.; Hennig, B.P.; Wendisch, D.; Pott, F.; Debnath, O.; Thürmann, L.; Kurth, F.; Völker, M.T.; et al. COVID-19 severity correlates with airway epithelium–immune cell interactions identified by single-cell analysis. *Nat. Biotechnol.* **2020**, *38*, 970–979. [CrossRef] [PubMed]
45. Huang, J.; Hume, A.J.; Abo, K.M.; Werder, R.B.; Villacorta-Martin, C.; Alysandratos, K.D.; Beermann, M.L.; Simone-Roach, C.; Lindstrom-Vautrin, J.; Olejnik, J.; et al. SARS-CoV-2 Infection of Pluripotent Stem Cell-Derived Human Lung Alveolar Type 2 Cells Elicits a Rapid Epithelial-Intrinsic Inflammatory Response. *Cell Stem Cell* **2020**, *27*, 962–973.e7. [CrossRef] [PubMed]
46. Hojyo, S.; Uchida, M.; Tanaka, K.; Hasebe, R.; Tanaka, Y.; Murakami, M.; Hirano, T. How COVID-19 induces cytokine storm with high mortality. *Inflamm. Regen.* **2020**, *40*, 37. [CrossRef] [PubMed]
47. Hirano, T.; Murakami, M. COVID-19: A New Virus, but a Familiar Receptor and Cytokine Release Syndrome. *Immunity* **2020**, *52*, 731–733. [CrossRef] [PubMed]
48. Hong, K.S.; Ahn, J.H.; Jang, J.G.; Lee, J.H.; Kim, H.N.; Kim, D.; Lee, W. GSK-LSD1, an LSD1 inhibitor, quashes SARS-CoV-2-triggered cytokine release syndrome in-vitro. *Signal Transduct. Target. Ther.* **2020**, *5*, 267. [CrossRef]
49. McAleavy, M.; Zhang, Q.; Ehmann, P.J.; Xu, J.; Wipperman, M.F.; Ajithdoss, D.; Pan, L.; Wakai, M.; Simonson, R.; Gadi, A.; et al. The Activin/FLRG Pathway Associates with Poor COVID-19 Outcomes in Hospitalized Patients. *Mol. Cell. Biol.* **2022**, *42*, e0046721. [CrossRef]
50. Hariharan, A.; Hakeem, A.R.; Radhakrishnan, S.; Reddy, M.S.; Rela, M. The Role and Therapeutic Potential of NF-kappa-B Pathway in Severe COVID-19 Patients. *Inflammopharmacology* **2021**, *29*, 91–100. [CrossRef]
51. Kircheis, R.; Haasbach, E.; Lueftenegger, D.; Heyken, W.T.; Ocker, M.; Planz, O. NF-κB Pathway as a Potential Target for Treatment of Critical Stage COVID-19 Patients. *Front. Immunol.* **2020**, *11*, 598444. [CrossRef] [PubMed]
52. Neufeldt, C.J.; Cerikan, B.; Cortese, M.; Frankish, J.; Lee, J.Y.; Plociennikowska, A.; Heigwer, F.; Prasad, V.; Joecks, S.; Burkart, S.S.; et al. SARS-CoV-2 infection induces a pro-inflammatory cytokine response through cGAS-STING and NF-κB. *Commun. Biol.* **2022**, *5*, 45. [CrossRef] [PubMed]
53. Lo, M.W.; Kemper, C.; Woodruff, T.M. COVID-19: Complement, Coagulation, and Collateral Damage. *J. Immunol.* **2020**, *205*, 1488–1495. [CrossRef] [PubMed]
54. Holter, J.C.; Pischke, S.E.; de Boer, E.; Lind, A.; Jenum, S.; Holten, A.R.; Tonby, K.; Barratt-Due, A.; Sokolova, M.; Schjalm, C.; et al. Systemic complement activation is associated with respiratory failure in COVID-19 hospitalized patients. *Proc. Natl. Acad. Sci. USA* **2020**, *117*, 25018–25025. [CrossRef] [PubMed]
55. Polycarpou, A.; Howard, M.; Farrar, C.A.; Greenlaw, R.; Fanelli, G.; Wallis, R.; Klavinskis, L.S.; Sacks, S. Rationale for targeting complement in COVID-19. *EMBO Mol. Med.* **2020**, *12*, e12642. [CrossRef] [PubMed]
56. Yu, J.; Yuan, X.; Chen, H.; Chaturvedi, S.; Braunstein, E.M.; Brodsky, R.A. Direct activation of the alternative complement pathway by SARS-CoV-2 spike proteins is blocked by factor D inhibition. *Blood* **2020**, *136*, 2080–2089. [CrossRef] [PubMed]
57. Fletcher-Sandersjöö, A.; Bellander, B.-M. Is COVID-19 associated thrombosis caused by overactivation of the complement cascade? A literature review. *Thromb. Res.* **2020**, *194*, 36–41. [CrossRef] [PubMed]
58. Mussbacher, M.; Salzmann, M.; Brostjan, C.; Hoesel, B.; Schoergenhofer, C.; Datler, H.; Hohensinner, P.; Basilio, J.; Petzelbauer, P.; Assinger, A.; et al. Cell Type-Specific Roles of NF-κB Linking Inflammation and Thrombosis. *Front. Immunol.* **2019**, *10*, 85. [CrossRef] [PubMed]
59. McCarthy, C.G.; Wilczynski, S.; Wenceslau, C.F.; Webb, R.C. A new storm on the horizon in COVID-19: Bradykinin-induced vascular complications. *Vasc. Pharmacol.* **2021**, *137*, 106826. [CrossRef]
60. Leppkes, M.; Knopf, J.; Naschberger, E.; Lindemann, A.; Singh, J.; Herrmann, I.; Stürzl, M.; Staats, L.; Mahajan, A.; Schauer, C.; et al. Vascular occlusion by neutrophil extracellular traps in COVID-19. *EBioMedicine* **2020**, *58*, 102925. [CrossRef]
61. Skendros, P.; Mitsios, A.; Chrysanthopoulou, A.; Mastellos, D.C.; Metallidis, S.; Rafailidis, P.; Ntinopoulou, M.; Sertaridou, E.; Tsironidou, V.; Tsigalou, C.; et al. Complement and tissue factor–enriched neutrophil extracellular traps are key drivers in COVID-19 immunothrombosis. *J. Clin. Investig.* **2020**, *130*, 6151–6157. [CrossRef] [PubMed]
62. Kvietys, P.R.; Fakhoury, H.M.A.; Kadan, S.; Yaqinuddin, A.; Al-Mutairy, E.; Al-Kattan, K. COVID-19: Lung-Centric Immunothrombosis. *Front. Cell. Infect. Microbiol.* **2021**, *11*, 679878. [CrossRef] [PubMed]

63. Gando, S.; Wada, T. Pathomechanisms Underlying Hypoxemia in Two COVID-19-Associated Acute Respiratory Distress Syndrome Phenotypes: Insights from Thrombosis and Hemostasis. *Shock* **2022**, *57*, 1–6. [CrossRef] [PubMed]
64. Gu, S.X.; Tyagi, T.; Jain, K.; Gu, V.W.; Lee, S.H.; Hwa, J.M.; Kwan, J.M.; Krause, D.S.; Lee, A.I.; Halene, S.; et al. Thrombocytopathy and endotheliopathy: Crucial contributors to COVID-19 thromboinflammation. *Nat. Rev. Cardiol.* **2021**, *18*, 194–209. [CrossRef] [PubMed]
65. Menter, T.; Haslbauer, J.D.; Nienhold, R.; Savic, S.; Hopfer, H.; Deigendesch, N.; Frank, S.; Turek, D.; Willi, N.; Pargger, H.; et al. Post-mortem examination of COVID19 patients reveals diffuse alveolar damage with severe capillary congestion and variegated findings of lungs and other organs suggesting vascular dysfunction. *Histopathology* **2020**, *77*, 198–209. [CrossRef] [PubMed]
66. Hulo, C.; de Castro, E.; Masson, P.; Bougueleret, L.; Bairoch, A.; Xenarios, I.; Le Mercier, P. ViralZone: A knowledge resource to understand virus diversity. *Nucleic Acids Res.* **2010**, *39*, D576–D582. [CrossRef] [PubMed]
67. Mannar, D.; Saville, J.W.; Zhu, X.; Srivastava, S.S.; Berezuk, A.M.; Tuttle, K.S.; Marquez, A.C.; Sekirov, I.; Subramaniam, S. SARS-CoV-2 Omicron variant: Antibody evasion and cryo-EM structure of spike protein-ACE2 complex. *Science* **2022**, *375*, 760–764. [CrossRef]
68. Barton, M.I.; MacGowan, S.A.; Kutuzov, M.A.; Dushek, O.; Barton, G.J.; van der Merwe, P.A. Effects of common mutations in the SARS-CoV-2 Spike RBD and its ligand, the human ACE2 receptor on binding affinity and kinetics. *Elife* **2021**, *10*, e70658. [CrossRef] [PubMed]
69. McCallum, M.; Czudnochowski, N.; Rosen, L.E.; Zepeda, S.K.; Bowen, J.E.; Walls, A.C.; Hauser, K.; Joshi, A.; Stewart, C.; Dillen, J.R.; et al. Structural basis of SARS-CoV-2 Omicron immune evasion and receptor engagement. *Science* **2022**, *375*, 864–868. [CrossRef]
70. Yuan, S.; Ye, Z.W.; Liang, R.; Tang, K.; Zhang, A.J.; Lu, G.; Ong, C.P.; Man Poon, V.K.; Chan, C.C.; Mok, B.W.; et al. Pathogenicity, transmissibility, and fitness of SARS-CoV-2 Omicron in Syrian hamsters. *Science* **2022**, *377*, 428–433. [CrossRef]
71. Thiruvengadam, R.; Rizvi, Z.A.; Raghavan, S.; Murugesan, D.R.; Gosain, M.; Dandotiya, J.; Ayushi; Samal, S.; Pandey, A.K.; Wadhwa, N.; et al. Clinical and experimental evidence suggest omicron variant of SARS-CoV-2 is inherently less pathogenic than delta variant independent of previous immunity. *Eur. J. Med. Res.* **2023**, *28*, 421. [CrossRef] [PubMed]
72. Ameratunga, R.; Leung, E.; Woon, S.-T.; Chan, L.; Steele, R.; Lehnert, K.; Longhurst, H. SARS-CoV-2 Omicron: Light at the End of the Long Pandemic Tunnel or Another False Dawn for Immunodeficient Patients? *J. Allergy Clin. Immunol. Pr.* **2022**, *10*, 2267–2273. [CrossRef]
73. Flisiak, R.; Rzymski, P.; Zarębska-Michaluk, D.; Ciechanowski, P.; Dobrowolska, K.; Rogalska, M.; Jaroszewicz, J.; Szymanek-Pasternak, A.; Rorat, M.; Kozielewicz, D.; et al. Variability in the Clinical Course of COVID-19 in a Retrospective Analysis of a Large Real-World Database. *Viruses* **2023**, *15*, 149. [CrossRef]
74. Sigal, A. Milder disease with Omicron: Is it the virus or the pre-existing immunity? *Nat. Rev. Immunol.* **2022**, *22*, 69–71. [CrossRef] [PubMed]
75. Menni, C.; Valdes, A.M.; Polidori, L.; Antonelli, M.; Penamakuri, S.; Nogal, A.; Louca, P.; May, A.; Figueiredo, J.C.; Hu, C.; et al. Symptom prevalence, duration, and risk of hospital admission in individuals infected with SARS-CoV-2 during periods of omicron and delta variant dominance: A prospective observational study from the ZOE COVID Study. *Lancet* **2022**, *399*, 1618–1624. [CrossRef]
76. Tamura, T.; Yamasoba, D.; Oda, Y.; Ito, J.; Kamasaki, T.; Nao, N.; Hashimoto, R.; Fujioka, Y.; Suzuki, R.; Wang, L.; et al. Comparative pathogenicity of SARS-CoV-2 Omicron subvariants including BA.1, BA.2, and BA.5. *Commun. Biol.* **2023**, *6*, 772. [CrossRef] [PubMed]
77. Bálint, G.; Vörös-Horváth, B.; Széchenyi, A. Omicron: Increased transmissibility and decreased pathogenicity. *Signal Transduct. Target. Ther.* **2022**, *7*, 151. [CrossRef]
78. Shervani, Z.; Jamal, N.; Qazi, U.Y.; Hasan, S.; Fatma, K.; Siddiquie, A.; Vuyyuru, V.P.S.R.; Ibrahim, A.; Kumar, P.D.M.; Khan, A.A.; et al. Prevalence and Pathogenicity of Omicron Varian. *Eur. J. Med. Health Sci.* **2022**, *4*, 125–132. [CrossRef]
79. Mohandas, S.; Shete, A.; Kumar, A.; Wakchaure, K.; Rai, V.; Mote, C.; Dighe, H.; Sarkale, P.; Gawande, P.; Yemul, J.; et al. Comparative pathogenicity of BA.2.12, BA.5.2 and XBB.1 with the Delta variant in Syrian hamsters. *Front. Microbiol.* **2023**, *14*, 1183763. [CrossRef] [PubMed] [PubMed Central]
80. WHO. EG.5 Updated Risk Evaluation. 21 September 2023. Available online: https://www.who.int/docs/default-source/coronaviruse/eg5-risk-evaluation.pdf?sfvrsn=6e9690e0_4 (accessed on 3 February 2024).
81. Arnaout, R.; Arnaout, R. Visualizing omicron: COVID-19 deaths vs. cases over time. *PLoS ONE* **2022**, *17*, e0265233. [CrossRef]
82. Korobova, Z.R.; Arsentieva, N.A.; Liubimova, N.E.; Batsunov, O.K.; Dedkov, V.G.; Gladkikh, A.S.; Sharova, A.A.; Adish, Z.; Chernykh, E.I.; Kaschenko, V.A.; et al. Cytokine Profiling in Different SARS-CoV-2 Genetic Variants. *Int. J. Mol. Sci.* **2022**, *23*, 14146. [CrossRef] [PubMed]
83. Barh, D.; Tiwari, S.; Gomes, L.G.R.; Pinto, C.H.R.; Andrade, B.S.; Ahmad, S.; Aljabali, A.A.A.; Alzahrani, K.J.; Banjer, H.J.; Hassan, S.S.; et al. SARS-CoV-2 Variants Show a Gradual Declining Pathogenicity and Pro-Inflammatory Cytokine Stimulation, an Increasing Antigenic and Anti-Inflammatory Cytokine Induction, and Rising Structural Protein Instability: A Minimal Number Genome-Based Approach. *Inflammation* **2023**, *46*, 297–312. [CrossRef] [PubMed]
84. Armando, F.; Beythien, G.; Kaiser, F.K.; Allnoch, L.; Heydemann, L.; Rosiak, M.; Becker, S.; Gonzalez-Hernandez, M.; Lamers, M.M.; Haagmans, B.L.; et al. SARS-CoV-2 Omicron variant causes mild pathology in the upper and lower respiratory tract of hamsters. *Nat. Commun.* **2022**, *13*, 3519. [CrossRef] [PubMed]

85. Shuai, H.; Chan, J.F.-W.; Hu, B.; Chai, Y.; Yuen, T.T.-T.; Yin, F.; Huang, X.; Yoon, C.; Hu, J.-C.; Liu, H.; et al. Attenuated replication and pathogenicity of SARS-CoV-2 B.1.1.529 Omicron. *Nature* **2022**, *603*, 693–699. [CrossRef] [PubMed]
86. Martins, M.; Nascimento, G.M.D.; Nooruzzaman, M.; Yuan, F.; Chen, C.; Caserta, L.C.; Miller, A.D.; Whittaker, G.R.; Fang, Y.; Diel, D.G. The Omicron Variant BA.1.1 Presents a Lower Pathogenicity than B.1 D614G and Delta Variants in a Feline Model of SARS-CoV-2 Infection. *J. Virol.* **2022**, *96*, e0096122. [CrossRef] [PubMed]
87. Khan, S.; Shafiei, M.S.; Longoria, C.; Schoggins, J.W.; Savani, R.C.; Zaki, H. SARS-CoV-2 spike protein induces inflammation via TLR2-dependent activation of the NF-κB pathway. *Elife* **2021**, *10*, e68563. [CrossRef] [PubMed]
88. Kim, S.; Kim, S.Y.; Pribis, J.P.; Lotze, M.; Mollen, K.P.; Shapiro, R.; Loughran, P.; Scott, M.J.; Billiar, T.R. Signaling of High Mobility Group Box 1 (HMGB1) through Toll-like Receptor 4 in Macrophages Requires CD14. *Mol. Med.* **2013**, *19*, 88–98. [CrossRef] [PubMed]
89. Turner, N.A. Inflammatory and fibrotic responses of cardiac fibroblasts to myocardial damage associated molecular patterns (DAMPs). *J. Mol. Cell. Cardiol.* **2016**, *94*, 189–200. [CrossRef] [PubMed]
90. Olejnik, J.; Hume, A.J.; Mühlberger, E. Toll-like receptor 4 in acute viral infection: Too much of a good thing. *PLoS Pathog.* **2018**, *14*, e1007390. [CrossRef]
91. Chow, J.C.; Young, D.W.; Golenbock, D.T.; Christ, W.J.; Gusovsky, F. Toll-like receptor-4 mediates lipopolysaccharide-induced signal transduction. *J. Biol. Chem.* **1999**, *274*, 10689–10692. [CrossRef]
92. Kuzmich, N.N.; Sivak, K.V.; Chubarev, V.N.; Porozov, Y.B.; Savateeva-Lyubimova, T.N.; Peri, F. TLR4 Signaling Pathway Modulators as Potential Therapeutics in Inflammation and Sepsis. *Vaccines* **2017**, *5*, 34. [CrossRef] [PubMed]
93. Ohto, U.; Fukase, K.; Miyake, K.; Shimizu, T. Structural basis of species-specific endotoxin sensing by innate immune receptor TLR4/MD-2. *Proc. Natl. Acad. Sci. USA* **2012**, *109*, 7421–7426. [CrossRef] [PubMed]
94. Schromm, A.B.; Brandenburg, K.; Loppnow, H.; Zähringer, U.; Rietschel, E.T.; Carroll, S.F.; Koch, M.H.J.; Kusumoto, S.; Seydel, U. The charge of endotoxin molecules influences their conformation and IL-6-inducing capacity. *J. Immunol.* **1998**, *161*, 5464–5471. [CrossRef] [PubMed]
95. Schromm, A.B.; Brandenburg, K.; Loppnow, H.; Moran, A.P.; Koch, M.H.J.; Rietschel, E.T.; Seydel, U. Biological activities of lipopolysaccharides are determined by the shape of their lipid A portion. *JBIC J. Biol. Inorg. Chem.* **2000**, *267*, 2008–2013. [CrossRef] [PubMed]
96. Shirato, K.; Kizaki, T. SARS-CoV-2 spike protein S1 subunit induces pro-inflammatory responses via toll-like receptor 4 signaling in murine and human macrophages. *Heliyon* **2021**, *7*, e06187. [CrossRef] [PubMed]
97. Wrobel, A.G.; Benton, D.J.; Xu, P.; Roustan, C.; Martin, S.R.; Rosenthal, P.B.; Skehel, J.J.; Gamblin, S.J. SARS-CoV-2 and bat RaTG13 spike glycoprotein structures inform on virus evolution and furin-cleavage effects. *Nat. Struct. Mol. Biol.* **2020**, *27*, 763–767. [CrossRef] [PubMed]
98. Park, B.S.; Song, D.H.; Kim, H.M.; Choi, B.-S.; Lee, H.; Lee, J.-O. The structural basis of lipopolysaccharide recognition by the TLR4–MD-2 complex. *Nature* **2009**, *458*, 1191–1195. [CrossRef] [PubMed]
99. Gobeil, S.M.; Henderson, R.; Stalls, V.; Janowska, K.; Huang, X.; May, A.; Speakman, M.; Beaudoin, E.; Manne, K.; Li, D.; et al. Structural diversity of the SARS-CoV-2 Omicron spike. *Mol. Cell* **2022**, *82*, 2050–2068.e6. [CrossRef]
100. Yan, Y.; Tao, H.; He, J.; Huang, S.-Y. The HDOCK server for integrated protein–protein docking. *Nat. Protoc.* **2020**, *15*, 1829–1852. [CrossRef]
101. Wauters, E.; Van Mol, P.; Garg, A.D.; Jansen, S.; Van Herck, Y.; Vanderbeke, L.; Bassez, A.; Boeckx, B.; Malengier-Devlies, B.; Timmerman, A.; et al. Discriminating mild from critical COVID-19 by innate and adaptive immune single-cell profiling of bronchoalveolar lavages. *Cell Res.* **2021**, *31*, 272–290. [CrossRef]
102. McMahon, S.; Grondin, F.; McDonald, P.P.; Richard, D.E.; Dubois, C.M. Hypoxia-enhanced expression of the proprotein convertase furin is mediated by hypoxia-inducible factor-1: Impact on the bioactivation of proproteins. *J. Biol. Chem.* **2005**, *280*, 6561–6569. [CrossRef]
103. Jiang, H.; Hu, R.; Sun, L.; Chai, D.; Cao, Z.; Li, Q. Critical role of toll-like receptor 4 in hypoxia-inducible factor 1alpha activation during trauma/hemorrhagic shocky induced acute lung injury after lymph infusion in mice. *Shock* **2014**, *42*, 271–278. [CrossRef]
104. Chakraborty, C.; Mallick, B.; Bhattacharya, M.; Byrareddy, S.N. SARS-CoV-2 Omicron Spike shows strong binding affinity and favourable interaction landscape with the TLR4/MD2 compared to other variants. *J. Genet. Eng. Biotechnol.* **2024**, *22*, 100347. [CrossRef]
105. Sahanic, S.; Hilbe, R.; Dünser, C.; Tymoszuk, P.; Löffler-Ragg, J.; Rieder, D.; Trajanoski, Z.; Krogsdam, A.; Demetz, E.; Yurchenko, M.; et al. SARS-CoV-2 activates the TLR4/MyD88 pathway in human macrophages: A possible correlation with strong pro-inflammatory responses in severe COVID-19. *Heliyon* **2023**, *9*, e21893. [CrossRef] [PubMed] [PubMed Central]
106. Li, J.; Wang, J.; Wang, H. Emerging Landscape of Preclinical Models for Studying COVID-19 Neurologic Diseases. *ACS Pharmacol. Transl. Sci.* **2023**, *6*, 1323–1339. [CrossRef] [PubMed]
107. Kaushik, D.; Bhandari, R.; Kuhad, A. TLR4 as a therapeutic target for respiratory and neurological complications of SARS-CoV-2. *Expert Opin. Ther. Targets* **2021**, *25*, 491–508. [CrossRef] [PubMed]
108. de la Torre, S.H.; Parlatini, V.; Cortese, S. Long-term central nervous system (CNS) consequences of COVID-19 in children. *Expert Rev. Neurother.* **2023**, *23*, 703–720. [CrossRef]

109. Rong, Z.; Mai, H.; Kapoor, S.; Puelles, V.G.; Czogalla, J.; Schädler, J.; Vering, J.; Delbridge, C.; Steinke, H.; Frenzel, H.; et al. SARS-CoV-2 Spike Protein Accumulation in the Skull-Meninges-Brain Axis: Potential Implications for Long-Term Neurological Complications in post-COVID-19. *bioRxiv* **2023**. [CrossRef]
110. Ordaya, E.E.; Razonable, R.R. Emerging anti-spike monoclonal antibodies against SARS-CoV-2. *Expert Opin. Biol. Ther.* **2024**, *accepted*. [CrossRef]
111. Feng, X.; Wang, H. Emerging Landscape of Nanobodies and Their Neutralizing Applications against SARS-CoV-2 Virus. *ACS Pharmacol. Transl. Sci.* **2023**, *6*, 925–942. [CrossRef]
112. Edesa Biotech Announces Positive Phase 2 Data of Its Monoclonal Antibody in Hospitalized COVID-19 Patients. TORONTO, ACCESSWIRE. 20 September 2021. Available online: https://feeds.issuerdirect.com/news-release.html?newsid=5635061214803143 (accessed on 6 May 2024).
113. Kircheis, R.; Planz, O. The Role of Toll-like Receptors (TLRs) and Their Related Signaling Pathways in Viral Infection and Inflammation. *Int. J. Mol. Sci.* **2023**, *24*, 6701. [CrossRef] [PubMed]
114. PDB 7LWU. Available online: https://www.wwpdb.org/pdb?id=pdb_00007lwu (accessed on 6 May 2024). [CrossRef]
115. PDB 7M8K, Map EMD-23718: EMDB EMDataResource. Available online: https://www.wwpdb.org/pdb?id=pdb_00007m8k (accessed on 6 May 2024). [CrossRef]
116. PDB 2W92, Map EMD-32359: EMDB EMDataResource. Available online: https://www.wwpdb.org/pdb?id=pdb_00002w92 (accessed on 7 May 2024). [CrossRef]
117. PDB 2Z7X. Available online: https://www.wwpdb.org/pdb?id=pdb_00002z7x (accessed on 6 May 2024). [CrossRef]
118. PDB 2A79. Available online: https://www.wwpdb.org/pdb?id=pdb_00002a79 (accessed on 6 May 2024). [CrossRef]
119. Uhlén, M.; Fagerberg, L.; Hallström, B.M.; Lindskog, C.; Oksvold, P.; Mardinoglu, A.; Sivertsson, Å.; Kampf, C.; Sjöstedt, E.; Asplund, A.; et al. Proteomics. Tissue-Based Map of the Human Proteome. *Science* **2015**, *347*, 1260419. [CrossRef] [PubMed]

International Journal of *Molecular Sciences*

MDPI

Review

An Update on Toll-like Receptor 2, Its Function and Dimerization in Pro- and Anti-Inflammatory Processes

Katrin Colleseli, Anna Stierschneider and Christoph Wiesner *

Department of Medical and Pharmaceutical Biotechnology, IMC University of Applied Sciences, 3500 Krems, Austria
* Correspondence: christoph.wiesner@fh-krems.ac.at

Abstract: While a certain level of inflammation is critical for humans to survive infection and injury, a prolonged inflammatory response can have fatal consequences. Pattern recognition Toll-like receptors (TLRs) are key players in the initiation of an inflammatory process. TLR2 is one of the most studied pattern recognition receptors (PRRs) and is known to form heterodimers with either TLR1, TLR4, TLR6, and TLR10, allowing it to recognize a wide range of pathogens. Although a large number of studies have been conducted over the past decades, there are still many unanswered questions regarding TLR2 mechanisms in health and disease. In this review, we provide an up-to-date overview of TLR2, including its homo- and heterodimers. Furthermore, we will discuss the pro- and anti-inflammatory properties of TLR2 and recent findings in prominent TLR2-associated infectious and neurodegenerative diseases.

Keywords: Toll-like receptor 2; dimerization; homodimerization; heterodimerization; inflammation; neuroinflammation; neurodegeneration

Citation: Colleseli, K.; Stierschneider, A.; Wiesner, C. An Update on Toll-like Receptor 2, Its Function and Dimerization in Pro- and Anti-Inflammatory Processes. *Int. J. Mol. Sci.* **2023**, *24*, 12464. https://doi.org/10.3390/ijms241512464

Academic Editors: Oliver Planz and Ralf Kircheis

Received: 9 July 2023
Revised: 1 August 2023
Accepted: 3 August 2023
Published: 5 August 2023

1. Introduction

Over the past few decades, a great deal of knowledge has been gained about human innate and adaptive immunity and its key players, including the pattern recognition receptors (PRRs). These germline-encoded receptors can rapidly identify the specific molecular structures on the surface of pathogens, so-called pathogen-associated molecular patterns (PAMPs) or endogenous damage-associated molecular patterns (DAMPs), linking non-specific immunity to specific immunity [1,2]. To date, two distinct classes of PRRs have been described: the membrane-bound receptors, including Toll-like receptors (TLRs) and C-type lectin receptors (CLRs), and the cytoplasmic proteins, such as nucleotide-binding oligomerization domain (NOD)-like receptors (NLRs), retinoic acid-inducible gene (RIG)-I-like receptors, absent in melanoma 2 (AIM2)-like receptors (ALRs), and DNA or RNA sensors, such as cyclic GMP-AMP synthase (cGAS) [3–5]. TLRs are type I transmembrane proteins with an extracellular domain containing leucine-rich repeats that facilitate recognition of PAMPs. They also have a transmembrane domain and an intracellular toll-interleukin-1 receptor (TIR) domain needed for downstream signaling [6]. So far, ten receptors (TLR1-10) have been identified in the human body, which can be divided into cell surface receptors such as TLR1, TLR2, TLR4-6, TLR10, and intracellular receptors including TLR3, and TLR7-9 [7,8]. TLRs are widely distributed on cells of the immune system, such as monocytes, as well as on non-immune cells, such as endothelial or epithelial cells [9]. Upon PAMP or DAMP recognition, homo- or heterodimerization of TLRs is triggered, leading to an intracellular signaling cascade that in most cases results in a pro-inflammatory response [10]. A prominent example of homodimerization is TLR4, which is induced after exposure to lipopolysaccharide (LPS) isolated from the outer membrane of Gram-negative bacteria such as *Escherichia coli*, whereas TLR2 is known to form heterodimers with either TLR1 or TLR6, depending on the ligand [11,12]. In addition to heterodimer formation, TLR2 can function as a single receptor and in a homodimer, albeit the latter is still under debate [12]. Although TLRs, especially TLR2, have been extensively studied, their role in health and disease is still very much in the spotlight, as evidenced by the large number of studies on TLRs and SARS-CoV-2 infection, for example [13–15].

In this review, we aim to provide an up-to-date overview of TLR2 dimerization, including homo- and heterodimerization, and propose a model to further investigate these dimerization mechanisms. Additionally, the expanded role of TLR2 with its pro- and anti-inflammatory capabilities, and the recent findings in prominent TLR2-associated (neuro) inflammatory diseases will be discussed.

2. TLR2 Ligands and Signaling

TLRs, which are the mammalian orthologues of the *Drosophila melanogaster* Toll receptor originally discovered in 1985, are well conserved and primarily associated with pattern recognition [12,16,17]. TLR2, along with TLR4, is one of the most studied pattern recognition receptors, and unlike certain other receptors in this family, TLR2 is known to form various heterodimers with TLR1, TLR4, TLR6, and TLR10 [18–20]. In recent years, a growing number of studies have shown that TLR2 may be the most versatile TLR, as this receptor recognizes a broad repertoire of ligands from a variety of pathogen sources and interacts with a large number of other receptors [12,21].

TLR2/1 heterodimers can sense triacylated lipopeptides (LPs) from Gram-negative bacteria or mycoplasma, such as lipoarabinomannans and lipomannans, whereas TLR2/TLR6 heterodimers recognize diacylated LPs, including lipoteichoic acid (LTA) from Gram-positive bacteria and mycoplasma [12,22–24]. Depending on the ligand, low-endotoxic atypical LPS can induce TLR2/TLR4 heterodimerization, and the TLR2/TLR10 heterodimer has been found to participate in *Helicobacter pylori* LPS recognition [25,26]. Additional co-receptors, such as clusters of differentiation 14 (CD14), have been described that bring CD14 and TLR2 and TLR1 into physical proximity by binding LPs to induce signaling [27]. Interestingly, the mRNA expression of TLR2, but not TLR1, TLR4, and TLR6, was strongly induced in alveolar macrophages by prolonged LPS treatment (24 h) [28]. Besides heterodimerization, some evidence also points to the existence of TLR2 homodimerization, but this is still controversially discussed [12,29]. Commonly used TLR2 ligands for in vitro and in vivo studies are synthetic di- and triacylated LPs such as Pam2CSK4, Pam3CSK4, isolated from *Escherichia coli*, and FSL-1, which represents a pathogen-associated molecular pattern of the LP secluded from *Mycoplasma fermentans* or *Mycoplasma salivarium* [12,30,31]. Moreover, endogenous ligands, danger signals for TLR2, have been identified, including heat shock proteins, human β-defensin-3, high mobility group box 1 protein (HMGB1), and hyaluronan fragments [32–35]. Other ligands of TLR2 that play a major role in neurodegenerative disorders are amyloid-β (Aβ) and α-synuclein (αSyn) [36,37]. Therefore, in addition to PAMPs, endogenous TLR2 activation by host-derived danger signals is a viable factor in the pathogenesis of inflammation and related diseases.

Ligand-induced dimerization is the starting point for TLR2 signal transduction. Dimerization brings the TIR domains of the cytoplasmic tails into close proximity, providing a platform for signaling by TIR domain-containing adaptor molecules [12]. TLR2 primarily relies on the myeloid differentiation factor 88 (MyD88) and the Toll-interleukin-1 receptor domain-containing adaptor protein (TIRAP) adaptors for signaling [38]. The MyD88-dependent pathway involves death-domain interactions mediating intracellular signaling in a stepwise manner. Specifically, MyD88 recruitment is indirect and mediated by TIRAP [39]. Active MyD88 then sequentially triggers the phosphorylation of interleukin-1 receptor-associated kinase 4 (IRAK4), IRAK1, and IRAK2 [40]. The IRAK complex engages tumor necrosis factor (TNF) receptor-associated factor 6 (TRAF6), which undergoes K63-linked autoubiquitination and ubiquitinates nuclear factor kappa B (NF-κB) essential modulator (NEMO). The complex of transforming growth factor-β-activated kinase-1 (TAK1), TAK1-binding protein 2 (TAB2), and TAB3 is then activated. Next, TAK1 phosphorylates IkappaB kinase alpha (IKKα) and IKKβ, and the IKKs phosphorylate IκB, are marked for degradation, which ultimately leads to the production of pro-inflammatory cytokines via NF-κB and activating protein-1 (AP-1), and activation of mitogen-activated protein kinase (MAPK), modulating cell proliferation and survival (Figure 1) [6,41]. Interestingly, the interaction of TIRAP with IRAK1 and IRAK4 can lead to the degradation of TIRAP following its phosphorylation and ubiquitination, resulting in the inhibition of TLR2 signaling [42].

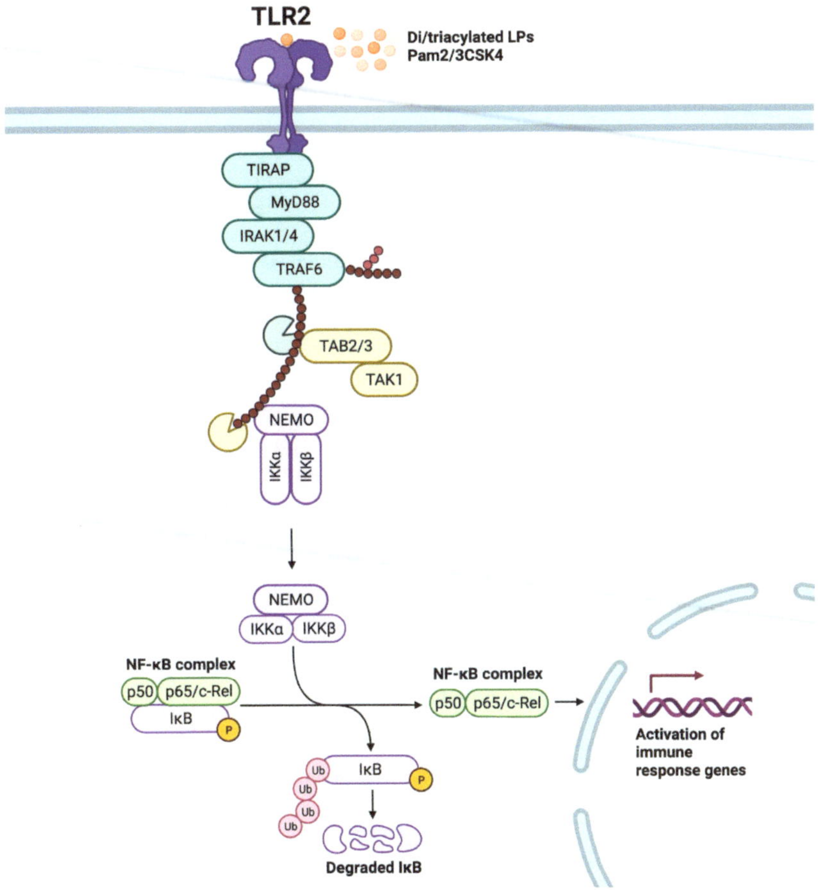

Figure 1. TLR2 MyD88-dependent signaling pathway. Dimerization is triggered by ligand binding (diacylated lipopeptides (LPs) or triacylated LPs, and synthetic Pam2CSK4 or Pam3CSK4), resulting in a signaling cascade that begins with Toll-interleukin-1 receptor domain-containing adaptor protein (TIRAP) binding to TLR2, which then leads to the recruitment of MyD88. Following phosphorylation of interleukin-1 receptor-associated kinase (IRAK) 4, IRAK1, and IRAK2, TNF receptor-associated factor 6 (TRAF6) undergoes K63-linked autoubiquitination and ubiquitinates nuclear factor kappa B (NF-κB) essential modulator (NEMO). The complex of transforming growth factor-β-activated kinase-1 (TAK1), TAK1-binding protein 2 (TAB2), and TAB3 complex is then activated to phosphorylate IkappaB kinase alpha (IKKα) and IKKβ, and the IKKs phosphorylate IκB, marking it for degradation and releasing the NF-κB complex (consisting of p50 and p65/c-Rel). This ultimately leads to the activation of immune response genes, including the production of pro-inflammatory cytokines via the transcription factor NF-κB. Adapted from "NF-KB signaling pathway", by BioRender.com (2023).

3. Dimerization and Additional Co-Receptors of TLR2

3.1. Homodimerization

The existence of TLR2 homodimers has been mentioned in a number of publications in recent years, however, the function of this homodimer is still controversial [12,43–45]. In vitro studies such as those of de Groot et al. suggest that TLR2 is functional as a homodimer, although it may require co-receptors to reach an active state. In particular, they were able to show that TLR2 signaling was reduced in the TLR1- and TLR6-deficient reporter systems, but was still evident in response to *Mycoplasma pneumoniae* and *Streptococcus pneumoniae* [46]. In a different investigation, another research group concluded that LTA-activated TLR2 via the TLR2 homodimer is likely to be less potent than that of LPs via the TLR2 heterodimer, which is limited to the induction of IRAK-M [47]. Su et al. identified diprovocim-1 as an effective inducer not only for the formation of TLR2/TLR1 heterodimers, but also for the formation of TLR2 homodimers in vitro [48]. Contrary studies show that the proline-proline-glutamic acid 18 (PPE18) protein of *Mycobacterium tuberculosis* induces TLR2 homodimerization, which triggers anti-inflammatory type responses that cause increased activation of the mammalian p38 MAPK [49]. Furthermore, by generating a CRISPR/Cas9-mediated knock-out of endogenous TLR6 in JE6-1 TLR2/6 reporter cells, it was recently shown that TLR2 homodimers are expressed on the cell surface. However, these reporter cells did not respond to any of the bacterial TLR2 agonists tested, nor could they be detected by diprovocim-1 induction. Activation of reporter cells occurred only in cells expressing TLR2/1 heterodimers [50]. Taken together, the recent results suggest that the function of a TLR2 homodimer is strongly dependent on ligands and co-receptors yet to be discovered. To address these unanswered questions, novel methods and strategies could help to further elucidate potential functions of the TLR2 homodimer.

3.2. TLR2 Heterodimers

The ability of TLR2 to dimerize not only with itself but also with a variety of TLRs, greatly expands the spectrum of detectable pathogens. Ozinsky et al. were among the first to discover that TLR2 relies on heterodimerization with either TLR1 or TLR6 to initiate adequate cell activation and a pro-inflammatory response [18,21]. Structural studies of the heterodimers have supported the importance of ligand binding to stabilize TLR2/1 and TLR2/6 dimerization for downstream signaling [51,52]. However, since the heterodimerization of TLR2 with TLR1 or TLR6 has already been extensively studied, we will focus on insights into other TLR2 heterodimerizations. There is accumulating evidence that TLR10 is capable of forming functional heterodimers with TLR2 [20,53]. This is plausible because in mammals, TLR10 is a member of the TLR1 subfamily and shares numerous paralogous genes with TLR1, 2, and 6 [54,55]. To assess whether the TLR2/10 heterodimer induces downstream signaling, Pachathundikandi et al. examined the mRNA expression levels of pro-inflammatory cytokines upon exposure to *Helicobacter pylori* or LPS and found a significant upregulation of interleukin-1β (IL-1β) in TLR2/10-transfected HEK293 cells [56]. In another study, they silenced TLR10 in the human colon adenocarcinoma cell line HT-29 and in monocytic THP-1 cells, which led to increased viability of *Listeria monocytogenes*, and they unveiled that the heterodimer of TLR2/TLR10 was the one behind the activation of NF-κB [57]. Nagashima et al. further confirmed this by using NC1-N87 gastric cells incubated with *Helicobacter pylori* and concluded that the TLR2/TLR10 heterodimer upregulates NF-κB activation to a greater extent than other TLR2 heterodimers [26]. A recent research report indicated that TLR10 regulates TLR2-induced cytokine production in monocytes isolated from Parkinson's disease patients [58]. However, the specific ligand(s) and function of the TLR2/10 heterodimer have not been fully elucidated. Interestingly, a report from 2014 suggested that the hemoglobin-induced TLR2/4 heterodimer mediates inflammatory injury in intracerebral hemorrhage in in vivo and in vitro models [19]. Furthermore, an in vitro study by Muniz-Bongers et al. showed that in HEK293, matrix metalloproteinase 2 (MMP2) binds to both TLR2 and TLR4 for signaling, suggesting that TLR2 and TLR4 may form a heterodimer and consequently trigger a MyD88-dependent signaling cascade [59].

The existence of functional TLR2/4 heterodimers was further confirmed by Francisco et al. who showed that low-endotoxic atypical LPS isolated from *Ochrobactrum intermedium* is a potent TLR2/TLR4-inducing agonist, which they additionally confirmed by molecular docking analysis and fluorescence resonance energy transfer [25]. Nevertheless, the detailed mechanism of these divergent TLR2 heterodimers is still in question, and additional research and specific ligands are needed to substantiate these findings.

3.3. Other Co-Receptors Supporting TLR2

Additional co-receptors, called accessory receptors, are often required to augment TLR2 ligand delivery, pattern recognition, and other functions (see Figure 2) [12,21,60]. Prominent adaptors include CD14 and CD36, which are glycoproteins that are expressed primarily on monocytes and macrophages and that promote TLR2-dependent inflammation [61–65]. By performing immunoprecipitation and sugar inhibition assays, the physical interaction of ArtinM, a TLR2 agonist, with TLR2 and CD14 was observed to be a prerequisite for ArtinM-induced M1 macrophage activation [66]. Another study demonstrated that treatment of the LPS-incubated human microglia with anti-CD36 regulated the inflammatory cytokine levels in the brains of newborn mice and resulted in a significant decrease in TLR2 expression levels, while TLR4 expression was not altered [67]. In addition to CD14 and CD36, the hyaluronan receptor CD44 was found to be significantly upregulated in THP-1 wild-type versus TLR2 knock-out cells using quantitative mass spectrometry [68]. A number of reports in the last few years have indicated an interaction between TLR2 and CD44 [69–72]. Another interesting group of TLR2 co-receptors are the heterodimeric integrin receptors, which are critical for facilitating cell–cell and cell–extracellular matrix adhesion during inflammation [73]. They are composed of an α and a β subunit, which allows them to recognize a variety of different ligands that are relevant to cell adhesion and migration [74]. Numerous research articles report that β1, β2, as well as β3, directly or indirectly interact with TLR2 to initiate signaling cascades that culminate in cytokine secretion [21,68]. An in vitro study from 2019 demonstrated that α- 1- acid glycoprotein potentiated TLR2-dependent β2-integrin-mediated cell adhesion of neutrophils [75]. On the other hand, De Baros et al. showed that Pam3CSK4 increased α3β1 integrin levels, which was inhibited by silencing TLR2. However, *Paracoccidioides brasiliensis* infection led to an increase in TLR2 and in parallel, α3β1 integrin levels were downregulated in human lung epithelial cells [76]. These results once again emphasize that TLR2 modulation is ligand- and cell-type dependent.

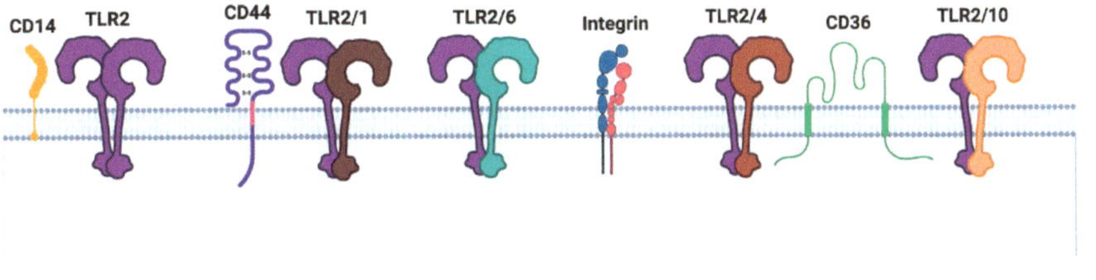

Figure 2. TLR2 homo- and heterodimers and additional co-receptors. Upon ligand binding, dimerization is initiated, often requiring additional co-receptors such as cluster of differentiation 14 (CD14), CD44 receptor, integrin receptors, or CD36. Created with BioRender.com.

3.4. Optogenetics, an Innovative Approach to Study TLR Dimerization and Signaling

Conventional techniques to further identify and characterize dimerization and signaling mechanisms, such as pharmacological control including ligand exposure, knock-out models, or biochemical assays, may not always be the most appropriate method. For example, models based on TLR2 knock-outs emphasize the importance of TLR2 but do not provide insight into the detailed signaling mechanism or interactions with other co-receptors. However, new experimental models, such as optogenetics, are emerging that do not rely on ligand binding to activate cellular receptors. Optogenetics was first introduced by Deisseroth in 2006 in the field of neuroscience, where light-sensitive ion channels were being utilized to remotely control action potentials and thereby neuronal networks [77]. In short, this biological technology uses genetic engineering to combine an effector protein with a light-sensitive protein domain of microbial, fungal, or plant photoreceptors to precisely control the activity of specific cells with light [78]. To enable the activation and inactivation of signaling pathways, a variety of optically controlled receptors have been developed in recent years, including G protein-coupled receptors (GPCRs) or receptor tyrosine kinases (RTKs) [79–82]. Grusch et al. have used the light–oxygen–voltage (LOV) sensing domain to induce RTK dimerization by light [80]. In a variety of eukaryotic and prokaryotic proteins, LOV domains bind a flavin chromophore to function as blue light sensors. In order to use these LOV domains for optogenetic approaches, they were isolated from different organisms, including *Arabidopsis thaliana* or *Chlamydomonas reinhardtii* [83,84]. Our research group has established light-inducible TLR4 cell models using human pancreatic adenocarcinoma and endothelial cell lines [85,86]. For this purpose, the LOV domain, isolated from the yellow-green alga *Vaucheria frigida aureochrome 1*, was C-terminally fused to the full-length TLR4. This innovative system allows precise on/off switching under temporal and spatial control, which may be an interesting approach to elucidate TLR2 homodimerization and signaling.

4. TLR2 Mediates Both Pro- and Anti-Inflammatory Responses

TLR2 activation can evoke pro-inflammatory and anti-inflammatory responses that are highly dependent on the cell type and the cell compartment in which it is expressed, the ligand, and the co-receptors. A complete overview of all cells expressing TLR2 and the functional impact of TLR2 activation is beyond the scope of this review. However, to illustrate the diverse and complex effects of TLR2 activation, a selection of recent findings will be mentioned in this section. While the well-known heterodimerization of TLR2 with TLR1 or TLR6 leads to the classical pro-inflammatory response in different cell types, the anti-inflammatory reaction serving as an immune regulator seems to be more heterogeneous. A study from 2021 showed that a high number and blockwise distribution of certain methyl esters of pectins were responsible for a TLR2/1-dependent anti-inflammatory effect [87]. Interestingly, the flavonoid baicalin isolated from *Scutellaria baicalensis Georgi* inhibited inflammation in rats with chronic obstructive pulmonary disease via the TLR2/MyD88/NF-κB pathway [88]. Furthermore, the anti-inflammatory properties of TLR2 were evident in response to polysaccharide A (PSA) from *Bacteroides fragilis* in B cells and T cells, resulting in the production of IL-10 and interferon-γ (IFN-γ) [89–91]. M2 macrophages from healthy donors or patients with rheumatoid arthritis showed reduced anti-inflammatory activity in the presence of abundant TLR2 [92]. Soluble TLR2 (sTLR2), which is generated by proteolytic cleavage of the TLR2 transmembrane protein, also known as ectodomain shedding, has been suggested as a promising biomarker of infections and systemic inflammation [93]. In this context, it has been shown that soluble and full-length TLR2 are released by macrophages under anti-inflammatory conditions, implicating that vesicle-bound full-length TLR2 exerts decoy properties that may be involved in immune suppression [28].

Although pattern recognition appears to be the most common activity of TLR2, it is not its only biological function. Our recent publication indicated that TLR2 also plays an important role in cell adhesion and migration of monocytes [68]. More specifically, we performed several functional cell-based assays using wild-type, TLR2 knock-out, and TLR2 knock-in THP-1 cells. We found that TLR2 enhances quicker and stronger adhesion of monocytes to the endothelium and a more pronounced endothelial barrier disruption after endothelial activation. The importance of ligand-activated TLR2 in cell adhesion and migration has been shown previously, but we pointed out that activation of endothelial cells is sufficient for a distinct effect of TLR2 expressed on monocytes [94–96]. In this context, several studies have confirmed the involvement of TLR2 in coagulation, resulting in increased monocyte adhesion, upregulation of tissue factor, increased extrinsic pathway coagulation including active recruitment of platelets, and formation of platelet–monocyte aggregates [97–99]. Interestingly, TLR2 was also found to be expressed in early endosomes, late endosomes/lysosomes, and Rab-11-positive compartments but not in the Golgi apparatus or endoplasmic reticulum of monocytes [100]. Recent data have provided evidence that IL-10 secretion in response to *Listeria monocytogenes* infection in vitro and in vivo (mouse model) is generally TLR2 dependent and that immune suppression by phagosome-confined bacteria in vivo is mostly dependent on endosomal TLRs [101]. Table 1 lists the most recent findings of pro- and anti-inflammatory responses induced by TLR2.

Table 1. TLR2-dependent pro- and anti-inflammatory responses.

TLR Dimer	Ligand	Cytokines/Chemokines	References
TLR2/1	Triacyl LPs Pam3CSK4	IL-6, IL-8, TNF α, IL-1α, IL-1β, IL-12, IL-13, IFN-α, IFN- β	[12,102,103]
TLR2/6	Diacyl LPs Pam2CSK4 FSL-1	IL-6, IL-8, IL-12, TNF-α, IFN-α, IFN- β, CXCL-10	[12,104,105]
TLR2/?	*Listeria monocytogenes*	IL-10	[101]
TLR2/?	*Bacteroides fragilis* PSA	IL-10, IFN-γ	[89–91]
TLR2/?	LPS	TNF, CXCL10, IL10	[28]
TLR2/4	Hemoglobin, atypical LPS	IL-12, IL-6, TNF-α	[25]
TLR2/10	*Helicobacter pylori*	TNF- α, IL-1β	[56]

5. The Role of TLR2 in Infectious and Neurodegenerative Diseases

TLR2 and its signaling have been implicated in a variety of inflammatory diseases such as rheumatoid arthritis, inflammatory bowel disease, asthma, and multiple sclerosis [106–109]. However, here we will discuss sepsis and COVID-19, both of which can lead to chronic inflammation and, in addition, neuroinflammation that may promote neurodegenerative diseases such as Alzheimer's disease (AD) or Parkinson's disease (PD) (see Figure 3).

5.1. Sepsis

Sepsis is described as a systemic dysregulated immune response of the body to an infectious reaction caused by pathogens, including bacteria, viruses, fungi, and parasites [110,111]. This life-threatening syndrome damages the body's tissues and organs, ultimately leading to organ failure and death, with mortality rates reaching 20–70% if left untreated [112,113]. Advances in medical health care have made it possible for most patients to survive the initial hyperinflammatory phase of sepsis [114]. However, the transition to the later immunosuppressive phase is still critical, with 30% of patients dying from secondary infections [115,116]. In recent years, TLR2 and TLR4, which are found on immune cells such as monocytes, endothelial cells, or platelets, have gained attention in the context of human sepsis [117,118]. Given the inherent role of TLR2 in acute and chronic inflammation, TLR2 needs to be strictly controlled to avoid a dysregulated host response. If this regulation fails, sepsis can be triggered by pathogens that activate platelets, which then interact with endothelial cells and immune cells, leading to disruption of the endothelial barrier, fluid leakage, and ultimately to tissue damage and organ failure [116]. Another scenario describes the activation of the endothelium by pathogens, resulting in the release of chemokines and cytokines (cytokine storm) that trigger the recruitment of platelets and immune cells, ultimately leading to a hyperinflammatory response that causes apoptosis, endothelial barrier breakdown, fluid leakage, and further tissue damage [119]. It has been shown that *Streptococcus pneumoniae* induces platelet activation via TLR2 and that its inhibition completely abolished platelet aggregation, implicating TLR2 in the thrombotic complications of sepsis [120]. Another research group studied 59 patients with *Staphylococcus aureus* bacteremia and found that TLR2 downregulation and high IL-6 and IL-10 levels, indicative of immune dysregulation during early bacteremia, may be linked to mortality [121]. Furthermore, Hoppstädter et al. showed that TLR2 is highly upregulated during the immunosuppressive phase of systemic inflammatory response syndrome (SIRS) and sepsis patients [28]. Consistent with previous data, using the experimental model of polymicrobial sepsis induced by cecal ligation and puncture, one group showed increased TLR2 levels in the kidney and intestine [122]. In the same sepsis mouse model, another group has recently demonstrated that TLR2-deficient mice have lower levels of IL-10 and reduced caspase-3 activation in the spleen [123]. Taken together, these findings underscore the multifaceted role of TLR2 in the hyperinflammatory state and also in sepsis-induced immunosuppression.

5.2. COVID-19

Severe acute respiratory syndrome coronavirus 2 (SARS-CoV-2), which has challenged human health and public safety in recent years, is a member of the *Coronaviridae* family of β-coronaviruses, which are positive-stranded, enveloped RNA viruses [124]. Clinical features include an increase in inflammatory monocytes and neutrophils, and a robust inflammatory milieu that includes cytokines such as IL-1β, IL-6, and TNF-α [125,126]. While these cytokines are an inherent part of innate immunity and essential for the elimination of pathogens, uncontrolled release can trigger hyperinflammation, leading to cell death and tissue damage [127,128]. Zheng et al. demonstrated that TLR2 can recognize the SARS-CoV-2 envelope (E) protein and is required for inflammatory cytokine release during β-coronavirus infection. In addition, they found that SARS-CoV-2 E protein can induce TLR2-dependent inflammation in the lungs of mice [129]. Consistent with this, another study also showed that E protein–TLR2 interaction leads to the downstream activation of the transcription factor NF-κB, which stimulates the production of IL-8 [130]. However, it has also been proposed that prophylactic intranasal administration of a TLR2/6 agonist reduces SARS-CoV-2 transmission and provides protection against COVID-19 [131]. Notably, sepsis and COVID-19 patients share many pathophysiological and clinical features, such as thrombocytopenia, coagulopathy, vascular micro thrombosis, high cytokine production, septic shock, multiorgan dysfunction syndrome, fever, leukopenia, hypotension, or leukocytosis [132]. These data highlight the importance of temporal dependency in the context of TLR2, which can lead to protective or harmful effects.

5.3. Neuroinflammation

TLR2 is expressed not only on a variety of immune cells, but also on neuronal and glial cells in the central, peripheral, and enteric nervous systems, enabling them to act as immune cells [133,134]. As evidenced by numerous publications in recent years, TLR2 appears to play a pivotal role in neurological and neurodegenerative disorders by influencing brain cells and plasticity. Neuroinflammation is characterized by functionally activated and proliferating microglia and astrocytes and increased levels of pro-inflammatory cytokines, chemokines, and cytotoxic molecules [135–137]. Ultimately, this leads to permeabilization of the blood–brain barrier, infiltration of peripheral immune cells, and neuronal cell death [137–139]. Although neuroinflammation is an essential tool to protect the central nervous system, a persistent inflammatory response, also known as chronic inflammation, can lead to the damage and death of neurons, predisposing to neurodegenerative diseases such as AD or PD [140,141].

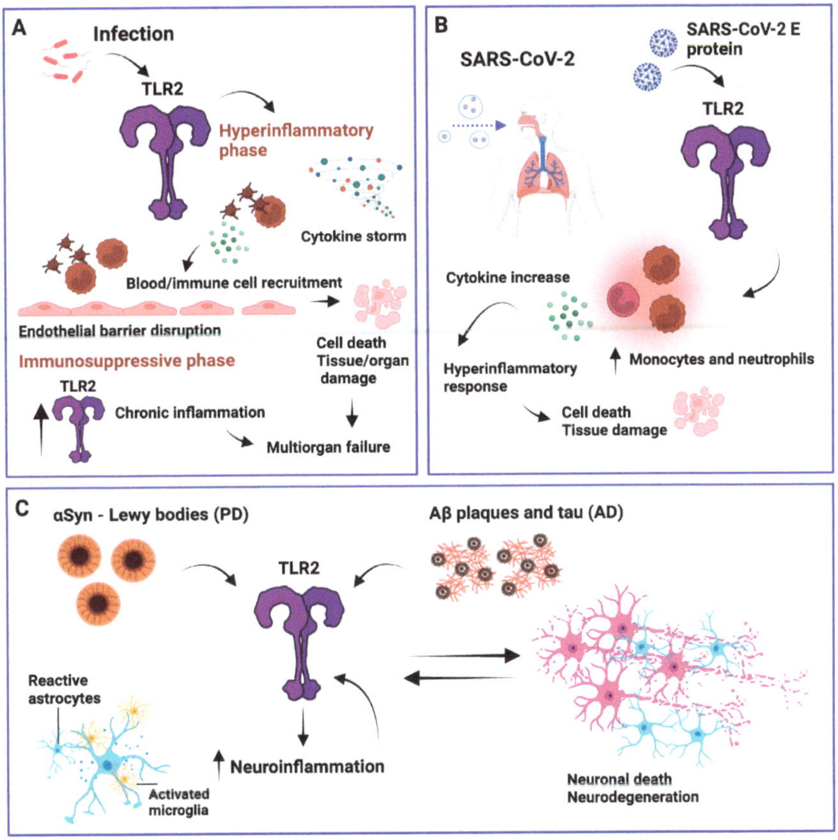

Figure 3. Simplified overview of TLR2 involvement in several relevant inflammatory diseases. (**A**) Sepsis. When immune regulation fails and a hyperinflammatory state is induced via elevated cytokine production (cytokine storm), sepsis can be triggered by TLR2, which further activates platelets and other immune cells that then interact with endothelial cells, leading to endothelial barrier disruption, cell death, and tissue and organ damage. In addition, TLR2 has been shown to be elevated during the immunosuppressive phase of sepsis which can lead to multiorgan failure. (**B**) Severe acute respiratory syndrome coronavirus 2 (SARS-CoV-2). TLR2 can recognize the SARS-CoV-2 envelope (E) protein, leading to an increase in inflammatory monocytes and neutrophils, inducing a hyperinflammatory response that can cause cell death and tissue damage. (**C**) Parkinson's disease (PD) and Alzheimer's disease (AD). TLR2 can recognize α-synuclein (αSyn) and amyloid-β (Aβ) plaques and tau, respectively. This promotes neuroinflammation characterized by reactive astrocytes and activated microglia, which further increase TLR2 levels, leading to a feedback loop of neuronal cell death, further TLR2 upregulation, and neuroinflammation. Created with BioRender.com.

5.3.1. Alzheimer's Disease

The most prevalent neurodegenerative disease in the aging population, AD, is pathologically characterized by the accumulation of Aβ plaques, hyperphosphorylated tau in neurofibrillary tangles, and consequently, enhanced neuroinflammation [142,143]. Since inflammation has a significant impact on the pathogenesis of AD, several inflammatory mediators have been proposed as AD markers, including IL-1β, TNF-α, NF-κB, and TLR2 [144]. A large body of research shows that inflammation can cause Aβ or tau overload, which then triggers inflammatory responses, creating a downward spiral of neuroinflammation and pathology [144–146]. In particular, microglia, the main cellular component of the innate immune system in the brain, express high levels of TLR2, making them a valuable target in the initiation, progression, and preservation of Alzheimer's disease [147,148]. Together with CD14 and TLR4, TLR2 is essential for fibrillar Aβ-stimulated microglial activation [149]. Rangasamy et al. showed that selectively targeting the TLR2–MyD88 interaction rescued AD pathology, including hippocampal glial activation and reduced Aβ burden [150]. However, in the context of TLR2, there is a fine line between acute, protective inflammation and chronic, neurotoxic inflammation, and the challenge is to find the right framework for therapeutic approaches.

5.3.2. Parkinson's Disease

PD is the most prominent form of α-synucleinopathies and the second most common neurodegenerative disease [151]. The pathological hallmark of PD is Lewy bodies, which are protein-containing inclusions in which misfolded and aggregated αSyn is the major component [152,153]. Similar to AD, the spread and process of αSyn misfolding remains elusive but in vivo approaches along with a variety of in vitro studies have provided strong evidence that environmental and genetic factors, as well as inflammation, are key contributors [154,155]. Since TLR2 can recognize so-called alarmins, including αSyn, which are released from damaged neuronal cells and lead to the production of cytokines, other immune cells, such as T cells and B cells, can be recruited to stimulate an adaptive immune response [156]. It has been suggested that TLR2 is upregulated in the pathogenesis of synucleinopathies in the brains of PD patients, thereby establishing TLR2 as a mediator of not only pro-inflammatory but also neurotoxic effects of extracellular α-synuclein aggregates. Kim et al. showed that blocking TLR2 reduced αSyn accumulation in neuronal and astroglial cells, neuroinflammation, neurodegeneration, and functional deficits in a PD mouse model [157]. Consistent with the previous study, it has recently been demonstrated that neuronal TLR2 activation acutely impairs the lysosomal autophagy pathway and amplifies αSyn pathology seeded with α-Syn preformed fibrils using human cell models. In addition, TLR2 knock-out in a human neuroblastoma cell line and the use of a small molecule TLR2 inhibitor (NPT1220-321) in induced pluripotent stem cell-derived neurons from a patient with PD have both demonstrated that αSyn pathology can be attenuated [158]. Another promising study has shown that specific inhibition of the TLR2 interaction domain of MyD88 and the NEMO binding domain can αSyn propagation in vitro and in vivo [159]. Taken together, several studies implicate TLR2 in the pathogenesis of PD but it is still unclear how exactly TLR2 affects the αSyn pathology [160].

6. Conclusions and Future Perspectives

One of the major discoveries in medicine over the last twenty years has been that the immune system and subsequent inflammatory processes are associated not only with a few specific diseases but also with a broad range of mental and physical health conditions [161,162]. In fact, chronic inflammatory disorders are now considered the leading cause of death worldwide, with more than 50% of all deaths due to inflammation-related diseases such as stroke, cancer, ischemic heart disease, autoimmune diseases, neurodegenerative diseases, etc. [163,164]. A key player in these inflammation-mediated events is TLR2, one of the most extensively studied pattern recognition receptors [12,18]. A considerable number of in vitro and in vivo studies have been conducted to fully elucidate the role of this protein. Among all known TLRs, TLR2 is rather exceptional in the sense that heterodimerization with multiple TLRs is possible, broadening the ligand spectrum and also exhibiting pro- and anti-inflammatory capacities [12,28]. Evidence has accumulated that TLR2 signaling is triggered by SARS-CoV-2, inducing robust pro-inflammatory cytokine expression that may contribute to severe COVID-19 [165]. Moreover, TLR2 has been identified as a key effector of neuroinflammation, which is highly relevant to the pathophysiology of neurodegenerative diseases [166]. Despite these advances, there are still many gaps in our knowledge of TLR2 mechanisms in health but particularly in disease. We have only scratched the surface of the mechanistic link between TLRs and neurodegenerative diseases, for example. Clearly, more studies, especially with patient-derived samples, are needed to further assess the complexity. In addition, emphasis should be placed on in silico techniques which provide attractive tools not only for drug design and discovery methodologies but also for cellular models of various pathologies [167,168]. There is an urgent need to fill these gaps in order to develop therapeutic strategies for inflammation-induced diseases, including neurodegenerative diseases, which will become an even greater burden as the population ages. As highlighted in a number of recent reports, neurological disorders pose a major challenge to the sustainability of global health care [169–171]. In this review, we have attempted to provide a current view of TLR2 focusing on dimerization, optogenetics as a novel experimental model, and involvement in common diseases which may help to push new and innovative ideas on its way.

Author Contributions: Conceptualization, K.C. and C.W.; writing—original draft preparation, K.C.; writing—review and editing, K.C., A.S. and C.W.; visualization, K.C.; supervision, C.W.; funding acquisition, C.W. All authors have read and agreed to the published version of the manuscript.

Funding: This work was supported by the Lower Austrian FTI Program, grant number K3-F-744/005-2019 and by the Austrian Research Promotion Agency (FFG), Project Bridge, grant number 35467434.

Institutional Review Board Statement: Not applicable.

Informed Consent Statement: Not applicable.

Data Availability Statement: Not applicable.

Acknowledgments: We would like to express our gratitude to Dipl.-Ing. Marie Ebeyer-Masotta for her thorough review of this paper.

Conflicts of Interest: The authors declare no conflict of interest.

References

1. Takeuchi, O.; Akira, S. Pattern Recognition Receptors and Inflammation. *Cell* **2010**, *140*, 805–820. [CrossRef]
2. Roh, J.S.; Sohn, D.H. Damage-Associated Molecular Patterns in Inflammatory Diseases. *Immune Netw.* **2018**, *18*, e27. [CrossRef]
3. Li, D.; Wu, M. Pattern Recognition Receptors in Health and Diseases. *Signal Transduct. Target. Ther.* **2021**, *6*, 291. [CrossRef] [PubMed]
4. Caneparo, V.; Landolfo, S.; Gariglio, M.; De Andrea, M. The Absent in Melanoma 2-Like Receptor IFN-Inducible Protein 16 as an Inflammasome Regulator in Systemic Lupus Erythematosus: The Dark Side of Sensing Microbes. *Front. Immunol.* **2018**, *9*, 1180. [CrossRef]
5. Hu, T.; Pan, M.; Yin, Y.; Wang, C.; Cui, Y.; Wang, Q. The Regulatory Network of Cyclic GMP-AMP Synthase-Stimulator of Interferon Genes Pathway in Viral Evasion. *Front. Microbiol.* **2021**, *12*, 790714. [CrossRef]
6. Kawai, T.; Akira, S. The Role of Pattern-Recognition Receptors in Innate Immunity: Update on Toll-like Receptors. *Nat. Immunol.* **2010**, *11*, 373–384. [CrossRef]
7. Takeda, K.; Akira, S. Toll-like Receptors in Innate Immunity. *Int. Immunol.* **2005**, *17*, 1–14. [CrossRef]
8. Lind, N.A.; Rael, V.E.; Pestal, K.; Liu, B.; Barton, G.M. Regulation of the Nucleic Acid-Sensing Toll-like Receptors. *Nat. Rev. Immunol.* **2022**, *22*, 224–235. [CrossRef]
9. Simpson, M.E.; Petri, W.A., Jr. TLR2 as a Therapeutic Target in Bacterial Infection. *Trends Mol. Med.* **2020**, *26*, 715–717. [CrossRef]
10. Moynagh, P.N. TLR Signalling and Activation of IRFs: Revisiting Old Friends from the NF-KB Pathway. *Trends Immunol.* **2005**, *26*, 469–476. [CrossRef]
11. Núñez Miguel, R.; Wong, J.; Westoll, J.F.; Brooks, H.J.; O'Neill, L.A.J.; Gay, N.J.; Bryant, C.E.; Monie, T.P. A Dimer of the Toll-like Receptor 4 Cytoplasmic Domain Provides a Specific Scaffold for the Recruitment of Signalling Adaptor Proteins. *PLoS ONE* **2007**, *2*, e788. [CrossRef] [PubMed]
12. Oliveira-Nascimento, L.; Massari, P.; Wetzler, L.M. The Role of TLR2 Ininfection and Immunity. *Front. Immunol.* **2012**, *18*, 79. [CrossRef]
13. Mantovani, S.; Oliviero, B.; Varchetta, S.; Renieri, A.; Mondelli, M.U. TLRs: Innate Immune Sentries against SARS-CoV-2 Infection. *Int. J. Mol. Sci.* **2023**, *24*, 8065. [CrossRef]
14. Jung, H.E.; Lee, H.K. Current Understanding of the Innate Control of Toll-like Receptors in Response to SARS-CoV-2 Infection. *Viruses* **2021**, *13*, 2132. [CrossRef] [PubMed]
15. Dai, J.; Wang, Y.; Wang, H.; Gao, Z.; Wang, Y.; Fang, M.; Shi, S.; Zhang, P.; Wang, H.; Su, Y.; et al. Toll-Like Receptor Signaling in Severe Acute Respiratory Syndrome Coronavirus 2-Induced Innate Immune Responses and the Potential Application Value of Toll-Like Receptor Immunomodulators in Patients With Coronavirus Disease 2019. *Front. Microbiol.* **2022**, *13*, 948770. [CrossRef] [PubMed]

16. Hashimoto, C.; Hudson, K.L.; Anderson, K. V The Toll Gene of Drosophila, Required for Dorsal-Ventral Embryonic Polarity, Appears to Encode a Transmembrane Protein. *Cell* **1988**, *52*, 269–279. [CrossRef]
17. Anderson, K.V.; Jürgens, G.; Nüsslein-Volhard, C. Establishment of Dorsal-Ventral Polarity in the Drosophila Embryo: Genetic Studies on the Role of the Toll Gene Product. *Cell* **1985**, *42*, 779–789. [CrossRef]
18. Ozinsky, A.; Underhill, D.M.; Fontenot, J.D.; Hajjar, A.M.; Smith, K.D.; Wilson, C.B.; Schroeder, L.; Aderem, A. The Repertoire for Pattern Recognition of Pathogens by the Innate Immune System Is Defined by Cooperation between Toll-like Receptors. *Proc. Natl. Acad. Sci. USA* **2000**, *97*, 13766–13771. [CrossRef]
19. Wang, Y.-C.; Zhou, Y.; Fang, H.; Lin, S.; Wang, P.-F.; Xiong, R.-P.; Chen, J.; Xiong, X.-Y.; Lv, F.-L.; Liang, Q.-L.; et al. Toll-like Receptor 2/4 Heterodimer Mediates Inflammatory Injury in Intracerebral Hemorrhage. *Ann. Neurol.* **2014**, *75*, 876–889. [CrossRef]
20. Su, S.-B.; Tao, L.; Deng, Z.-P.; Chen, W.; Qin, S.-Y.; Jiang, H.-X. TLR10: Insights, Controversies and Potential Utility as a Therapeutic Target. *Scand. J. Immunol.* **2021**, *93*, e12988. [CrossRef]
21. van Bergenhenegouwen, J.; Plantinga, T.S.; Joosten, L.A.B.; Netea, M.G.; Folkerts, G.; Kraneveld, A.D.; Garssen, J.; Vos, A.P. TLR2 & Co: A Critical Analysis of the Complex Interactions between TLR2 and Coreceptors. *J. Leukoc. Biol.* **2013**, *94*, 885–902. [CrossRef] [PubMed]
22. Keshavarz-Fathi, M.B.T.-V.; Rezaei, N. *Chapter 3—Vaccines, Adjuvants, and Delivery Systems*; Academic Press: Cambridge, MA, USA, 2019; pp. 45–59, ISBN 978-0-12-814039-0.
23. Quesniaux, V.J.; Nicolle, D.M.; Torres, D.; Kremer, L.; Guérardel, Y.; Nigou, J.; Puzo, G.; Erard, F.; Ryffel, B. Toll-like Receptor 2 (TLR2)-Dependent-Positive and TLR2-Independent-Negative Regulation of Proinflammatory Cytokines by Mycobacterial Lipomannans. *J. Immunol.* **2004**, *172*, 4425–4434. [CrossRef] [PubMed]
24. Long, E.M.; Millen, B.; Kubes, P.; Robbins, S.M. Lipoteichoic Acid Induces Unique Inflammatory Responses When Compared to Other Toll-like Receptor 2 Ligands. *PLoS ONE* **2009**, *4*, e5601. [CrossRef]
25. Francisco, S.; Billod, J.-M.; Merino, J.; Punzón, C.; Gallego, A.; Arranz, A.; Martin-Santamaria, S.; Fresno, M. Induction of TLR4/TLR2 Interaction and Heterodimer Formation by Low Endotoxic Atypical LPS. *Front. Immunol.* **2021**, *12*, 748303. [CrossRef] [PubMed]
26. Nagashima, H.; Iwatani, S.; Cruz, M.; Jiménez Abreu, J.A.; Uchida, T.; Mahachai, V.; Vilaichone, R.-K.; Graham, D.Y.; Yamaoka, Y. Toll-like Receptor 10 in Helicobacter Pylori Infection. *J. Infect. Dis.* **2015**, *212*, 1666–1676. [CrossRef]
27. Manukyan, M.; Triantafilou, K.; Triantafilou, M.; Mackie, A.; Nilsen, N.; Espevik, T.; Wiesmüller, K.-H.; Ulmer, A.J.; Heine, H. Binding of Lipopeptide to CD14 Induces Physical Proximity of CD14, TLR2 and TLR1. *Eur. J. Immunol.* **2005**, *35*, 911–921. [CrossRef]
28. Hoppstädter, J.; Dembek, A.; Linnenberger, R.; Dahlem, C.; Barghash, A.; Fecher-Trost, C.; Fuhrmann, G.; Koch, M.; Kraegeloh, A.; Huwer, H.; et al. Toll-Like Receptor 2 Release by Macrophages: An Anti-Inflammatory Program Induced by Glucocorticoids and Lipopolysaccharide. *Front. Immunol.* **2019**, *10*, 1634. [CrossRef]
29. Zheng, H.; Tan, Z.; Zhou, T.; Zhu, F.; Ding, Y.; Liu, T.; Wu, Y.; Xu, W. The TLR2 Is Activated by Sporozoites and Suppresses Intrahepatic Rodent Malaria Parasite Development. *Sci. Rep.* **2015**, *5*, 18239. [CrossRef]
30. Oosenbrug, T.; van de Graaff, M.J.; Ressing, M.E.; van Kasteren, S.I. Chemical Tools for Studying TLR Signaling Dynamics. *Cell Chem. Biol.* **2017**, *24*, 801–812. [CrossRef]
31. Kurkjian, C.J.; Guo, H.; Montgomery, N.D.; Cheng, N.; Yuan, H.; Merrill, J.R.; Sempowski, G.D.; Brickey, W.J.; Ting, J.P.-Y. The Toll–Like Receptor 2/6 Agonist, FSL–1 Lipopeptide, Therapeutically Mitigates Acute Radiation Syndrome. *Sci. Rep.* **2017**, *7*, 17355. [CrossRef]
32. Scheibner, K.A.; Lutz, M.A.; Boodoo, S.; Fenton, M.J.; Powell, J.D.; Horton, M.R. Hyaluronan Fragments Act as an Endogenous Danger Signal by Engaging TLR21. *J. Immunol.* **2006**, *177*, 1272–1281. [CrossRef] [PubMed]
33. Funderburg, N.; Lederman, M.M.; Feng, Z.; Drage, M.G.; Jadlowsky, J.; Harding, C.V.; Weinberg, A.; Sieg, S.F. Human β-Defensin-3 Activates Professional Antigen-Presenting Cells via Toll-like Receptors 1 and 2. *Proc. Natl. Acad. Sci. USA* **2007**, *104*, 18631–18635. [CrossRef] [PubMed]
34. Erridge, C. Endogenous Ligands of TLR2 and TLR4: Agonists or Assistants? *J. Leukoc. Biol.* **2010**, *87*, 989–999. [CrossRef] [PubMed]
35. Marks, K.E.; Cho, K.; Stickling, C.; Reynolds, J.M. Toll-like Receptor 2 in Autoimmune Inflammation. *Immune Netw.* **2021**, *21*, e18. [CrossRef]
36. Liu, S.; Liu, Y.; Hao, W.; Wolf, L.; Kiliaan, A.J.; Penke, B.; Rübe, C.E.; Walter, J.; Heneka, M.T.; Hartmann, T.; et al. TLR2 Is a Primary Receptor for Alzheimer's Amyloid β Peptide to Trigger Neuroinflammatory Activation. *J. Immunol.* **2012**, *188*, 1098–1107. [CrossRef]
37. Kim, C.; Ho, D.-H.; Suk, J.-E.; You, S.; Michael, S.; Kang, J.; Joong Lee, S.; Masliah, E.; Hwang, D.; Lee, H.-J.; et al. Neuron-Released Oligomeric α-Synuclein Is an Endogenous Agonist of TLR2 for Paracrine Activation of Microglia. *Nat. Commun.* **2013**, *4*, 1562. [CrossRef]
38. Yamamoto, M.; Sato, S.; Hemmi, H.; Sanjo, H.; Uematsu, S.; Kaisho, T.; Hoshino, K.; Takeuchi, O.; Kobayashi, M.; Fujita, T.; et al. Essential Role for TIRAP in Activation of the Signalling Cascade Shared by TLR2 and TLR4. *Nature* **2002**, *420*, 324–329. [CrossRef]
39. Rajpoot, S.; Wary, K.K.; Ibbott, R.; Liu, D.; Saqib, U.; Thurston, T.L.M.; Baig, M.S. TIRAP in the Mechanism of Inflammation. *Front. Immunol.* **2021**, *12*, 697588. [CrossRef]

40. Lin, S.-C.; Lo, Y.-C.; Wu, H. Helical Assembly in the MyD88–IRAK4–IRAK2 Complex in TLR/IL-1R Signalling. *Nature* **2010**, *465*, 885–890. [CrossRef]
41. Kawasaki, T.; Kawai, T. Toll-like Receptor Signaling Pathways. *Front. Immunol.* **2014**, *5*, 461. [CrossRef]
42. Dunne, A.; Carpenter, S.; Brikos, C.; Gray, P.; Strelow, A.; Wesche, H.; Morrice, N.; O'Neill, L.A.J. IRAK1 and IRAK4 Promote Phosphorylation, Ubiquitination, and Degradation of MyD88 Adaptor-like (Mal). *J. Biol. Chem.* **2016**, *291*, 24802. [CrossRef] [PubMed]
43. Piao, W.; Ru, L.W.; Toshchakov, V.Y. Differential Adapter Recruitment by TLR2 Co-Receptors. *Pathog. Dis.* **2016**, *74*, ftw043. [CrossRef] [PubMed]
44. Sahoo, B.R.; Basu, M.; Swain, B.; Dikhit, M.R.; Jayasankar, P.; Samanta, M. Elucidation of Novel Structural Scaffold in Rohu TLR2 and Its Binding Site Analysis with Peptidoglycan, Lipoteichoic Acid and Zymosan Ligands, and Downstream MyD88 Adaptor Protein. *Biomed Res. Int.* **2013**, *2013*, 185282. [CrossRef]
45. Shukla, S.; Richardson, E.T.; Drage, M.G.; Boom, W.H.; Harding, C. V Mycobacterium Tuberculosis Lipoprotein and Lipoglycan Binding to Toll-Like Receptor 2 Correlates with Agonist Activity and Functional Outcomes. *Infect. Immun.* **2018**, *86*, e00450-18. [CrossRef] [PubMed]
46. de Groot, R.C.A.; Zhu, H.; Hoogenboezem, T.; de Bruijn, A.C.J.M.; Eenjes, E.; Jong, A.E.J.'t.; Belo, A.I.; Estevão, S.C.; Bajramovic, J.J.; Rottier, R.J.; et al. Mycoplasma Pneumoniae Compared to Streptococcus Pneumoniae Avoids Induction of Proinflammatory Epithelial Cell Responses despite Robustly Inducing TLR2 Signaling. *Infect. Immun.* **2022**, *90*, e00129-22. [CrossRef] [PubMed]
47. Im, J.; Baik, J.E.; Lee, D.; Park, O.-J.; Park, D.H.; Yun, C.-H.; Han, S.H. Bacterial Lipoproteins Induce BAFF Production via TLR2/MyD88/JNK Signaling Pathways in Dendritic Cells. *Front. Immunol.* **2020**, *11*, 564699. [CrossRef] [PubMed]
48. Su, L.; Wang, Y.; Wang, J.; Mifune, Y.; Morin, M.D.; Jones, B.T.; Moresco, E.M.Y.; Boger, D.L.; Beutler, B.; Zhang, H. Structural Basis of TLR2/TLR1 Activation by the Synthetic Agonist Diprovocim. *J. Med. Chem.* **2019**, *62*, 2938–2949. [CrossRef] [PubMed]
49. Udgata, A.; Qureshi, R.; Mukhopadhyay, S. Transduction of Functionally Contrasting Signals by Two Mycobacterial PPE Proteins Downstream of TLR2 Receptors. *J. Immunol.* **2016**, *197*, 1776–1787. [CrossRef]
50. Radakovics, K.; Battin, C.; Leitner, J.; Geiselhart, S.; Paster, W.; Stöckl, J.; Hoffmann-Sommergruber, K.; Steinberger, P. A Highly Sensitive Cell-Based TLR Reporter Platform for the Specific Detection of Bacterial TLR Ligands. *Front. Immunol.* **2022**, *12*, 817604. [CrossRef]
51. Jin, M.S.; Kim, S.E.; Heo, J.Y.; Lee, M.E.; Kim, H.M.; Paik, S.-G.; Lee, H.; Lee, J.-O. Crystal Structure of the TLR1-TLR2 Heterodimer Induced by Binding of a Tri-Acylated Lipopeptide. *Cell* **2007**, *130*, 1071–1082. [CrossRef]
52. Kang, J.Y.; Nan, X.; Jin, M.S.; Youn, S.-J.; Ryu, Y.H.; Mah, S.; Han, S.H.; Lee, H.; Paik, S.-G.; Lee, J.-O. Recognition of Lipopeptide Patterns by Toll-like Receptor 2-Toll-like Receptor 6 Heterodimer. *Immunity* **2009**, *31*, 873–884. [CrossRef] [PubMed]
53. Fore, F.; Indriputri, C.; Mamutse, J.; Nugraha, J. TLR10 and Its Unique Anti-Inflammatory Properties and Potential Use as a Target in Therapeutics. *Immune Netw.* **2020**, *20*, e21. [CrossRef] [PubMed]
54. Koponen, P.; Vuononvirta, J.; Nuolivirta, K.; Helminen, M.; He, Q.; Korppi, M. The Association of Genetic Variants in Toll-like Receptor 2 Subfamily with Allergy and Asthma after Hospitalization for Bronchiolitis in Infancy. *Pediatr. Infect. Dis. J.* **2014**, *33*, 463–466. [CrossRef]
55. Huang, Y.; Temperley, N.D.; Ren, L.; Smith, J.; Li, N.; Burt, D.W. Molecular Evolution of the Vertebrate TLR1 Gene Family--a Complex History of Gene Duplication, Gene Conversion, Positive Selection and Co-Evolution. *BMC Evol. Biol.* **2011**, *11*, 149. [CrossRef] [PubMed]
56. Pachathundikandi, S.K.; Backert, S. Differential Expression of Interleukin 1β During Helicobacter Pylori Infection of Toll-like Receptor 2 (TLR2) and TLR10-Expressing HEK293 Cell Lines. *J. Infect. Dis.* **2016**, *214*, 166–167. [CrossRef]
57. Regan, T.; Nally, K.; Carmody, R.; Houston, A.; Shanahan, F.; Macsharry, J.; Brint, E. Identification of TLR10 as a Key Mediator of the Inflammatory Response to Listeria Monocytogenes in Intestinal Epithelial Cells and Macrophages. *J. Immunol.* **2013**, *191*, 6084–6092. [CrossRef] [PubMed]
58. da Rocha Sobrinho, H.M.; Saar Gomes, R.; da Silva, D.J.; Quixabeira, V.B.L.; Joosten, L.A.B.; Ribeiro de Barros Cardoso, C.; Ribeiro-Dias, F. Toll-like Receptor 10 Controls TLR2-Induced Cytokine Production in Monocytes from Patients with Parkinson's Disease. *J. Neurosci. Res.* **2021**, *99*, 2511–2524. [CrossRef]
59. Muniz-Bongers, L.R.; McClain, C.B.; Saxena, M.; Bongers, G.; Merad, M.; Bhardwaj, N. MMP2 and TLRs Modulate Immune Responses in the Tumor Microenvironment. *JCI Insight* **2021**, *6*, e144913. [CrossRef]
60. Behzadi, P.; García-Perdomo, H.A.; Karpiński, T.M. Toll-Like Receptors: General Molecular and Structural Biology. *J. Immunol. Res.* **2021**, *2021*, 9914854. [CrossRef]
61. Raby, A.-C.; Holst, B.; Le Bouder, E.; Diaz, C.; Ferran, E.; Conraux, L.; Guillemot, J.-C.; Coles, B.; Kift-Morgan, A.; Colmont, C.S.; et al. Targeting the TLR Co-Receptor CD14 with TLR2-Derived Peptides Modulates Immune Responses to Pathogens. *Sci. Transl. Med.* **2013**, *5*, 185ra64. [CrossRef]
62. Janot, L.; Secher, T.; Torres, D.; Maillet, I.; Pfeilschifter, J.; Quesniaux, V.F.J.; Landmann, R.; Ryffel, B.; Erard, F. CD14 Works with Toll-Like Receptor 2 to Contribute to Recognition and Control of Listeria Monocytogenes Infection. *J. Infect. Dis.* **2008**, *198*, 115–124. [CrossRef] [PubMed]
63. Chávez-Sánchez, L.; Garza-Reyes, M.G.; Espinosa-Luna, J.E.; Chávez-Rueda, K.; Legorreta-Haquet, M.V.; Blanco-Favela, F. The Role of TLR2, TLR4 and CD36 in Macrophage Activation and Foam Cell Formation in Response to OxLDL in Humans. *Hum. Immunol.* **2014**, *75*, 322–329. [CrossRef] [PubMed]

64. Granucci, F.; Zanoni, I. Role of CD14 in Host Protection against Infections and in Metabolism Regulation. *Front. Cell. Infect. Microbiol.* **2013**, *3*, 32. [CrossRef]
65. Sheedy, F.J.; Grebe, A.; Rayner, K.J.; Kalantari, P.; Ramkhelawon, B.; Carpenter, S.B.; Becker, C.E.; Ediriweera, H.N.; Mullick, A.E.; Golenbock, D.T.; et al. CD36 Coordinates NLRP3 Inflammasome Activation by Facilitating Intracellular Nucleation of Soluble Ligands into Particulate Ligands in Sterile Inflammation. *Nat. Immunol.* **2013**, *14*, 812–820. [CrossRef]
66. da Silva, T.A.; Zorzetto-Fernandes, A.L.V.; Cecílio, N.T.; Sardinha-Silva, A.; Fernandes, F.F.; Roque-Barreira, M.C. CD14 Is Critical for TLR2-Mediated M1 Macrophage Activation Triggered by N-Glycan Recognition. *Sci. Rep.* **2017**, *7*, 7083. [CrossRef]
67. Gadagkar, S.G.; Lalancette-Hébert, M.; Thammisetty, S.S.; Vexler, Z.S.; Kriz, J. CD36 Neutralisation Blunts TLR2-IRF7 but Not IRF3 Pathway in Neonatal Mouse Brain and Immature Human Microglia Following Innate Immune Challenge. *Sci. Rep.* **2023**, *13*, 2304. [CrossRef]
68. Colleselli, K.; Ebeyer-Masotta, M.; Neuditschko, B.; Stierschneider, A.; Pollhammer, C.; Potocnjak, M.; Hundsberger, H.; Herzog, F.; Wiesner, C. Beyond Pattern Recognition: TLR2 Promotes Chemotaxis, Cell Adhesion, and Migration in THP-1 Cells. *Cells* **2023**, *12*, 1425. [CrossRef]
69. Kawana, H.; Karaki, H.; Higashi, M.; Miyazaki, M.; Hilberg, F.; Kitagawa, M.; Harigaya, K. CD44 Suppresses TLR-Mediated Inflammation. *J. Immunol.* **2008**, *180*, 4235–4245. [CrossRef]
70. Liang, J.; Jiang, D.; Griffith, J.; Yu, S.; Fan, J.; Zhao, X.; Bucala, R.; Noble, P.W. CD44 Is a Negative Regulator of Acute Pulmonary Inflammation and Lipopolysaccharide-TLR Signaling in Mouse Macrophages. *J. Immunol.* **2007**, *178*, 2469–2475. [CrossRef]
71. Ezz, M.A.; Mansouri, A.; Akthar, I.; Yousef, M.S.; Kowsar, R.; Miyamoto, A. Hyaluronan Regulates Sperm-Induced Inflammatory Response by Enhancing Sperm Attachment to Bovine Endometrial Epithelial Cells via CD44: In-Silico and in-Vitro Approaches. *Front. Endocrinol.* **2023**, *14*, 1134868. [CrossRef]
72. D'Ascola, A.; Scuruchi, M.; Ruggeri, R.M.; Avenoso, A.; Mandraffino, G.; Vicchio, T.M.; Campo, S.; Campo, G.M. Hyaluronan Oligosaccharides Modulate Inflammatory Response, NIS and Thyreoglobulin Expression in Human Thyrocytes. *Arch. Biochem. Biophys.* **2020**, *694*, 108598. [CrossRef]
73. Zhang, Y.; Wang, H. Integrin Signalling and Function in Immune Cells. *Immunology* **2012**, *135*, 268–275. [CrossRef] [PubMed]
74. Sales, A.; Ende, K.; Diemer, J.; Kyvik, A.R.; Veciana, J.; Ratera, I.; Kemkemer, R.; Spatz, J.P.; Guasch, J. Cell Type-Dependent Integrin Distribution in Adhesion and Migration Responses on Protein-Coated Microgrooved Substrates. *ACS Omega* **2019**, *4*, 1791–1800. [CrossRef]
75. Sumanth, M.S.; Abhilasha, K.V.; Jacob, S.P.; Chaithra, V.H.; Basrur, V.; Willard, B.; McIntyre, T.M.; Prabhu, K.S.; Marathe, G.K. Acute Phase Protein, α-1-Acid Glycoprotein (AGP-1), Has Differential Effects on TLR-2 and TLR-4 Mediated Responses. *Immunobiology* **2019**, *224*, 672–680. [CrossRef]
76. de Barros, B.C.S.C.; Almeida, B.R.; Suzuki, E. Paracoccidioides Brasiliensis Downmodulates A3 Integrin Levels in Human Lung Epithelial Cells in a TLR2-Dependent Manner. *Sci. Rep.* **2020**, *10*, 19483. [CrossRef] [PubMed]
77. Deisseroth, K.; Feng, G.; Majewska, A.K.; Miesenböck, G.; Ting, A.; Schnitzer, M.J. Next-Generation Optical Technologies for Illuminating Genetically Targeted Brain Circuits. *J. Neurosci.* **2006**, *26*, 10380. [CrossRef]
78. Joshi, J.; Rubart, M.; Zhu, W. Optogenetics: Background, Methodological Advances and Potential Applications for Cardiovascular Research and Medicine. *Front. Bioeng. Biotechnol.* **2020**, *7*, 466. [CrossRef]
79. Kolar, K.; Weber, W. Synthetic Biological Approaches to Optogenetically Control Cell Signaling. *Curr. Opin. Biotechnol.* **2017**, *47*, 112–119. [CrossRef]
80. Grusch, M.; Schelch, K.; Riedler, R.; Reichhart, E.; Differ, C.; Berger, W.; Inglés-Prieto, Á.; Janovjak, H. Spatio-temporally Precise Activation of Engineered Receptor Tyrosine Kinases by Light. *EMBO J.* **2014**, *33*, 1713–1726. [CrossRef]
81. Tichy, A.-M.; So, W.L.; Gerrard, E.J.; Janovjak, H. Structure-Guided Optimization of Light-Activated Chimeric G-Protein-Coupled Receptors. *Structure* **2022**, *30*, 1075–1087.e4. [CrossRef]
82. Abreu, N.; Levitz, J. Optogenetic Techniques for Manipulating and Sensing G Protein-Coupled Receptor Signaling. *Methods Mol. Biol.* **2020**, *2173*, 21–51. [CrossRef] [PubMed]
83. Huang, K.; Merkle, T.; Beck, C.F. Isolation and Characterization of a Chlamydomonas Gene That Encodes a Putative Blue-Light Photoreceptor of the Phototropin Family. *Physiol. Plant.* **2002**, *115*, 613–622. [CrossRef] [PubMed]
84. Kinoshita, T.; Doi, M.; Suetsugu, N.; Kagawa, T.; Wada, M.; Shimazaki, K. Phot1 and Phot2 Mediate Blue Light Regulation of Stomatal Opening. *Nature* **2001**, *414*, 656–660. [CrossRef] [PubMed]
85. Stierschneider, A.; Neuditschko, B.; Colleselli, K.; Hundsberger, H.; Herzog, F.; Wiesner, C. Comparative and Temporal Characterization of LPS and Blue-Light-Induced TLR4 Signal Transduction and Gene Expression in Optogenetically Manipulated Endothelial Cells. *Cells* **2023**, *12*, 697. [CrossRef] [PubMed]
86. Stierschneider, A.; Grünstäudl, P.; Colleselli, K.; Atzler, J.; Klein, C.T.; Hundsberger, H.; Wiesner, C. Light-Inducible Spatio-Temporal Control of TLR4 and NF-KB-Gluc Reporter in Human Pancreatic Cell Line. *Int. J. Mol. Sci.* **2021**, *22*, 9232. [CrossRef]
87. Beukema, M.; Jermendi, É.; van den Berg, M.A.; Faas, M.M.; Schols, H.A.; de Vos, P. The Impact of the Level and Distribution of Methyl-Esters of Pectins on TLR2-1 Dependent Anti-Inflammatory Responses. *Carbohydr. Polym.* **2021**, *251*, 117093. [CrossRef]
88. Ju, J.; Li, Z.; Shi, Q. Baicalin Inhibits Inflammation in Rats with Chronic Obstructive Pulmonary Disease by the TLR2/MYD88/NF-κBp65 Signaling Pathway. *Evid.-Based Complement. Altern. Med.* **2022**, *2022*, 7273387. [CrossRef]
89. Ramakrishna, C.; Kujawski, M.; Chu, H.; Li, L.; Mazmanian, S.K.; Cantin, E.M. Bacteroides Fragilis Polysaccharide A Induces IL-10 Secreting B and T Cells That Prevent Viral Encephalitis. *Nat. Commun.* **2019**, *10*, 2153. [CrossRef]

90. Round, J.L.; Mazmanian, S.K. Inducible Foxp3+ Regulatory T-Cell Development by a Commensal Bacterium of the Intestinal Microbiota. *Proc. Natl. Acad. Sci. USA* **2010**, *107*, 12204–12209. [CrossRef]
91. Dasgupta, S.; Erturk-Hasdemir, D.; Ochoa-Reparaz, J.; Reinecker, H.-C.; Kasper, D.L. Plasmacytoid Dendritic Cells Mediate Anti-Inflammatory Responses to a Gut Commensal Molecule via Both Innate and Adaptive Mechanisms. *Cell Host Microbe* **2014**, *15*, 413–423. [CrossRef]
92. Quero, L.; Hanser, E.; Manigold, T.; Tiaden, A.N.; Kyburz, D. TLR2 Stimulation Impairs Anti-Inflammatory Activity of M2-like Macrophages, Generating a Chimeric M1/M2 Phenotype. *Arthritis Res. Ther.* **2017**, *19*, 245. [CrossRef] [PubMed]
93. Holst, B.; Szakmany, T.; Raby, A.-C.; Hamlyn, V.; Durno, K.; Hall, J.E.; Labéta, M.O. Soluble Toll-like Receptor 2 Is a Biomarker for Sepsis in Critically Ill Patients with Multi-Organ Failure within 12 h of ICU Admission. *Intensive Care Med. Exp.* **2017**, *5*, 2. [CrossRef]
94. Chung, K.-J.; Mitroulis, I.; Wiessner, J.R.; Zheng, Y.Y.; Siegert, G.; Sperandio, M.; Chavakis, T. A Novel Pathway of Rapid TLR-Triggered Activation of Integrin-Dependent Leukocyte Adhesion That Requires Rap1 GTPase. *Mol. Biol. Cell* **2014**, *25*, 2948–2955. [CrossRef] [PubMed]
95. Harokopakis, E.; Albzreh, M.H.; Martin, M.H.; Hajishengallis, G. TLR2 Transmodulates Monocyte Adhesion and Transmigration via Rac1- and PI3K-Mediated inside-out Signaling in Response to Porphyromonas Gingivalis Fimbriae. *J. Immunol.* **2006**, *176*, 7645–7656. [CrossRef] [PubMed]
96. McGarry, T.; Veale, D.J.; Gao, W.; Orr, C.; Fearon, U.; Connolly, M. Toll-like Receptor 2 (TLR2) Induces Migration and Invasive Mechanisms in Rheumatoid Arthritis. *Arthritis Res. Ther.* **2015**, *17*, 153. [CrossRef] [PubMed]
97. Williams, B.; Neder, J.; Cui, P.; Suen, A.; Tanaka, K.; Zou, L.; Chao, W. Toll-like Receptors 2 and 7 Mediate Coagulation Activation and Coagulopathy in Murine Sepsis. *J. Thromb. Haemost.* **2019**, *17*, 1683–1693. [CrossRef]
98. Foley, J.H.; Conway, E.M. Cross Talk Pathways Between Coagulation and Inflammation. *Circ. Res.* **2016**, *118*, 1392–1408. [CrossRef]
99. Marín Oyarzún, C.P.; Glembotsky, A.C.; Goette, N.P.; Lev, P.R.; De Luca, G.; Baroni Pietto, M.C.; Moiraghi, B.; Castro Ríos, M.A.; Vicente, A.; Marta, R.F.; et al. Platelet Toll-Like Receptors Mediate Thromboinflammatory Responses in Patients With Essential Thrombocythemia. *Front. Immunol.* **2020**, *11*, 705. [CrossRef]
100. Nilsen, N.J.; Deininger, S.; Nonstad, U.; Skjeldal, F.; Husebye, H.; Rodionov, D.; von Aulock, S.; Hartung, T.; Lien, E.; Bakke, O.; et al. Cellular Trafficking of Lipoteichoic Acid and Toll-like Receptor 2 in Relation to Signaling: Role of CD14 and CD36. *J. Leukoc. Biol.* **2008**, *84*, 280–291. [CrossRef]
101. Nguyen, B.N.; Chávez-Arroyo, A.; Cheng, M.I.; Krasilnikov, M.; Louie, A.; Portnoy, D.A. TLR2 and Endosomal TLR-Mediated Secretion of IL-10 and Immune Suppression in Response to Phagosome-Confined Listeria Monocytogenes. *PLoS Pathog.* **2020**, *16*, e1008622. [CrossRef]
102. Thwaites, R.S.; Unterberger, S.; Chamberlain, G.; Walker-Bone, K.; Davies, K.A.; Sacre, S. TLR1/2 and 5 Induce Elevated Cytokine Levels from Rheumatoid Arthritis Monocytes Independent of ACPA or RF Autoantibody Status. *Rheumatology* **2020**, *59*, 3533–3539. [CrossRef]
103. Agrawal, S.; Gupta, S. TLR1/2, TLR7, and TLR9 Signals Directly Activate Human Peripheral Blood Naive and Memory B Cell Subsets to Produce Cytokines, Chemokines, and Hematopoietic Growth Factors. *J. Clin. Immunol.* **2011**, *31*, 89–98. [CrossRef]
104. Mauldin, I.S.; Wang, E.; Deacon, D.H.; Olson, W.C.; Bao, Y.; Slingluff, C.L.J. TLR2/6 Agonists and Interferon-Gamma Induce Human Melanoma Cells to Produce CXCL10. *Int. J. cancer* **2015**, *137*, 1386–1396. [CrossRef] [PubMed]
105. Flynn, C.M.; Garbers, Y.; Lokau, J.; Wesch, D.; Schulte, D.M.; Laudes, M.; Lieb, W.; Aparicio-Siegmund, S.; Garbers, C. Activation of Toll-like Receptor 2 (TLR2) Induces Interleukin-6 Trans-Signaling. *Sci. Rep.* **2019**, *9*, 7306. [CrossRef]
106. McGarry, T.; Biniecka, M.; Gao, W.; Cluxton, D.; Canavan, M.; Wade, S.; Wade, S.; Gallagher, L.; Orr, C.; Veale, D.J.; et al. Resolution of TLR2-Induced Inflammation through Manipulation of Metabolic Pathways in Rheumatoid Arthritis. *Sci. Rep.* **2017**, *7*, 43165. [CrossRef] [PubMed]
107. Zakeri, A.; Russo, M. Dual Role of Toll-like Receptors in Human and Experimental Asthma Models. *Front. Immunol.* **2018**, *9*, 1027. [CrossRef] [PubMed]
108. Cari, L.; Rosati, L.; Leoncini, G.; Lusenti, E.; Gentili, M.; Nocentini, G.; Riccardi, C.; Migliorati, G.; Ronchetti, S. Association of GILZ with MUC2, TLR2, and TLR4 in Inflammatory Bowel Disease. *Int. J. Mol. Sci.* **2023**, *24*, 2235. [CrossRef]
109. Wasko, N.J.; Nichols, F.; Clark, R.B. Multiple Sclerosis, the Microbiome, TLR2, and the Hygiene Hypothesis. *Autoimmun. Rev.* **2020**, *19*, 102430. [CrossRef]
110. Gyawali, B.; Ramakrishna, K.; Dhamoon, A.S. Sepsis: The Evolution in Definition, Pathophysiology, and Management. *SAGE Open Med.* **2019**, *7*, 2050312119835043. [CrossRef]
111. Singer, M.; Deutschman, C.S.; Seymour, C.; Shankar-Hari, M.; Annane, D.; Bauer, M.; Bellomo, R.; Bernard, G.R.; Chiche, J.D.; Coopersmith, C.M.; et al. The Third International Consensus Definitions for Sepsis and Septic Shock (Sepsis-3). *JAMA J. Am. Med. Assoc.* **2016**, *315*, 801–810. [CrossRef]
112. Angus, D.C.; Linde-Zwirble, W.T.; Lidicker, J.; Clermont, G.; Carcillo, J.; Pinsky, M.R. Epidemiology of Severe Sepsis in the United States: Analysis of Incidence, Outcome, and Associated Costs of Care. *Crit. Care Med.* **2001**, *29*, 1303–1310. [CrossRef] [PubMed]
113. Fujishima, S. Organ Dysfunction as a New Standard for Defining Sepsis. *Inflamm. Regen.* **2016**, *36*, 24. [CrossRef] [PubMed]
114. Brady, J.; Horie, S.; Laffey, J.G. Role of the Adaptive Immune Response in Sepsis. *Intensive Care Med. Exp.* **2020**, *8*, 20. [CrossRef] [PubMed]

115. Liu, D.; Huang, S.-Y.; Sun, J.-H.; Zhang, H.-C.; Cai, Q.-L.; Gao, C.; Li, L.; Cao, J.; Xu, F.; Zhou, Y.; et al. Sepsis-Induced Immunosuppression: Mechanisms, Diagnosis and Current Treatment Options. *Mil. Med. Res.* **2022**, *9*, 56. [CrossRef] [PubMed]
116. Jarczak, D.; Kluge, S.; Nierhaus, A. Sepsis-Pathophysiology and Therapeutic Concepts. *Front. Med.* **2021**, *8*, 628302. [CrossRef]
117. Savva, A.; Roger, T. Targeting Toll-like Receptors: Promising Therapeutic Strategies for the Management of Sepsis-Associated Pathology and Infectious Diseases. *Front. Immunol.* **2013**, *4*, 387. [CrossRef]
118. Hamzeh-Cognasse, H.; Berthelot, P.; Tardy, B.; Pozzetto, B.; Bourlet, T.; Laradi, S.; Garraud, O.; Cognasse, F. Platelet Toll-like Receptors Are Crucial Sensors of Infectious Danger Moieties. *Platelets* **2018**, *29*, 533–540. [CrossRef]
119. Kerrigan, S.W.; Devine, T.; Fitzpatrick, G.; Thachil, J.; Cox, D. Early Host Interactions That Drive the Dysregulated Response in Sepsis. *Front. Immunol.* **2019**, *10*, 1748. [CrossRef]
120. Keane, C.; Tilley, D.; Cunningham, A.; Smolenski, A.; Kadioglu, A.; Cox, D.; Jenkinson, H.F.; Kerrigan, S.W. Invasive Streptococcus Pneumoniae Trigger Platelet Activation via Toll‐like Receptor 2. *J. Thromb. Haemost.* **2010**, *8*, 2757–2765. [CrossRef]
121. Kim, N.-H.; Sung, J.Y.; Choi, Y.J.; Choi, S.-J.; Ahn, S.; Ji, E.; Kim, M.; Kim, C.J.; Song, K.-H.; Choe, P.G.; et al. Toll-like Receptor 2 Downregulation and Cytokine Dysregulation Predict Mortality in Patients with Staphylococcus Aureus Bacteremia. *BMC Infect. Dis.* **2020**, *20*, 901. [CrossRef]
122. Krivan, S.; Kapelouzou, A.; Vagios, S.; Tsilimigras, D.I.; Katsimpoulas, M.; Moris, D.; Aravanis, C.V.; Demesticha, T.D.; Schizas, D.; Mavroidis, M.; et al. Increased Expression of Toll-like Receptors 2, 3, 4 and 7 MRNA in the Kidney and Intestine of a Septic Mouse Model. *Sci. Rep.* **2019**, *9*, 4010. [CrossRef]
123. Wang, X.; Li, D.; Qin, Y.-Y.; Gong, J.; Zou, L.; Chao, W.; Gong, Y. Toll-like Receptor 2 Deficiency Relieves Splenic Immunosuppression during Sepsis. *Immunobiology* **2023**, *228*, 152374. [CrossRef] [PubMed]
124. Hu, B.; Guo, H.; Zhou, P.; Shi, Z.-L. Characteristics of SARS-CoV-2 and COVID-19. *Nat. Rev. Microbiol.* **2021**, *19*, 141–154. [CrossRef] [PubMed]
125. Giamarellos-Bourboulis, E.J.; Netea, M.G.; Rovina, N.; Akinosoglou, K.; Antoniadou, A.; Antonakos, N.; Damoraki, G.; Gkavogianni, T.; Adami, M.-E.; Katsaounou, P.; et al. Complex Immune Dysregulation in COVID-19 Patients with Severe Respiratory Failure. *Cell Host Microbe* **2020**, *27*, 992–1000.e3. [CrossRef]
126. Lucas, C.; Wong, P.; Klein, J.; Castro, T.B.R.; Silva, J.; Sundaram, M.; Ellingson, M.K.; Mao, T.; Oh, J.E.; Israelow, B.; et al. Longitudinal Analyses Reveal Immunological Misfiring in Severe COVID-19. *Nature* **2020**, *584*, 463–469. [CrossRef] [PubMed]
127. Karki, R.; Sharma, B.R.; Tuladhar, S.; Williams, E.P.; Zalduondo, L.; Samir, P.; Zheng, M.; Sundaram, B.; Banoth, B.; Malireddi, R.K.S.; et al. Synergism of TNF-α and IFN-γ Triggers Inflammatory Cell Death, Tissue Damage, and Mortality in SARS-CoV-2 Infection and Cytokine Shock Syndromes. *Cell* **2021**, *184*, 149–168.e17. [CrossRef]
128. Jose, R.J.; Manuel, A. COVID-19 Cytokine Storm: The Interplay between Inflammation and Coagulation. *Lancet. Respir. Med.* **2020**, *8*, e46–e47. [CrossRef] [PubMed]
129. Zheng, M.; Karki, R.; Williams, E.P.; Yang, D.; Fitzpatrick, E.; Vogel, P.; Jonsson, C.B.; Kanneganti, T.-D. TLR2 Senses the SARS-CoV-2 Envelope Protein to Produce Inflammatory Cytokines. *Nat. Immunol.* **2021**, *22*, 829–838. [CrossRef]
130. Planès, R.; Bert, J.-B.; Tairi, S.; BenMohamed, L.; Bahraoui, E. SARS-CoV-2 Envelope (E) Protein Binds and Activates TLR2 Pathway: A Novel Molecular Target for COVID-19 Interventions. *Viruses* **2022**, *14*, 999. [CrossRef]
131. Proud, P.C.; Tsitoura, D.; Watson, R.J.; Chua, B.Y.; Aram, M.J.; Bewley, K.R.; Cavell, B.E.; Cobb, R.; Dowall, S.; Fotheringham, S.A.; et al. Prophylactic Intranasal Administration of a TLR2/6 Agonist Reduces Upper Respiratory Tract Viral Shedding in a SARS-CoV-2 Challenge Ferret Model. *EBioMedicine* **2021**, *63*, 103153. [CrossRef]
132. Olwal, C.; Nganyewo, N.; Tapela, K.; Lindsey, A.; Zune, D.; Owoicho, O.; Bediako, Y.; Duodu, S. Parallels in Sepsis and COVID-19 Conditions: Implications for Managing Severe COVID-19. *Front. Immunol.* **2021**, *12*, 602848. [CrossRef] [PubMed]
133. Rietdijk, C.D.; van Wezel, R.J.A.; Garssen, J.; Kraneveld, A.D. Neuronal Toll-like Receptors and Neuro-Immunity in Parkinson's Disease, Alzheimer's Disease and Stroke. *Neuroimmunol. Neuroinflamm.* **2016**, *3*, 27–37. [CrossRef]
134. Jack, C.S.; Arbour, N.; Manusow, J.; Montgrain, V.; Blain, M.; McCrea, E.; Shapiro, A.; Antel, J.P. TLR Signaling Tailors Innate Immune Responses in Human Microglia and Astrocytes. *J. Immunol.* **2005**, *175*, 4320–4330. [CrossRef] [PubMed]
135. Chitnis, T.; Weiner, H.L. CNS Inflammation and Neurodegeneration. *J. Clin. Investig.* **2017**, *127*, 3577–3587. [CrossRef]
136. Radenovic, L.; Nenadic, M.; Ułamek-Kozioł, M.; Januszewski, S.; Czuczwar, S.J.; Andjus, P.R.; Pluta, R. Heterogeneity in Brain Distribution of Activated Microglia and Astrocytes in a Rat Ischemic Model of Alzheimer's Disease after 2 Years of Survival. *Aging* **2020**, *12*, 12251–12267. [CrossRef]
137. Moyse, E.; Krantic, S.; Djellouli, N.; Roger, S.; Angoulvant, D.; Debacq, C.; Leroy, V.; Fougere, B.; Aidoud, A. Neuroinflammation: A Possible Link Between Chronic Vascular Disorders and Neurodegenerative Diseases. *Front. Aging Neurosci.* **2022**, *14*, 827263. [CrossRef]
138. Dansokho, C.; Ait Ahmed, D.; Aid, S.; Toly-Ndour, C.; Chaigneau, T.; Calle, V.; Cagnard, N.; Holzenberger, M.; Piaggio, E.; Aucouturier, P.; et al. Regulatory T Cells Delay Disease Progression in Alzheimer-like Pathology. *Brain* **2016**, *139*, 1237–1251. [CrossRef]
139. Skaper, S.D.; Facci, L.; Zusso, M.; Giusti, P. An Inflammation-Centric View of Neurological Disease: Beyond the Neuron. *Front. Cell. Neurosci.* **2018**, *12*, 72. [CrossRef]
140. Wang, L.; Yang, H.-Y.; Zang, C.-X.; Shang, J.-M.; Liu, H.; Zhang, Z.-H.; Yuan, F.-Y.; Ju, C.; Li, F.-Y.; Bao, X.-Q.; et al. TLR2 Potentiates SR-Marco-Mediated Neuroinflammation by Interacting with the SRCR Domain. *Mol. Neurobiol.* **2021**, *58*, 5743–5755. [CrossRef]

141. Ferrucci, L.; Fabbri, E. Inflammageing: Chronic Inflammation in Ageing, Cardiovascular Disease, and Frailty. *Nat. Rev. Cardiol.* **2018**, *15*, 505–522. [CrossRef]
142. Ferrari, C.; Sorbi, S. The Complexity of Alzheimer's Disease: An Evolving Puzzle. *Physiol. Rev.* **2021**, *101*, 1047–1081. [CrossRef]
143. Song, C.; Shi, J.; Zhang, P.; Zhang, Y.; Xu, J.; Zhao, L.; Zhang, R.; Wang, H.; Chen, H. Immunotherapy for Alzheimer's Disease: Targeting β-Amyloid and Beyond. *Transl. Neurodegener.* **2022**, *11*, 18. [CrossRef] [PubMed]
144. Novoa, C.; Salazar, P.; Cisternas, P.; Gherardelli, C.; Vera-Salazar, R.; Zolezzi, J.M.; Inestrosa, N.C. Inflammation Context in Alzheimer's Disease, a Relationship Intricate to Define. *Biol. Res.* **2022**, *55*, 39. [CrossRef] [PubMed]
145. Kinney, J.W.; Bemiller, S.M.; Murtishaw, A.S.; Leisgang, A.M.; Salazar, A.M.; Lamb, B.T. Inflammation as a Central Mechanism in Alzheimer's Disease. *Alzheimer's Dement. Transl. Res. Clin. Interv.* **2018**, *4*, 575–590. [CrossRef]
146. Pascoal, T.A.; Benedet, A.L.; Ashton, N.J.; Kang, M.S.; Therriault, J.; Chamoun, M.; Savard, M.; Lussier, F.Z.; Tissot, C.; Karikari, T.K.; et al. Microglial Activation and Tau Propagate Jointly across Braak Stages. *Nat. Med.* **2021**, *27*, 1592–1599. [CrossRef]
147. Dallas, M.L.; Widera, D. TLR2 and TLR4-Mediated Inflammation in Alzheimer's Disease: Self-Defense or Sabotage? *Neural Regen. Res.* **2021**, *16*, 1552–1553. [CrossRef] [PubMed]
148. Wang, Z.; Weaver, D.F. Microglia and Microglial-Based Receptors in the Pathogenesis and Treatment of Alzheimer's Disease. *Int. Immunopharmacol.* **2022**, *110*, 109070. [CrossRef]
149. Reed-Geaghan, E.G.; Savage, J.C.; Hise, A.G.; Landreth, G.E. CD14 and Toll-like Receptors 2 and 4 Are Required for Fibrillar A{beta}-Stimulated Microglial Activation. *J. Neurosci.* **2009**, *29*, 11982–11992. [CrossRef] [PubMed]
150. Rangasamy, S.B.; Jana, M.; Roy, A.; Corbett, G.T.; Kundu, M.; Chandra, S.; Mondal, S.; Dasarathi, S.; Mufson, E.J.; Mishra, R.K.; et al. Selective Disruption of TLR2-MyD88 Interaction Inhibits Inflammation and Attenuates Alzheimer's Pathology. *J. Clin. Investig.* **2018**, *128*, 4297–4312. [CrossRef]
151. Henderson, M.X.; Trojanowski, J.Q.; Lee, V.M.-Y. α-Synuclein Pathology in Parkinson's Disease and Related α-Synucleinopathies. *Neurosci. Lett.* **2019**, *709*, 134316. [CrossRef]
152. Spillantini, M.G.; Schmidt, M.L.; Lee, V.M.-Y.; Trojanowski, J.Q.; Jakes, R.; Goedert, M. α-Synuclein in Lewy Bodies. *Nature* **1997**, *388*, 839–840. [CrossRef] [PubMed]
153. Marotta, N.P.; Ara, J.; Uemura, N.; Lougee, M.G.; Meymand, E.S.; Zhang, B.; Petersson, E.J.; Trojanowski, J.Q.; Lee, V.M.-Y. Alpha-Synuclein from Patient Lewy Bodies Exhibits Distinct Pathological Activity That Can Be Propagated in Vitro. *Acta Neuropathol. Commun.* **2021**, *9*, 188. [CrossRef]
154. Luna, E.; Luk, K.C. Bent out of Shape: α-Synuclein Misfolding and the Convergence of Pathogenic Pathways in Parkinson's Disease. *FEBS Lett.* **2015**, *589*, 3749–3759. [CrossRef] [PubMed]
155. Forloni, G. Alpha Synuclein: Neurodegeneration and Inflammation. *Int. J. Mol. Sci.* **2023**, *24*, 5914. [CrossRef] [PubMed]
156. Mazzotta, G.M.; Ceccato, N.; Conte, C. Synucleinopathies Take Their Toll: Are TLRs a Way to Go? *Cells* **2023**, *12*, 1231. [CrossRef]
157. Kim, C.; Spencer, B.; Rockenstein, E.; Yamakado, H.; Mante, M.; Adame, A.; Fields, J.A.; Masliah, D.; Iba, M.; Lee, H.-J.; et al. Immunotherapy Targeting Toll-like Receptor 2 Alleviates Neurodegeneration in Models of Synucleinopathy by Modulating α-Synuclein Transmission and Neuroinflammation. *Mol. Neurodegener.* **2018**, *13*, 43. [CrossRef]
158. Chedid, J.; Labrador-Garrido, A.; Zhong, S.; Gao, J.; Zhao, Y.; Perera, G.; Kim, W.S.; Halliday, G.M.; Dzamko, N. A Small Molecule Toll-like Receptor Antagonist Rescues α-Synuclein Fibril Pathology. *J. Biol. Chem.* **2022**, *298*, 102260. [CrossRef]
159. Dutta, D.; Jana, M.; Majumder, M.; Mondal, S.; Roy, A.; Pahan, K. Selective Targeting of the TLR2/MyD88/NF-KB Pathway Reduces α-Synuclein Spreading in Vitro and in Vivo. *Nat. Commun.* **2021**, *12*, 5382. [CrossRef]
160. Caplan, I.F.; Maguire-Zeiss, K.A. Toll-Like Receptor 2 Signaling and Current Approaches for Therapeutic Modulation in Synucleinopathies. *Front. Pharmacol.* **2018**, *9*, 417. [CrossRef]
161. Bennett, J.M.; Reeves, G.; Billman, G.E.; Sturmberg, J.P. Inflammation-Nature's Way to Efficiently Respond to All Types of Challenges: Implications for Understanding and Managing "the Epidemic" of Chronic Diseases. *Front. Med.* **2018**, *5*, 316. [CrossRef]
162. Netea, M.G.; Balkwill, F.; Chonchol, M.; Cominelli, F.; Donath, M.Y.; Giamarellos-Bourboulis, E.J.; Golenbock, D.; Gresnigt, M.S.; Heneka, M.T.; Hoffman, H.M.; et al. A Guiding Map for Inflammation. *Nat. Immunol.* **2017**, *18*, 826–831. [CrossRef]
163. Roth, G.A.; Abate, D.; Abate, K.H.; Abay, S.M.; Abbafati, C.; Abbasi, N.; Abbastabar, H.; Abd-Allah, F.; Abdela, J.; Abdelalim, A.; et al. Global, Regional, and National Age-Sex-Specific Mortality for 282 Causes of Death in 195 Countries and Territories, 1980-2017: A Systematic Analysis for the Global Burden of Disease Study 2017. *Lancet* **2018**, *392*, 1736–1788. [CrossRef] [PubMed]
164. Furman, D.; Campisi, J.; Verdin, E.; Carrera-Bastos, P.; Targ, S.; Franceschi, C.; Ferrucci, L.; Gilroy, D.W.; Fasano, A.; Miller, G.W.; et al. Chronic Inflammation in the Etiology of Disease across the Life Span. *Nat. Med.* **2019**, *25*, 1822–1832. [CrossRef] [PubMed]
165. Sariol, A.; Perlman, S. SARS-CoV-2 Takes Its Toll. *Nat. Immunol.* **2021**, *22*, 801–802. [CrossRef] [PubMed]
166. Wang, Y.; Lv, S.; Zhou, X.; Niu, X.; Chen, L.; Yang, Z.; Peng, D. Identification of TLR2 as a Key Target in Neuroinflammation in Vascular Dementia. *Front. Genet.* **2022**, *13*, 860122. [CrossRef]
167. Chung, B.K.-S.; Dick, T.; Lee, D.-Y. In Silico Analyses for the Discovery of Tuberculosis Drug Targets. *J. Antimicrob. Chemother.* **2013**, *68*, 2701–2709. [CrossRef]
168. Macalino, S.J.Y.; Billones, J.B.; Organo, V.G.; Carrillo, M.C.O. In Silico Strategies in Tuberculosis Drug Discovery. *Molecules* **2020**, *25*, 665. [CrossRef]

169. Feigin, V.L.; Abajobir, A.A.; Abate, K.H.; Abd-Allah, F.; Abdulle, A.M.; Abera, S.F.; Abyu, G.Y.; Ahmed, M.B.; Aichour, A.N.; Aichour, I.; et al. Global, Regional, and National Burden of Neurological Disorders, 1990–2016: A Systematic Analysis for the Global Burden of Disease Study 2016. *Lancet. Neurol.* **2019**, *18*, 459–480. [CrossRef]
170. Ding, C.; Wu, Y.; Chen, X.; Chen, Y.; Wu, Z.; Lin, Z.; Kang, D.; Fang, W.; Chen, F. Global, Regional, and National Burden and Attributable Risk Factors of Neurological Disorders: The Global Burden of Disease Study 1990–2019. *Front. Public Health* **2022**, *10*, 952161. [CrossRef]
171. Van Schependom, J.; D'haeseleer, M. Advances in Neurodegenerative Diseases. *J. Clin. Med.* **2023**, *12*, 1709. [CrossRef]

International Journal of *Molecular Sciences*

MDPI

Review

From Immunity to Neurogenesis: Toll-like Receptors as Versatile Regulators in the Nervous System

Daniela Melissa Abarca-Merlin [1], J. Abigail Martínez-Durán [1], J. David Medina-Pérez [1], Guadalupe Rodríguez-Santos [1] and Lourdes Alvarez-Arellano [1,2,*]

1 Laboratorio de Investigación en Neurociencias, Hospital Infantil de México Federico Gómez, Av. Dr. Márquez 162. Colonia Doctores, Mexico City 06720, Mexico; melyabarca@gmail.com (D.M.A.-M.)

2 CONAHCYT-Hospital Infantil de México Federico Gómez, Mexico City 06720, Mexico

* Correspondence: malvarez@himfg.edu.mx

Abstract: Toll-like receptors (TLRs) are among the main components of the innate immune system. They can detect conserved structures in microorganisms and molecules associated with stress and cellular damage. TLRs are expressed in resident immune cells and both neurons and glial cells of the nervous system. Increasing evidence is emerging on the participation of TLRs not only in the immune response but also in processes of the nervous system, such as neurogenesis and cognition. Below, we present a review of the literature that evaluates the expression and role of TLRs in processes such as neurodevelopment, behavior, cognition, infection, neuroinflammation, and neurodegeneration.

Keywords: TLR; brain; neurogenesis; blood–brain barrier; behavior; cognition; neurodegeneration; neuroinflammation

check for updates

Citation: Abarca-Merlin, D.M.; Martínez-Durán, J.A.; Medina-Pérez, J.D.; Rodríguez-Santos, G.; Alvarez-Arellano, L. From Immunity to Neurogenesis: Toll-like Receptors as Versatile Regulators in the Nervous System. *Int. J. Mol. Sci.* **2024**, *25*, 5711. https://doi.org/10.3390/ijms25115711

Academic Editors: Irmgard Tegeder, Oliver Planz and Ralf Kircheis

Received: 26 March 2024
Revised: 10 May 2024
Accepted: 11 May 2024
Published: 24 May 2024

1. Introduction

The activation of immune system cells is based on the recognition of exogenous molecules from organisms called microorganism-associated molecular patterns (MAMPs), as well as the recognition of endogenous molecules considered warning signals called damage-associated molecular patterns (DAMPs). The recognition of MAMPs and DAMPs is accomplished by germline-encoded receptors called pattern recognition receptors (PRRs), and this family of receptors includes Toll-like receptors (TLRs) [1]. TLRs are type I transmembrane glycoproteins that are characterized by an extracellular domain with leucine-rich repetitive regions (LRRs); MAMPs and DAMPs are recognized through this domain. They also have an intracellular domain similar to that of the interleukin I receptor (IL-1R) family called Toll/IL-1R (TIR), through which they initiate downstream signaling processes. At present, 13 members of this family have been identified in mammals, 10 of which are functional in humans (TLR1–TLR10) and 12 in mice (TLR1–TLR9 and TLR11–TLR13). TLRs can be subdivided based on their predominant cellular localization. The group of TLRs expressed on the cell membrane includes TLR1, TLR2, TLR4, TLR5, TLR6, and TLR10, while TLR3, TLR7, TLR8, and TLR9 are mainly expressed in membranes of intracellular compartments, such as endosomes, lysosomes and the endoplasmic reticulum [1].

TLR2 forms heterodimers with TLR1, TLR6, and TLR10, and in conjunction, these recognize a wide variety of MAMPs and DAMPs, such as lipopeptides, glycolipids, peptidoglycans, lipoteichoic acid, and zymosan. Among the most important described TLR3 agonists are double-stranded RNA (dsRNA) and polyinosinic-polycytidylic acid [poly(I:C), a synthetic agonist]. TLR4 recognizes lipopolysaccharide (LPS), heat shock proteins, and components of the extracellular matrix, and TLR5 recognizes flagellin. TLR7 recognizes single-stranded RNA (ssRNA), microRNAs (miRNAs), small interfering RNAs (siRNAs), and synthetic agonists, such as imidazoquinoline derivatives, including imiquimod and resiquimod (R848) and guanine analogs, such as loxoribine. TLR8 is structurally similar to TLR7 and preferentially recognizes bacterial RNA, viral dsRNA, and synthetic agonists, such as R848. TLR9 recognizes DNA with unmethylated cytosine–guanine (CpG) motifs, synthetic CpG oligonucleotides (ODNs), and immunoglobulin–DNA complexes, and TLR10 has been reported to be activated by proteins of a viral origin [1–3].

Following the recognition of their agonists, TLRs trigger a signaling cascade (Figure 1). The canonical signaling pathway induced by TLRs (other than TLR3) begins with the recruitment of an adapter protein called MyD88. Subsequently, the IRAK4 kinase is recruited to the MyD88–receptor complex, resulting in the phosphorylation and activation of IRAK1, which allows the recruitment of TRAF6. TRAF6 then activates the TAK1/Tab1/2/3 complex, which, in turn, phosphorylates and activates the IKK complex; ultimately, NF-κB is released from its inhibitor IκB. NF-κB is translocated to the nucleus, where it activates the expression of genes encoding molecules that participate in processes such as the immune response, survival, migration, proliferation, and apoptosis, among others. TLR4 can also lead to the activation of TRIF, culminating in the activation of transcription factor IRF-3, which, in turn, promotes transcription mainly of type I interferons. For its part, TLR3 also signals through TRAF3 but without the intervention of MyD88 [3–5].

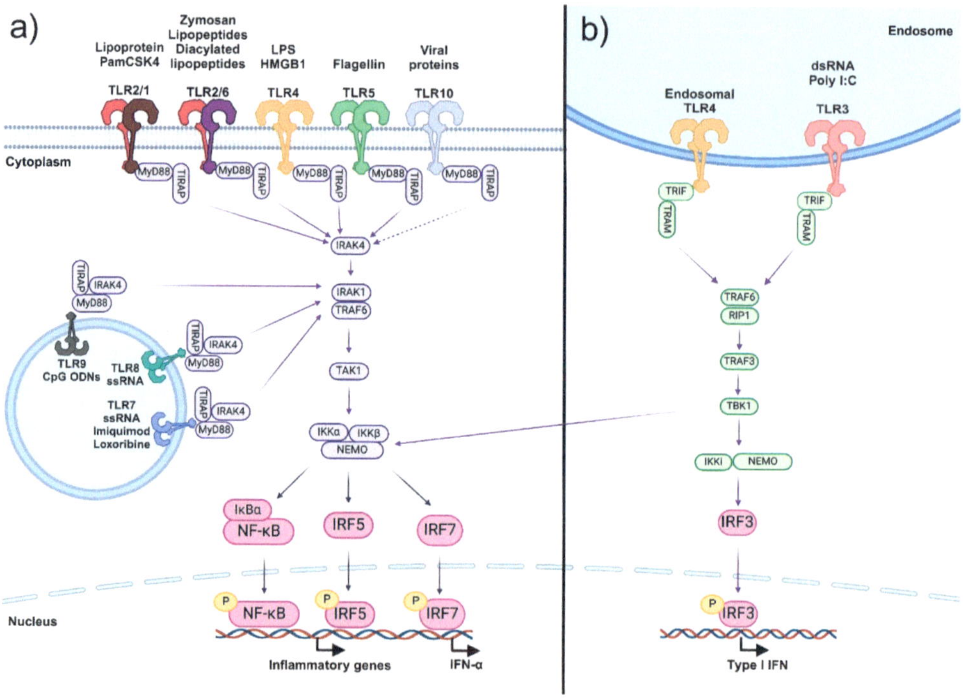

Figure 1. TLR-mediated signaling. Activation of TLRs can induce different signaling pathways. The most common signaling pathway induced by TLRs is the MyD88-dependent pathway (**a**). Although TLR10 has been shown to interact with MyD88, the complete signaling pathway has not been demonstrated experimentally. The MyD88-independent pathway is induced only by the activation of TLR4 and TLR3 (**b**). The signaling cascade culminates with the activation of transcription factors that, in turn, induce the expression of genes involved in various processes, including the immune response. Created with BioRender.com.

Under homeostatic and pathological conditions, TLRs play important roles in nervous system processes, such as neurodevelopment, neurogenesis, cellular migration, differentiation, the neuroinflammatory response, and neurodegeneration [3,5]. TLRs are found in virtually all resident cells of the nervous system. TLRs are expressed on glial cells (microglia, astrocytes, and oligodendrocytes), neuronal progenitors, mature neurons, and cancer cells. Below, we present an overview of the expression of TLRs in the main cell types found in the nervous system and a brief description of the current evidence of their contribution to the function and maintenance of central nervous system (CNS) homeostasis.

2. TLR Expression and Function in Glial Cells and Neurons

A significant proportion of the brain mass is composed of nonneuronal cells called glial cells, which include microglia, astrocytes, and oligodendrocytes. In addition to providing neuronal support, some of their functions include regulating the proliferation, migration, and neurogenesis of neuroblasts; promoting axonal remodeling; secreting the extracellular matrix; modulating synaptic transmission by controlling neurotransmitter availability; and maintaining osmotic, ionic, and pH homeostasis. Additionally, these nonneuronal cells contribute to the provision of energetic and metabolic support to the cellular microenvironment, among other functions [6].

2.1. Microglia

Microglia are resident macrophages of the CNS that specialize in the host defense response through the detection of MAMPs and DAMPs by PRRs, the secretion of proinflammatory cytokines (IL-1β, TNF-α, IL-6, and IL-18) and chemokines, the phagocytosis of cellular debris and dead cells, and the maintenance of myelin [7]. Some evidence indicates that human microglia express TLR1–TLR9, and their activation induces different responses in these cells [8].

Microglia can acquire different functional phenotypes depending on the brain microenvironment. Two phenotypes are traditionally recognized: M1 (proinflammatory) and M2 (neuroprotective). When activated by PAMPs (such as LPS) or inflammatory factors (such as IFN-γ), microglia induce the M1 phenotype, characterized by the production of IL-1β, TNF-α, IL-6, IL-12, metalloproteases (MMPs), and inducible nitric oxide synthase (iNOS), among others, which contribute to the inflammatory environment and neurotoxicity. The M2 phenotype is induced by IL-4, IL-10, IL-13, and TGF-β and promotes the expression of factors involved in tissue repair and cellular homeostasis, such as FIZZ1, arginase 1 (Arg-1), CD206, and insulin-like growth factor 1 (IGF-1) [9,10]. Moreover, a study revealed a novel state in which microglia, which are activated simultaneously by proinflammatory (LPS) and anti-inflammatory (IL-4/IL-13) triggers, can generate a regulatory phenotype through immune pathways [11]. The functional phenotype or polarization of microglia is reversible, unlike that of other terminally differentiated body cells [12]. Recently, a study of M2-polarized microglia revealed that TLR9 activation using a conjugate of ODN 1826 with hyaluronic acid could alter the expression of immunomodulatory cytokines, reversing the antitumor M1 phenotype and improving the survival of an orthotopic glioblastoma model [13].

On the other hand, the activation of TLR4 by LPS has been found to suppress autophagic activity in primary cultures of murine microglia by activating the PI3K-AKT1 pathway and subsequently inhibiting FOXO3, contrary to the autophagic activity triggered by TLR4 activation in bone marrow macrophages [14]. The activation of TLR5 by flagellin induces the release of proinflammatory molecules, chemotaxis, and increased phagocytic activity, while coculture with neurons induces neuronal loss and apoptosis [15].

The activation of TLR7 by the let7 miRNA family revealed that members with a specific nucleotide sequence (UUGU) induced greater secretion of proinflammatory cytokines and the expression of molecules involved in antigen presentation (MHC-I and CD45) and chemotaxis in primary cultures of mouse microglia [16]. This finding is very interesting because it is one of the few reports of the activation of TLRs with endogenous agonists in the context of the CNS.

TLR10 has also been shown to participate in the polarization of microglia to the immunoregulatory M2 phenotype. A study demonstrated that treatment with the active metabolite of vitamin D3 in a human microglia cell line considerably increased the expres sion of TLR10 and typical cytokines of M2, such as IL-10. This effect was sustained even when the cells were simultaneously stimulated with LPS, suggesting the anti-inflammatory role of TLR10 [17].

2.2. *Astrocytes*

Astrocytes are the most abundant glial cells in the CNS and support neural activity. Their functions are the recycling of neurotransmitters, ionic balance, maintenance, and formation of the blood–brain barrier (BBB), as well as communication with other cell types in the CNS [18]. Like microglia, astrocytes adopt the A1/A2 nomenclature for the functional phenotypes they present. The A1 phenotype is involved in neuroinflammatory processes, and the A2 phenotype is considered neuroprotective. Classically, microglia can induce the A1 phenotype through IL-1β [19].

Astrocytes contribute to the immune response against viral infections in the CNS by expressing TLRs and producing cytokines and chemokines. Human astrocytes in embryonic culture have been shown to express TLR1–TLR10 mRNAs in response to ligands of TLRs. TLR3 activation with poly(I:C) in astrocytes leads to the production of cytokines, such as IL-6 and TNF-α, as well as chemokines, such as CCL2, CCL5, CCL20, and CXCL10. These cytokines and chemokines recruit and activate cells of the immune system, such as T lymphocytes, B lymphocytes, monocytes, and dendritic cells [20].

Several studies have analyzed the regulation of astrocyte function mediated by TLRs. Most of these studies focused on the expression or release of inflammatory mediators, such as cytokines and chemokines. For example, in a rat model of intracerebral hemorrhage (ICH), researchers found that aquaporin-2 (AQP2), an astrocyte water channel, is a positive regulator of astrocyte activation that promotes IL-1β expression via TLR4/NF-κB [21]. Another group performed studies with mouse spinal cord astrocytes focusing on the inhibition of TLR9 through its antagonist ODN 2088 and reported that in vitro stimulation with ODN 2088 caused a decrease in the proliferation and migration of astrocytes and modulated the release of chemoattractants, such as CCL1 and CCL2, and chemokines, such as CCL9, whereas in a murine model, treatment with ODN 2088 caused a decrease in reactive astrocytes and induced a neuroprotective environment [22,23].

2.3. *Oligodendrocytes*

The cells responsible for myelinating axons of the CNS are oligodendrocytes; these make the myelin sheath extend from the plasma membrane and supply energy to the axon [24]. The myelinating equivalent of these cells in the peripheral nervous system is Schwann cells, although these cells have been shown to invade the CNS under pathological demyelinating conditions as a rescue mechanism [25].

Most research on TLRs in oligodendrocytes has focused on their activation and myelinating function [26]. The TLR2 and STAT3/SOCS3 signaling pathways are associated with low expression of myelin genes, such as MOG (a protein related to myelin deposits of oligodendrocytes) [27]. TLR2 overexpression is also associated with cuprizone-induced demyelinating damage in a murine model, where a decrease in mature oligodendrocytes in the corpus callosum was also found after cuprizone intoxication [28]. In addition, the activation of TLR3 with poly(I:C) in primary cultures of mature murine oligodendrocytes decreased the expression of myelin genes (PLP1 and MOG) and increased the expression of CCL2 and CXCL10 [29].

The proliferation of oligodendrocyte progenitor cells (OPCs) can also be altered by the activation of TLRs. Blocking the interaction of human endogenous retrovirus type W with TLR4 rescued the maturation of OPCs into mature myelinating oligodendrocytes and decreased the secretion of proinflammatory factors (iNOS and IL-6) [30]. A similar effect was observed on traumatic brain injury (TBI), in which the postinjury release of HMGB1 caused a decrease in OPC proliferation through TLR2/TLR4, revealing the participation of these receptors in the regulation of OPCs [31].

However, the evidence has shown that Schwann cells not only participate in myelination but also promote immunomodulatory responses through TLRs. For example, a study of sciatic nerve injury in a murine model revealed that such injury results in the recruitment of Schwann cells and the secretion of proinflammatory factors (TNF-α, IL-1β, and IL-6) via TLR4/MD2 [32]. Similarly, the administration of 1-methyl-4-phenyl-1,2,3,6-tetrahydropyridine (MPTP) in a murine model of Parkinson's disease causes the formation of p-α-synuclein deposits and an increase in TLR2 expression in enteric Schwann cells, promoting the production of proinflammatory factors (IL-1β, TNF-α, and NLRP3) through the TLR2/MyD88/NF-κB pathway, and the recovery of intestinal activity was achieved using a TLR2 inhibitor [33,34]. Despite the progress to date, more studies are required that address the function and expression of all members of the TLR family in the context of oligodendrocytes.

2.4. Neurons

Neurons are the functional unit of the nervous system. Their participation is not limited to information processing, as they also intervene in other processes, such as angiogenesis, the inflammatory response, proliferation, and neurogenesis, through the secretion of different factors and the modulation of different signaling pathways [35]. Interestingly, neurons express all TLRs in mice and humans, although their function has not yet been fully explored [36–42]. TLRs can trigger very diverse effects depending on the type of neuron and the context in which they are found. For example, TLR4 can modulate the feeding behavior of mice, as demonstrated by a TLR4 conditional knockdown mouse study on dopaminergic neurons of the ventral tegmental area, in which these mice reduced food reward behavior and suppressed the expression of tyrosine hydroxylase and dopamine at the neuronal level [43]. A study in primary cultures of murine hippocampal neurons revealed that activation of the TLR3/TRIF pathway reduced excitability by inhibiting Na^+ channels and synaptic activity by reducing AMPA receptor expression; these results were similar to those in cultures of neurons derived from human pluripotent stem cells [44]. TLR9 can modulate the survival and fate of neurons, as shown in an in vivo model of sepsis-associated encephalopathy, in which mice in the treated group exhibited increased neurodegeneration and TLR9 expression in cortical neurons. The TLR9/p38 MAPK/ERK axis was shown to be associated with the activation of PANoptosis, while treatment with the TLR9 inhibitor ODN 2088 improved the survival rate of the mice [45].

In a model of peripheral nerve damage, murine dorsal root ganglion neurons (mDRGs) increased TLR7 expression after injury, and silencing TLR7 decreased hypersensitivity and chronic pain in mice [46]. Interestingly, TLR10 expression has also been demonstrated in human neurons. TLR10 mRNA expression is higher in primary cultures of neurons derived from fetal brains compared with neuronal cell lines, but its function has been poorly studied [47]. However, more studies are necessary to clarify its function.

3. Role of TLRs in the Blood–Brain Barrier

The blood–brain barrier (BBB) is an essential biological barrier for the functioning of the CNS. The BBB is a highly specialized and selective structure found in the brain that controls the exchange of substances between circulating blood and nervous tissue, limiting the access of potent harmful or toxic substances circulating in the blood, which allows the maintenance of homeostasis of the CNS microenvironment [48,49]. The BBB consists of a layer of endothelial cells (ECs) connected by tight junctions lining the blood vessels of the brain, pericytes, astrocytes, and other cellular components, including microglia, neurons, and the vascular basement membrane, which, in turn, constitute the structure of the neurovascular unit (NVU) [48,50]. The vascular basement membrane is a network formed by different extracellular matrix proteins, including laminin, an important protein related to the degree of vascular integrity of the BBB. Additionally, pericytes around blood vessels that are embedded within the vascular basement membrane and astrocytic end-feet that also cover the vascular basement membrane, connecting the rest of their cell bodies to neurons and microglia within the brain parenchyma, are responsible for the induction and maintenance of the integrity of the BBB in the NVU (Figure 2) [48,51].

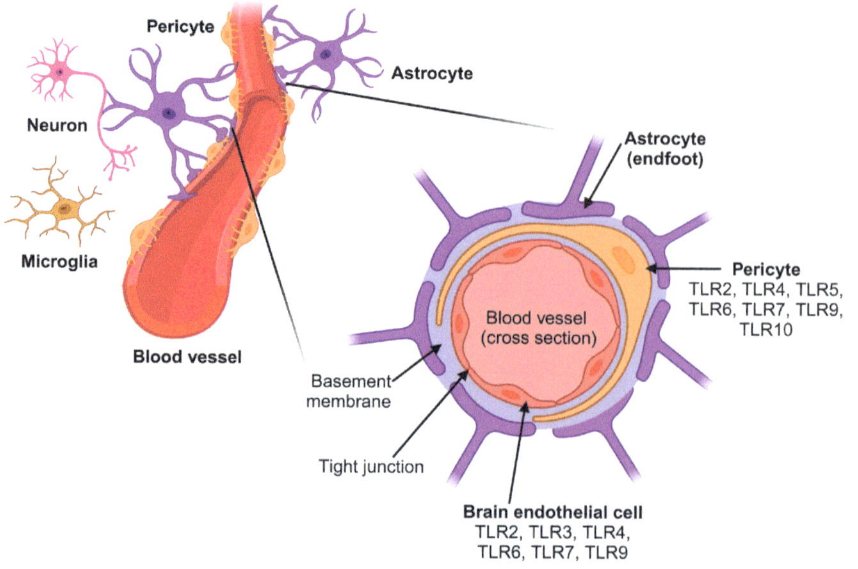

Figure 2. Schematic representation of the BBB located in the central part of the NVU. A cross-section of a blood vessel of the blood–brain barrier is depicted, showing the three main cell types that compose it (endothelial cells, pericytes, and astrocytes), other cell types of the neurovascular unit (microglia and neurons), and their respective TLRs that are expressed. The main function of the neurovascular unit is the formation of the blood–brain barrier and neurovascular coupling. Created with BioRender.com.

3.1. Brain Endothelial Cells

Brain endothelial cells (BECs) differ from other vascular endothelial cells in the human body because they act as selective transport barriers composed of intimate cell junctions that determine their permeability [49,50,52]. BECs are involved in several physiological processes, including the regulation of blood flow, maintenance of vascular integrity, and immune responses in the CNS [53]. Therefore, BECs are at the forefront of CNS protection and respond to stress factors, such as inflammatory, ischemic, and infectious events, in which TLRs play an important role [49,52]. Oxidative stress and TNF-α positively regulate the expression of TLR2, TLR3, TLR4, and TLR6 in human brain endothelial cells (hCMEC/D3) and mouse brain endothelial cells. Moreover, TLR2 and TLR6 activation increases the permeability of the BBB by downregulating adherens junction proteins [52]. The expression of TLR3, TLR4, and TLR7 is increased upon activation with the ligands poly(I:C), LPS, and imiquimod, respectively, in human cerebral microvascular endothelial cells (hCMVECs). Furthermore, activation of TLR3 and TLR4 induces a proinflammatory response; in contrast, activation of TLR7 with imiquimod has an anti-inflammatory effect [54]. The expression of TLR2, TLR4, and TLR9 mRNAs was described in endothelial cells and brain tissue from a murine model after experimental cerebral ischemic injury [55]. Specifically, the most highly expressed gene was TLR2, and a global gene expression analysis was used to determine the regulation of TLRs after cerebral ischemia. Additionally, TLR2 knockout in mice has been shown to have a protective effect at the neuronal level by reducing the cerebral infarct volume. TLR2 signaling may contribute to ischemic brain injury by mediating the production of proinflammatory cytokines, such as TNF-α and IL-1β, through the NF-κB signaling pathway. Therefore, TLR2 inhibition contributes to neuroprotection, and TLR2 signaling promotes ischemic brain injury [56].

On the other hand, the activation of TLR3 and RIG-I (retinoic acid-inducible gene-I) with synthetic ligands induces the expression of IFN-β and IFN-λ in BECs. In turn, the release of interferons into the extracellular medium decreases HIV (human immunodeficiency virus) replication in infected macrophage cultures. HIV infection is widely known to compromise the integrity and permeability of the BBB, but BECs mount an innate immune response to HIV infection in the CNS through TLRs and the RIG-I receptor [57].

3.2. Pericytes

Pericytes are cells with elongations that extend around the vascular wall and are closely related to endothelial cells. Pericytes, also called mural cells, are specialized cells that provide structural support to help maintain the integrity and stability of blood vessels. They are involved in angiogenesis, vascular homeostasis, vessel maturation, and blood flow regulation [58,59]. Little is known about the expression of TLRs in pericytes; the mRNA expression of TLR2, TLR4, TLR5, TLR6, and TLR10 was detected under basal conditions, and TLR9 mRNA expression was detected after TNF-α treatment and IFN-γ and IL-1β induction in human brain vascular pericyte cells [60]. Furthermore, TLR2 expression was observed in the postischemic vasculature of aged rats [61].

TLRs have a multifaceted immunological function in pericytes due to their secretion of several types of chemokines and cytokines, as well as their ability to regulate immune cell trafficking and the expression of adhesion molecules [58]. For example, the activation of the TLR4 receptor upon stimulation with LPS and high mobility group box 1 protein (HMGB1) in human pericytes triggers proinflammatory and proangiogenic factors through the NF-κB signaling pathway [58]. Another study analyzed the inflammatory response of brain pericytes after infection with the Japanese encephalitis virus (JEV). This study detected the expression of TLR7 in a pericyte cell line infected with JEV. TLR7 signaling in pericytes after JEV infection, as well as that induced by the TLR7 agonist gardiquimod, contributed to the production of IL-6, RANTES, and PGE2, compromising the integrity of the endothelial barrier and leukocyte chemotaxis [62]. However, more research is needed to understand the roles of TLRs expressed in pericytes in the function of the BBB.

3.3. End-Feet of Perivascular Astrocytes

The terminal feet of astrocytes are essential for the formation and maintenance of the BBB; these layers cover the perivascular structure of the NVU and, in turn, most of the cerebrovascular surface by enveloping endothelial cells and pericytes with their terminal feet [63]. Astrocytes perform several physiological functions to maintain the integrity of the BBB, such as regulating cerebral blood flow by releasing vasoactive factors, such as nitric oxide (NO) and arachidonic acid, which act on smooth muscle cells in blood vessels; regulating permeability via tight junction proteins, such as claudin-5; removing waste and toxins from the brain through phagocytosis and active transport; maintaining ionic balance within the brain by removing excess potassium ions released during neuronal activity; and by taking up and recycling neurotransmitters released at synapses [63,64].

TLR2 mRNA expression is increased in hemorrhagic tissues in two different mouse models of intracerebral hemorrhage. The volume of brain lesions and neurological deficits were reduced in TLR2 knockout mice compared to the wild-type control in both models. Moreover, in TLR2 knockout mice, there was less damage to the integrity of the BBB, neutrophil infiltration, and proinflammatory gene expression in injured brain tissue compared to wild-type mice. These mechanisms indicate that TLR2 plays a principal role in ICH-induced brain injury [65]. Another finding in the context of ischemic stroke is the expression of MMP9 induced by HMGB1 via TLR4 in astrocytes and the mouse brain after cerebral ischemia [66]. TLR4 protein expression was significantly increased in an astrocyte oxygen–glucose deprivation model, which evaluated the neuroinflammatory effects of Z-guggulsterone after cerebral ischemic injury [67]. Moreover, TLR4 expression in astrocytes increased after middle cerebral artery occlusion (MCAO). TLR4-deficient mice exhibited a reduction in infarct size and improvements in neurological and behavioral outcomes after MCAO compared to those of control mice [68].

3.4. Other Cell Types of the NVU

Microglia and neurons also express TLRs, which play important roles in the function of the cerebral vasculature. Microglia play a dual role in promoting both protection and injury in brain injury caused by subarachnoid hemorrhage (SAH). Microglial activation after SAH is mediated by microglia-associated receptors, such as TLR4, the receptor for advanced glycation end products (RAGE), and myeloid cell-expressed receptor-1 (TREM-1). These receptors are activated by DAMPs released during SAH, which trigger the production of proinflammatory mediators and the release of cytokines, such as TNF-α, IL-1β, and IL-12. These inflammatory cytokines aggravate brain injury by triggering neuronal apoptosis, the disruption of the BBB, and brain edema [69]. It has been shown that inhibiting TLR4 activation could reduce the inflammatory response and mitigate the detrimental effects of activated microglia after SAH in mice [70]. In focal cerebral ischemia, TLR2 knockout mice exhibited reduced CNS damage compared to wild-type mice. These findings suggest that TLR2 in microglia contributes to stroke-induced CNS injury [71]. Together, these findings suggest that the regulatory pathways of responses induced by TLRs could be therapeutic targets for reducing the damage generated during cerebral vascular disease. However, more studies are needed to elucidate the cellular and molecular mechanisms mediated by TLRs expressed in the BBB.

4. Regulation of Neurogenesis by TLRs

Neurogenesis is the process by which new neurons are generated and integrated into the neural circuit. The evidence shows that neurogenesis occurs mostly during embryogenesis and decreases during early postnatal development and adulthood [72]. However, there is considerable evidence that neurogenesis occurs in the adult human brain in specific regions, like the dentate gyrus of the hippocampus and the lateral subventricular zone [72,73]. Neurogenesis in the adult phase has also been documented in various species, including rats, mice, songbirds, and nonhuman primates [72,74]. Neurogenesis has been postulated to play a role in memory and learning processes, as it is also a protective mechanism against stress-induced attrition of the brain [75].

In humans, neurogenesis starts midway through the first trimester in the ventricular zone, where these cells eventually migrate to their destination [72]. Neural progenitor cells (NPCs) divide to contribute to the generation of additional progenitor cells and differentiated neurons [76,77]. This intricate process occurs across various proliferative zones, culminating in the migration of neurons to their final destinations for long-term maintenance [77]. Concurrent sets of progenitors that later generate oligodendrocytes and astrocytes are maintained [76–78]. Notably, microglia are not derived from neural progenitors but rather from the hematopoietic lineage [78]. Neurodevelopmental processes include several key processes, such as neurogenesis, gliogenesis, cell death, synaptic pruning, and myelination, which collectively shape the intricate architecture of the developing brain [77]. TLRs participate in different brain and plasticity-related processes, including neurogenesis [79]. Different studies have shown that TLRs are involved in various steps of successful neurogenesis, such as NPC proliferation, migration, differentiation, maturation, and the integration of newly formed neuronal networks [79,80]. In mice, during the early stages of neurodevelopment, the activation of TLRs mainly affects neurogenesis, and in postnatal stages, it becomes involved in neuronal function, including memory, learning, and behavior. Furthermore, these effects have been shown to be related to the development of behavioral disorders and psychiatric disorders, such as seizures or schizophrenia-like behaviors [81–83].

Studies have shown that the activation of TLR2 contributes to the differentiation of NPCs into neurons via MyD88-dependent signaling and induces their differentiation into astrocytes but has no effect on cell proliferation or cell survival [84]. It has recently been discovered that neurogenesis in the adult brain is promoted by TLR2 in the neural stem cells (NSC) of the hippocampal dentate gyrus in rats [85]. Overexpression of TLR2 promotes the cell proliferation of NSC and increases cell numbers [85]. In the same way, neuronal activation via a heat-sensitive cation channel called TrpA1 also increases proliferation, and this effect depends on Toll-2 in Drosophila [86]. The deletion of TLR2 in NPCs results in heightened self-renewal and apoptosis, as well as a delay in neural differentiation and increased proliferation. Moreover, the absence of TLR4 in NPCs has been found to enhance neuronal differentiation without affecting NPC proliferation.

Notably, a significant decrease in TLR1 expression, which similarly impedes the maturation of oligodendrocyte precursor cells from NPCs, was induced by the absence of TLR2 and TLR4 [87]. Additionally, a potential correlation between TLR2 and the proliferation of neural stem cells (NSCs) after traumatic brain injury (TBI) has been suggested. In a rat model of TBI, TLR2 expression in the dentate gyrus increased post trauma, and on the third day, TLR2 expression peaked. Later, on the seventh day post trauma, its expression decreased, but it was increased compared with that in the sham group. In addition, increased proliferation of NSCs was observed post trauma [85]. In contrast, during mouse brain development, embryonic NPC proliferation is negatively regulated by TLR3 [88]. Heightened neuronal proliferation and differentiation were seen in the absence of TLR4, with the extent varying with sex and age. Specifically, young female TLR4 knockout mice showed increased neuronal production in both the ventral and dorsal hippocampus, whereas in young males, this was seen primarily in the ventral hippocampus [89]. Inhibiting TLR4 with neutralizing antibodies increases NPC self-renewal [84]. Furthermore, TLR4 has a crucial role in both neurodevelopment and the regulation of NPC proliferation and differentiation in disease settings [90,91]. TLR4 was shown to play an important role in neurogenesis by promoting neuroblast migration and increasing the number of new cortical neurons after stroke, despite its negative effect on proliferation in the subventricular zone [91]. Consistent with another experimental model of stroke, the evidence demonstrated that HMGB1-mediated TLR4 activation is necessary to maintain NSC proliferation and promote cell differentiation into neuroblasts, promoting their migration [92]. These findings indicate the importance of TLR4 as a therapeutic target in the modulation of neurogenesis after stroke. Moreover, stimulation of TLR3 and TLR4 curtails NPC's capacity for self-renewal, yet deficiency promotes NPC proliferation [91–93].

Recent research unveiled the role of TLR5 in regulating the cell cycle by inhibiting proliferation in adult hippocampal NSCs and promoting neural differentiation [41]. Additionally, deficits of TLR5 in mice lead to impaired fear conditioning memory performance, indicating its vital role in the modulation of neurogenesis and suggesting it as a potential target for treating neurological disorders. TLR5 regulates contextual fear conditioning memory associated with adult hippocampal neurogenesis [41]. Neuronal activation of TLR3 with poly(I:C), TLR7 with CL075, and TLR8 with CL075/polydT negatively modulates neurite outgrowth, alters synapse formation in a cell-autonomous manner, and regulates neuronal differentiation [94]. TLRs can regulate slow cellular processes through the canonical MyD88 and ERK signaling pathways and gene expression, including cell renewal and NPC differentiation to neurons [84,88]. Moreover, enhanced NPC proliferation in the subgranular zone of the dentate gyrus was seen in the absence of MyD88, highlighting the essential role of TLR signaling components in NPC proliferation, even without exogenous TLR ligands. Similar to their impact on NPCs in the subependymal and subventricular zones, as well as in the hippocampus, TLRs also influence the proliferation of retinal progenitor cells [84,88,93,95]. In newborn MyD88 knockout mice, the cortical neuron density was increased, the neocortical thickness was reduced, and the numbers of microglia, astrocytes, oligodendrocytes, and proliferating cells were similar in these mice compared to wild-type mice [96]. In adult MyD88 knockout mice, the neuronal density in the neocortex and hippocampus was increased, while the neocortical thickness did not differ [96].

Cells with diminished activation or expression of TLR3, TLR7, and TLR8 display aberrant neuronal maturation [94,97]. Different experiments have revealed that TLR3 stimuli with poly(I:C) in cultured mouse and cortical and hippocampal neurons lead to reduced dendrite length and the modulation of dendrite morphology. These effects were TLR3-dependent since they were not observed when neuronal TLR3 was knocked down. In this experimental paradigm, TLR3 signaling was mediated by the MyD88 pathway rather than by TRIF [97,98]. Furthermore, through the manipulation of TLR7 expression levels, neuronal morphology is altered, and the addition of ssRNA and synthetic TLR7 ligands, such as CL075 and loxoribine, but not imiquimod, to cultured neurons specifically restricts dendrite growth via TLR7 [97]. The authors subsequently revealed, using MyD88-, IL-6-, and TNF-deficient neurons, the importance of IL-6 and MyD88 in the TLR7 track to restrict dendrite growth [97]. In comparison, the stimulation of TLR8 with a synthetic agonist inhibited neurite outgrowth and induced the death of cortical neurons in vitro. These effects of the agonist were not mediated through the canonical TLR-MyD88-NF-κB signaling pathway [99]. However, decreases in IRAK4 and IκBα, which are involved in TLR-MyD88-NF-κB downstream signaling, coincided with the inhibition of neurite outgrowth and neuronal apoptosis [99]. It is important to highlight that within mammalian brains, under homeostatic conditions (the absence of MAMPs and DAMPs), TLRs take part in the regulation of NPC proliferation, dendrite and axon growth, synapse formation, and neuronal morphology [77,94].

In summary, TLRs in the human brain regulate the proliferation and differentiation of neural progenitors and stem cells into neurons or glial cells. They also control cell survival, cell death, neurite growth and retraction, synaptogenesis, and the modulation of spine density and size (Figure 3).

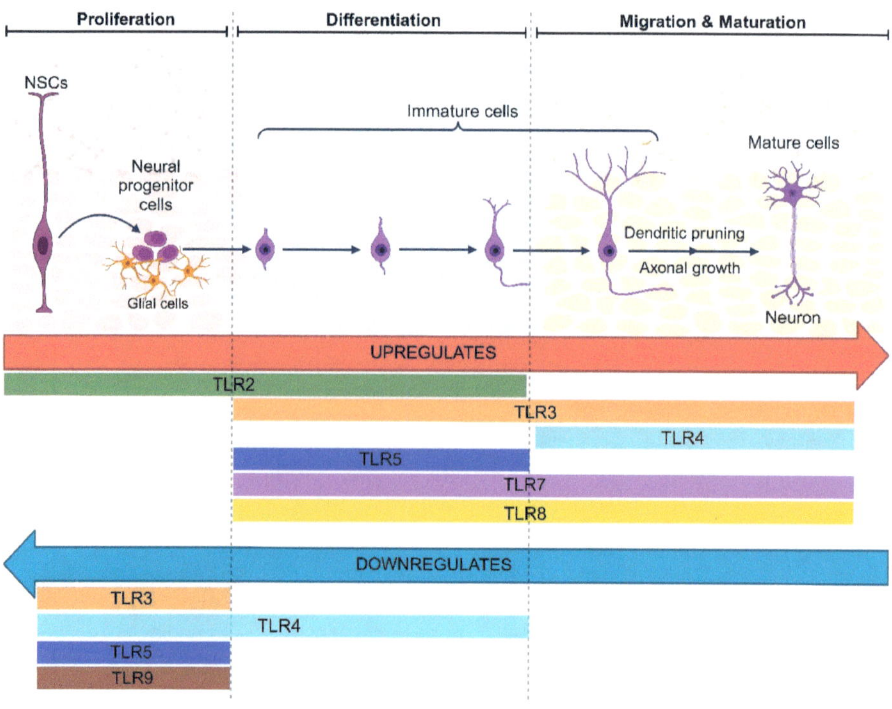

Figure 3. Roles and expression of TLRs in neurogenesis. Schematic representation of the principal stages of neurogenesis and the Toll-like receptors involved in the up- and downregulation of each process. Created with BioRender.com.

5. Effects of TLR Misregulation on Behavior and Cognitive Function

TLRs have shown functional importance in behavior and cognitive function [96,100]. The absence of some TLRs results in behavioral abnormalities. For example, TLR2 knockout mice exhibit reduced anxiety, impaired sociability, aggression, and cognitive defects [100]. Similarly, the deficiency of MyD88 in mice results in specific behavioral traits, such as decreased locomotor activity and increased anxiety-like behavior. This highlights the diverse effects of TLR signaling pathways on behavior and cognitive function in animal models [96].

Another study has described TLR2 as a metabolic regulator of hypothalamic neurons that modulates feeding behavior and metabolic regulation [101]. Also, TLR3 knockout mice are less anxious than wild-type mice. TLR3-deficient mice, however, exhibit enhanced memory functions linked to the hippocampus [89]. Meanwhile, knockout TLR4 mice demonstrated improved spatial memory, which was particularly notable in young males and aged females, suggesting that it not only enhances memory but also modulates hippocampal inflammation, with effects varying by sex and age [102]. Interestingly, TLR4 deficiency was found to protect against sevoflurane-induced cognitive impairment and inflammation in mice [103]. Furthermore, TLR4 has been shown to play a significant role in cancer-associated cognitive impairment. It was found that the activation of TLR4 through cisplatin is a key mechanism leading to sphingosine-1-phosphate (S1P) production, resulting in cognitive dysfunction. S1P is a regulatory signaling molecule of vascular permeability, inflammation, angiogenesis, cancer growth, metastasis, and brain and cardiac development that contributes to the development of cancer-related cognitive impairment [104].

Recent investigations in organotypic slice cultures of rat hippocampus have revealed that the activation of microglia through the simultaneous or individual administration of TLR2, TLR3, and TLR4 agonists can induce diverse levels of dysfunction within neuronal networks [105]. Similarly, another study found that the activation of microglia, marked by increased proinflammatory cytokines, as a consequence of repeated social defeat stress was linked to social avoidance behavior, which was mitigated in mice lacking TLR2/TLR4. These findings highlight the significance of innate immunity in the medial prefrontal cortex during repeated environmental stress-induced behavioral alterations [106]. Correspondingly, the use of an antagonist (OxPAPC) to block TLR2 and TLR4 effectively prevented the escalation of inflammatory responses in the hippocampus, triggered by a 24 h challenge with LPS in vivo [107]. In addition, TLR signaling is involved in the inflammatory pathways that contribute to mediating depressogenic effects in the context of major depressive disorder [108].

TLR7-deficient mice exhibit reduced exploratory behaviors, decreased anxiety, reduced aggression, increased olfaction, and impaired contextual fear memory [82]. In contrast, prenatal TLR7 activation in mice with imiquimod induced a phenotype of reduced anxiety-like behavior and fragmented social behavior, including social partners, circadian cues, and gonadal hormone fluctuations, but normal locomotor activity. All of these effects were accompanied by decreases in microglial density but increases in ramifications [109].

On the other hand, TLR9 knockout mice exhibited hyper-responsive sensory and motor phenotypes, with no alterations in learning or memory [110]. Furthermore, seizure-induced cognitive decline was intensified by TLR9 deficiency [111]. The brains of adult MyD88 knockout mice showed specific behavioral traits, suggesting that MyD88 is a key factor in neurodevelopment. Morphologic and cellular changes in the mouse brain, as well as altered natural and specific behaviors, result from its absence [96].

TLRs play key roles in a range of mental health conditions, including depression, stress, and schizophrenia, and in addictions such as alcohol and drug consumption. For instance, in a rat model of stress-induced depression called chronic mild stress (CMS), it was demonstrated that by reducing the gastrointestinal microflora, the elevated expression of TLR4/MD2 mRNA and protein and the elevated levels of inflammatory mediators were reversed in the prefrontal cortex [112]. Subsequently, investigations by the same research group revealed that CMS resulted in increased intestinal permeability, enhanced migration of bacteria into tissues, and activation of TLR-associated transcription factors, along with reduced antioxidant activity [113]. These findings confirm the importance of the intestinal–neuroimmune axis interaction.

Recently, chronic social defeat stress (CSDS) in mice led to depressive-like behaviors accompanied by a significant increase in hippocampal TLR4 protein expression. Treatment with fluoxetine, an antidepressant, reversed depressive-like behaviors and increased the expression of TLR4 and TNF-α in the hippocampus. Inhibiting TLR4 with TAK-242 blocked CSDS-induced behavioral despair without affecting social avoidance or anxiety-like behaviors. TLR4 knockout also prevented CSDS-induced behavioral despair and increased the expression of TNF-α in the hippocampus. These findings suggest that TLR4 plays a role in behavioral despair, and its inhibition may be a potential treatment for individuals experiencing depressive-like behaviors [114].

Moreover, individuals with stable chronic schizophrenia showed evidence of overactivation of the monocytic TLR4 signaling pathway. TLR4 hyperactivity was related to cognitive impairment and reduced white matter integrity. Additionally, the study identified a sluggish response of the TLR4/NF-κB/IL-1β signaling pathway to LPS stimulation in patients with stable chronic schizophrenia compared to healthy controls [115].

TLR7 expression was heightened in postmortem samples of the brains of individuals (humans) with alcohol use disorder [116]. The stimulation of TLR7 with R848 in mice resulted in a reduction in both preference and ethanol consumption, effects that persisted even after removing R848 [116]. Conversely, the pre-administration of R848 before drinking led to an increase in ethanol intake and preference. Notably, specific alterations in innate immune transcript levels across different brain regions, including increased expression of MyD88 and TRIF-dependent signaling, were triggered after acute TLR7 activation. The study underscores the relationship between maladaptive behaviors, in this case, excessive ethanol consumption, and the innate immune system [116]. Previous research has also demonstrated a direct correlation between signaling TLR pathways and alcohol intake. For example, LPS TLR4-activation was directly linked to increased alcohol consumption, whereas inhibition of the downstream signaling component IKKβ (MyD88-dependent) using siRNA resulted in reduced alcohol intake [117,118]. Similarly, the activation of TLR3 increases alcohol intake in mice. Repeated exposure to poly(I:C) and ethanol leads to a persistent increase in alcohol intake, which is dependent on TRIF signaling. This study identified TLR3 as a potential therapeutic target to reduce excessive alcohol consumption and treat substance abuse disorders [119]. In addition, TLR4 signaling has also been demonstrated to induce neuroinflammation and have a direct impact on the onset of addictions to substances such as alcohol and opioids. Blocking TLR4 or its downstream processes has been shown to diminish the reinforcing effects of opioids and alcohol, with implications for withdrawal and relapse-like behaviors [120,121]. These findings highlight the close and complex interaction between innate immunity and diseases of neuronal origin, emphasizing the potential of TLRs as targets for future studies on therapeutic approaches.

6. Involvement of TLRs in Neuroinflammation and Neurodegeneration

Inflammation is a physiological process by which the immune system responds to injury and infection. Neuroinflammation is characterized by the presence of reactive glia and infiltration of immune system cells, in addition to the release of several inflammatory mediators, including cytokines and free radicals. This response has the objective of protecting and repairing neuronal tissue, but persistent neuroinflammatory conditions and excessive oxidative stress lead to a decrease or loss of neuronal function (neurodegeneration), cell death, and the inhibition of regeneration [122,123].

During neuroinflammation, microglia and astrocytes that can produce cytokines (such as IL-1β, IL-6, TNF-α, and IFN-γ), chemokines, prostaglandins, and ROS are activated and proliferate. Subsequently, peripheral immune cells, such as monocytes, macrophages, and T and B cells, infiltrate the circulation [124,125].

Below, we analyze the role of TLR-mediated neuroinflammation in nervous system diseases.

6.1. TLRs in Bacterial and Viral Infections of the Nervous System

Persistent viral infection in the CNS can produce inflammation in different regions of the nervous system, for example, meningitis (meninges) and encephalitis (brain), caused by the production of various inflammatory factors, including cytokines. The exacerbated release of cytokines constitutes what is called a "cytokine storm", which can result in neurodegeneration and, consequently, behavioral and neurological diseases [126,127]. Recent studies have highlighted the roles of TLRs in CNS viral infection. For example, in BV2 microglial cells infected with JEV, the activation of TLR2 and TLR6 is key to the production of inflammatory factors, such as TNF-α, CCL3, IL-1β, and iNOS, via the PI3K-AKT pathway [128]. Moreover, JEV induces let7a/b expression in N9 microglia, and let7a/b can activate the NOTCH-TLR7 pathway and induce NF-κB activation and cytokine production [129].

It has been shown that in a mouse model of experimental autoimmune encephalomyelitis (EAE), neuroinflammation increased via the TLR2/MyD88 pathway, in addition to increasing M1 microglial polarization and the production of proinflammatory cytokines, such as TNF-α, IL-6, IL-1β, and MCP-1 [130]. Subcutaneous administration of the TLR2/TLR4 antagonist naltrexone to mice with EAE has been shown to reduce the mRNA levels of IL-1β, NLRP3, IkBα, TNF-α, and IL-7 in the dorsal lumbar spinal cord [131]. Consistent with these findings, another study of EAE mice revealed that the attenuated expression of TLR4 and HMGB1 in astrocytes and microglia reduced inflammation and demyelination [132]. On the other hand, TLR3 deficiency in humans may predispose to the autoimmune and demyelinating manifestations that occur in acute disseminated encephalitis triggered by the herpes simplex virus (HSV) [133]. Furthermore, complications, such as encephalitis, that occur in some patients with Puumala hantavirus and HSV may be due to a heterozygous mutation in the TLR3 gene [134].

The envelope (E) protein of SARS-CoV-2 in a mouse model can trigger neuroinflammation and depression-like behaviors mediated by the activation of microglia and astrocytes through TLR2 in the CNS, while blocking TLR2 can alleviate E-induced neuroinflammation [135]. In murine models, exposure to the SARS-CoV-2 spike protein affects cognitive function and synaptic loss via microglial activation and neuroinflammation mediated by TLR4 activation [136]. Polymorphisms of the GG genotype of TLR4 2604G>A (rs10759931) are associated with a greater risk of developing cognitive impairment following SARS-CoV-2 infection than the GA genotype [136].

In a microglial model of the human immunodeficiency virus (HIV), TLR3 activation mediated by poly(I:C) effectively and dose-dependently inhibited HIV infection by triggering the IFN signaling pathway, inducing the expression of ISGs (IFN-stimulated genes; restriction factors involved in anti-HIV activities), and the production of chemokines (RANTES, MIP-1α and MIP-1β) [137].

On the other hand, polymorphisms of TLRs are involved in the susceptibility to infections; in this context, the TLR2 2477G/A polymorphism contributes to an increased risk of pneumococcal meningitis in the Caucasian population [138]. In Angolan children, the single nucleotide polymorphism TLR4 rs4986790 increases the susceptibility to bacterial meningitis caused by Haemophilus influenzae, the risk of ataxia, and neurological sequelae, while TLR9 rs187084 decreases the risk [139]. A meta-analysis revealed that the TLR9 polymorphism rs352140 is associated with a decreased risk of bacterial meningitis [140].

In a model of pneumococcal meningitis, the TLR2/TLR4 heterodimer regulates inflammatory and antibacterial mechanisms; in mice lacking both TLR2 and TLR4 inoculated intracerebroventricularly with Streptococcus pneumoniae, disease severity is increased, survival is significantly impaired, and the proinflammatory cytokines IL-1β, TNF-α, and IL-6 are decreased, while IFN-γ increases and might lead to long-term neurological problems [141]. Likewise, pretreatment with the intraperitoneal administration of TLR9 CpG

ODN has been shown to have a protective effect on mice with Streptococcus pneumoniae infection, prolong survival, decrease the bacterial concentration in the systemic circulation and CNS, and promote major microglial proliferation [142]. In contrast, in patients and mice with pneumococcal meningitis, TLR7 was overexpressed. The TLR7 overexpression-induced model showed NF-κB activation and proinflammatory cytokine production, while TLR7 silencing suppressed inflammatory reactions and apoptosis and promoted cell viability, reducing brain damage caused by meningitis in mice [143].

6.2. Neurodegenerative Diseases and TLRs

Many studies have reported the participation of TLRs in the most common neurodegenerative diseases; a few examples are mentioned below. Alzheimer's disease (AD) is a progressive neurodegenerative disease characterized by the accumulation of amyloid-beta (Aβ) peptides and phosphorylated tau in the medial temporal lobe and neocortical structures, in addition to neuroinflammation and oxidative stress [144]. In murine models of AD, TLR1 and TLR2 are overexpressed [145,146]. In addition, human brain tissue samples from patients with AD expressed TLR3 in microglia and endothelial cells associated with Aβ plaques [147]. Tau fibrils in AD mouse models induced the activation of TLR2/MyD88/NF-κB in microglia and stimulated neuroinflammation [148]. Systemic administration of the TLR2 agonist zymosan and lipoteichoic acid in an AD mouse model induced microglia-mediated neuronal death; therefore, microbial infections may facilitate neurodegeneration [149]. The activation of TLR3 by the agonist poly(I:C) can modulate early life stress-induced working memory impairments in adolescent mice. The intraperitoneal injection of poly(I:C) increased the mRNA levels of IL-6 and NF-κB but not oxidative stress and contributed to neuroinflammation, consequently decreasing working memory performance [150]. TLR2 inhibition in mice with tauopathy results in improved learning and spatial memory retention, and thus, TLR2 is a potential therapeutic target for AD [148]. Blocking TLR2 with an antibody increased Aβ phagocytosis by microglia and inhibited the inflammasome, suggesting that inflammatory microglia are less phagocytic [151]. Selective inhibition of the TLR2 pathway decreased neuronal apoptosis, inhibited the hippocampal activation of NF-κB and microglial inflammation, and protected cognitive function in an AD mouse model [152].

However, existing studies on the role of TLR2 in AD have yielded conflicting results; the deletion of TLR2 impaired learning ability and memory function, aggravated increased anxiety and depression, and enhanced the activation of astrocytes in AD mice [153]. Pretreatment with Pam3Cys and the monophosphorylated lipid A ligands TLR2 and TLR4 in a rat model of AD increased the release of TNF-α and CCL3, improved cognitive function, restored synaptic plasticity, and decreased Aβ deposits in the brains of patients treated with Aβ to the control level [154]. Pretreatment with an intraperitoneal injection of poly(I:C) in mice with AD reduced anxiety and depression, increased spontaneous activity, and improved memory deficits. In addition, poly(I:C) eliminated Aβ deposits, attenuated neuronal loss, and reduced the activation of microglia in the hippocampus of AD mice [155]. The administration of a TLR9 agonist before and after AD in mice eliminated short-term memory deficits and decreased Aβ deposits through the phagocytic activity of microglia [156].

Parkinson's disease (PD) is the second most common neurodegenerative condition, and it is characterized by the presence of Lewy bodies and substantia nigra neuronal loss [157,158]. There are several studies that address the role of TLRs in PD; however, they are still inconclusive and, in some cases, contradictory. Increased expression of TLR4, which can promote neuroinflammation, has been observed in patients with PD [159]. Also, in patients with PD, neuroinflammation activated by microglia upregulated the expression of neuronal TLR2 and TLR4 to activate the p38/JNK pathway, leading to an increase in phosphorylation, the accumulation of α-synuclein proteins, and the shortening of neurites [160].

In a mouse model of PD, TLR2 was found to be overexpressed, and its depletion had a neuroprotective effect by reducing α-synuclein expression and neuroinflammation;

therefore, the loss of dopaminergic neurons was reduced [161]. The absence of TLR4 reduces neuroinflammation and prevents striatal dopaminergic neuron loss and dopamine depletion associated with a PD mouse model [162]. In contrast, a recent study found that the absence of TLR4 resulted in a decrease in lysosomal autophagy (CD68 and p62) in microglia and facilitated the propagation of α-synuclein aggregates and the progression of the disease [163].

Similarly, in MPTP-treated mice, TLR7 and TLR8 were upregulated, and TLR8 deletion prevented the loss of locomotor agility and stability and ameliorated the emotional state of the mice. In addition, the double knockout of TLR7 and TLR8 reduced α-synuclein aggregate formation, T-cell activation and dendritic cell numbers, and, accordingly, reduced neuroinflammation and neurodegeneration [164].

The TLR3 rs3775290 polymorphism and TLR9 rs352140 polymorphism seemed to reduce PD susceptibility and may be protective factors in Chinese Han and Iranian populations, respectively [165,166]. In contrast, the TLR4 rs4986791 and rs1153889 polymorphisms were associated with an increased risk of PD, while the interaction between rs1927914 and alcohol consumption was associated with a decreased risk of PD in the Chinese Han population [167]. Further studies are needed to clarify the participation of TLRs in the pathogenesis of PD.

7. Conclusions and Perspectives

Beyond their functions in responding to infections or damage, TLRs play an important role in processes such as neurogenesis, cognition, and behavior, revealing the close connections that exist between the nervous system and the immune system (Table 1). However, despite the large number of published works, many of the mechanisms are unknown, or the results are controversial and, therefore, not decisive. More studies are needed to determine the expression and involvement of TLRs in nervous system functionality in the context of humans specifically. There are both differences and coincidences in the expression, activation mechanisms, and signaling of TLRs between humans and other mammals, which complicates the interpretation of the results obtained and means that the findings of animal models need to be interpreted with caution.

Finally, it is important to highlight the lack of knowledge about the participation of endogenous agonists of TLRs that are not associated with damage or infections during neurodevelopment, which is a great area of opportunity for future research. Endogenous ligands could be the key to both the diagnosis and treatment of different neurodevelopmental and neurodegenerative diseases.

Table 1. Summary of TLR participation in the nervous system.

Receptor	Outcome	Related Disease	Model	Refs.
TLR1	Increased receptor expression.	Epilepsy	Mouse model	[36]
TLR2	Negatively regulated maturation of oligodendrocyte precursor cells.	Multiple sclerosis	KO mouse	[26]
	Receptor deficiency decreased neuronal death, inflammation, and brain injury.	Cerebral ischemia	KO mouse	[37,56,71]
	Increased receptor expression.	Amyotrophic lateral sclerosis	Postmortem human spinal cord sample	[38]
	Receptor activation increased transendothelial permeability.	Not	Rat and human cerebral endothelial cells	[52]
	Receptor deficiency reduced MMP-9 activation in astrocytes.	Intracerebral hemorrhage	KO mouse	[65]
	Receptor deficiency induced schizophrenia-like symptoms.	Schizophrenia	KO mouse	[100]
	Regulated metabolic processes.	Obesity	Mouse neuron cell line	[101]
	Administration of antagonist decreased proinflammatory cytokines.	Multiple sclerosis	EAE mouse model	[131]
	Regulated neuroinflammation, depression-like behaviors, and dysosmia.	COVID-19	Mouse	[135]
	Receptor activation with tau fibrils induced stimulated microglial inflammation.	Tauopathy	Mouse	[148]
	Induced death of neurons and inflammation.	AD	AD mouse model	[149]
	Increased accumulation of α-synuclein proteins and the shortening of neurites.	PD	Dopaminergic neuronal cell line	[160]
TLR3	Increased receptor expression.	Multiple sclerosis	Human postmortem brain tissue	[8]
	Receptor activation decreased myelin expression and increased cytokine and chemokine production.	Maternal/fetal infection	Mouse oligodendrocyte precursor cells	[29]
	Receptor activation induced IL-1β secretion.	Viral infection	Human neurons	[39]
	Increased receptor expression.	Viral encephalitis	Human postmortem brain tissue	[40]
	Induced interferon expression.	HIV infection	Microvascular endothelial cells	[57]
	Negatively regulated NSPC proliferation.	Not	KO mouse	[88]
	Receptor activation increased alcohol intake.	Alcohol use disorder	Mouse	[119]
	Receptor activation in early stage decreased neuron loss and improved neurobehavioral functions.	AD	AD mouse model	[155]

Table 1. *Cont.*

Receptor	Outcome	Related Disease	Model	Refs.
TLR4	Increased receptor expression.	Multiple sclerosis	Human postmortem brain tissue	[8]
	Receptor activation inhibited autophagy and phagocytic function.	Not	Mouse microglial cells	[14]
	Induced neuronal death.	Cerebral ischemia	KO mouse	[37]
	Receptor deficiency decreased infarct size, improvements in neurological and behavioral outcomes, and inflammation.	Cerebral ischemia	TLR4-deficient mutant mouse	[66,68]
	Inhibited spatial memory and induced inflammation.	Not	KO mouse	[102,103]
	Receptor activation through cisplatin resulted in cognitive dysfunction.	Cancer	KO mouse	[104]
	Receptor deficiency decreased ethanol-induced impairment of BBB.	Alcohol use disorder	KO mouse	[121]
	Receptor blockade prevented detrimental effects on synapse and memory.	COVID-19	Mouse	[136]
	Receptor deficiency decreased inflammation, dopaminergic neuron loss, and autophagy and induced α-synuclein aggregate formation.	PD	KO mouse	[162,163]
TLR5	Induced inflammatory factors and neuronal apoptosis.	Not	Murine microglial and neuronal cultures	[15]
	Increased neuronal differentiation.	Not	KO mouse	[41]
TLR6	Activation of TLR2/TLR6 heterodimer increased interferon production.	HSV-1 infection	Mouse neuronal cell line	[42]
TLR7	Receptor activation induced microglial inflammatory cytokine production and antitumor effect.	Glioblastoma	Mouse glioblastoma	[16]
	Induced chemotaxis and prostaglandin production.	JEV infection	Pericyte cell line	[62]
	Receptor deficiency reduced anxiety, aggression, and contextual memory.	Not	KO mouse	[82]
	Acute activation of receptor reduced ethanol intake. Chronic activation induced tolerance to the adverse effects and an increase in ethanol consumption.	Alcohol use disorder	Mouse	[116]
	eceptor overexpression promoted meningitis.	Meningitis	Mouse	[143]
TLR8	TLR7 and TLR8 deficiency reduced astrogliosis, microgliosis, α-synuclein aggregate formation, and T-cell activation.	PD	TLR7/TLR8 KO mouse	[164]

Table 1. *Cont.*

Receptor	Outcome	Related Disease	Model	Refs.
TLR9	Induced antitumor effect.	Glioblastoma	Mouse microglial cell line	[13]
	Receptor inhibition decreased neurodegeneration.	Sepsis-associated encephalopathy	Sepsis rat model	[45]
	Induced TNF-α production in microglia and inhibition of seizure-induced aberrant neurogenesis.	Epilepsy	KO mouse	[111]
	Administration of ODN decreased bacterial burden in the cerebellum and bacteremia.	Meningitis	Mouse	[142]
	Receptor activation induced cognitive improvements and amyloid angiopathy reduction.	AD	AD mouse model	[156]
TLR10	Metabolite of vitamin D increased the expression of receptor.	Not	Human microglial cell line	[17]

KO, knockout; MMP-9, matrix metalloproteinase-9; EAE, experimental autoimmune encephalomyelitis; AD, Alzheimer's disease; PD, Parkinson's disease; HIV, human immunodeficiency virus; NSPC, neural stem and progenitor cells; HSV-1, herpes simplex virus-1; JEV, Japanese encephalitis virus.

Author Contributions: All authors contributed to the writing, review, and editing; D.M.A.-M. and L.A.-A. finalized the manuscript for submission; L.A.-A., conceptualization and funding acquisition. All authors have read and agreed to the published version of the manuscript.

Funding: This work was supported by Hospital Infantil de México Federico Gómez HIM-2017-09-SSA-1304.

Acknowledgments: G.R.-S. was a student from Posgrado en Ciencias Biológicas UNAM and received a fellowship from CONAHCYT.

Conflicts of Interest: The authors declare no conflicts of interest.

References

1. Li, D.; Wu, M. Pattern Recognition Receptors in Health and Diseases. *Signal Transduct. Target. Ther.* **2021**, *6*, 291. [CrossRef] [PubMed]
2. Su, S.B.; Tao, L.; Deng, Z.P.; Chen, W.; Qin, S.Y.; Jiang, H.X. TLR10: Insights, Controversies and Potential Utility as a Therapeutic Target. *Scand. J. Immunol.* **2021**, *93*, e12988. [CrossRef] [PubMed]
3. Vidya, M.K.; Kumar, V.G.; Sejian, V.; Bagath, M.; Krishnan, G.; Bhatta, R. Toll-like Receptors: Significance, Ligands, Signaling Pathways, and Functions in Mammals. *Int. Rev. Immunol.* **2018**, *37*, 20–36. [CrossRef] [PubMed]
4. Anthoney, N.; Foldi, I.; Hidalgo, A. Toll and Toll-like Receptor Signalling in Development. *Development* **2018**, *145*, 156018. [CrossRef] [PubMed]
5. Li, X.; Jiang, S.; Tapping, R.I. Toll-like Receptor Signaling in Cell Proliferation and Survival. *Cytokine* **2010**, *49*, 1–9. [CrossRef] [PubMed]
6. Starkey, J.; Horstick, E.J.; Ackerman, S.D. Glial Regulation of Critical Period Plasticity. *Front. Cell. Neurosci.* **2023**, *17*, 1247335. [CrossRef] [PubMed]
7. Prinz, M.; Jung, S.; Priller, J. Microglia Biology: One Century of Evolving Concepts. *Cell* **2019**, *179*, 292–311. [CrossRef] [PubMed]
8. Bsibsi, M.; Ravid, R.; Gveric, D.; Van Noort, J.M. Broad Expression of Toll-like Receptors in the Human Central Nervous System. *J. Neuropathol. Exp. Neurol.* **2002**, *61*, 1013–1021. [CrossRef] [PubMed]
9. Guo, S.; Wang, H.; Yin, Y. Microglia Polarization from M1 to M2 in Neurodegenerative Diseases. *Front. Aging Neurosci.* **2022**, *14*, 815347. [CrossRef]
10. Kwon, H.S.; Koh, S.H. Neuroinflammation in Neurodegenerative Disorders: The Roles of Microglia and Astrocytes. *Transl. Neurodegener.* **2020**, *9*, 42. [CrossRef]
11. Mishra, M.K.; Rawji, K.S.; Keough, M.B.; Kappen, J.; Dowlatabadi, R.; Vogel, H.J.; Chopra, S.; Distéfano-Gagné, F.; Dufour, A.; Gosselin, D.; et al. Harnessing the Benefits of Neuroinflammation: Generation of Macrophages/Microglia with Prominent Remyelinating Properties. *J. Neurosci.* **2021**, *41*, 3366–3385. [CrossRef] [PubMed]
12. Ren, C.; Li, D.; Zhou, Q.; Hu, X. Mitochondria-Targeted TPP-MoS2 with Dual Enzyme Activity Provides Efficient Neuroprotection through M1/M2 Microglial Polarization in an Alzheimer's Disease Model. *Biomaterials* **2020**, *232*, 119752. [CrossRef] [PubMed]
13. Catania, G.; Rodella, G.; Vanvarenberg, K.; Préat, V.; Malfanti, A. Combination of Hyaluronic Acid Conjugates with Immunogenic Cell Death Inducer and CpG for Glioblastoma Local Chemo-Immunotherapy Elicits an Immune Response and Induces Long-Term Survival. *Biomaterials* **2023**, *294*, 122006. [CrossRef] [PubMed]
14. Lee, J.W.; Nam, H.; Kim, L.E.; Jeon, Y.; Min, H.; Ha, S.; Lee, Y.; Kim, S.Y.; Lee, S.J.; Kim, E.K.; et al. TLR4 (Toll-like Receptor 4) Activation Suppresses Autophagy through Inhibition of FOXO3 and Impairs Phagocytic Capacity of Microglia. *Autophagy* **2019**, *15*, 753–770. [CrossRef] [PubMed]
15. Ifuku, M.; Hinkelmann, L.; Kuhrt, L.D.; Efe, I.E.; Kumbol, V.; Buonfiglioli, A.; Krüger, C.; Jordan, P.; Fulde, M.; Noda, M.; et al. Activation of Toll-like Receptor 5 in Microglia Modulates Their Function and Triggers Neuronal Injury. *Acta Neuropathol. Commun.* **2020**, *8*, 159. [CrossRef] [PubMed]
16. Buonfiglioli, A.; Efe, I.E.; Guneykaya, D.; Ivanov, A.; Huang, Y.; Orlowski, E.; Krüger, C.; Deisz, R.A.; Markovic, D.; Flüh, C.; et al. Let-7 MicroRNAs Regulate Microglial Function and Suppress Glioma Growth through Toll-Like Receptor 7. *Cell Rep.* **2019**, *29*, 3460–3471.e7. [CrossRef] [PubMed]
17. Verma, R.; Kim, J.Y. 1,25-Dihydroxyvitamin D3 Facilitates M2 Polarization and Upregulates TLR10 Expression on Human Microglial Cells. *Neuroimmunomodulation* **2016**, *23*, 75–80. [CrossRef] [PubMed]
18. Lee, H.G.; Wheeler, M.A.; Quintana, F.J. Function and Therapeutic Value of Astrocytes in Neurological Diseases. *Nat. Rev. Drug Discov.* **2022**, *21*, 339–358. [CrossRef] [PubMed]
19. Rupareliya, V.P.; Singh, A.A.; Butt, A.M.; Hariharan, A.; Kumar, H. The "Molecular Soldiers" of the CNS: Astrocytes, a Comprehensive Review on Their Roles and Molecular Signatures. *Eur. J. Pharmacol.* **2023**, *959*, 176048. [CrossRef]
20. Farina, C.; Krumbholz, M.; Giese, T.; Hartmann, G.; Aloisi, F.; Meinl, E. Preferential Expression and Function of Toll-like Receptor 3 in Human Astrocytes. *J. Neuroimmunol.* **2005**, *159*, 12–19. [CrossRef]
21. Deng, S.; Chen, X.; Lei, Q.; Lu, W. AQP2 Promotes Astrocyte Activation by Modulating the TLR4/NFκB-P65 Pathway following Intracerebral Hemorrhage. *Front. Immunol.* **2022**, *13*, 847360. [CrossRef]

22. Li, L.; Ni, L.; Eugenin, E.A.; Heary, R.F.; Elkabes, S. Toll-like Receptor 9 Antagonism Modulates Astrocyte Function and Preserves Proximal Axons following Spinal Cord Injury. *Brain Behav. Immun.* **2019**, *80*, 328–343. [CrossRef] [PubMed]
23. Li, L.; Ni, L.; Heary, R.F.; Elkabes, S. Astroglial TLR9 Antagonism Promotes Chemotaxis and Alternative Activation of Macrophages via Modulation of Astrocyte-Derived Signals: Implications for Spinal Cord Injury. *J. Neuroinflamm.* **2020**, *17*, 73. [CrossRef]
24. Kuhn, S.; Gritti, L.; Crooks, D.; Dombrowski, Y. Oligodendrocytes in Development, Myelin Generation and Beyond. *Cells* **2019**, *8*, 1424. [CrossRef] [PubMed]
25. Garcia-Diaz, B.; Baron-Van Evercooren, A. Schwann Cells: Rescuers of Central Demyelination. *Glia* **2020**, *68*, 1945–1956. [CrossRef]
26. Sloane, J.A.; Batt, C.; Ma, Y.; Harris, Z.M.; Trapp, B.; Vartanian, T. Hyaluronan Blocks Oligodendrocyte Progenitor Maturation and Remyelination through TLR2. *Proc. Natl. Acad. Sci. USA* **2010**, *107*, 11555–11560. [CrossRef] [PubMed]
27. Gyetvai, G.; Roe, C.; Heikal, L.; Ghezzi, P.; Mengozzi, M. Leukemia Inhibitory Factor Inhibits Erythropoietin-Induced Myelin Gene Expression in Oligodendrocytes. *Mol. Med.* **2018**, *24*, 51. [CrossRef] [PubMed]
28. Esser, S.; Göpfrich, L.; Bihler, K.; Kress, E.; Nyamoya, S.; Tauber, S.C.; Clarner, T.; Stope, M.B.; Pufe, T.; Kipp, M.; et al. Toll-Like Receptor 2-Mediated Glial Cell Activation in a Mouse Model of Cuprizone-Induced Demyelination. *Mol. Neurobiol.* **2018**, *55*, 6237–6249. [CrossRef] [PubMed]
29. Boccazzi, M.; Van Steenwinckel, J.; Schang, A.L.; Faivre, V.; Le Charpentier, T.; Bokobza, C.; Csaba, Z.; Verderio, C.; Fumagalli, M.; Mani, S.; et al. The Immune-Inflammatory Response of Oligodendrocytes in a Murine Model of Preterm White Matter Injury: The Role of TLR3 Activation. *Cell Death Dis.* **2021**, *12*, 166. [CrossRef]
30. Göttle, P.; Schichel, K.; Reiche, L.; Werner, L.; Zink, A.; Prigione, A.; Küry, P. TLR4 Associated Signaling Disrupters as a New Means to Overcome HERV-W Envelope-Mediated Myelination Deficits. *Front. Cell. Neurosci.* **2021**, *15*, 777542. [CrossRef]
31. Ved, R.; Sharouf, F.; Harari, B.; Muzaffar, M.; Manivannan, S.; Ormonde, C.; Gray, W.P.; Zaben, M. Disulfide HMGB1 Acts via TLR2/4 Receptors to Reduce the Numbers of Oligodendrocyte Progenitor Cells after Traumatic Injury in Vitro. *Sci. Rep.* **2021**, *11*, 6181. [CrossRef] [PubMed]
32. Xu, D.; Liang, J.; Cui, M.; Zhang, L.; Ren, S.; Zheng, W.; Dong, X.; Zhang, B. Saturated Fatty Acids Activate the Inflammatory Signalling Pathway in Schwann Cells: Implication in Sciatic Nerve Injury. *Scand. J. Immunol.* **2020**, *92*, e12896. [CrossRef] [PubMed]
33. Jiang, W.; Cheng, Y.; Wang, Y.; Wu, J.; Rong, Z.; Sun, L.; Zhou, Y.; Zhang, K. Involvement of Abnormal P-α-Syn Accumulation and TLR2-Mediated Inflammation of Schwann Cells in Enteric Autonomic Nerve Dysfunction of Parkinson's Disease: An Animal Model Study. *Mol. Neurobiol.* **2023**, *60*, 4738–4752. [CrossRef] [PubMed]
34. Sun, L.; Jiang, W.W.; Wang, Y.; Yuan, Y.S.; Rong, Z.; Wu, J.; Fan, Y.; Lu, M.; Zhang, K.Z. Phosphorylated α-Synuclein Aggregated in Schwann Cells Exacerbates Peripheral Neuroinflammation and Nerve Dysfunction in Parkinson's Disease through TLR2/NF-KB Pathway. *Cell Death Discov.* **2021**, *7*, 289. [CrossRef] [PubMed]
35. Schiera, G.; Di Liegro, C.M.; Di Liegro, I. Cell-to-Cell Communication in Learning and Memory: From Neuro- and Glio-Transmission to Information Exchange Mediated by Extracellular Vesicles. *Int. J. Mol. Sci.* **2019**, *21*, 266. [CrossRef] [PubMed]
36. Wang, F.X.; Liu, S.Y.; Zheng, X.; Chen, X.; Lu, L.X.; Chen, B.; Xiong, X.Y.; Shu, H.F.; Yang, Q.W.; Yang, H. TLR1 Expression in Mouse Brain Was Increased in a KA-Induced Seizure Model. *Inflamm. Res.* **2015**, *64*, 487–495. [CrossRef] [PubMed]
37. Tang, S.C.; Arumugam, T.V.; Xu, X.; Cheng, A.; Mughal, M.R.; Dong, G.J.; Lathia, J.D.; Siler, D.A.; Chigurupati, S.; Ouyang, X.; et al. Pivotal Role for Neuronal Toll-like Receptors in Ischemic Brain Injury and Functional Deficits. *Proc. Natl. Acad. Sci. USA* **2007**, *104*, 13798–13803. [CrossRef] [PubMed]
38. Casula, M.; Iyer, A.M.; Spliet, W.G.M.; Anink, J.J.; Steentjes, K.; Sta, M.; Troost, D.; Aronica, E. Toll-like Receptor Signaling in Amyotrophic Lateral Sclerosis Spinal Cord Tissue. *Neuroscience* **2011**, *179*, 233–243. [CrossRef] [PubMed]
39. Préhaud, C.; Mégret, F.; Lafage, M.; Lafon, M. Virus Infection Switches TLR-3-Positive Human Neurons To Become Strong Producers of Beta Interferon. *J. Virol.* **2005**, *79*, 12893. [CrossRef]
40. Jackson, A.C.; Rossiter, J.P.; Lafon, M. Expression of Toll-like Receptor 3 in the Human Cerebellar Cortex in Rabies, Herpes Simplex Encephalitis, and Other Neurological Diseases. *J. Neurovirol.* **2006**, *12*, 229–234. [CrossRef]
41. Seong, K.J.; Choi, S.; Lee, H.G.; Rhee, J.H.; Lee, J.H.; Koh, J.T.; Kim, S.H.; Choi, W.S.; Jung, J.Y.; Kim, W.J. Toll-Like Receptor 5 Promotes the Neurogenesis From Embryonic Stem Cells and Adult Hippocampal Neural Stem Cells in Mice. *Stem Cells* **2022**, *40*, 303–317. [CrossRef] [PubMed]
42. Bansode, Y.D.; Chattopadhyay, D.; Saha, B. Transcriptomic Analysis of Interferon Response in Toll-Like Receptor 2 Ligand-Treated and Herpes Simplex Virus 1-Infected Neurons and Astrocytes. *Viral Immunol.* **2021**, *34*, 256–266. [CrossRef] [PubMed]
43. Li, Y.; Chen, L.; Zhao, W.; Sun, L.; Zhang, R.; Zhu, S.; Xie, K.; Feng, X.; Wu, X.; Sun, Z.; et al. Food Reward Depends on TLR4 Activation in Dopaminergic Neurons. *Pharmacol. Res.* **2021**, *169*, 105659. [CrossRef] [PubMed]
44. Ritchie, L.; Tate, R.; Chamberlain, L.H.; Robertson, G.; Zagnoni, M.; Sposito, T.; Wray, S.; Wright, J.A.; Bryant, C.E.; Gay, N.J.; et al. Toll-like Receptor 3 Activation Impairs Excitability and Synaptic Activity via TRIF Signalling in Immature Rat and Human Neurons. *Neuropharmacology* **2018**, *135*, 1–10. [CrossRef] [PubMed]
45. Zhou, R.; Ying, J.; Qiu, X.; Yu, L.; Yue, Y.; Liu, Q.; Shi, J.; Li, X.; Qu, Y.; Mu, D.; et al. A New Cell Death Program Regulated by Toll-like Receptor 9 through P38 Mitogen-Activated Protein Kinase Signaling Pathway in a Neonatal Rat Model with Sepsis Associated Encephalopathy. *Chin. Med. J.* **2022**, *135*, 1474–1485. [CrossRef] [PubMed]

46. He, L.; Han, G.; Wu, S.; Du, S.; Zhang, Y.; Liu, W.; Jiang, B.; Zhang, L.; Xia, S.; Jia, S.; et al. Toll-like Receptor 7 Contributes to Neuropathic Pain by Activating NF-KB in Primary Sensory Neurons. *Brain Behav. Immun.* **2020**, *87*, 840–851. [CrossRef] [PubMed]
47. Zhou, Y.; Ye, L.; Wan, Q.; Zhou, L.; Wang, X.; Li, J.; Hu, S.; Zhou, D.; Ho, W. Activation of Toll-like Receptors Inhibits Herpes Simplex Virus-1 Infection of Human Neuronal Cells. *J. Neurosci. Res.* **2009**, *87*, 2916–2925. [CrossRef]
48. Kadry, H.; Noorani, B.; Cucullo, L. A Blood-Brain Barrier Overview on Structure, Function, Impairment, and Biomarkers of Integrity. *Fluids Barriers CNS* **2020**, *17*, 69. [CrossRef]
49. Chen, Z.; Li, G. Immune Response and Blood-Brain Barrier Dysfunction during Viral Neuroinvasion. *Innate Immun.* **2021**, *27*, 109–117. [CrossRef]
50. Wu, D.; Chen, Q.; Chen, X.; Han, F.; Chen, Z.; Wang, Y. The Blood-Brain Barrier: Structure, Regulation, and Drug Delivery. *Signal Transduct. Target. Ther.* **2023**, *8*, 217. [CrossRef]
51. Halder, S.K.; Sapkota, A.; Milner, R. The Importance of Laminin at the Blood-Brain Barrier. *Neural Regen. Res.* **2023**, *18*, 2557–2563. [CrossRef] [PubMed]
52. Nagyoszi, P.; Wilhelm, I.; Farkas, A.E.; Fazakas, C.; Dung, N.T.K.; Haskó, J.; Krizbai, I.A. Expression and Regulation of Toll-like Receptors in Cerebral Endothelial Cells. *Neurochem. Int.* **2010**, *57*, 556–564. [CrossRef] [PubMed]
53. Krüger-Genge, A.; Blocki, A.; Franke, R.P.; Jung, F. Vascular Endothelial Cell Biology: An Update. *Int. J. Mol. Sci.* **2019**, *20*, 4411. [CrossRef]
54. Johnson, R.H.; Kho, D.T.; O'Carroll, S.J.; Angel, C.E.; Graham, E.S. The Functional and Inflammatory Response of Brain Endothelial Cells to Toll-Like Receptor Agonists. *Sci. Rep.* **2018**, *8*, 10102. [CrossRef] [PubMed]
55. Zhao, Y.; Zhang, X.; Chen, X.; Wei, Y. Neuronal Injuries in Cerebral Infarction and Ischemic Stroke: From Mechanisms to Treatment (Review). *Int. J. Mol. Med.* **2022**, *49*, 15. [CrossRef] [PubMed]
56. Ziegler, G.; Harhausen, D.; Schepers, C.; Hoffmann, O.; Röhr, C.; Prinz, V.; König, J.; Lehrach, H.; Nietfeld, W.; Trendelenburg, G. TLR2 Has a Detrimental Role in Mouse Transient Focal Cerebral Ischemia. *Biochem. Biophys. Res. Commun.* **2007**, *359*, 574–579. [CrossRef] [PubMed]
57. Li, J.; Wang, Y.; Wang, X.; Ye, L.; Zhou, Y.; Persidsky, Y.; Ho, W. Immune Activation of Human Brain Microvascular Endothelial Cells Inhibits HIV Replication in Macrophages. *Blood* **2013**, *121*, 2934–2942. [CrossRef] [PubMed]
58. Guijarro-Muñoz, I.; Compte, M.; Álvarez-Cienfuegos, A.; Álvarez-Vallina, L.; Sanz, L. Lipopolysaccharide Activates Toll-like Receptor 4 (TLR4)-Mediated NF-KB Signaling Pathway and Proinflammatory Response in Human Pericytes. *J. Biol. Chem.* **2014**, *289*, 2457–2468. [CrossRef] [PubMed]
59. Dabravolski, S.A.; Andreeva, E.R.; Eremin, I.I.; Markin, A.M.; Nadelyaeva, I.I.; Orekhov, A.N.; Melnichenko, A.A. The Role of Pericytes in Regulation of Innate and Adaptive Immunity. *Biomedicines* **2023**, *11*, 600. [CrossRef]
60. Nyúl-Tóth, Á.; Kozma, M.; Nagyőszi, P.; Nagy, K.; Fazakas, C.; Haskó, J.; Molnár, K.; Farkas, A.E.; Végh, A.G.; Váró, G.; et al. Expression of Pattern Recognition Receptors and Activation of the Non-Canonical Inflammasome Pathway in Brain Pericytes. *Brain Behav. Immun.* **2017**, *64*, 220–231. [CrossRef]
61. Zhang, L.; Chopp, M.; Liu, X.; Teng, H.; Tang, T.; Kassis, H.; Zhang, Z.G. Combination Therapy with VELCADE and Tissue Plasminogen Activator Is Neuroprotective in Aged Rats after Stroke and Targets MicroRNA-146a and the Toll-like Receptor Signaling Pathway. *Arterioscler. Thromb. Vasc. Biol.* **2012**, *32*, 1856–1864. [CrossRef] [PubMed]
62. Chang, C.Y.; Li, J.R.; Ou, Y.C.; Lin, S.Y.; Wang, Y.Y.; Chen, W.Y.; Hu, Y.H.; Lai, C.Y.; Chang, C.J.; Chen, C.J. Interplay of Inflammatory Gene Expression in Pericytes following Japanese Encephalitis Virus Infection. *Brain Behav. Immun.* **2017**, *66*, 230–243. [CrossRef] [PubMed]
63. Yue, Q.; Hoi, M.P.M. Emerging Roles of Astrocytes in Blood-Brain Barrier Disruption upon Amyloid-Beta Insults in Alzheimer's Disease. *Neural Regen. Res.* **2023**, *18*, 1890. [CrossRef] [PubMed]
64. Macvicar, B.A.; Newman, E.A. Astrocyte Regulation of Blood Flow in the Brain. *Cold Spring Harb. Perspect. Biol.* **2015**, *7*, a020388. [CrossRef] [PubMed]
65. Min, H.; Hong, J.; Cho, I.H.; Jang, Y.H.; Lee, H.; Kim, D.; Yu, S.W.; Lee, S.; Lee, S.J. TLR2-Induced Astrocyte MMP9 Activation Compromises the Blood Brain Barrier and Exacerbates Intracerebral Hemorrhage in Animal Models. *Mol. Brain* **2015**, *8*, 23. [CrossRef] [PubMed]
66. Qiu, J.; Xu, J.; Zheng, Y.; Wei, Y.; Zhu, X.; Lo, E.H.; Moskowitz, M.A.; Sims, J.R. High-Mobility Group Box 1 Promotes Metalloproteinase-9 Upregulation through Toll-like Receptor 4 after Cerebral Ischemia. *Stroke* **2010**, *41*, 2077–2082. [CrossRef] [PubMed]
67. Liu, T.; Liu, M.; Zhang, T.; Liu, W.; Xu, H.; Mu, F.; Ren, D.; Jia, N.; Li, Z.; Ding, Y.; et al. Z-Guggulsterone Attenuates Astrocytes-Mediated Neuroinflammation after Ischemia by Inhibiting Toll-like Receptor 4 Pathway. *J. Neurochem.* **2018**, *147*, 803–815. [CrossRef] [PubMed]
68. Caso, J.R.; Pradillo, J.M.; Hurtado, O.; Lorenzo, P.; Moro, M.A.; Lizasoain, I. Toll-like Receptor 4 Is Involved in Brain Damage and Inflammation after Experimental Stroke. *Circulation* **2007**, *115*, 1599–1608. [CrossRef] [PubMed]
69. Wang, X.Y.; Wu, F.; Zhan, R.Y.; Zhou, H.J. Inflammatory Role of Microglia in Brain Injury Caused by Subarachnoid Hemorrhage. *Front. Cell. Neurosci.* **2022**, *16*, 956185. [CrossRef]
70. Okada, T.; Kawakita, F.; Nishikawa, H.; Nakano, F.; Liu, L.; Suzuki, H. Selective Toll-Like Receptor 4 Antagonists Prevent Acute Blood-Brain Barrier Disruption after Subarachnoid Hemorrhage in Mice. *Mol. Neurobiol.* **2019**, *56*, 976–985. [CrossRef]

71. Lehnardt, S.; Lehmann, S.; Kaul, D.; Tschimmel, K.; Hoffmann, O.; Cho, S.; Krueger, C.; Nitsch, R.; Meisel, A.; Weber, J.R. Toll-like Receptor 2 Mediates CNS Injury in Focal Cerebral Ischemia. *J. Neuroimmunol.* **2007**, *190*, 28–33. [CrossRef]
72. Kumar, A.; Pareek, V.; Faiq, M.A.; Ghosh, S.K.; Kumari, C. Adult Neurogenesis in Humans: A Review of Basic Concepts, History, Current Research, and Clinical Implications. *Innov. Clin. Neurosci.* **2019**, *16*, 30.
73. Kempermann, G.; Gage, F.H.; Aigner, L.; Song, H.; Curtis, M.A.; Thuret, S.; Kuhn, H.G.; Jessberger, S.; Frankland, P.W.; Cameron, H.A.; et al. Human Adult Neurogenesis: Evidence and Remaining Questions. *Cell Stem Cell* **2018**, *23*, 25–30. [CrossRef]
74. Gage, F.H. Neurogenesis in the Adult Brain. *J. Neurosci.* **2002**, *22*, 612. [CrossRef]
75. Bergmann, O.; Spalding, K.L.; Frisén, J. Adult Neurogenesis in Humans. *Cold Spring Harb. Perspect. Biol.* **2015**, *7*, a018994. [CrossRef]
76. Stiles, J.; Jernigan, T.L. The Basics of Brain Development. *Neuropsychol. Rev.* **2010**, *20*, 327–348. [CrossRef]
77. Zengeler, K.E.; Lukens, J.R. Innate Immunity at the Crossroads of Healthy Brain Maturation and Neurodevelopmental Disorders. *Nat. Rev. Immunol.* **2021**, *21*, 454–468. [CrossRef]
78. Ransohoff, R.M.; Cardona, A.E. The Myeloid Cells of the Central Nervous System Parenchyma. *Nature* **2010**, *468*, 253–262. [CrossRef]
79. Barak, B.; Feldman, N.; Okun, E. Toll-like Receptors as Developmental Tools That Regulate Neurogenesis during Development: An Update. *Front. Neurosci.* **2014**, *8*, 91968. [CrossRef]
80. Sakamoto, M.; Kageyama, R.; Imayoshi, I. The Functional Significance of Newly Born Neurons Integrated into Olfactory Bulb Circuits. *Front. Neurosci.* **2014**, *8*, 87675. [CrossRef]
81. Shen, Y.; Qin, H.; Chen, J.; Mou, L.; He, Y.; Yan, Y.; Zhou, H.; Lv, Y.; Chen, Z.; Wang, J.; et al. Postnatal Activation of TLR4 in Astrocytes Promotes Excitatory Synaptogenesis in Hippocampal Neurons. *J. Cell Biol.* **2016**, *215*, 719. [CrossRef]
82. Hung, Y.F.; Chen, C.Y.; Li, W.C.; Wang, T.F.; Hsueh, Y.P. Tlr7 Deletion Alters Expression Profiles of Genes Related to Neural Function and Regulates Mouse Behaviors and Contextual Memory. *Brain Behav. Immun.* **2018**, *72*, 101–113. [CrossRef]
83. Hui, C.W.; St-Pierre, A.; El Hajj, H.; Remy, Y.; Hébert, S.S.; Luheshi, G.N.; Srivastava, L.K.; Tremblay, M.È. Prenatal Immune Challenge in Mice Leads to Partly Sex-Dependent Behavioral, Microglial, and Molecular Abnormalities Associated with Schizophrenia. *Front. Mol. Neurosci.* **2018**, *11*, 13. [CrossRef]
84. Rolls, A.; Shechter, R.; London, A.; Ziv, Y.; Ronen, A.; Levy, R.; Schwartz, M. Toll-like Receptors Modulate Adult Hippocampal Neurogenesis. *Nat. Cell Biol.* **2007**, *9*, 1081–1088. [CrossRef]
85. Zhang, X.; Hei, Y.; Bai, W.; Huang, T.; Kang, E.; Chen, H.; Kong, C.; Yang, Y.; Ye, Y.; He, X. Toll-Like Receptor 2 Attenuates Traumatic Brain Injury-Induced Neural Stem Cell Proliferation in Dentate Gyrus of Rats. *Neural Plast.* **2020**, *2020*, 9814978. [CrossRef]
86. Li, G.; Forero, M.G.; Wentzell, J.; Durmus, I.; Wolf, R.; Anthoney, N.; Parker, M.; Jiang, R.; Hasenauer, J.; Strausfeld, N.; et al. A Toll-Receptor Map Underlies Structural Brain Plasticity. *Elife* **2020**, *9*, e52743. [CrossRef]
87. Sanchez-Petidier, M.; Guerri, C.; Moreno-Manzano, V. Toll-like Receptors 2 and 4 Differentially Regulate the Self-Renewal and Differentiation of Spinal Cord Neural Precursor Cells. *Stem Cell Res. Ther.* **2022**, *13*, 117. [CrossRef]
88. Lathia, J.D.; Okun, E.; Tang, S.C.; Griffioen, K.; Cheng, A.; Mughal, M.R.; Laryea, G.; Selvaraj, P.K.; Ffrench-Constant, C.; Magnus, T.; et al. Toll-like Receptor 3 Is a Negative Regulator of Embryonic Neural Progenitor Cell Proliferation. *J. Neurosci.* **2008**, *28*, 13978–13984. [CrossRef]
89. Okun, E.; Griffioen, K.; Barak, B.; Roberts, N.J.; Castro, K.; Pita, M.A.; Cheng, A.; Mughal, M.R.; Wan, R.; Ashery, U.; et al. Toll-like Receptor 3 Inhibits Memory Retention and Constrains Adult Hippocampal Neurogenesis. *Proc. Natl. Acad. Sci. USA* **2010**, *107*, 15625–15630. [CrossRef]
90. Connolly, M.G.; Yost, O.L.; Potter, O.V.; Giedraitis, M.E.; Kohman, R.A. Toll-like Receptor 4 Differentially Regulates Adult Hippocampal Neurogenesis in an Age- and Sex-Dependent Manner. *Hippocampus* **2020**, *30*, 958–969. [CrossRef]
91. Moraga, A.; Pradillo, J.M.; Cuartero, M.I.; Hernández-Jiménez, M.; Oses, M.; Moro, M.A.; Lizasoain, I. Toll-like Receptor 4 Modulates Cell Migration and Cortical Neurogenesis after Focal Cerebral Ischemia. *FASEB J.* **2014**, *28*, 4710–4718. [CrossRef]
92. Palma-Tortosa, S.; Hurtado, O.; Pradillo, J.M.; Ferreras-Martín, R.; García-Yébenes, I.; García-Culebras, A.; Moraga, A.; Moro, M.Á.; Lizasoain, I. Toll-like Receptor 4 Regulates Subventricular Zone Proliferation and Neuroblast Migration after Experimental Stroke. *Brain Behav. Immun.* **2019**, *80*, 573–582. [CrossRef]
93. Shechter, R.; Ronen, A.; Rolls, A.; London, A.; Bakalash, S.; Young, M.J.; Schwartz, M. Toll-like Receptor 4 Restricts Retinal Progenitor Cell Proliferation. *J. Cell Biol.* **2008**, *183*, 393. [CrossRef]
94. Hung, Y.F.; Chen, C.Y.; Shih, Y.C.; Liu, H.Y.; Huang, C.M.; Hsueh, Y.P. Endosomal TLR3, TLR7, and TLR8 Control Neuronal Morphology through Different Transcriptional Programs. *J. Cell Biol.* **2018**, *217*, 2727–2742. [CrossRef]
95. Okun, E.; Griffioen, K.J.; Mattson, M.P. Toll-like Receptor Signaling in Neural Plasticity and Disease. *Trends Neurosci.* **2011**, *34*, 269. [CrossRef]
96. Schroeder, P.; Rivalan, M.; Zaqout, S.; Krüger, C.; Schüler, J.; Long, M.; Meisel, A.; Winter, Y.; Kaindl, A.M.; Lehnardt, S. Abnormal Brain Structure and Behavior in MyD88-Deficient Mice. *Brain Behav. Immun.* **2021**, *91*, 181–193. [CrossRef]
97. Liu, H.Y.; Hong, Y.F.; Huang, C.M.; Chen, C.Y.; Huang, T.N.; Hsueh, Y.P. TLR7 Negatively Regulates Dendrite Outgrowth through the Myd88–c-Fos–IL-6 Pathway. *J. Neurosci.* **2013**, *33*, 11479. [CrossRef]
98. Chen, C.; Liu, H.; Hsueh, Y. TLR3 Downregulates Expression of Schizophrenia Gene Disc1 via MYD88 to Control Neuronal Morphology. *EMBO Rep.* **2017**, *18*, 169–183. [CrossRef]

99. Ma, Y.; Li, J.; Chiu, I.; Wang, Y.; Sloane, J.A.; Lü, J.; Kosaras, B.; Sidman, R.L.; Volpe, J.J.; Vartanian, T. Toll-like Receptor 8 Functions as a Negative Regulator of Neurite Outgrowth and Inducer of Neuronal Apoptosis. *J. Cell Biol.* **2006**, *175*, 209. [CrossRef]
100. Park, S.J.; Lee, J.Y.; Kim, S.J.; Choi, S.Y.; Yune, T.Y.; Ryu, J.H. Toll-like Receptor-2 Deficiency Induces Schizophrenia-like Behaviors in Mice. *Sci. Rep.* **2015**, *5*, 8502. [CrossRef]
101. Shechter, R.; London, A.; Kuperman, Y.; Ronen, A.; Rolls, A.; Chen, A.; Schwartz, M. Hypothalamic Neuronal Toll-like Receptor 2 Protects against Age-Induced Obesity. *Sci. Rep.* **2013**, *3*, 1254. [CrossRef]
102. Potter, O.V.; Giedraitis, M.E.; Johnson, C.D.; Cox, M.N.; Kohman, R.A. Young and Aged TLR4 Deficient Mice Show Sex-Dependent Enhancements in Spatial Memory and Alterations in Interleukin-1 Related Genes. *Brain Behav. Immun.* **2019**, *76*, 37–47. [CrossRef]
103. Fei, X.; Wang, J.X.; Wu, Y.; Dong, N.; Sheng, Z.Y. Sevoflurane-Induced Cognitive Decline in Aged Mice: Involvement of Toll-like Receptors 4. *Brain Res. Bull.* **2020**, *165*, 23–29. [CrossRef]
104. Squillace, S.; Niehoff, M.L.; Doyle, T.M.; Green, M.; Esposito, E.; Cuzzocrea, S.; Arnatt, C.K.; Spiegel, S.; Farr, S.A.; Salvemini, D. Sphingosine-1-Phosphate Receptor 1 Activation in the Central Nervous System Drives Cisplatin-Induced Cognitive Impairment. *J. Clin. Investig.* **2022**, *132*, e157738. [CrossRef]
105. Schilling, S.; Chausse, B.; Dikmen, H.O.; Almouhanna, F.; Hollnagel, J.O.; Lewen, A.; Kann, O. TLR2- and TLR3-Activated Microglia Induce Different Levels of Neuronal Network Dysfunction in a Context-Dependent Manner. *Brain Behav. Immun.* **2021**, *96*, 80–91. [CrossRef]
106. Nie, X.; Kitaoka, S.; Tanaka, K.; Segi-Nishida, E.; Imoto, Y.; Ogawa, A.; Nakano, F.; Tomohiro, A.; Nakayama, K.; Taniguchi, M.; et al. The Innate Immune Receptors TLR2/4 Mediate Repeated Social Defeat Stress-Induced Social Avoidance through Prefrontal Microglial Activation. *Neuron* **2018**, *99*, 464–479.e7. [CrossRef]
107. Weber, M.D.; Frank, M.G.; Sobesky, J.L.; Watkins, L.R.; Maier, S.F. Blocking Toll-like Receptor 2 and 4 Signaling during a Stressor Prevents Stress-Induced Priming of Neuroinflammatory Responses to a Subsequent Immune Challenge. *Brain Behav. Immun.* **2013**, *32*, 112–121. [CrossRef]
108. Figueroa-Hall, L.K.; Paulus, M.P.; Savitz, J. Toll-Like Receptor Signaling in Depression. *Psychoneuroendocrinology* **2020**, *121*, 104843. [CrossRef]
109. Missig, G.; Robbins, J.O.; Mokler, E.L.; McCullough, K.M.; Bilbo, S.D.; McDougle, C.J.; Carlezon, W.A. Sex-Dependent Neurobiological Features of Prenatal Immune Activation via TLR7. *Mol. Psychiatry* **2020**, *25*, 2330–2341. [CrossRef]
110. Khariv, V.; Pang, K.; Servatius, R.J.; David, B.T.; Goodus, M.T.; Beck, K.D.; Heary, R.F.; Elkabes, S. Toll-like Receptor 9 Deficiency Impacts Sensory and Motor Behaviors. *Brain Behav. Immun.* **2013**, *32*, 164–172. [CrossRef]
111. Matsuda, T.; Murao, N.; Katano, Y.; Juliandi, B.; Kohyama, J.; Akira, S.; Kawai, T.; Nakashima, K. TLR9 Signalling in Microglia Attenuates Seizure-Induced Aberrant Neurogenesis in the Adult Hippocampus. *Nat. Commun.* **2015**, *6*, 6514. [CrossRef]
112. Gárate, I.; García-Bueno, B.; Madrigal, J.L.M.; Bravo, L.; Berrocoso, E.; Caso, J.R.; Micó, J.A.; Leza, J.C. Origin and Consequences of Brain Toll-like Receptor 4 Pathway Stimulation in an Experimental Model of Depression. *J. Neuroinflamm.* **2011**, *8*, 151. [CrossRef]
113. Martín-Hernández, D.; Caso, J.R.; Bris, Á.G.; Maus, S.R.; Madrigal, J.L.M.; García-Bueno, B.; MacDowell, K.S.; Alou, L.; Gómez-Lus, M.L.; Leza, J.C. Bacterial Translocation Affects Intracellular Neuroinflammatory Pathways in a Depression-like Model in Rats. *Neuropharmacology* **2016**, *103*, 122–133. [CrossRef]
114. Zhang, K.; Lin, W.; Zhang, J.; Zhao, Y.; Wang, X.; Zhao, M. Effect of Toll-like Receptor 4 on Depressive-like Behaviors Induced by Chronic Social Defeat Stress. *Brain Behav.* **2020**, *10*, e01525. [CrossRef]
115. Li, H.; Chen, W.; Gou, M.; Li, W.; Tong, J.; Zhou, Y.; Xie, T.; Yu, T.; Feng, W.; Li, Y.; et al. The Relationship between TLR4/NF-KB/IL-1β Signaling, Cognitive Impairment, and White-Matter Integrity in Patients with Stable Chronic Schizophrenia. *Front. Psychiatry* **2022**, *13*, 966657. [CrossRef]
116. Grantham, E.K.; Warden, A.S.; McCarthy, G.S.; DaCosta, A.; Mason, S.; Blednov, Y.; Mayfield, R.D.; Harris, R.A. Role of Toll-like Receptor 7 (TLR7) in Voluntary Alcohol Consumption. *Brain Behav. Immun.* **2020**, *89*, 423–432. [CrossRef]
117. Blednov, Y.A.; Benavidez, J.M.; Geil, C.; Perra, S.; Morikawa, H.; Harris, R.A. Activation of Inflammatory Signaling by Lipopolysaccharide Produces a Prolonged Increase of Voluntary Alcohol Intake in Mice. *Brain Behav. Immun.* **2011**, *25* (Suppl. S1), S92–S105. [CrossRef]
118. Truitt, J.M.; Blednov, Y.A.; Benavidez, J.M.; Black, M.; Ponomareva, O.; Law, J.; Merriman, M.; Horani, S.; Jameson, K.; Lasek, A.W.; et al. Inhibition of IKKβ Reduces Ethanol Consumption in C57BL/6J Mice. *eNeuro* **2016**, *3*, ENEURO.0256-16.2016. [CrossRef]
119. Warden, A.S.; Azzam, M.; DaCosta, A.; Mason, S.; Blednov, Y.A.; Messing, R.O.; Mayfield, R.D.; Harris, R.A. Toll-like Receptor 3 Activation Increases Voluntary Alcohol Intake in C57BL/6J Male Mice. *Brain Behav. Immun.* **2019**, *77*, 55–65. [CrossRef]
120. Gabr, M.M.; Saeed, I.; Miles, J.A.; Ross, B.P.; Shaw, P.N.; Hollmann, M.W.; Parat, M.O. Interaction of Opioids with TLR4-Mechanisms and Ramifications. *Cancers* **2021**, *13*, 5274. [CrossRef]
121. Rubio-Araiz, A.; Porcu, F.; Pérez-Hernández, M.; García-Gutiérrez, M.S.; Aracil-Fernández, M.A.; Gutierrez-López, M.D.; Guerri, C.; Manzanares, J.; O'Shea, E.; Colado, M.I. Disruption of Blood-Brain Barrier Integrity in Postmortem Alcoholic Brain: Preclinical Evidence of TLR4 Involvement from a Binge-like Drinking Model. *Addict. Biol.* **2017**, *22*, 1103–1116. [CrossRef] [PubMed]
122. Shabab, T.; Khanabdali, R.; Moghadamtousi, S.Z.; Kadir, H.A.; Mohan, G. Neuroinflammation Pathways: A General Review. *Int. J. Neurosci.* **2017**, *127*, 624–633. [CrossRef] [PubMed]
123. Kaur, N.; Chugh, H.; Sakharkar, M.K.; Dhawan, U.; Chidambaram, S.B.; Chandra, R. Neuroinflammation Mechanisms and Phytotherapeutic Intervention: A Systematic Review. *ACS Chem. Neurosci.* **2020**, *11*, 3707–3731. [CrossRef]

124. DiSabato, D.J.; Quan, N.; Godbout, J.P. Neuroinflammation: The Devil Is in the Details. *J. Neurochem.* **2016**, *139* (Suppl. S2), 136–153. [CrossRef]
125. Moyse, E.; Krantic, S.; Djellouli, N.; Roger, S.; Angoulvant, D.; Debacq, C.; Leroy, V.; Fougere, B.; Aidoud, A. Neuroinflammation: A Possible Link between Chronic Vascular Disorders and Neurodegenerative Diseases. *Front. Aging Neurosci.* **2022**, *14*, 827263. [CrossRef] [PubMed]
126. Singh, H.; Koury, J.; Kaul, M. Innate Immune Sensing of Viruses and Its Consequences for the Central Nervous System. *Viruses* **2021**, *13*, 170. [CrossRef] [PubMed]
127. Gern, O.L.; Mulenge, F.; Pavlou, A.; Ghita, L.; Steffen, I.; Stangel, M.; Kalinke, U. Toll-like Receptors in Viral Encephalitis. *Viruses* **2021**, *13*, 2065. [CrossRef] [PubMed]
128. Zhao, G.; Gao, Y.; Zhang, J.; Zhang, H.; Xie, C.; Nan, F.; Feng, S.; Ha, Z.; Li, C.; Zhu, X.; et al. Toll-like Receptor 2 Signaling Pathway Activation Contributes to a Highly Efficient Inflammatory Response in Japanese Encephalitis Virus-Infected Mouse Microglial Cells by Proteomics. *Front. Microbiol.* **2022**, *13*, 989183. [CrossRef] [PubMed]
129. Mukherjee, S.; Akbar, I.; Kumari, B.; Vrati, S.; Basu, A.; Banerjee, A. Japanese Encephalitis Virus-Induced Let-7a/b Interacted with the NOTCH-TLR7 Pathway in Microglia and Facilitated Neuronal Death via Caspase Activation. *J. Neurochem.* **2019**, *149*, 518–534. [CrossRef]
130. Shen, Y.; Yang, R.; Zhao, J.; Chen, M.; Chen, S.; Ji, B.; Chen, H.; Liu, D.; Li, L.; Du, G. The Histone Deacetylase Inhibitor Belinostat Ameliorates Experimental Autoimmune Encephalomyelitis in Mice by Inhibiting TLR2/MyD88 and HDAC3/NF-KB P65-Mediated Neuroinflammation. *Pharmacol. Res.* **2022**, *176*, 105969. [CrossRef]
131. Kwilasz, A.J.; Green Fulgham, S.M.; Duran-Malle, J.C.; Schrama, A.E.W.; Mitten, E.H.; Todd, L.S.; Patel, H.P.; Larson, T.A.; Clements, M.A.; Harris, K.M.; et al. Toll-like Receptor 2 and 4 Antagonism for the Treatment of Experimental Autoimmune Encephalomyelitis (EAE)-Related Pain. *Brain Behav. Immun.* **2021**, *93*, 80–95. [CrossRef]
132. Chu, Y.; Jing, Y.; Zhao, X.; Wang, M.; Zhang, M.; Ma, R.; Ma, W.; Lv, Y.; Zhu, L. Modulation of the HMGB1/TLR4/NF-KB Signaling Pathway in the CNS by Matrine in Experimental Autoimmune Encephalomyelitis. *J. Neuroimmunol.* **2021**, *352*, 577480. [CrossRef]
133. Vitturi, B.K.; Rosemberg, S.; Arita, F.N.; da Rocha, A.J.; Forte, W.C.N.; Tilbery, C.P. Multiphasic Disseminated Encephalomyelitis Associated with Herpes Virus Infection in a Patient with TLR3 Deficiency. *Mult. Scler. Relat. Disord.* **2019**, *36*, 101379. [CrossRef] [PubMed]
134. Partanen, T.; Chen, J.; Lehtonen, J.; Kuismin, O.; Rusanen, H.; Vapalahti, O.; Vaheri, A.; Anttila, V.J.; Bode, M.; Hautala, N.; et al. Heterozygous TLR3 Mutation in Patients with Hantavirus Encephalitis. *J. Clin. Immunol.* **2020**, *40*, 1156–1162. [CrossRef]
135. Su, W.; Ju, J.; Gu, M.; Wang, X.; Liu, S.; Yu, J.; Mu, D. SARS-CoV-2 Envelope Protein Triggers Depression-like Behaviors and Dysosmia via TLR2-Mediated Neuroinflammation in Mice. *J. Neuroinflamm.* **2023**, *20*, 110. [CrossRef]
136. Fontes-Dantas, F.L.; Fernandes, G.G.; Gutman, E.G.; De Lima, E.V.; Antonio, L.S.; Hammerle, M.B.; Mota-Araujo, H.P.; Colodeti, L.C.; Araújo, S.M.B.; Froz, G.M.; et al. SARS-CoV-2 Spike Protein Induces TLR4-Mediated Long-Term Cognitive Dysfunction Recapitulating Post-COVID-19 Syndrome in Mice. *Cell Rep.* **2023**, *42*, 112189. [CrossRef]
137. Wang, P.; Liu, J.B.; Wang, X.; Meng, F.Z.; Xiao, Q.H.; Liu, L.; Zhu, J.; Hu, W.H.; Ho, W.Z. Activation of Toll-like Receptor 3 Inhibits HIV Infection of Human IPSC-Derived Microglia. *J. Med. Virol.* **2023**, *95*, e29217. [CrossRef]
138. Jin, X.; Yin, S.; Zhang, Y.; Chen, X. Association between TLR2 + 2477G/A Polymorphism and Bacterial Meningitis: A Meta-Analysis. *Epidemiol. Infect.* **2018**, *146*, 642–647. [CrossRef] [PubMed]
139. Tenhu, E.; Teräsjärvi, J.; Cruzeiro, M.L.; Savonius, O.; Rugemalira, E.; Roine, I.; He, Q.; Pelkonen, T. Gene Polymorphisms of TLR4 and TLR9 and Haemophilus Influenzae Meningitis in Angolan Children. *Genes* **2020**, *11*, 1099. [CrossRef]
140. Xue, H.; Peng, H.; Li, J.; Li, M.; Lu, S. TLR9 Rs352140 Polymorphism Contributes to a Decreased Risk of Bacterial Meningitis: Evidence from a Meta-Analysis. *Epidemiol. Infect.* **2020**, *148*, e294. [CrossRef]
141. Too, L.K.; Yau, B.; Baxter, A.G.; McGregor, I.S.; Hunt, N.H. Double Deficiency of Toll-like Receptors 2 and 4 Alters Long-Term Neurological Sequelae in Mice Cured of Pneumococcal Meningitis. *Sci. Rep.* **2019**, *9*, 16189. [CrossRef] [PubMed]
142. Ribes, S.; Zacke, L.; Nessler, S.; Saiepour, N.; Avendaño-Guzmán, E.; Ballüer, M.; Hanisch, U.K.; Nau, R. Oligodeoxynucleotides Containing Unmethylated Cytosine-Guanine Motifs Are Effective Immunostimulants against Pneumococcal Meningitis in the Immunocompetent and Neutropenic Host. *J. Neuroinflamm.* **2021**, *18*, 39. [CrossRef] [PubMed]
143. He, Y.; Zhang, X.; Sun, Y.; Gong, P.; Yu, H. Promotion Properties of TLR7 in Pediatric Meningitis via the NF-KB Pathway. *J. Bioenerg. Biomembr.* **2021**, *53*, 39–48. [CrossRef] [PubMed]
144. Scheltens, P.; De Strooper, B.; Kivipelto, M.; Holstege, H.; Chételat, G.; Teunissen, C.E.; Cummings, J.; van der Flier, W.M. Alzheimer's Disease. *Lancet* **2021**, *397*, 1577–1590. [CrossRef] [PubMed]
145. Zhang, J.; Zheng, Y.; Zhao, Y.; Zhang, Y.; Liu, Y.; Ma, F.; Wang, X.; Fu, J. Andrographolide Ameliorates Neuroinflammation in APP/PS1 Transgenic Mice. *Int. Immunopharmacol.* **2021**, *96*, 107808. [CrossRef] [PubMed]
146. Rangaraju, S.; Dammer, E.B.; Raza, S.A.; Rathakrishnan, P.; Xiao, H.; Gao, T.; Duong, D.M.; Pennington, M.W.; Lah, J.J.; Seyfried, N.T.; et al. Identification and Therapeutic Modulation of a Pro-Inflammatory Subset of Disease-Associated-Microglia in Alzheimer's Disease. *Mol. Neurodegener.* **2018**, *13*, 24. [CrossRef] [PubMed]
147. Walker, D.G.; Tang, T.M.; Lue, L.F. Increased Expression of Toll-like Receptor 3, an Anti-Viral Signaling Molecule, and Related Genes in Alzheimer's Disease Brains. *Exp. Neurol.* **2018**, *309*, 91–106. [CrossRef] [PubMed]

148. Dutta, D.; Jana, M.; Paidi, R.K.; Majumder, M.; Raha, S.; Dasarathy, S.; Pahan, K. Tau Fibrils Induce Glial Inflammation and Neuropathology via TLR2 in Alzheimer's Disease–Related Mouse Models. *J. Clin. Investig.* **2023**, *133*, e161987. [CrossRef] [PubMed]
149. Lax, N.; Fainstein, N.; Nishri, Y.; Ben-Zvi, A.; Ben-Hur, T. Systemic Microbial TLR2 Agonists Induce Neurodegeneration in Alzheimer's Disease Mice. *J. Neuroinflamm.* **2020**, *17*, 55. [CrossRef]
150. Viola, T.W.; Creutzberg, K.C.; Zaparte, A.; Kestering-Ferreira, É.; Tractenberg, S.G.; Centeno-Silva, A.; Orso, R.; Lumertz, F.S.; Brietzke, E.; Wearick-Silva, L.E.; et al. Acute Neuroinflammation Elicited by TLR-3 Systemic Activation Combined with Early Life Stress Induces Working Memory Impairments in Male Adolescent Mice. *Behav. Brain Res.* **2019**, *376*, 112221. [CrossRef]
151. Rubio-Araiz, A.; Finucane, O.M.; Keogh, S.; Lynch, M.A. Anti-TLR2 Antibody Triggers Oxidative Phosphorylation in Microglia and Increases Phagocytosis of β-Amyloid. *J. Neuroinflamm.* **2018**, *15*, 247. [CrossRef] [PubMed]
152. Rangasamy, S.B.; Jana, M.; Roy, A.; Corbett, G.T.; Kundu, M.; Chandra, S.; Mondal, S.; Dasarathi, S.; Mufson, E.J.; Mishra, R.K.; et al. Selective Disruption of TLR2-MyD88 Interaction Inhibits Inflammation and Attenuates Alzheimer's Pathology. *J. Clin. Investig.* **2018**, *128*, 4297–4312. [CrossRef]
153. Zhou, C.; Sun, X.; Hu, Y.; Song, J.; Dong, S.; Kong, D.; Wang, Y.; Hua, X.; Han, J.; Zhou, Y.; et al. Genomic Deletion of TLR2 Induces Aggravated White Matter Damage and Deteriorated Neurobehavioral Functions in Mouse Models of Alzheimer's Disease. *Aging* **2019**, *11*, 7257–7273. [CrossRef]
154. Pourbadie, H.G.; Sayyah, M.; Khoshkholgh-Sima, B.; Choopani, S.; Nategh, M.; Motamedi, F.; Shokrgozar, M.A. Early Minor Stimulation of Microglial TLR2 and TLR4 Receptors Attenuates Alzheimer's Disease-Related Cognitive Deficit in Rats: Behavioral, Molecular, and Electrophysiological Evidence. *Neurobiol. Aging* **2018**, *70*, 203–216. [CrossRef]
155. Wang, S.; Zhu, T.; Ni, W.; Zhou, C.; Zhou, H.; Lin, L.; Hu, Y.; Sun, X.; Han, J.; Zhou, Y.; et al. Early Activation of Toll-like Receptor-3 Reduces the Pathological Progression of Alzheimer's Disease in APP/PS1 Mouse. *Alzheimer's Res. Ther.* **2023**, *15*, 33. [CrossRef]
156. Scholtzova, H.; Do, E.; Dhakal, S.; Sun, Y.; Liu, S.; Mehta, P.D.; Wisniewski, T. Innate Immunity Stimulation via Toll-Like Receptor 9 Ameliorates Vascular Amyloid Pathology in Tg-SwDI Mice with Associated Cognitive Benefits. *J. Neurosci.* **2017**, *37*, 936–959. [CrossRef] [PubMed]
157. Ye, H.; Robak, L.A.; Yu, M.; Cykowski, M.; Shulman, J.M. Genetics and Pathogenesis of Parkinson's Syndrome. *Annu. Rev. Pathol.* **2023**, *18*, 95–121. [CrossRef]
158. Marogianni, C.; Sokratous, M.; Dardiotis, E.; Hadjigeorgiou, G.M.; Bogdanos, D.; Xiromerisiou, G. Neurodegeneration and Inflammation-An Interesting Interplay in Parkinson's Disease. *Int. J. Mol. Sci.* **2020**, *21*, 8421. [CrossRef] [PubMed]
159. Conte, C.; Ingrassia, A.; Breve, J.; Bol, J.J.; Timmermans-Huisman, E.; van Dam, A.M.; Beccari, T.; van de Berg, W.D.J. Toll-like Receptor 4 Is Upregulated in Parkinson's Disease Patients and Co-Localizes with PSer129αSyn: A Possible Link with the Pathology. *Cells* **2023**, *12*, 1368. [CrossRef]
160. Chung, L.Y.R.; Lin, Y.T.; Liu, C.; Tai, Y.C.; Lin, H.Y.; Lin, C.H.; Chen, C.C. Neuroinflammation Upregulated Neuronal Toll-Like Receptors 2 and 4 to Drive Synucleinopathy in Neurodegeneration. *Front. Pharmacol.* **2022**, *13*, 845930. [CrossRef]
161. He, Y.; Zhao, J.; Dong, H.; Zhang, X.; Duan, Y.; Ma, Y.; Yu, M.; Fei, J.; Huang, F. TLR2 Deficiency Is Beneficial at the Late Phase in MPTP-Induced Parkinson' Disease Mice. *Life Sci.* **2023**, *333*, 122171. [CrossRef] [PubMed]
162. Campolo, M.; Paterniti, I.; Siracusa, R.; Filippone, A.; Esposito, E.; Cuzzocrea, S. TLR4 Absence Reduces Neuroinflammation and Inflammasome Activation in Parkinson's Diseases in Vivo Model. *Brain Behav. Immun.* **2019**, *76*, 236–247. [CrossRef] [PubMed]
163. Venezia, S.; Kaufmann, W.A.; Wenning, G.K.; Stefanova, N. Toll-like Receptor 4 Deficiency Facilitates α-Synuclein Propagation and Neurodegeneration in a Mouse Model of Prodromal Parkinson's Disease. *Park. Relat. Disord.* **2021**, *91*, 59–65. [CrossRef] [PubMed]
164. Campolo, M.; Filippone, A.; Biondo, C.; Mancuso, G.; Casili, G.; Lanza, M.; Cuzzocrea, S.; Esposito, E.; Paterniti, I. TLR7/8 in the Pathogenesis of Parkinson's Disease. *Int. J. Mol. Sci.* **2020**, *21*, 9384. [CrossRef] [PubMed]
165. Wang, J.; Liu, Y.; Liu, Y.; Zhu, K.; Xie, A. The Association between TLR3 Rs3775290 Polymorphism and Sporadic Parkinson's Disease in Chinese Han Population. *Neurosci. Lett.* **2020**, *728*, 135005. [CrossRef] [PubMed]
166. Miri, N.S.; Saadat, P.; Azadmehr, A.; Oladnabi, M.; Daraei, A. Toll-Like Receptor (TLR)-9 Rs352140 Polymorphism Is an Immunopathology Protective Factor in Parkinson's Disease in the Northern Iranian Population. *Iran. J. Immunol.* **2020**, *17*, 313–323. [CrossRef]
167. Li, Z.; Song, A.; Yu, H. Interaction between Toll-like Receptor 4 (TLR4) Gene and Alcohol Drinking on Parkinson's Disease Risk in Chinese Han Population. *J. Clin. Neurosci.* **2019**, *62*, 128–132. [CrossRef]

International Journal of *Molecular Sciences*

MDPI

Article

Synergistic Targeting of Innate Receptors TLR7 and NOD2 for Therapeutic Intervention in Multiple Sclerosis

Magdalena Dubik, Joanna Marczynska-Grzelak, Michael Zaucha Sørensen, Ruthe Storgaard Dieu, Dominika Rusin, Eydís Sigurdardóttir Schiöth, Bita Ramazani, Rouhin Belal, Bhavya Ojha, Jonathan Krieger, Dina S. Arengoth, Agnieszka Wlodarczyk, Trevor Owens and Reza Khorooshi *

Department of Neurobiology Research, Institute of Molecular Medicine, University of Southern Denmark, Campusvej 55, DK-5000 Odense C, Denmark

* Correspondence: rkhorooshi@health.sdu.dk; Tel.: +45-30519922

Abstract: Regulation of neuroinflammation is critical for maintaining central nervous system (CNS) homeostasis and holds therapeutic promise in autoimmune diseases such as multiple sclerosis (MS). Previous studies have highlighted the significance of selective innate signaling in triggering anti-inflammatory mechanisms, which play a protective role in an MS-like disease, experimental autoimmune encephalomyelitis (EAE). However, the individual intra-CNS administration of specific innate receptor ligands or agonists, such as for toll-like receptor 7 (TLR7) and nucleotide-binding oligomerization-domain-containing protein 2 (NOD2), failed to elicit the desired anti-inflammatory response in EAE. In this study, we investigated the potential synergistic effect of targeting both TLR7 and NOD2 simultaneously to prevent EAE progression. Our findings demonstrate that simultaneous intrathecal administration of NOD2- and TLR7-agonists led to synergistic induction of Type I IFN (IFN I) and effectively suppressed EAE in an IFN I-dependent manner. Suppression of EAE was correlated with a significant decrease in the infiltration of monocytes, granulocytes, and natural killer cells, reduced demyelination, and downregulation of *IL-1β*, *CCL2*, and *IFNγ* gene expression in the spinal cord. These results underscore the therapeutic promise of concurrently targeting the TLR7 and NOD2 pathways in alleviating neuroinflammation associated with MS, paving the way for novel and more efficacious treatment strategies.

Keywords: TLR7; NOD2; synergy; type I IFN; experimental autoimmune encephalomyelitis

Citation: Dubik, M.; Marczynska-Grzelak, J.; Sørensen, M.Z.; Dieu, R.S.; Rusin, D.; Schiöth, E.S.; Ramazani, B.; Belal, R.; Ojha, B.; Krieger, J.; et al. Synergistic Targeting of Innate Receptors TLR7 and NOD2 for Therapeutic Intervention in Multiple Sclerosis. *Int. J. Mol. Sci.* **2024**, *25*, 7462. https://doi.org/10.3390/ijms25137462

Academic Editors: Oliver Planz and Ralf Kircheis

Received: 7 June 2024
Revised: 28 June 2024
Accepted: 4 July 2024
Published: 7 July 2024

1. Introduction

Multiple sclerosis (MS) is a chronic autoimmune disease characterized by inflammation and demyelination within the central nervous system (CNS), leading to neurological dysfunction and disability [1]. Despite advancements in understanding the pathophysiology of MS, current treatment options remain limited and often associated with significant side effects. Neuroinflammation, orchestrated by dysregulated immune responses, plays a pivotal role in MS progression and pathogenesis. Thus, modulating immune responses and restoring CNS homeostasis represent attractive therapeutic approaches for MS management. Recent studies have highlighted the significance of selective innate signaling in triggering anti-inflammatory mechanisms, including the production of IFN I, which play a protective role in the MS-like disease experimental autoimmune encephalomyelitis (EAE) [2,3]. However, the individual intra-CNS administration of specific innate receptor ligands or agonists, such as toll-like receptor 7 (TLR7) was insufficient to elicit an anti-inflammatory response in EAE [4]. Similarly, in this study, we found that the stimulation of nucleotide-binding oligomerization domain-containing protein 2 (NOD2) had no beneficial effect on EAE when given at the time of disease onset.

NOD2 is a cytosolic receptor primarily responsible for detecting bacterial peptidoglycan fragments (bacterial muramyl dipeptide, MDP), while TLR7 is an endosomal receptor that responds to viral single-stranded RNA [5,6]. Co-stimulation with NOD2 and TLR7 agonists has been reported to synergistically enhance cytokine production in peripheral blood mononuclear cells and macrophages [7,8]. Additionally, simultaneous targeting of both receptors produces super-additive effects on the development of immune responses in vivo [9].

Recent studies have demonstrated that the administration of an immunomodulatory microparticle that activates TLR9 and NOD2 signaling is protective against EAE [10,11]. To investigate the potential synergistic effect of targeting CNS innate receptors TLR7 and NOD2 in preventing EAE, we employed intrathecal administration of specific agonists for TLR7 and NOD2 in combination. Our results indicate that while individual administration of TLR7 or NOD2 agonists failed to induce an anti-inflammatory program in EAE, simultaneous activation of both receptors resulted in a synergistic induction of IFN I. Furthermore, intrathecal administration of combined NOD2 and TLR7 agonists effectively reduced infiltration of immune cells into the CNS and suppressed EAE progression in an IFN I-dependent manner.

2. Results

2.1. While Individual Intrathecal Delivery of Imiquimod or MDP Failed to Suppress EAE, Their Combination Had Disease-Modifying Activity

Previously, we reported that the stimulation of TLR7 within the CNS had no beneficial effects on EAE [4]. Similarly, in this study, stimulation of NOD2 by MDP within the CNS did not alter the progression of EAE (Supplementary Figure S1). We and others have shown that the delivery of a bispecific immunomodulatory microparticle that activates TLR9 and NOD2 signaling reduces the severity of EAE [10,11]. Together, these findings prompted us to examine the potential synergistic impact of concurrently targeting both TLR7 and NOD2 on EAE progression.

Following EAE induction in mice, upon the manifestation of initial symptoms (floppy tail, defined as day 0), a single intrathecal injection of combined imiquimod and MDP (hereafter called MI27) was administered. Subsequent disease progression was monitored over 4 days. Intrathecal administration of MI27 resulted in significant reduction in disease severity compared to controls. The therapeutic effects of MI27 were apparent as early as day one post-treatment and persisted through day three (Figure 1A).

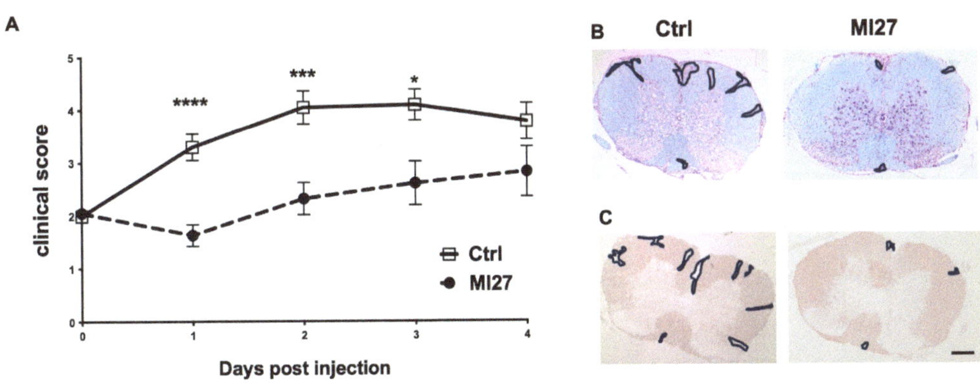

Figure 1. Intrathecal MI27 suppressed EAE. (**A**) C57BL/6 mice were immunized with MOGp35-55 to induce EAE, and at disease onset, they received intrathecal MI27 or PBS, and clinical signs were scored daily. At the onset of EAE (day 0), mice that received MI27 showed marked attenuation of disease compared to control mice (n = 15–18 in each group). The data are pooled from three independent studies. Data are presented as mean ± SEM. Results were analyzed using the two-tailed Mann–Whitney u-test. * $p < 0.05$, *** $p < 0.001$, **** $p < 0.0001$. (**B**,**C**) Micrographs of spinal cord sections from mice with EAE that received MI27 or PBS intrathecally, stained with LFB and cresyl

violet and with anti-MOG antibody. (**B**) In the control group, cresyl violet staining showed cell infiltration into the parenchyma of the spinal cord with extensive loss of LFB (marked area) and (**C**) MOG staining in corresponding areas (marked area). Mice with EAE that were treated with intrathecal MI27 showed cell accumulation in the meninges and reduced loss of LFB- and MOG staining. Scale bar. 200 μm.

As expected, histological analysis of the spinal cord white matter in control animals showed increased nuclear density, indicative of leukocyte infiltration (marked regions in Figure 1B) and reduced MOG staining indicative of demyelination (marked regions in Figure 1C). Notably, treatment with MI27 led to a decrease in immune cell infiltration, primarily confined to the extraparenchymal regions within the meninges (Figure 1B). Additionally, MI27 treatment mitigated myelin loss in the spinal cord (Figure 1C).

2.2. *Intrathecal MI27 Induced IFN Beta Response in the CNS and Suppressed EAE in an IFN I Dependent Manner*

Interferon beta (IFNβ) has demonstrated therapeutic efficacy in the treatment of MS. Given that activation of either TLR7 or NOD2 pathways can induce IFN I expression, we aimed to investigate whether intrathecal administration of MI27 induces IFNβ expression in the steady state CNS. For this purpose, we utilized mice expressing a luciferase gene under the control of the IFNβ promoter [12]. Mice received intrathecal injections of imiquimod, MDP, MDP-ctrl combined with imiquimod, or MI27. Luciferase activity indicative of IFNβ expression was measured 4, 24, and 48 h post-treatment. In vivo imaging revealed that a single intrathecal injection of MI27 significantly increased IFNβ expression within the CNS at 4 h post-injection (Figure 2A,B). In contrast, intrathecal administration of imiquimod, MDP, or MDP-ctrl + imiquimod failed to induce IFNβ expression (Figure 2B).

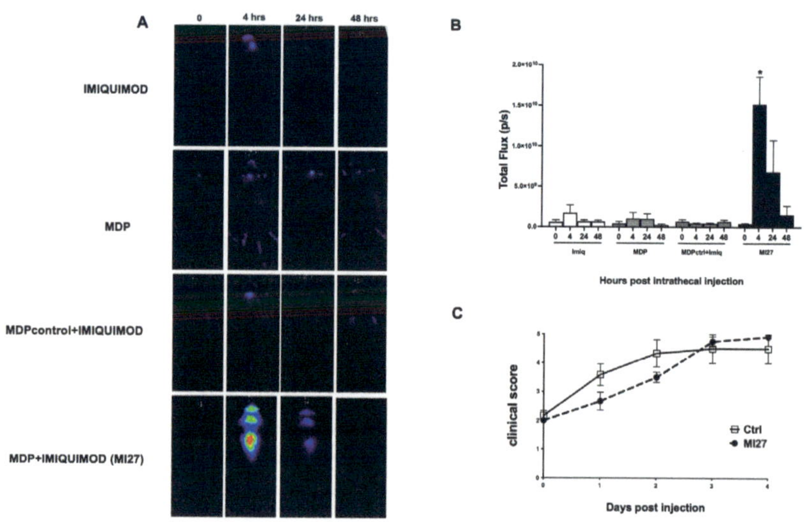

Figure 2. Intrathecal MI27 induced IFNβ and suppressed EAE in an IFNAR-dependent manner. (**A**,**B**) In vivo imaging of IFNβ/luciferase reporter mice that received intrathecal imiquimod, MDP, MDP control + imiquimod, or MDP control + imiquimod (MI27). The level of IFNβ was significantly increased at 4- and 24-h post-injection. $n = 3$ in each group. (**C**) IFNAR1-deficient mice were immunized with MOGp35-55 to induce EAE and at disease onset they received intrathecal MI27 or PBS, and clinical signs were scored daily. The protective effect of intrathecal MI27 was abrogated in IFNAR1-KO mice. ($n = 6$ per group). Data are presented as mean ± SEM. Results were analyzed using paired Student's *t*-test. * $p < 0.05$.

The robust induction of IFNβ expression observed in mice treated with MI27 strongly suggested a pivotal role for IFNAR signaling in mediating the effects of MI27. To further explore this, we investigated the potential synergistic impact of concurrently targeting both TLR7 and NOD2 on EAE progression in mice lacking IFN I signaling. IFNAR1-deficient mice were immunized to induce EAE, and upon disease onset, they received a single intrathecal injection of MI27. Disease progression was monitored for 4 days. Disease severity remained unaffected in IFNAR1-deficient mice treated with intrathecal MI27 (Figure 2C). This finding indicates that the therapeutic efficacy of MI27 is dependent on IFN I signaling pathways.

2.3. *Intrathecal MI27 Treatment Reduced CNS Recruitment of Selected Immune Cells into the Spinal Cord*

The results from histology suggested that the modulation of immune cell infiltration within the CNS can serve as a potential mechanism underlying the efficacy of MI27 treatment. To further characterize infiltrating immune cells in the CNS in response to intrathecal MI27, we conducted flow cytometry analysis in mice with EAE (Figure 3A). The results showed that the total number of $CD45^{high}$ cells was significantly reduced in MI27-treated mice compared to control mice (Figure 3B). Further analysis of $CD45^{high}$ cells demonstrated a decrease in the total numbers of infiltrating myeloid cell populations, including monocytes ($CD45^{high}CD11b^{high}GR1^{int}MHCII^{+}$) and granulocytes ($CD45^{high}CD11b^{high}GR1^{high}MHCII^{-}$). Additionally, the total number of NK cells ($CD45^{high}CD11b^{int}TCRb^{-}NK1.1^{+}$) was significantly lower (Figure 3B), whereas the number of T and B lymphocytes was not affected in MI27-treated mice.

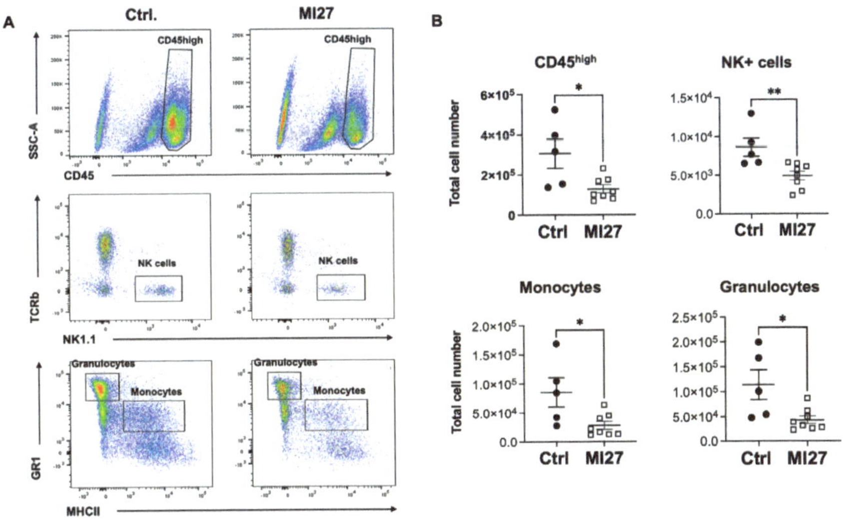

Figure 3. Intrathecal MI27 inhibited infiltration of immune cells to the CNS of mice with EAE. (**A**) Flow cytometric gating strategy to distinguish $CD45^{high}$ leukocytes from $CD45^{dim}$ microglia. Monocytes ($CD45^{high}CD11b^{high}GR1^{int}MHCII^{+}$), granulocytes ($CD45^{high}CD11b^{high}GR1^{high}MHCII^{-}$), and NK cells ($CD45^{high}CD11b^{int}TCRb\text{-}NK1.1^{+}$) were gated from $CD45^{high}$. (**B**) The total number of $CD45^{high}$ cells was reduced in spinal cords from MI27 treated mice. Monocytes and granulocytes, as well as NK cells, were significantly reduced in the spinal cord tissues of mice upon intrathecal MI27 treatment (n = 5–8 per group). Data are presented as mean ± SEM. Results were analyzed using the non-parametric Mann-Whitney test. * $p < 0.05$, ** $p < 0.01$.

2.4. *Intrathecal MI27 Altered Inflammatory Programs in the CNS of Mice with EAE*

Flow cytometry results suggested that MI27 reduced the recruitment of CD45+ cells into the CNS. To further examine the distribution of CD45+ cells, we performed immunofluorescent staining on spinal cord sections from EAE mice intrathecally treated with PBS or MI27. At one day post-MI27 administration, CD45+ cells were primarily located in the leptomeningeal space (Figure 4A). In contrast, CD45+ cells were more abundant in the spinal cord parenchyma of control mice with EAE (Figure 4A).

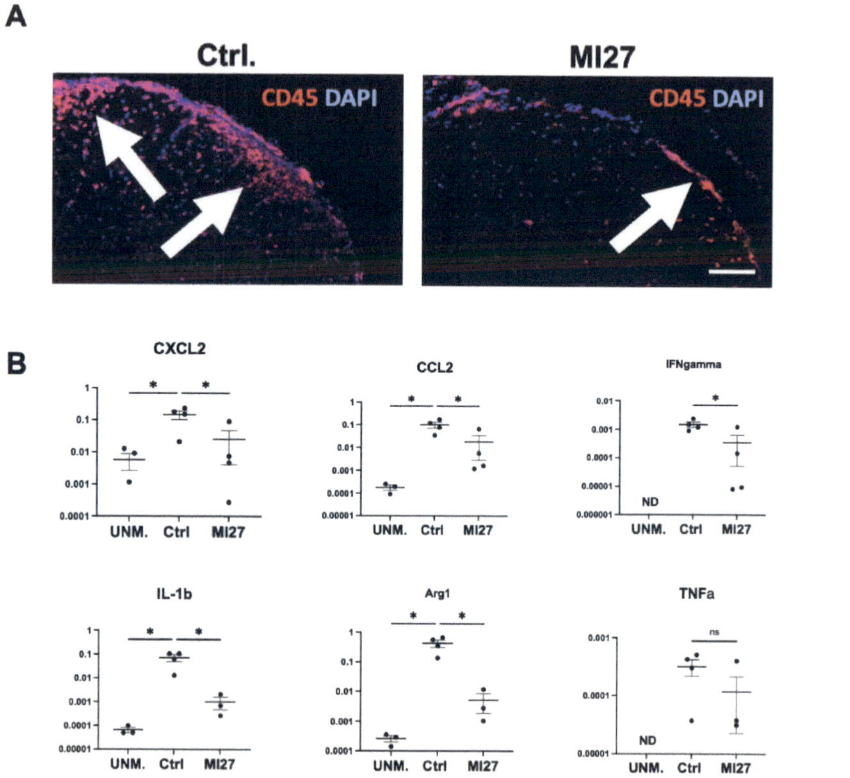

Figure 4. Intrathecal MI27 in mice with EAE-altered localization of CD45+ cells to the CNS. (**A**) Micrographs of spinal cord sections from mice with EAE that received MI27 or PBS intrathecally, stained with anti-CD45 antibody. Mice with EAE that were treated with intrathecal MI27 showed CD45+ cell accumulation in the meninges (arrow), whereas in control mice (ctrl), CD45+ cells were found in the spinal cord parenchyma (arrows). Scale bar: 100 μm. (**B**) RT-qPCR analysis of spinal cords showed chemokines, including *CCL2* and *CXCL2* as well as cytokines *IFNγ*, *IL-1β*, and *Arg1*, were significantly reduced upon intrathecal MI27 treatment at 1-day post-dose (n = 4 in each group). Unm = unmanipulated. ns = not significant. ND = not detected. Data are presented as mean ± SEM. Results were analyzed using the two-tailed Mann–Whitney u-test. * $p < 0.05$.

To assess the impact of simultaneous activation of TLR7 and NOD2 downstream signaling on CNS inflammatory responses in mice with EAE, we analyzed the expression of inflammation-associated mediators using RT-qPCR. Intrathecal MI27 treatment resulted in significant downregulation of *CCL2* and *CXCL2* mRNA at one day post-dose (Figure 4B). Moreover, we found that treatment with intrathecal MI27 significantly reduced the levels of *IL-1β*, *IFN-γ*, and *Arg1* mRNA (Figure 4B). The levels of *CSF1*, *IFNα*, and *IL-6* and the levels of transcription factors including nuclear factor-kappa B (NF-κB), interferon regulatory factors 7 (*IRF7*) and *IRF3* were not affected, at one day post-treatment (Supplementary Figure S2).

3. Discussion

The present study highlights the potential therapeutic efficacy of synergistic targeting of innate receptors TLR7 and NOD2 in mitigating neuroinflammation and suppressing EAE progression. By harnessing the joint effects of these receptors, we induced robust anti-inflammatory responses characterized by the production of IFN I, which exerted protective effects in the context of EAE. MI27 treatment led to reduced recruitment of CD45 high cells, including monocytes, granulocytes, and NK cells, into the spinal cord. The reduced cellular infiltration was associated with the downregulation of chemokines including *CCL2*, and *CXCL2* as well as cytokines including *IL-1β*, and *IFNγ*, whereas MI27 did not affect the expression of *CSF1*, *CXCL10*, *IL-6*, *TNFα*, *IFNα*, *NF-κB*, *IRF3*, and *IRF7*.

Previous studies have shown that peripheral administration of imiquimod, a TLR7 ligand, to mice immunized for EAE improved symptoms by inducing IFN I production [13]. Conversely, research with TLR7 knockout mice indicated that TLR7 signaling exacerbates EAE [14]. In our previous study, we demonstrated that intrathecal imiquimod had minimal effects on IFN I induction, immune cell infiltration, and did not suppress EAE [4]. Similarly, stimulation of NOD2 has been found to promote EAE progression, with NOD2-deficient mice being resistant to EAE and displaying reduced numbers of T cells and activated dendritic cells in the CNS [15]. In the present study, we found that intrathecal administration of NOD2 ligand did not induce an optimal IFN I response and failed to suppress EAE. Interestingly, peripheral administration of MDP in EAE exhibited a significant beneficial effect both before and after disease onset [16]. Using NOD2 deficient mice, the study demonstrated that NOD2 signaling is crucial for MDP-dependent immune modulation and resistance to EAE [16]. Whether these differential effects of NOD2 stimulation is due to the timing of treatment or the rout of MDP administration needs further investigation. Interestingly, the combination of imiquimod and MDP resulted in robust IFN I production, suppression of EAE, and a reduction in the pro-inflammatory response. These results suggest that the distinct impact of intrathecally administered MI27 on EAE may be attributed to its synergistic effects, highlighting the importance of combined receptor stimulation for effective modulation of EAE progression.

Although several scenarios can be proposed, the molecular mechanisms behind this synergy remain poorly understood. One possible scenario is a direct interaction of signaling pathways downstream of NOD2 and TLR7. It is pertinent to consider the potential involvement of intracellular signaling pathways, such as NF-κB and IRFs, including IRF3 and IRF7, in mediating the anti-inflammatory effects of combined TLR7 and NOD2 activation. Both TLR7 and NOD2 signaling pathways converge on these transcription factors, which play critical roles in regulating immune responses and gene expression of inflammatory mediators, including IFN I production [17–19]. Although no changes in the levels of *IRF7*, *IRF3*, and *NF-κB* in response to MI27 were observed one day post-treatment, we cannot exclude the possibility that the synergistic activation of TLR7 and NOD2 leads to activation of these pathways at earlier time points. This is plausible, as we observed strong but transient IFN I induction at 4 h that contributed to the observed IFN I-dependent suppression of neuroinflammation. Given the intricate crosstalk between TLR7, NOD2, NF-κB, and IRF signaling pathways, it is plausible that the therapeutic effects of combined receptor activation observed in our study may involve complex interactions at the level of intracellular signaling cascades. Future investigations into the specific roles of NF-κB and IRFs in mediating the anti-inflammatory and immunomodulatory effects of synergistic receptor targeting will provide valuable mechanistic insights and may identify additional therapeutic targets for MS and related autoimmune disorders.

It is unclear whether the therapeutic effect of MI27 is achieved through the simultaneous activation of NOD2 and TLR7 on a single cell or whether these two ligands act separately, targeting different cells. It is therefore important to study the cellular target of MI27. This would require the usage of covalently conjugated NOD2/TLR7 agonists or fluorescent-labeled TLR7- and NOD2- agonists [7,8,20].

The reduction in cytokines, including *IL-1β* and *IFNγ*, and chemokines, including *CCL2* and *CXCL2* in mice with EAE treated with MI27, may be attributed to two potential mechanisms. First, MI27 may decrease the infiltration of immune cells into the CNS tissue. This might be due to the robust induction of IFN I by MI27, which has been demonstrated to decrease immune cell infiltration into the CNS as well as reduction of *IFNγ* and *FoxP3* expression in mice with EAE [21], which was also downregulated in our study. Second, MI27 may induce an anti-inflammatory state by preventing the induction of proinflammatory mediators within the CNS. The current study suggests that MI27 blocks the infiltration of proinflammatory/encephalitogenic immune cells and reduces proinflammatory responses. In this regard, monocytes, granulocytes, and NK cells are among the cells that were significantly reduced in the CNS of mice treated with MI27, consistent with the chemokine and cytokine profile. Monocytes and granulocytes play a critical role in MS-like disease, and their recruitment to the CNS is essential for EAE progression [22,23]. Our findings are in agreement with these studies and suggest that MI27 reduces infiltrating monocytes and granulocytes, thereby preventing EAE progression. NK cells are a specialized population of innate lymphoid cells that help control local immune responses. Several studies demonstrated that the NK cells play a detrimental role in EAE, and the presence of NK cells, and more specifically NK cell-derived IFNγ, are critical for the development of EAE pathology [24–26]. Our findings support these notions. However, others have suggested a protective role for NK cells [27]. Future studies may clarify which NK cells subtypes are involved in MI27-mediated response.

In conclusion, our study demonstrates that simultaneous targeting of innate receptors TLR7 and NOD2 leads to synergistic induction of IFN I and effectively suppresses EAE progression in a murine model of MS. These findings support the therapeutic potential of combinatorial approaches aimed at modulating neuroinflammation and restoring CNS homeostasis in MS. Future studies focusing on elucidating the precise mechanisms underlying the synergistic effects of TLR7 and NOD2 signaling pathways will facilitate the development of novel treatment strategies for MS and related autoimmune disorders.

4. Materials and Methods

4.1. Mice

Female albino IFNβ+/Δβ-luc mice (IFNβ/luciferase reporter mice) [12] and IFNAR1-KO mice (C57BL/6 background) were specifically bred and maintained under standardized conditions at the Biomedical Laboratory, University of Denmark. Female C57BL/6j mice were purchased from Taconic Europe A/S (Lille Skensved, Denmark). All procedures involving animals were conducted in strict accordance with the guidelines provided by the Danish Animal Experiments Inspectorate, and the study was conducted under the approved protocol number 2020-15-0201-00652.

4.2. EAE Induction

Experimental autoimmune encephalomyelitis (EAE) was induced in both C57BL/6 and IFNAR1-KO mice following established protocols [2]. Briefly, mice were immunized with a 100 μL emulsion containing 100 μg of myelin oligodendrocyte glycoprotein (MOG)p35-55 (TAG Copenhagen A/S, Copenhagen, Denmark) in complete Freund's adjuvant (BD Biosciences, Lyngby, Denmark), supplemented with 200 μg of heat-inactivated Mycobacterium tuberculosis (BD Biosciences), administered subcutaneously into each hind flank. Furthermore, mice received an intraperitoneal injection of Bordetella pertussis toxin (0.3 μg, Sigma-Aldrich, Søborg, Denmark) at the time of immunization and again 1 day post-immunization. To monitor disease progression, mice were weighed daily and evaluated for EAE symptoms, graded on a scale ranging from 0 to 5 based on predefined criteria. The EAE grades were defined as follows: grade 0, no signs of disease; grade 1, weak or hooked tail; grade 2, floppy tail, indicating complete loss of tonus; grade 3, floppy tail and hind limb paresis, grade 4: floppy tail and unilateral hind limb paralysis; grade 5, floppy tail and bilateral hind limb paralysis. Due to ethical reasons, mice were sacrificed when they reached grade 5 or if hind limb paralysis persisted for 2 days.

4.3. Intrathecal Injection

Intrathecal injections were performed under anesthesia induced by inhalation of isoflurane (Abbott Laboratories, Chicago, IL, USA). A 30-gauge needle, bent at a 55° angle with a 2 mm tip, attached to a 50 μL Hamilton syringe, was used for precise delivery into the intrathecal space. Mice received injections of either 12.5 μg imiquimod (R837, Invivogen, San Diego, CA, USA), 25 μg Muramyl dipeptide (MDP, L-D isomer, active, Invivogen), 25 μg MDP control (L-L isomer, inactive, Invivogen) + imiquimod (12.5 μg), MI27 (25 μg MDP + 12.5 μg imiquimod), or phosphate-buffered saline (PBS). These doses were selected based on preliminary experiments demonstrating optimal effects and were consistently used throughout the study.

4.4. In Vivo Imaging

To visualize and quantify luciferase activity, in vivo imaging was performed according to established protocols. IFN-β+/Δβ-luc mice were injected intraperitoneally with D-luciferin (150 mg/kg) and subsequently monitored at 4-, 24-, and 48 h post dose, using an IVIS 200 imaging system (CaliperLS) (DaMBIC. Photon flux emitted from luciferase activity was quantified using Living Image 4.4 software (CaliperLS).

4.5. Tissue Processing

Upon completion of experiments, mice were euthanized using an overdose of sodium pentobarbital (100 mg/kg, Glostrup Apotek, Glostrup, Denmark) and subsequently perfused with ice-cold PBS. For subsequent analysis, CNS tissue was either placed in ice-cold PBS for flow cytometry or post-fixed with 4% paraformaldehyde (PFA), followed by immersion in 30% sucrose in PBS for histological processing. Tissue sections (16 μm thick) were obtained using a cryostat (Leica). Additionally, CNS tissues, intended for reverse transcriptase–quantitative polymerase chain reaction (RT-qPCR) analysis, were stored in Trizol Reagent (Ambion, Austin, TX, USA) at −80 °C until RNA extraction.

4.6. Flow Cytometry

A single-cell suspension was obtained by forcing the CNS tissue through a 70 μm cell strainer (Falcon, Miami, FL, USA) with Hank's buffered salt solution (HBSS, Gibco, Waltham, MA, USA) supplemented with 2% fetal bovine serum (FBS, Merck, Germany). Myelin was cleared by resuspending cells in 37% Percoll (GE Healthcare Bio-sciences AB, Chicago, IL, USA) in a buffer consisting of PBS pH 7.2. The myelin layer was removed, and the cell pellet was washed and incubated in a blocking solution containing HBSS, 2% FBS, anti-CD16/32 antibody (1 μg/mL, Clone 2.4G2, BD Biosciences, San Jose, CA, USA), and Syrian hamster IgG (50 μg/mL, Jackson ImmunoResearch Laboratories Inc., West Grove, PA, USA). The cells were then labeled with fluorophore-conjugated antibodies (BioLegend, San Diego, CA, USA), including anti-CD45 (clone 30-F11), CD11b (M1/70), GR-1 (RB6-8C5), NK1.1 (PK136), and TCRβ (H57-597), MHC II (M5/114.15.2). Fluorescence was measured using an LSRII flow cytometer (BD Biosciences) with FACS Diva^{TM} software version 6.1.2 (BD Biosciences) and analyzed with Flow logic (Inivai Technologies, Victoria, Australia) and FlowJo (Becton, Dickinson & Company, Franklin Lakes, NJ, USA) as described previously [2,28].

4.7. Histology

Luxol Fast Blue (LFB) and cresyl violet staining were performed as described [2]. To investigate glial response, frozen spinal cord sections from C57BL/6 mice with EAE, intrathecally injected with MI27 or PBS respectively, were stained with anti-IBA1- or CY3-conjugated anti-glial fibrillary acidic protein (GFAP, C9205-2ML, Sigma, St. Louis, MO,

USA), anti-MOG antibody, anti-MBP antibody and PE- conjugated anti-CD45 antibody as described previously [2,4,28]. Images were acquired using an Olympus DP71 digital camera mounted on an Olympus BX51 microscope (Olympus, Ballerup, Denmark). Images were combined using Adobe Photoshop CS3 (Adobe Systems Denmark A/S, København, Denmark).

4.8. RNA Isolation and RT-qPCR

RNA extraction from spinal cords was performed using TRIzol reagent in accordance with the manufacturer's protocol. RNA was converted into cDNA using a high-capacity cDNA reverse transcription kit (Applied Biosystems, Waltham, MA, USA). RT-qPCR was performed using an ABI Prism 7300 sequence detection system (Applied Biosystems), using primers and probes described earlier [2,4,28] Ct values were determined, and the results presented as genes of interest relative to 18sRNA ($2^{\Delta CT}$ method).

4.9. Statistical Analysis

Outliers were identified and excluded using the ROUT method prior to further statistical analysis. Comparisons between two groups were conducted using the non-parametric Mann–Whitney test. For comparisons among three groups, one-way ANOVA followed by a multiple comparison test was utilized. For analyzing IFNβ induction in IFNβ/luciferase reporter mice, a paired Student's *t*-test was performed. All statistical analysis was performed using GraphPad Prism version 10 (Graphpad Software Inc., La Jolla, CA, USA). Results are presented as means ± SEM. Values of $p < 0.05$ were considered significant.

Supplementary Materials: The following supporting information can be downloaded at: https://www.mdpi.com/article/10.3390/ijms25137462/s1.

Author Contributions: Conceptualization, R.K. and T.O.; methodology, M.D., J.M.-G., R.S.D., M.Z.S., D.R., E.S.S., B.O., B.R., R.B., J.K., D.S.A., A.W. and R.K.; writing—original draft preparation, R.K., M.D. and A.W. All authors have read and agreed to the published version of the manuscript.

Funding: This research was supported by grants from the Danish Research Council (DFF, 4183-00198A and 8020-00157B), the Danish Multiple Sclerosis Society, the European Committee for Treatment and Research in Multiple Sclerosis (ECTRIMS) Postdoctoral Research Fellowship Programme, A.P. Møller Fonden, Direktør Ejnar Jonasson kaldet Johnsen og hustrus mindelegat, and EU.MSCA (Marie Sklodowska–Curie Individual Actions Agreement No. 813263) ITN PMSMatTrain.

Institutional Review Board Statement: The animal study protocol was approved by the Danish Animal Experiments Inspectorate, approval number: 2020-15-0201-0065.

Data Availability Statement: The data will be available upon request.

Acknowledgments: We thank Pia Nyborg Nielsen and Dorte Overgaard Brorsen for excellent technical assistance. We also thank Lars Vitved (Department of Cancer and Inflammation Research, IMM, SDU) for assistance with flow cytometry at the FACS core facility.

Conflicts of Interest: The authors declare no conflicts of interest.

References

1. Thompson, A.J.; Baranzini, S.E.; Geurts, J.; Hemmer, B.; Ciccarelli, O. Multiple sclerosis. *Lancet* **2018**, *391*, 1622–1636. [CrossRef] [PubMed]
2. Kronborg Hansen, A.K.; Dubik, M.; Marczynska, J.; Ojha, B.; Nistal-Villan, E.; Gonzalez Aseguinolaza, G.; Arengoth, D.S.; Owens, T.; Khorooshi, R. Targeting Signaling Pathway Downstream of RIG-I/MAVS in the CNS Stimulates Production of Endogenous Type I IFN and Suppresses EAE. *Int. J. Mol. Sci.* **2022**, *23*, 11292. [CrossRef]
3. Owens, T.; Benmamar-Badel, A.; Wlodarczyk, A.; Marczynska, J.; Morch, M.T.; Dubik, M.; Arengoth, D.S.; Asgari, N.; Webster, G.; Khorooshi, R. Protective roles for myeloid cells in neuroinflammation. *Scand. J. Immunol.* **2020**, *92*, e12963. [CrossRef]
4. Dieu, R.S.; Wais, V.; Sorensen, M.Z.; Marczynska, J.; Dubik, M.; Kavan, S.; Thomassen, M.; Burton, M.; Kruse, T.; Khorooshi, R.; et al. Central Nervous System-Endogenous TLR7 and TLR9 Induce Different Immune Responses and Effects on Experimental Autoimmune Encephalomyelitis. *Front. Neurosci.* **2021**, *15*, 685645. [CrossRef] [PubMed]
5. Girardin, S.E.; Boneca, I.G.; Viala, J.; Chamaillard, M.; Labigne, A.; Thomas, G.; Philpott, D.J.; Sansonetti, P.J. Nod2 is a general sensor of peptidoglycan through muramyl dipeptide (MDP) detection. *J. Biol. Chem.* **2003**, *278*, 8869–8872. [CrossRef] [PubMed]

6. Pandey, S.; Kawai, T.; Akira, S. Microbial sensing by Toll-like receptors and intracellular nucleic acid sensors. *Cold Spring Harb. Perspect. Biol.* **2014**, *7*, a016246. [CrossRef]
7. Gutjahr, A.; Papagno, L.; Vernejoul, F.; Lioux, T.; Jospin, F.; Chanut, B.; Perouzel, E.; Rochereau, N.; Appay, V.; Verrier, B.; et al. New chimeric TLR7/NOD2 agonist is a potent adjuvant to induce mucosal immune responses. *EBioMedicine* **2020**, *58*, 102922. [CrossRef] [PubMed]
8. Guzelj, S.; Weiss, M.; Slutter, B.; Frkanec, R.; Jakopin, Z. Covalently Conjugated NOD2/TLR7 Agonists Are Potent and Versatile Immune Potentiators. *J. Med. Chem.* **2022**, *65*, 15085–15101. [CrossRef]
9. Passlick, D.; Piradashvili, K.; Bamberger, D.; Li, M.; Jiang, S.; Strand, D.; Wich, P.R.; Landfester, K.; Bros, M.; Grabbe, S.; et al. Delivering all in one: Antigen-nanocapsule loaded with dual adjuvant yields superadditive effects by DC-directed T cell stimulation. *J. Control. Release* **2018**, *289*, 23–34. [CrossRef]
10. White, M.P.J.; Webster, G.; Leonard, F.; La Flamme, A.C. Innate IFN-gamma ameliorates experimental autoimmune encephalomyelitis and promotes myeloid expansion and PDL-1 expression. *Sci. Rep.* **2018**, *8*, 259. [CrossRef]
11. Khorooshi, R.; Marczynska, J.; Dieu, R.S.; Wais, V.; Hansen, C.R.; Kavan, S.; Thomassen, M.; Burton, M.; Kruse, T.; Webster, G.A.; et al. Innate signaling within the central nervous system recruits protective neutrophils. *Acta Neuropathol. Commun.* **2020**, *8*, 2. [CrossRef]
12. Lienenklaus, S.; Cornitescu, M.; Zietara, N.; Lyszkiewicz, M.; Gekara, N.; Jablonska, J.; Edenhofer, F.; Rajewsky, K.; Bruder, D.; Hafner, M.; et al. Novel reporter mouse reveals constitutive and inflammatory expression of IFN-beta in vivo. *J. Immunol.* **2009**, *183*, 3229–3236. [CrossRef]
13. O'Brien, K.; Fitzgerald, D.; Rostami, A.; Gran, B. The TLR7 agonist, imiquimod, increases IFN-beta production and reduces the severity of experimental autoimmune encephalomyelitis. *J. Neuroimmunol.* **2010**, *221*, 107–111. [CrossRef]
14. Lalive, P.H.; Benkhoucha, M.; Tran, N.L.; Kreutzfeldt, M.; Merkler, D.; Santiago-Raber, M.L. TLR7 signaling exacerbates CNS autoimmunity through downregulation of Foxp3+ Treg cells. *Eur. J. Immunol.* **2014**, *44*, 46–57. [CrossRef]
15. Shaw, P.J.; Lamkanfi, M.; Kanneganti, T.D. NOD-like receptor (NLR) signaling beyond the inflammasome. *Eur. J. Immunol.* **2010**, *40*, 624–627. [CrossRef]
16. Fani Maleki, A.; Cisbani, G.; Laflamme, N.; Prefontaine, P.; Plante, M.M.; Baillargeon, J.; Rangachari, M.; Gosselin, J.; Rivest, S. Selective Immunomodulatory and Neuroprotective Effects of a NOD2 Receptor Agonist on Mouse Models of Multiple Sclerosis. *Neurotherapeutics* **2021**, *18*, 889–904. [CrossRef]
17. Fritz, J.H.; Ferrero, R.L.; Philpott, D.J.; Girardin, S.E. Nod-like proteins in immunity, inflammation and disease. *Nat. Immunol.* **2006**, *7*, 1250–1257. [CrossRef]
18. Kawai, T.; Akira, S. The role of pattern-recognition receptors in innate immunity: Update on Toll-like receptors. *Nat. Immunol.* **2010**, *11*, 373–384. [CrossRef]
19. Liu, T.; Zhang, L.; Joo, D.; Sun, S.C. NF-kappaB signaling in inflammation. *Signal Transduct. Target. Ther.* **2017**, *2*, 17023. [CrossRef] [PubMed]
20. Butchi, N.B.; Woods, T.; Du, M.; Morgan, T.W.; Peterson, K.E. TLR7 and TLR9 trigger distinct neuroinflammatory responses in the CNS. *Am. J. Pathol.* **2011**, *179*, 783–794. [CrossRef]
21. Vasquez, M.; Consuegra-Fernandez, M.; Aranda, F.; Jimenez, A.; Tenesaca, S.; Fernandez-Sendin, M.; Gomar, C.; Ardaiz, N.; Di Trani, C.A.; Casares, N.; et al. Treatment of Experimental Autoimmune Encephalomyelitis by Sustained Delivery of Low-Dose IFN-alpha. *J. Immunol.* **2019**, *203*, 696–704. [CrossRef] [PubMed]
22. Ajami, B.; Bennett, J.L.; Krieger, C.; McNagny, K.M.; Rossi, F.M. Infiltrating monocytes trigger EAE progression, but do not contribute to the resident microglia pool. *Nat. Neurosci.* **2011**, *14*, 1142–1149. [CrossRef] [PubMed]
23. Rumble, J.M.; Huber, A.K.; Krishnamoorthy, G.; Srinivasan, A.; Giles, D.A.; Zhang, X.; Wang, L.; Segal, B.M. Neutrophil-related factors as biomarkers in EAE and MS. *J. Exp. Med.* **2015**, *212*, 23–35. [CrossRef]
24. Shi, F.D.; Takeda, K.; Akira, S.; Sarvetnick, N.; Ljunggren, H.G. IL-18 directs autoreactive T cells and promotes autodestruction in the central nervous system via induction of IFN-gamma by NK cells. *J. Immunol.* **2000**, *165*, 3099–3104. [CrossRef]
25. Vollmer, T.L.; Liu, R.; Price, M.; Rhodes, S.; La Cava, A.; Shi, F.D. Differential effects of IL-21 during initiation and progression of autoimmunity against neuroantigen. *J. Immunol.* **2005**, *174*, 2696–2701. [CrossRef]
26. Winkler-Pickett, R.; Young, H.A.; Cherry, J.M.; Diehl, J.; Wine, J.; Back, T.; Bere, W.E.; Mason, A.T.; Ortaldo, J.R. In vivo regulation of experimental autoimmune encephalomyelitis by NK cells: Alteration of primary adaptive responses. *J. Immunol.* **2008**, *180*, 4495–4506. [CrossRef]
27. Yang, Y.; Day, J.; Souza-Fonseca Guimaraes, F.; Wicks, I.P.; Louis, C. Natural killer cells in inflammatory autoimmune diseases. *Clin. Transl. Immunol.* **2021**, *10*, e1250. [CrossRef] [PubMed]
28. Dubik, M.; Marczynska, J.; Morch, M.T.; Webster, G.; Jensen, K.N.; Wlodarczyk, A.; Khorooshi, R.; Owens, T. Innate Signaling in the CNS Prevents Demyelination in a Focal EAE Model. *Front. Neurosci.* **2021**, *15*, 682451. [CrossRef]

International Journal of
Molecular Sciences

MDPI

Article

n-3 Polyunsaturated Fatty Acids Modulate LPS-Induced ARDS and the Lung–Brain Axis of Communication in Wild-Type versus Fat-1 Mice Genetically Modified for Leukotriene B4 Receptor 1 or Chemerin Receptor 23 Knockout

Jessica Hernandez [1,†], Julia Schäffer [1,2,†], Christiane Herden [3], Fabian Johannes Pflieger [1], Sylvia Reiche [2], Svenja Körber [3], Hiromu Kitagawa [4], Joelle Welter [1], Susanne Michels [5], Carsten Culmsee [5,6], Jens Bier [2], Natascha Sommer [2], Jing X. Kang [7], Konstantin Mayer [8], Matthias Hecker [2,*] and Christoph Rummel [1,6,*]

1 Institute of Veterinary Physiology and Biochemistry, Justus Liebig University Giessen, 35392 Giessen, Germany; jessica.hernandez-2@vetmed.uni-giessen.de (J.H.); julia.schaeffer@vetmed.uni-giessen.de (J.S.)
2 Excellence Cluster Cardio-Pulmonary Institute (CPI), Universities of Giessen and Marburg Lung Center (UGMLC), Member of the German Center for Lung Research (DZL), Justus Liebig University Giessen, 35392 Giessen, Germany; jens.bier@innere.med.uni-giessen.de (J.B.); natascha.sommer@innere.med.uni-giessen.de (N.S.)
3 Institute of Veterinary Pathology, Justus Liebig University Giessen, 35392 Giessen, Germany; christiane.herden@vetmed.uni-giessen.de (C.H.); svenja.koerber@vetmed.uni-giessen.de (S.K.)
4 Department of Biomedical Engineering, Osaka Institute of Technology, Omiya, Osaka 535-8585, Japan
5 Institute of Pharmacology and Clinical Pharmacy, Philipps University of Marburg, 35032 Marburg, Germany; culmsee@staff.uni-marburg.de (C.C.)
6 Center for Mind Brain and Behavior, Universities Giessen and Marburg, 35032 Marburg, Germany
7 Laboratory for Lipid Medicine and Technology, Department of Medicine, Massachusetts General Hospital and Harvard Medical, Boston, MA 02129, USA
8 Department of Internal Medicine, Justus Liebig University Giessen, 35392 Giessen, Germany; konstantin.mayer@innere.med.uni-giessen.de
* Correspondence: matthias.hecker@innere.med.uni-giessen.de (M.H.); christoph.d.rummel@vetmed.uni-giessen.de (C.R.)
† These authors contributed equally to this work.

Citation: Hernandez, J.; Schäffer, J.; Herden, C.; Pflieger, F.J.; Reiche, S.; Körber, S.; Kitagawa, H.; Welter, J.; Michels, S.; Culmsee, C.; et al. n-3 Polyunsaturated Fatty Acids Modulate LPS-Induced ARDS and the Lung–Brain Axis of Communication in Wild-Type versus Fat-1 Mice Genetically Modified for Leukotriene B4 Receptor 1 or Chemerin Receptor 23 Knockout. *Int. J. Mol. Sci.* **2023**, *24*, 13524. https://doi.org/10.3390/ijms241713524

Academic Editors: Oliver Planz and Ralf Kircheis

Received: 28 July 2023
Revised: 25 August 2023
Accepted: 27 August 2023
Published: 31 August 2023

Abstract: Specialized pro-resolving mediators (SPMs) and especially Resolvin E1 (RvE1) can actively terminate inflammation and promote healing during lung diseases such as acute respiratory distress syndrome (ARDS). Although ARDS primarily affects the lung, many ARDS patients also develop neurocognitive impairments. To investigate the connection between the lung and brain during ARDS and the therapeutic potential of SPMs and its derivatives, *fat-1* mice were crossbred with RvE1 receptor knockout mice. ARDS was induced in these mice by intratracheal application of lipopolysaccharide (LPS, 10 µg). Mice were sacrificed at 0 h, 4 h, 24 h, 72 h, and 120 h post inflammation, and effects on the lung, liver, and brain were assessed by RT-PCR, multiplex, immunohistochemistry, Western blot, and LC-MS/MS. Protein and mRNA analyses of the lung, liver, and hypothalamus revealed LPS-induced lung inflammation increased inflammatory signaling in the hypothalamus despite low signaling in the periphery. Neutrophil recruitment in different brain structures was determined by immunohistochemical staining. Overall, we showed that immune cell trafficking to the brain contributed to immune-to-brain communication during ARDS rather than cytokines. Deficiency in RvE1 receptors and enhanced omega-3 polyunsaturated fatty acid levels (*fat-1* mice) affect lung–brain interaction during ARDS by altering profiles of several inflammatory and lipid mediators and glial activity markers.

Keywords: immune-to-brain communication; lung–brain axis; lipopolysaccharide; acute respiratory distress syndrome; cytokines; lipid mediators; omega-3 fatty acids; resolvin E1 receptors

1. Introduction

The omega (n)-3 polyunsaturated fatty acids (PUFA) α-linolenic acid (ALA), eicosapentaenoic acid (EPA), and docosahexaenoic acid (DHA) are essential PUFAs that have demonstrated several anti-inflammatory effects [1–6]. Indeed, diets enriched in n-3 PUFAs have been beneficial in human clinical trials [7]. Derivatives of EPA and DHA such as resolvins (Rv), maresins, and protectins belong to a group of specialized pro-resolving mediators (SPMs), which actively terminate inflammation and promote healing as previously demonstrated in animals and humans [8–23]. All SPMs have been shown to inhibit neutrophil trans-endothelial migration, enhance macrophage phagocytosis of apoptotic neutrophils [8,9,19], and decrease cytokine production [11–15]. Some derivatives of DHA, such as the D-series of Rv such as RvD1, can reduce the deterioration of tight junctions during inflammation and, therefore, decrease edema in the lungs in a model of lipopolysaccharide (LPS)-induced acute lung injury (ALI) [17]. D-series Rv also play an important role in the development of the brain [18,20] and have been shown to reduce cytokine production in in vitro cultures of microglia cells following treatment with LPS [14]. Moreover, EPA-derived E-series Rv also promote the resolution of inflammation by reducing the migration of immune cells such as neutrophils and enhancing the clearance of apoptotic neutrophils by macrophages, as has been shown in different inflammatory models such as during peritonitis or allergic airway inflammation [8,12,21–24]. Of the E-series Rv, particular attention has been placed on Resolvin E1 (RvE1), which is derived from EPA [12]. Interestingly, RvE1 has a high potential for the treatment of several lung and brain-inflammatory diseases such as LPS-induced depressive behaviors [25,26].

RvE1 predominantly acts through two receptors, the G-protein coupled receptor chemerin receptor 23 (CR) and the leukotriene B4 (LTB_4) receptor 1 (LR) [21]. The LR is mainly found on leukocytes and neutrophils but CR is expressed by a variety of other tissues [12,27–31]. Notably, CR is found in several organs, such as the liver, and has, thus far, been identified on many cell types including dendritic cells, monocytes, macrophages, endothelial cells, and to a small degree on neutrophils [12,27–31]. RvE1 has been shown to inhibit tumor necrosis factor (TNF) α-induced activation of the inflammatory nuclear factor (NF) κB pathway through CR while also acting as an antagonist on the LR to dampen pro-inflammatory signals [12,21,32–34]. Moreover, Rv are antagonists for the LR and inhibit LTB_4-induced calcium and NFκB signaling to mediate the resolution of inflammation during murine peritonitis [21,35].

Indeed, previous studies have shown that SPMs such as Rv can be beneficial during neuronal damage by promoting brain cell survival and blocking glial cell cytokine production [14,36,37]. For lung inflammation, SPMs and especially Rv have been well studied during acute respiratory distress syndrome (ARDS) and the restoration of homeostasis [26,38–40]. ARDS is a type of acute diffuse lung injury, characterized by inflammation leading to increased pulmonary vascular permeability [41]. Mice expressing the *fat-1* gene and therefore endogenously synthesizing n-3 PUFAs, or mice receiving a fish oil infusion, demonstrated reduced leukocyte invasion, protein leakage, and decreased levels of several pro-inflammatory chemokines and cytokines in the lung tissue during LPS-induced ARDS when compared to solvent-treated or wild-type (WT) ARDS controls [38,39]. Notably, the n-3 PUFA derivative RvE1 increased alveolar fluid clearance and decreased inflammatory signaling in a model for bacterial pneumonia and LPS-induced ARDS [26,38–40].

In addition to the lung, other organs such as the kidneys, liver, and even the brain can be affected by ARDS [42,43]. In mice, LPS-induced ARDS can impair the blood–brain barrier (BBB) by reducing tight junction protein expression in endothelial cells, while simultaneously impairing cerebral blood flow [44,45]. Moreover, ARDS patients showed impaired cognitive outcomes due to brain damage assessed by S-100 protein and altered glucose metabolism in the brain [46,47]. Neurocognitive impairments such as compromised memory, attention and concentration, and/or decreased mental processing speed have even been reported in some patients several months or years after recovery from ARDS [48,49].

However, the molecular mechanisms responsible for the lung and brain connection during ARDS, i.e., lung-to-brain signaling, remain to be further clarified.

Since n-3 PUFA and their derivatives have demonstrated anti-inflammatory effects on the brain and given the impact of ARDS on the brain, we were particularly interested to see if n-3 PUFAs reduced inflammation during LPS-induced ARDS. We previously found that genetic n-3 PUFA enrichment accelerates recovery from central nervous system-controlled sickness responses such as fever after i.t LPS installation [39]. Because of its role as a regulator in the resolution of inflammation, particular attention was placed on endogenous RvE1, or more precisely, on its two prominent receptors: CR and LR. The high prevalence of both receptors on immune cells as well as the expression of the CR in the lung ensures their presence during ARDS [50]. In our present experiment, we sought to replicate a model of ARDS using LPS-induced local lung inflammation while assessing the anti-inflammatory roles of n-3 PUFAs with a focus on RvE1 actions through both receptors. To investigate the connection between the lung and brain during ARDS and the therapeutic potential of n-3 PUFAs and its derivatives, genetically n-3 enriched *fat-1* (Fat) versus control (WT) mice with unmodified (Norm) or deficient Rv receptors (CR KO or LR KO mice) were used in this study (Figure 1). Fat mice convert n-6 PUFAs to n-3 PUFAs due to the expression of an n-3 PUFA desaturase [51] and are hence a popular model to investigate n-3 PUFAs. To determine underlying signaling pathways of lung-to-brain communication and its alterations by deficient RvE1 signaling, we assessed the lung inflammatory response by multiplex analyses of lung inflammatory marker proteins. Similarly, the systemic inflammatory response was determined in the liver. In the brain, mRNA expression levels of a large panel of inflammatory marker proteins served to screen the effects of ARDS on the brain and its modulation by lipid mediators. The hypothalamus was chosen as this brain structure contains pivotal hubs in immune-to-brain communication and sickness signaling during inflammatory peripheral insults [52]. These include so-called sensory circumventricular organs (CVO) such as the vascular organ of lamina terminalis (OVLT) with a leaky BBB prone to detecting circulating mediators [53,54]. Changes in brain and lung lipids, brain inflammatory cytokines, brain oxidative stress, genomic NF-IL6 brain cell activation, and neutrophil recruitment to the brain were further determined through LC-MS/MS, protein carbonyl content (oxidative stress), Western blot, or immunofluorescence staining, as previously reported [52,55,56].

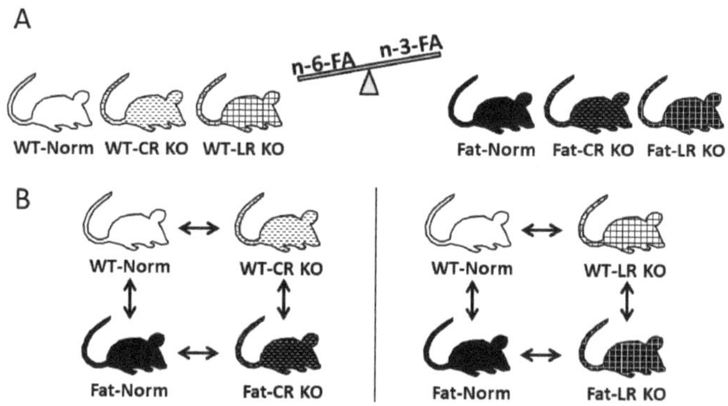

Figure 1. Schematic for genetically modified mouse strains. The breeding strategy outlines the mice used in the experiments (**A**). Genetically enriched n-3 fatty acid producing *fat-1* mice (Fat; black) were compared to wild-type control mice (WT; white). Both WT and Fat mice were cross bred with knockout (KO) mice deficient in the resolvin receptor chemerin receptor 23 (CR KO; spotted) or leukotriene B4 receptor (LR KO; checkered) to generate KO mice on WT and Fat backgrounds. Mice that maintained intact resolvin receptors were designated as Norm. Comparisons were made between Norm and receptor-deficient mice on both WT and Fat backgrounds but CR KO and LR KO mice were not compared to each other (**B**).

2. Results

Previous studies have already shown the dynamics of LPS-induced lung inflammation [57,58] Therefore, we focused on changes in the inflammatory response by genetic n-3 enrichment, i.e., Fat background compared to WT counterparts. Moreover, we investigated how RvE1 receptor deficiency altered this response, i.e., CR or LR KO mice.

2.1. Protein Analyses of Inflammatory Mediators in the Lung: n-3 PUFAs and RvE1 Receptors Altered LPS-Induced Pro-Inflammatory Cytokines, Anti-Inflammatory IL-10, and Neutrophil Markers at 0 h, 24 h, and 72 h p.i. in CR and LR KO Mice

Lung cytokine and neutrophil chemoattractant levels assessed by multiplex assays were used to screen for inflammatory modulation by n-3 PUFAs and the RvE1 receptors during LPS-induced ARDS at 0 h, 24 h, and 72 h post inflammation (p.i.). Descriptively, inflammation peaked 24 h after LPS installation into the lung. We found that the main effects due to n-3 PUFAs and deficiencies of both RvE1 receptors were already present at basal levels and persisted until the 72 h time point (Figure 2, Supplementary Table S1).

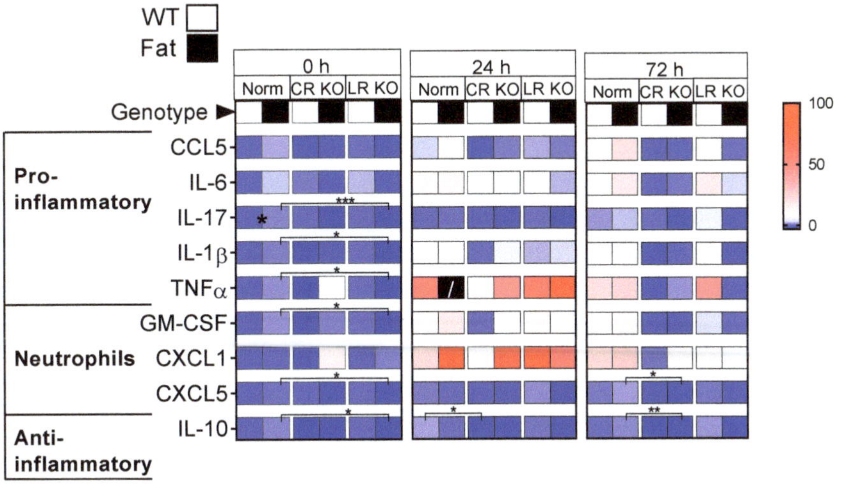

Figure 2. Changes in mediators of lung inflammation following intratracheal LPS-induced ARDS. Multiplex cytokine measurements from homogenized lung tissue for pro-inflammatory cytokines: chemokine (C-C motif) ligand 5 (CCL5), interleukin (IL)-6, IL-17, IL-1β, and tumor necrosis factor (TNF)α; neutrophil chemoattractants: granulocyte-macrophage colony-stimulating factor (GM-CSF), chemokine (C-X-C motif) ligand (CXCL)1 and CXCL5; and the anti-inflammatory cytokine: IL-10. Mice deficient in chemerin receptor 23 (CR KO) or leukotriene B4 receptor (LR KO) as well as unmodified mice (Norm) bred on wild-type (WT) or transgenic omega-3 (n-3) synthesizing *fat-1* (Fat) background received an intratracheal (i.t.) instillation with lipopolysaccharide (LPS, 10 μg) and were sacrificed at 0 h, 24 h, or 72 h p.i. Impacts of n-3 PUFAs and the RvE1 receptors CR and LR are shown at each time point to different extents. Indeed, significant alterations from n-3 PUFAs and the CR are detected for IL-10. The average fold change for each sample set was normalized to WT-Norm 0 h and is presented as a value on a heat map. A black box with a white diagonal line indicates a value that exceeds the scaled range. n = 3 per group for all groups except Fat-CR KO 0 h; n = 2; therefore, this group was excluded from statistical analyses but visually kept as descriptive information. * $p < 0.05$, ** $p < 0.01$, *** $p < 0.001$.

At 0 h, LR KO mice genetically enriched in n-3 PUFAs (Fat background) showed reduced expression of most inflammatory mediators compared to their Norm counterpart. This included pro-inflammatory cytokines (IL-17 ($p = 0.0010$), IL-1β ($p = 0.0366$), TNFα ($p = 0.0266$)), neutrophil chemoattractants (GM-CSF ($p = 0.0481$), CXCL5 ($p = 0.0447$)), and anti-inflammatory IL-10 ($p = 0.0328$). Although inflammation seemed highest at

24 h p.i., specific changes were not observed at this time point save for a CR KO-dependent reduction in IL-10 compared to Norm controls on the WT background (p = 0.0125). By 72 h, nearly all differences in inflammatory mediators had ceased and the only reductions could be found in CXCL5 (p = 0.0463) and IL-10 (p = 0.0079) for the genetically enriched in n-3 PUFAs (Fat background) CR KO group compared to their Norm counterpart. Minor impacts of n-3 PUFAs (i.e., Fat background) were observed with reduced protein levels of the neutrophil markers GM-CSF (0 h, Norm and CR KO), CXCL1 (24 h, Norm and CR KO), and CXCL5 (24 h, Norm and LR KO). However, n-3 PUFAs did not significantly affect other inflammatory mediators in Norm mice, save for an increase in IL-17 (p = 0.0267) at basal levels, or RvE1 receptor-deficient mice (Supplementary Table S1).

Overall, at basal levels, genetic n-3 PUFA enrichment (Fat background) reduced several inflammatory mediators in mice deficient in the LR compared to Norm controls such as IL-17, IL-1β, TNFα, GM-CSF, CXCL5, and IL-10. This effect of the LR did not extend past the 0 h time point and, during inflammation; the deficiency in the CR became more relevant as seen at 72 h p.i. At the late stage of inflammation, a robust effect of the CR emerged where CR deficiency decreased expression of IL-17, IL-1β, TNFα, GM-CSF, CXCL5, and IL-10 compared to Norm controls (Supplementary Table S1), although significant differences between groups were limited. Given the low numbers of animals per group and a high degree of variation, differences could not always be detected. However, the most consistent impact of the CR observed was for the anti-inflammatory cytokine IL-10 where a prolonged reduction in the CR was determined at the 24 h and 72 h time points for this important anti-inflammatory cytokine.

2.2. *Analyses of Inflammatory Marker Proteins in the Liver: n-3 PUFAs and RvE1 Receptors Altered LPS-Induced Pro-Inflammatory Cytokines, Anti-Inflammatory IL-10, and Neutrophil Markers at 24 h p.i. in Norm, CR, and LR KO Mice*

As a proxy for systemic inflammatory spillover of LPS-induced ARDS, changes in cytokine and neutrophil chemoattractant protein levels were also analyzed in the liver. LPS-induced alterations in the production of inflammatory mediators were less pronounced and the ability for genetic n-3 PUFA enrichment to reduce inflammatory mediators was remarkably weaker in the liver than in the lung. Significant differences were restricted to 24 h p.i. and, overall, only a small systemic inflammatory response was detected, which was modestly impacted by n-3 PUFAs and RvE1 receptors (Figure S1, Supplementary Table S2).

2.3. *mRNA Analyses of Hypothalamic Inflammatory Marker Proteins: n-3 PUFAs and RvE1 Receptors Altered LPS-Induced Glial Activation, Neutrophil Markers, Pro-Inflammatory IL-1ß, Anti-Inflammatory IL-1ra and IL-10, Signaling Pathways, and the Enzyme Precursor for Prostaglandin Synthesis mPGES in Norm, CR, and LR KO Mice*

Having observed only minor changes in inflammatory marker protein levels in the periphery (i.e., lung and liver) due to LPS-induced ARDS and its modulation by n-3 PUFAs enrichment and RvE1 receptor deficiencies, our next aim was to assess markers for lung-to-brain signaling. To this end, mRNA expression for markers of brain oxidative stress, immune cell trafficking, brain inflammatory responses (cytokines and signaling, enzyme of prostaglandin (PG) synthesis), and markers of neuroplasticity and glial activation were assessed in the hypothalamus (Figure 3). The glial activation marker CD68 did not show any significant changes in the first screening attempt and, thus, further analyses were not performed for microglia (Supplementary Figure S2). Overall, LPS-induced ARDS also increased markers of inflammation in the brain with a peak at 24 h after stimulation (e.g., IκBα, elastase). Genetic n-3 PUFA enrichment reduced some but not all of the analyzed markers. As previously observed in the lung, we found that the main effects due to n-3 PUFAs and deficiencies of both RvE1 receptors were already present at basal levels and persisted until the 72 h time point (Supplementary Table S3).

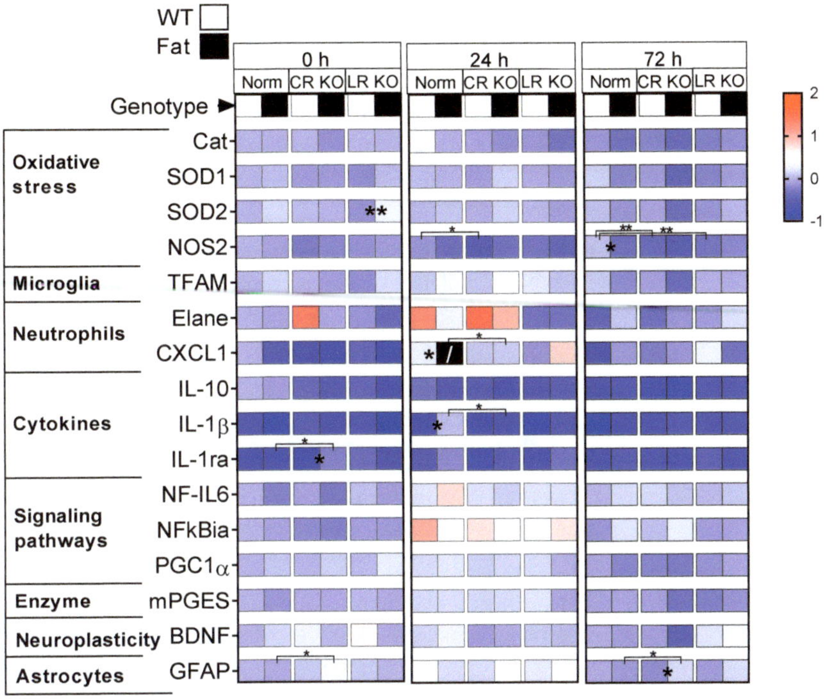

Figure 3. Changes in hypothalamic inflammatory mediator mRNA expression following intratracheal LPS-induced ARDS. The hypothalamus was analyzed for expression of markers for oxidative stress: catalase (Cat), superoxide dismutase (SOD)1, SOD2 and nitric oxide synthase (NOS2); mitochondrial biogenesis: transcription factor A, mitochondrial (TFAM); neutrophil trafficking: elastase, neutrophil expressed (Elane), and chemokine (C-X-C motif) ligand (CXCL)1; cytokines: interleukin (IL)10, IL-1β, and IL-1ra; signaling pathways: nuclear factor (NF)-IL6, NFκB via IκBα (ia), and peroxisome proliferator-activated receptor-α coactivator (PGC)1α; enzymes involved in prostaglandin expression: microsomal prostaglandin E synthase (mPGES); neuroplasticity: brain-derived neurotrophic factor (BDNF); and astrocyte activation: glial fibrillary acidic protein (GFAP). Mice deficient in chemerin receptor 23 (CR KO) or leukotriene B4 receptor (LR KO) as well as unmodified mice (Norm) bred on wild-type (WT) or transgenic omega-3 (n-3) synthesizing *fat-1* (Fat) background received an intratracheal (i.t.) instillation with lipopolysaccharide (LPS, 10 μg) and were sacrificed at 0 h, 24 h, or 72 h p.i. An impact of n-3 PUFAs and the RvE1 receptors CR and LR are evident at each time point but for different markers. The average fold change for each sample set was normalized to WT-Norm 0 h and presented as a value on a heat map. A black box with a white diagonal line indicates a value that exceeds the scaled range. n = 3–6 per group after outlier exclusions and undetermined values. Only IL-10 had fewer samples (n = 1–5) for which statistics were not performed but visually kept as descriptive information. * $p < 0.05$, ** $p < 0.01$.

Analysis of markers for oxidative stress included catalase (Cat), superoxide dismutase (SOD)1, SOD2, and nitric oxide synthase (NOS)2. Over the course of inflammation, there were significant alterations by n-3 PUFAs and RvE1 receptor deficiencies (Figure 3, Supplementary Table S3). In particular, at 72 h p.i., n-3 PUFA enrichment (Fat background) reduced all markers of oxidative stress when comparing the Norm and CR KO groups regardless of RvE1 receptor expression. However, it was only at basal levels that genetic n-3 PUFA enrichment (Fat background) increased the expression of SOD2 in LR KO mice ($p = 0.0016$). During the late stage of inflammation (72 h), n-3 PUFAs also decreased the production of NOS2 in Norm mice ($p = 0.0120$). Additional comparisons were made between

Norm and receptor-deficient mice on either the WT or Fat background for each time point. At 24 h p.i., CR KO mice on a WT background had decreased levels of NOS2 ($p = 0.0133$), while at 72 h p.i. expression levels were decreased for both CR KO ($p = 0.0049$) and LR KO mice ($p = 0.0068$). Further analysis of transcription factor A, mitochondrial (TFAM), a marker of mitochondrial biogenesis and antioxidant effects [59,60], also showed moderate n-3 PUFA-dependent reductions between Norm and LR KO (0 h) and Norm and CR KO (24 h, 72 h) regardless of RvE1 receptor expression (Supplementary Table S3).

Together, these results showed a modest reduction in the expression of markers of oxidative stress due to n-3 PUFAs alone at basal levels and again at 72 h p.i., particularly for SOD2, NOS2, and TFAM. Receptor KO-related differences to Norm mice were restricted to NOS2 and here we saw an earlier impact of CR at 24 h p.i. and CR and LR at 72 h p.i.

Neutrophils have been highlighted as an important immune-to-brain signaling pathway during systemic inflammation; therefore, we used the neutrophil marker elastase, neutrophil expressed (Elane) [61], as well as the neutrophil-specific chemokine CXCL1 to gain further insights into their recruitment to the hypothalamus (Figure 3) [55]. Indeed, some main effects of the RvE1 receptor deficiencies and n-3 PUFAs were found in the neutrophil markers (Supplementary Table S3). The majority of these effects observed on CXCL1 occurred at 24 h p.i. when impacts of the CR, LR, and n-3 PUFA enrichment were all present. Between Norm mice, at 24 h p.i. n-3 PUFA enrichment did increase the expression of CXCL1 ($p = 0.0248$). This n-3 PUFA-dependent increase was absent in both RvE1 receptor-deficient lines. In fact, in comparison to Fat-CR KO mice, the expression of CXCL1 was significantly lower than in Fat-Norm ($p = 0.0111$), but this was not the case for Fat-LR KO mice. Although there was significant variation in Elane expression between samples, genetic n-3 PUFA enrichment did not seem to have an effect, and the only impact observed was at 24 h p.i. for the LR. The results seen in the neutrophil markers indicate n-3 PUFAs influenced the expression of CXCL1 and deficiency of RvE1 receptors decreased the expression of Elane (LR KO) and CXCL1 (CR KO, LR KO).

Macrophages have also been found in the brain perivascular space and choroid plexus during neuroinflammation [62]. Using the macrophage marker CD163 (perivascular macrophages) [63,64], a preliminary analysis was also carried out at 24 h and 72 h p.i. (Supplementary Figure S2). During late-stage inflammation, n-3 PUFA enrichment did increase the expression of CD163 in Norm mice but there were no effects of RvE1 receptors; therefore, further analysis at 0 h p.i. was not performed.

Cytokine expression in the hypothalamus was determined using a panel of pro- and anti-inflammatory cytokines known to be influential during ARDS, namely, anti-inflammatory IL-10, pro-inflammatory IL-1β and its antagonist IL-1ra (Figure 3), as well as pro-inflammatory IL-6 (Supplementary Figure S2). It is important to note that the expression of IL-10 was difficult to detect and several samples had to be excluded due to 'undetermined' expression values, which could explain a failure to identify significant differences. The main effects of n-3 PUFA enrichment and RvE1 receptor deficiencies were restricted to the IL-1 family (Supplementary Table S3). Among the IL-1 family, n-3 PUFA enrichment did increase the expression of IL-1β in Norm mice ($p = 0.0147$) at 24 h p.i. but neither CR nor LR KO mice showed the same response and Fat-CR KO mice even presented significantly reduced expression compared to Fat-Norm ($p = 0.0191$). A similar spike was not seen for IL-1ra although, at the basal level, expression was elevated by n-3 PUFA enrichment in CR KO mice ($p = 0.0221$), significantly higher than the Fat-Norm mice ($p = 0.0307$). Preliminary investigation of IL-6 at 24 h and 72 h did not show significant differences between groups and was, therefore, not analyzed at 0 h p.i. (Supplementary Figure S2).

Inflammatory signaling pathways can be activated by cytokines such as IL-1β and IL-6 via several inflammatory transcription factors such as NF-IL6, NFκB, and signal transducer and activator of transcription (STAT)3 [65–67]. In addition, lipid mediators can signal via the anti-inflammatory PPARγ pathway, i.e., the peroxisome proliferator-activated receptor-α coactivator (PGC) 1α, which has been shown to increase in the brain during

LPS-induced inflammation [68]. Finally, the enzyme microsomal prostaglandin E synthase (mPGES) is yet another important target gene analyzed due to its role in PG synthesis and fever development [69]. We assessed the expression of NF-IL6, NFκB (via IκBα), PGC1α (Figure 3), and STAT3 (via suppressor of cytokine signaling (SOCS)3) as previously shown during preliminary analyses at 24 h and 72 h p.i. (Supplementary Figure S2) [68]. Moderate changes due to genetic n-3 PUFA enrichment were detected when comparing Norm and CR KO mice for NF-IL6 (decreased) and Norm and LR KO mice for PGC1α (increased) at the basal levels regardless of RvE1 receptor expression (Supplementary Table S3). Moreover, a modest increase in SOCS3 expression was revealed during late-stage inflammation in the LR KO mice regardless of n-3 PUFA status (Supplementary Figure S2). Together, the apparent peak in NFκbiα at 24 h p.i. is indicative of an increased inflammatory response in the brain. In the absence of significant differences between groups at basal levels, n-3 PUFA enrichment still had a modest influence on mRNA expression of NF-IL6 and PGC1α while mPGES remained unchanged.

Neuroplasticity and astrocytes were assessed to identify activation and function within the hypothalamus, expression of brain-derived neurotrophic factor (BDNF), astrocytic marker protein glial fibrillary acidic protein (GFAP) (Figure 3), and microglial activation marker CD68 [70,71] (Supplementary Figure S2) were analyzed. No effects of either genetic n-3 PUFA enrichment or RvE1 receptors on BDNF or CD68 could be detected but GFAP was altered at the basal level by CR (Figure 3 and Figure S2). Specifically, in n-3 PUFA genetically enriched (Fat background) mice the expression of GFAP was increased in Fat-CR KO mice compared to Fat-Norm at 0 h p.i. ($p = 0.0162$) and again at 72 h p.i. ($p = 0.0480$). Moreover, at 72 h p.i. n-3 PUFAs increased the expression of GFAP in CR KO mice ($p = 0.0146$). We did not detect any changes in CD68 mRNA expression at 24 h or 72 h p.i., thus, we omitted such assessments for further analyses (Supplementary Figure S2). Based on these results, it appears that genetic n-3 PUFA enrichment increases the mRNA expression of GFAP but only in mice deficient in CR while BDNF remains unaffected.

Overall, in the hypothalamus during LPS-induced ARDS, we found that n-3 PUFA enrichment exerted a broader effect on inflammatory markers and while the effects of the RvE1 receptors were not as present they still occurred in most groups. Although some basal effects of the receptors were detected, the strongest impacts of the receptor deficiencies occurred at 24 h p.i., particularly for the CR KO mice, which had reduced expression of NOS2, CXCL1, and IL-1β compared to Norm controls. At the late stage of inflammation (72 h p.i.), minimal effects were still observed and these were predominantly due to n-3 PUFA enrichment effects on markers for oxidative stress and GFAP expression in CR KO mice.

2.4. Protein Analyses of Hypothalamic Oxidative Stress Markers and Immunohistochemical Detection of Total Astrocyte Area in the Paraventricular Nucleus: RvE1 Receptor Deficiency Increased Astrocytic Activation at 24 h p.i but Did Not Alter Mediators for Oxidative Stress in Norm or CR KO Mice

With the observed changes in the mRNA expression of markers for oxidative stress and astrocyte activation, mediator levels in the hypothalamus were further analyzed by Western blot for oxidative stress and immunohistochemistry for total astrocyte area. Since we did not detect major changes in oxidative stress for WT-LR KO or Fat-LR KO mice at 72 h p.i. on mRNA levels during preliminary experiments (Figure 3), we decided to focus on samples from Norm and CR KO mice at 0 h and 24 h p.i. However, we did not reveal significant changes in the content of protein carbonylation, protein levels of catalase and mitochondrial SOD2 in response to genetic n-3 PUFA enrichment (Fat background) or RvE1 receptor deficiency (Supplementary Figure S3).

As with the oxidative stress markers, a comprehensive analysis of the total astrocyte area was performed focusing on the Norm and CR KO mice. The hypothalamic structure known as the paraventricular nucleus (PVN) was selected for analysis and changes in total astrocyte area were shown over time at 0 h, 24 h, and 72 h p.i., but comparisons were only made at 24 h p.i. (Supplementary Figure S4). Interestingly, at 24 h p.i. a deficiency in CR changed the percentage of GFAP expression in the PVN where the CR KO groups covered a higher percentage of the PVN in comparison to Norm mice regardless of n-3 PUFA status. The evaluation of astrocyte morphology through GFAP staining in the PVN confirms the mRNA effect we first observed in the hypothalamus (Figure 3) that deficiency in CR increases astrocyte activation.

2.5. Immunohistochemical Detection of Inflammatory Transcription Factor Expression at the OVLT, a Brain Structure with a Leaky BBB, and Bifurcation, a Brain Structure with a Complete BBB: Preliminary Results Indicate That n-3 PUFAs and RvE1 Receptors May Alter LPS-Induced NF-IL6 Immunoreactivity at the BIF in Norm and CR KO Mice

To further assess genomic brain cell activation as a marker of humoral immune-to-brain signaling [52], immunohistochemical detection of nuclear NF-IL6 signals was used. Since we did not have enough animals in our WT-Norm control group or the Fat-CR KO group for all time points, the analyses served as preliminary data and were carried out at the level of the OVLT, which is a brain structure with a leaky BBB known to detect circulating mediators [72]. However, semi-quantitative evaluations of NF-IL6 immunoreactivity in the OVLT did not reveal differences between time points (Supplementary Figure S5) or in groups (Supplementary Figures S6 and S7) possibly related to very low levels of genomic activation per section. Overall, we did not detect significant changes in NF-IL6 activation in the OVLT by n-3 PUFA enrichment or the RvE1 receptor deficiency. Similar to the very modest systemic inflammatory response detected in the liver, these results suggest that circulating cytokines may not play a major role in immune-to-brain communication at the level of the OVLT in the ARDS model applied here.

In addition to the OVLT, a brain structure with a complete BBB was also analyzed, namely the bifurcation (BIF), for differences in NF-IL6 (Figure 4 and Figure S8). A semi-quantitative evaluation was used to assess immunoreactivity levels of NF-IL6 at the BIF. In contrast to the OVLT, at 4 h p.i. NF-IL6 immunoreactivity seemed to peak in WT-Norm and Fat-Norm mice. RvE1 receptor deficiency may affect this process in the CR KO mice since they did not appear to display the same trend in NF-IL6 immunoreactivity, however, in the LR KO mice, there appeared a similar progression in NF-IL6 as seen in the Norm controls. Given that n-3 PUFA enrichment caused modest reductions of NF-IL6 mRNA expression in CR KO mice (Figure 3), immunofluorescence showed that CR but not genetic n-3 PUFA enrichment altered the immunoreactivity at 24 h p.i. ($p = 0.0197$) in the BIF (Figure 5). This effect could not be observed in LR KO mice, where RvE1 receptor-dependent alterations were only significant at 0 h p.i. ($p = 0.0098$) (Supplementary Figure S5).

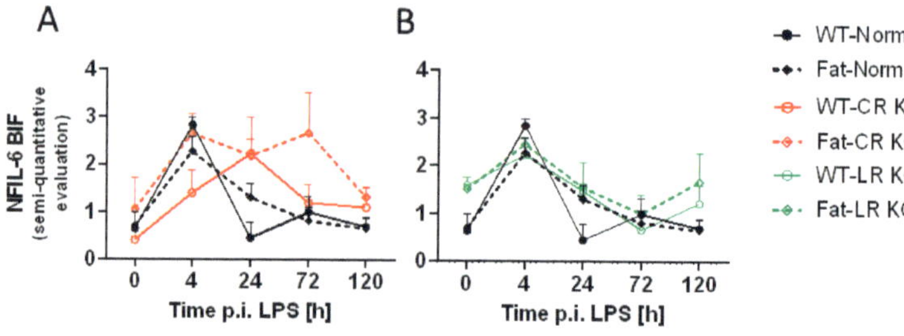

Figure 4. NF-IL6 immunoreactivity in Norm, CR KO, and LR KO mice at the level of the BIF over time after intratracheal LPS-induced ARDS. Sections of the brain analyzed at the level of the bifurcation (BIF) for nuclear factor interleukin 6 (NF-IL6) immunoreactivity on a scale ranging from 0 to 4 (**A**,**B**). Unmodified mice (Norm) were compared to mice deficient in chemerin receptor 23 (CR KO; (**A**)) or leukotriene B4 receptor (LR KO; (**B**)) bred on a wild-type (WT) or transgenic omega-3 (n-3) synthesizing *fat-1* (Fat) background received an intratracheal (i.t.) instillation with lipopolysaccharide (LPS, 10 µg) and were sacrificed at 0 h, 4 h, 24 h, 72 h, and 120 h p.i. Sections were analyzed by immunofluorescent staining using an antibody for NF-IL6. Immunoreactivity was assessed using semi-quantitative evaluation. n = 2–5 per group. Statistics were not performed due to low 'n' numbers.

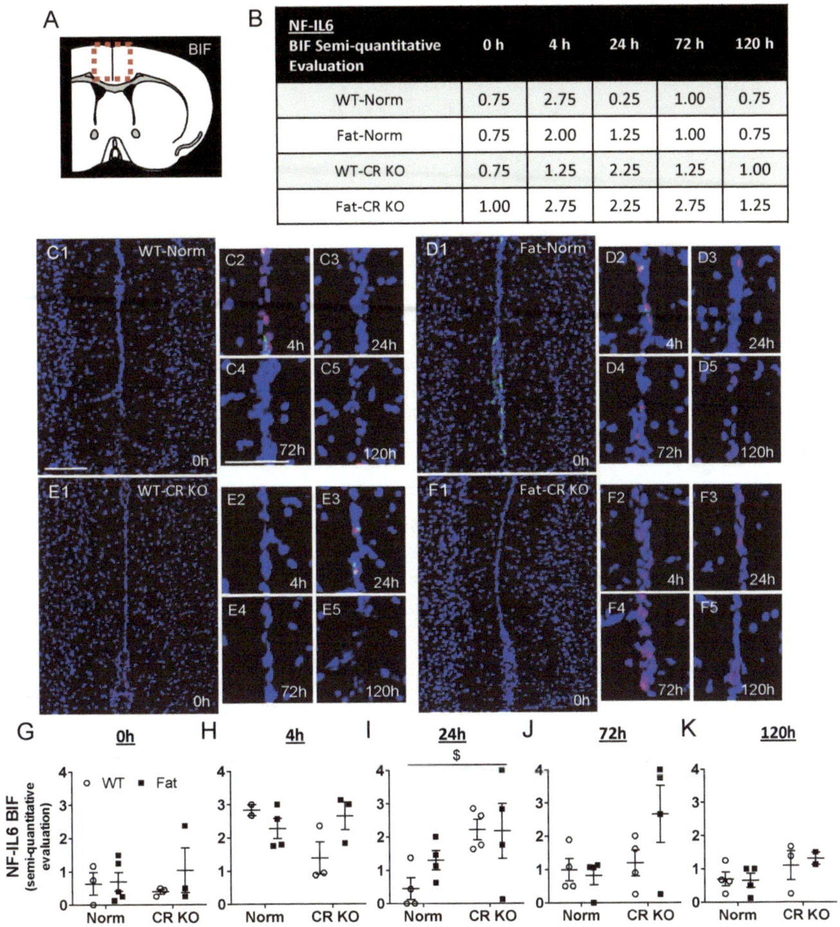

Figure 5. Intratracheal LPS-induced ARDS NF-IL6 immunoreactivity at the level of the BIF in Norm compared to CR KO mice is significantly altered at 24 h p.i. Sections of the brain were analyzed at the level of the bifurcation (BIF; (**A**) indicated by dashed square) by semi-quantitative evaluation for nuclear factor interleukin 6 (NF-IL6; red, (**C–F**)) immunoreactivity on a scale ranging from 0 to 4 (**B**). Unmodified mice (Norm; (**C**)) were compared to mice deficient in the chemerin receptor 23 (CR KO; (**E**,**F**)) bred on a wild-type (WT; (**C**,**E**)) or transgenic omega-3 (n-3) synthesizing *fat-1* (Fat; (**D**,**F**)) background received an intratracheal (i.t.) instillation with lipopolysaccharide (LPS, 10 µg) and were sacrificed at 0 h, 4 h, 24 h, 72 h, and 120 h p.i. WT-/Fat-Norm groups were compared to WT-/Fat-CR KO groups at each time point (**G–K**). At 24 h p.i with LPS CR deficiency increased NF-IL6 immunoreactivity compared to Norm. Von Willebrand factor (green; (**C–F**)) depicts brain vasculature. DAPI (blue; (**C–F**)) visualizes the surrounding tissue. n = 2–5 per group. Statistics were only performed when n = >3 per group. $ main effect Norm vs. CR KO. Scale bar in (**C1**) = 100 µm and is representative of (**C1**,**D1**,**E1**,**F1**); Scale bar in (**C4**) = 50 µm and is representative of (**C2–C5**,**D2–D5**,**E2–E5**,**F2–F5**). $ $p < 0.05$.

Overall, these preliminary descriptive data suggest a modest increase in NF-IL6 activation at 4 h after the lung insult for the BIF, suggestive of some moderate but low systemic inflammatory response (Figure 4) leading to a few humoral signals to the brain in the current model of LPS-induced ARDS.

2.6. Analyses of Immune Cell Recruitment to the OVLT and Bifurcation: Preliminary Results Indicate That LPS Stimulated Neutrophil Recruitment in CR KO Mice May Be Altered in the OVLT by n-3 PUFAs at 4 h p.i. and by CR at 0 h and 120 h p.i. in the Bifurcation. LPS Stimulated Neutrophil Recruitment in CR KO Mice Were Altered by n-3 PUFAs at 0 h and 72 h p.i. and by CR at 0 h, 24 h and 72 h p.i.

To further investigate data on hypothalamic mRNA expression for markers of neutrophil recruitment to the brain during LPS-induced ARDS (Figure 3), we performed immunohistochemical myeloperoxidase (MPO) staining for neutrophils in the OVLT (Figure 6). During early systemic inflammation, such staining reflects neutrophil recruitment to the brain [55]. Once again, the low 'n' for our WT-Norm control group and Fat-CR KO group meant that these analyses served to gather preliminary descriptive data. Here, the neutrophil trafficking into the brain appeared to peak at 4 h p.i. in WT-Norm controls but was decidedly absent in the Fat-Norm mice. While mice in the LR KO groups closely mirrored the response of WT-Norm controls, the CR KO groups appeared to have the opposite response in comparison. In the CR KO groups the peak seemed to be present in the n-3 PUFA-enriched (Fat background) CR KO mice while the WT-CR KO mice maintained relatively constant neutrophil levels.

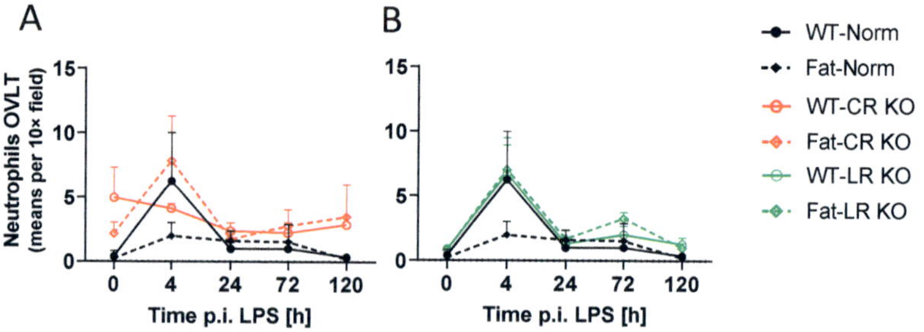

Figure 6. Neutrophil recruitment in Norm, CR KO, and LR KO mice at the level of the OVLT over time after intratracheal LPS-induced ARDS. Sections of the brain were analyzed at the level of the vascular organ of lamina terminalis (OVLT) for neutrophil trafficking to the brain (**A**,**B**). Unmodified mice (Norm) were compared to mice deficient in chemerin receptor 23 (CR KO; (**A**)) or leukotriene B4 receptor (LR KO; (**B**)) bred on wild-type (WT) or transgenic omega-3 (n-3) synthesizing *fat-1* (Fat) background received an intratracheal (i.t.) instillation with lipopolysaccharide (LPS, 10 µg) and were sacrificed at 0 h, 4 h, 24 h, 72 h, and 120 h p.i. Sections were analyzed by immunofluorescence staining using an antibody for myeloperoxidase. Images were taken and neutrophils were counted per 10× field of view. n = 2–5 per group. Statistics were not performed due to low 'n' numbers.

Although neutrophil trafficking to the OVLT occurred at low levels, their recruitment may peak 4 h after LPS stimulation. This response was abolished in Fat-Norm mice and partially recovered with RvE1 receptor deficiency. Significant differences were observed between Norm and CR KO at 0 h p.i. (p = 0.0340) (Supplementary Figure S9) but not LR KO (Supplementary Figure S10). Alterations between Norm and CR KO mice suggest a role for RvE1 receptors in neutrophil recruitment to the OVLT.

For the BIF, neutrophil recruitment to the brain also appeared to peak at 4 h p.i. in WT-Norm controls while the other groups showed consistently low levels of neutrophils or higher variation per time point (Figure 7). Again, this response was abolished in Fat-Norm mice and partially recovered with RvE1 receptor deficiency. Indeed, the 4 h peaks observed in WT-Norm and Fat-LR KO were absent in Fat-Norm and WT-LR KO mice over the course of the experiment.

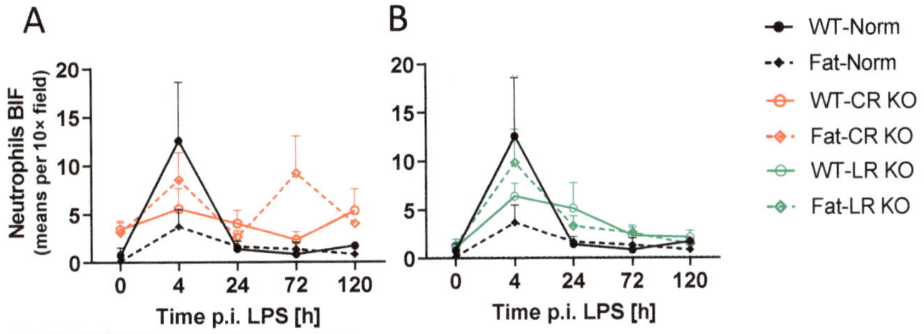

Figure 7. Neutrophil recruitment in Norm, CR KO, and LR KO mice at the level of the BIF over time after intratracheal LPS-induced ARDS. Sections of the brain were analyzed at the level of the bifurcation (BIF) for neutrophil trafficking to the brain (**A**,**B**). Unmodified mice (Norm) were compared to mice deficient in chemerin receptor 23 (CR KO; (**A**)) or leukotriene B4 receptor (LR KO; (**B**)) bred on a wild-type (WT) or transgenic omega-3 (n-3) synthesizing *fat-1* (Fat) background received an intratracheal (i.t.) instillation with lipopolysaccharide (LPS, 10 µg) and were sacrificed at 0 h, 4 h, 24 h, 72 h, and 120 h p.i. Sections were analyzed by immunofluorescence staining using an antibody for myeloperoxidase. Images were taken and neutrophils were counted per 10× field of view. n = 2–6 per group. Statistics were not performed due to low 'n' numbers.

Upon further analysis at the level of the BIF, n-3 PUFAs did not significantly affect neutrophil recruitment over time; however, at 0 h ($p = 0.0441$) and 72 h p.i. ($p = 0.0274$) n-3 PUFA enrichment (Fat background) in CR KO mice did show higher levels of neutrophils compared to Fat-Norm controls (Figure 8). Moreover, CR KO mice had an altered response in neutrophil recruitment compared to Norm at 0 h ($p = 0.0025$), 24 h ($p = 0.0339$), and 72 h ($p = 0.0216$). Interestingly, these results were not evident in LR KO mice, which showed no significant impact of n-3 PUFAs or RvE1 receptors (Supplementary Figure S11).

Similar to the OVLT, n-3 PUFAs alone may inhibit the ability of neutrophils to be recruited to the BIF during ARDS. However, the combination of LR KO and n-3 PUFA expression may enhance recruitment since mice deficient in the LR appeared to have a stronger peak in the n-3 PUFA enriched group. Overall, the observed impact of the CR on neutrophil recruitment to the BIF suggests that modulation of neutrophil recruitment by Rv may be mediated via CR and not LR.

2.7. Analyses of Lipid Mediators in the Lung and Brain: n-3 PUFAs and RvE1 Receptors Altered Lipid Mediators at 0 h, 24 h, and 72 h p.i. in CR and LR KO Mice

The lipid mediators in the lung and the brain were analyzed with LC-MS/MS to uncover their potential contributions to observed effects on inflammatory signaling and lung-to-brain communication pathways. Comparisons were made between Norm and receptor-deficient mice on either the WT or Fat background for each time point.

In the lung, n-3 PUFAs and RvE1 receptors had an impact on the levels of lipid mediators in the Norm and KO groups (Supplementary Table S4). These main effects were present at 0 h, 24 h, and 72 h p.i. due to LPS-induced ARDS. In general, genetic n-3 PUFA enrichment (Fat background) increased levels of lipid mediators during all three time points (Figure 9A). Interestingly, deficiency in RvE1 receptors had strong impacts increasing LTB_4 and EPA levels, suggesting potential negative feedback mechanisms.

A

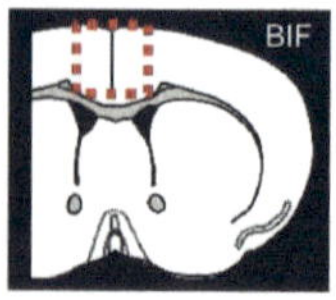

B

Neutrophils BIF Means per 10× field	0 h	4 h	24 h	72 h	120 h
WT-Norm	0.8	12.6	1.4	0.8	1.7
Fat-Norm	0.1	3.7	2.0	1.3	0.8
WT-CR KO	3.4	5.6	4.1	2.3	5.4
Fat-CR KO	3.1	8.6	2.6	9.2	4.0

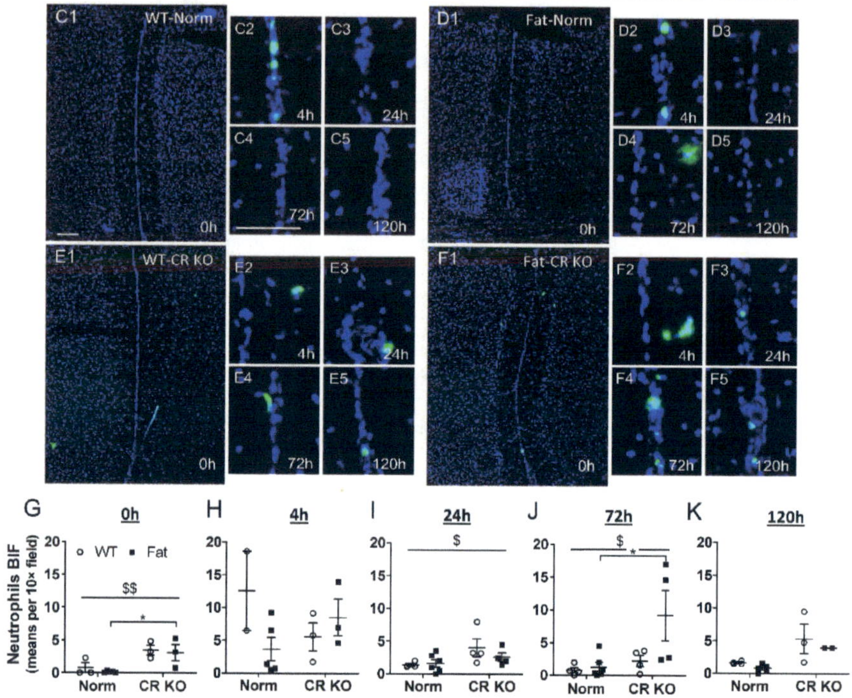

Figure 8. LPS-induced ARDS neutrophil recruitment at the level of the BIF in Norm compared to CR KO mice is significantly altered at 0 h, 24 h, 72 h, and 120 h p.i. Sections of the brain were analyzed at the level of the bifurcation (BIF; (**A**) indicated by the dashed square) and neutrophils (green, (**C–F**)) were counted per 10× field of view (**B**). Unmodified mice (Norm; (**C**)) were compared to mice deficient in the chemerin receptor 23 (CR KO; (**E**)) bred on wild-type (WT; (**C**,**E**)) or transgenic omega-3 (n-3) synthesizing *fat-1* (Fat; (**D**,**F**)) background received an intratracheal (i.t.) instillation with lipopolysaccharide (LPS, 10 µg) and were sacrificed at 0 h, 4 h, 24 h, 72 h, and 120 h p.i. WT-/Fat-Norm groups were compared to WT-/Fat-CR KO groups at each time point (**G–K**). At 0 h, 24 h, 72 h and 120 h p.i with LPS CR KO had altered neutrophil recruitment compared to Norm. DAPI (blue; (**C–F**)) visualizes the surrounding tissue. n = 2–6 per group. Statistics were only performed when n = >3 per group. $ main effect Norm vs. CR KO. Scale bar in (**C1**) = 100 µm and is representative of (**C1**,**D1**,**E1**,**F1**); Scale bar in (**C4**) = 50 µm and is representative of (**C2–C5**,**D2–D5**,**E2–E5**,**F2–F5**). * $p < 0.05$, \$ $p < 0.05$, \$\$ $p < 0.01$.

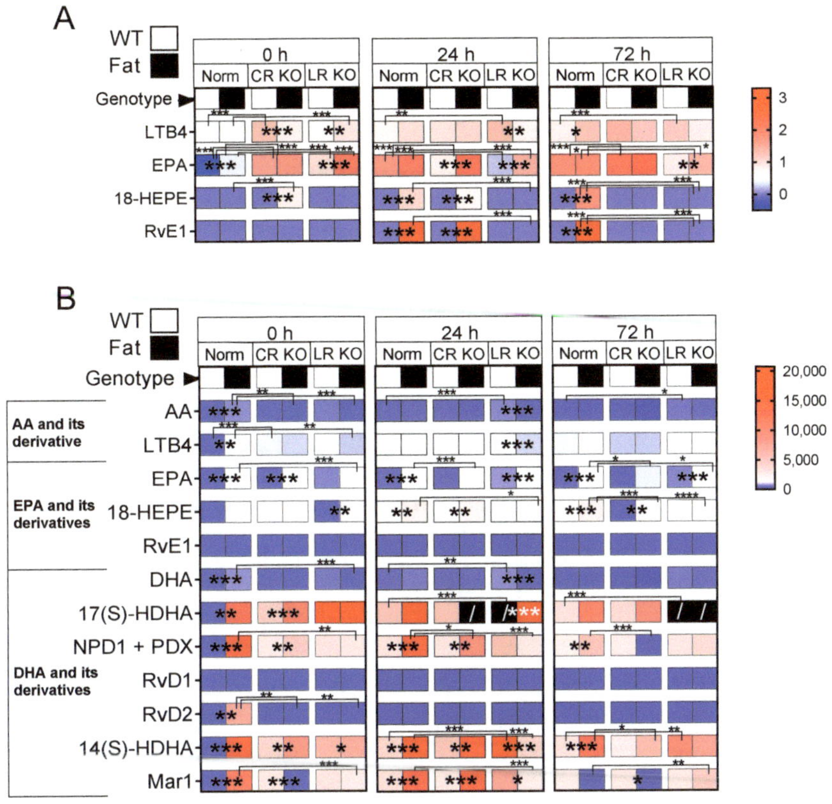

Figure 9. Differences in the lipid mediators in the (**A**) lung and (**B**) brain following intratracheal LPS-induced ARDS. Lipid mediators were measured from homogenized lung (**A**) and brain tissue (**B**) with LC-MS/MS. Mice deficient in chemerin receptor 23 (CR KO) or leukotriene B4 receptor (LR KO) as well as unmodified mice (Norm) bred on wild-type (WT) or transgenic omega-3 (n-3) synthesizing *fat-1* (Fat) background received an intratracheal (i.t.) instillation with lipopolysaccharide (LPS, 10 μg) and were sacrificed at 0 h, 24 h, or 72 h p.i. Notably, impacts of genetic n-3 PUFA enrichment and RvE1 receptor deficiencies (CR or LR) are observed in the lung in leukotriene B4 (LTB_4), eicosapentaenoic acid (EPA), 18R-hydroxy eicosapentaenoic acid (18-HEPE) and resolvin E1 (RvE1) levels. In the hypothalamus, several polyunsaturated fatty acids (PUFA) and their derivatives including arachidonic acid (AA), LTB_4, EPA, 18-HEPE, RvE1, docosahexaenoic acid (DHA), 17(S)-hydroxy docosahexaenoic acid (17(S)-HDHA), neuroprotectin D1 (NPD1) + protectin DX (PDX), resolvin (RV)D1, RvD2, 14(S)-hydroxy docosahexaenoic acid (14(S)-HDHA), and maresin 1 (Mar1) were altered. The average fold change for each sample set is normalized to WT-Norm 0 h and is presented as a value on a heat map. A black box with a white diagonal line indicates a value that exceeds the scaled range. n = 3–6 per group after outlier exclusions and undetermined values. n = 5 per group for all groups except at the 0 h time point Fat-CR KO and at 72 h WT-Norm (n = 4). * $p < 0.05$, ** $p < 0.01$, *** $p < 0.001$.

Levels of the n-6 PUFA AA derivative LTB_4, which binds to the LR, were elevated in CR KO compared to Norm mice ($p \leq 0.0001$) at basal levels. LTB_4 decreased in mice with genetic n-3 PUFA enrichment and CR receptor deficiency ($p \leq 0.0001$). After 24 h, n-3 PUFA enrichment decreased LTB_4 expression in the LR KO groups ($p \leq 0.0001$) but, in comparison to WT-Norm mice, levels were elevated in WT-LR KO mice ($p \leq 0.0001$) (Figure 9A). At 72 h p.i., the levels of LTB_4 were greatly increased in WT-LR KO ($p = 0.0073$)

and Fat-Norm ($p = 0.0001$) compared to WT-Norm mice. Genetic n-3 PUFA enrichment in CR KO and LR KO mice led to a decrease in LTB_4 (CR KO ($p = 0.0016$) and LR KO ($p = 0.0104$)) (Figure 9A).

The n-3 PUFA EPA levels increased significantly with genetic n-3 PUFA enrichment in LR KO in the lung at basal levels ($p = 0.0024$). At 24 h p.i., EPA was increased in mice with genetic n-3 PUFA enrichment and CR deficiency ($p \leq 0.0001$) and in Norm mice compared to CR KO ($p = 0.0013$) and LR KO ($p = 0.0008$). On the contrary, 72 h p.i. EPA was decreased in LR KO mice with and without genetic n-3 enrichment compared to Norm mice (Fat-Norm vs. Fat-LR KO, $p = 0.0003$; WT-Norm vs. WT-LR KO, $p = 0.0003$) (Figure 9A).

The EPA derivatives 18-HEPE and RvE1 showed very similar changes in the lung during all time points with almost no differences between the groups at basal levels. At 24 h p.i., 18-hydroxy eicosapentaenoic acid (HEPE) and RvE1 levels increased in Norm groups (18-HEPE, $p = 0.0451$; RvE1, $p \leq 0.0001$) with genetic n-3 PUFA enrichment compared to mice on the WT background. These elevations of n-3 PUFA EPA derivatives in Norm mice remained until 72 h p.i. (both, $p \leq 0.0001$). In n-3 genetically enriched CR KO mice, on the one hand, 18-HEPE levels remained mostly the same. RvE1, on the other hand, increased at 24 h p.i. in Fat-CR KO compared to WT-CR KO ($p \leq 0.0001$) and went back to baseline levels by 72 h p.i (Figure 9A).

Significant differences in lipid mediators in the brain were revealed between the groups at all time points (Figure 9B). Both genetic n-3 PUFA enrichment and RvE1 receptor deficiencies had an impact on the levels of lipid mediators in the Norm and KO groups (Supplementary Table S5). Due to LPS-induced ARDS, more main effects were present at 24 h and 72 h p.i. In most groups, the amount of lipid mediators increased after the LPS challenge. In general, n-3 PUFAs impacted the levels of lipid mediators to a higher extent than RvE1 receptors. Overall, genetic n-3 PUFA enrichment caused higher levels of lipid mediators during all three investigated time points in Norm and CR KO mice. Interestingly, these effects were reversed in LR KO mice. With the genetic enrichment of n-3 PUFAs in LR KO groups, the levels of lipid mediators were, for the most part, decreased. However, two exceptions were observed in LR KO mice with genetic n-3 enrichment; EPA was increased at 24 h ($p \leq 0.0001$) and 72 h p.i. ($p \leq 0.0001$) and 18-HEPE at 0 h p.i. ($p = 0.0023$) (Figure 9B).

Interestingly the amount of n-6 PUFA AA and its derivative LTB_4 were higher in Norm mice with genetic n-3 enrichment at 0 h p.i. (AA, $p \leq 0.0001$; LTB_4, $p = 0.0002$). Over time, the effect changed such that at 24 h and 72 h p.i. levels increased in Norm and decreased in Fat mice. In response to LPS in WT-LR KO mice, the levels of AA were increased compared to WT-Norm mice at 24 h and 72 h p.i. (24 h, $p = 0.0004$; 72 h $p = 0.0489$) and the levels of LTB_4 decreased with genetic n-3 enrichment 24 h p.i ($p = 0.0006$) (Figure 9B).

The n-3 PUFA EPA is increased in genetic n-3 enriched Norm, CR KO, and LR KO mice at different time points. Genetic n-3 enriched Norm mice demonstrate higher levels of EPA during all three time points (0 h, 24 h, $p \leq 0.0001$; 72 h, $p = 0.0211$). On the one hand, genetic n-3 enrichment in CR KO mice caused an increase in EPA at basal levels ($p \leq 0.0001$), on the other hand, in LR KO mice it led to an increase at 24 h ($p \leq 0.0001$) and 72 h p.i. ($p \leq 0.0001$). **Its derivative, 18-HEPE,** was increased in genetic n-3-enriched Norm and CR KO mice at 24 h (both, $p \leq 0.0001$) and 72 h p.i. (Norm, $p \leq 0.0001$; CR KO, $p = 0.0023$). As previously mentioned, the 18-HEPE derivative can be further converted to RvE1; however, levels of RvE1 were not detectable in any brain sample (Figure 9B).

The levels of DHA remained mostly unchanged over time. **The DHA derivative 17(S)-hydroxy docosahexaenoic acid (17(S)-HDHA)** was increased in Norm ($p = 0.0028$) and CR KO ($p \leq 0.0001$) mice with genetic n-3 enrichment at 0 h p.i. Interestingly, 17(S)-HDHA increased in WT-LR KO mice at 24 h p.i. compared to Fat-LR KO and WT-Norm mice (both, $p \leq 0.0001$) (Figure 9B).

The 17(S)-HDHA derivative can be further converted to **neuroprotectin D1 (NPD1) and protectin DX (PDX)** (both mediators cannot be distinguished by LC-MS/MS). Overall, these protectins were increased in Norm and CR KO mice at basal levels with genetic n-3

PUFA enrichment (Norm, $p \leq 0.0001$; CR KO, $p = 0.0003$), 24 h p.i. (Norm, $p \leq 0.0001$; CR KO, $p = 0.0078$) and also at 72 h p.i. for Norm mice alone ($p = 0.0025$). Moreover, 17(S)-HDHA can be converted to **RvD1 and RvD2** but only RvD2 was found to be increased in Fat-Norm mice at 0h p.i. (Figure 9B).

Interestingly, the levels of **14(S)-hydroxy docosahexaenoic acid (14(S)-HDHA)**, another derivative of DHA, were changed in a similar way to 17(S)-HDHA derivatives NPD1 and PDX. The derivate 14(S)-HDHA can be further converted to **Maresin 1** (Mar1) [9]. As such, Mar 1 showed similar alterations as NPD1 and PDX as well. In general, it was increased in most groups with n-3 enrichment at 24 h p.i. but decreased again at 72 h p.i. It is nonetheless interesting to note that genetic n-3 enrichment in CR KO mice led to a very high increase of Mar 1 at 24 h p.i. ($p \leq 0.0001$) (Figure 9B).

Overall, our data show that EPA and its derivatives are more abundant in the lung than in the brain [73,74] whereas DHA and its derivatives are found at higher levels in the brain (Figure 9), confirming previous reports by others [75–77].

3. Discussion

In the present study, we are the first to report underlying mechanisms of lung-to-brain signaling during LPS-induced ARDS with respect to the role of cytokines, lipid mediators, and trafficking of neutrophils to the brain. Indeed, LPS-induced lung inflammation did induce signs of brain inflammatory signaling such as NFκB. Overall, hallmarks of low-grade inflammation and other surrogate parameters such as genomic NF-IL6 activation in brain structures with and without complete BBB (i.e., BIF and OVLT) were not detected at levels to allow for a major impact in the lung-to-brain axis. Interestingly, neutrophil trafficking may contribute to underlying mechanisms of lung–brain communication. While descriptive, our data show an LPS-induced increase in neutrophil recruitment at 4 h p.i. at the level of the BIF, which was abolished by genetic n-3 PUFA enrichment in a CR-dependent manner. Moreover, we revealed that during LPS-induced ARDS genetic n-3 PUFA enrichment and the absence of RvE1 receptors, mediators of inflammation in the lung, liver, and hypothalamus were altered. Overall, pro-inflammatory cytokines such as IL-17, IL-1β, and TNFα, as well as the anti-inflammatory cytokine IL-10, were reduced by n-3 PUFAs in RvE1 receptor-deficient mice in the periphery. Additionally, the neutrophil markers GM-CSF and CXCL5 were similarly impacted. Within the brain, TFAM, NF-IL6, and PGC1α were seen to be altered by genetic n-3 PUFA enrichment while CXCL1, IL-1β, IL-1ra, and GFAP were affected by both n-3 PUFA enrichment and the deficient RvE1 receptors.

Specifically, deficiency of the LR had basal effects in the lung but deficiency of CR showed more significant effects than the LR on pro-inflammatory cytokines and neutrophil markers in the lung during later stages of inflammation. In the brain, impacts of the CR were especially evident for the BIF, where NF-IL6 immunoreactivity and neutrophil recruitment were both altered in mice lacking CR and, in the case of neutrophils, elevated recruitment by n-3 PUFAs was observed in CR but not LR-deficient mice. Moreover, our new insights into differences in lung and brain lipid mediators provide evidence for potential candidates such as 18-HEPE, protectins, RvD2, and Mar1 to explain n-3 PUFA-dependent alterations in lung–brain communication.

During lung inflammation, a consequence of the enhanced leukocyte accumulation and activation is high levels of cytokine synthesis [78–81]. Notably, IL-1β is one of the key mediators of the lung inflammatory process [81]. A bidirectional interaction exists between the lung and the brain, where insults or injuries to either organ can have impacts on the other [79,80,82,83]. One study performed by Sahu and colleagues (2018) found that following a "two-hit" model for ALI, where mice received an i.t. instillation of hydrochloric acid and LPS, the animals experienced cognitive impairments and elevated levels of systemic pro-inflammatory cytokines [80]. By transmitting information via cytokines and immune cell trafficking, a circuit of communication may be formed between the lung and the brain as has been previously shown with neutrophils and neutrophilic NADPH oxidase, for example, in a mouse model of LPS-induced ALI [82,84,85]. Here, we show only some minor contributions of low-grade systemic inflammation as evidenced in the liver and genomic NF-IL6 brain cell activation as a marker of humoral immune-to-brain communication.

During ARDS, the release of inflammatory mediators from the lung can influence levels of circulating mediators in the periphery and even reach the brain [84]. By investigating a range of cytokines, we hoped to reveal effects of the RvE1 receptors and/or n-3 PUFA enrichment on lung–brain communication pathways. To assess lung–brain interactions, we first analyzed the overall inflammatory response in the primary site of LPS-induced inflammation in the lung before analyzing the liver, which would provide a systemic cytokine profile. Previous studies have already identified upregulation of the pro-inflammatory cytokines IL-1β, TNFα, and IL-6 in bronchoalveolar lavage fluid (BALF) in ARDS patients [86]. These pro-inflammatory cytokines are able to cross the BBB, especially when permeability increases such as during inflammation [87–89].

In our present study investigating organ-specific inflammatory responses, nearly all effects were localized to the lung tissue. Fat mice deficient in the LR showed reduced basal levels of pro-inflammatory IL-1β and TNFα along with a reduction in the anti-inflammatory IL-10. Over the course of the experiment, mice deficient in the CR had a consistent reduction in IL-10 suggesting a complex regulation of the pro- and anti-inflammatory balance. On the one hand, even though RvE1 acts as an antagonist for the LR against LTB_4, the LTB_4-BLT1 complex can still signal and exacerbate inflammation as has, for example, been observed in models of myocardial infarction [90]. On the other hand, in addition to RvE1, chemerin can also bind the CR to encourage viral clearance and moderate levels of inflammation to the extent that CR KO mice infected with the mouse pneumonia virus had higher mortality rates than the immune-competent controls [50]. Therefore, in addition to the benefits of RvE1s interaction, the pro- and anti-inflammatory capacities of the LR and CR, respectively, likely contribute to the significance of each KO. In the absence of Rv receptors, however, with the lack of anti-inflammatory Rv signaling, we would have expected a dampened anti-inflammatory response, as previously shown by Arita (2007) in a mouse model of peritonitis and an enhanced inflammatory response, as shown by Bondue (2011) in the mouse model of pneumonia [21,50]. Indeed, reduced lung IL-10 levels support the notion of dampened anti-inflammatory signaling in mice with defective Rv signaling in our present study [91]. Yet, reduced lung IL-1β and TNFα levels seemingly do not fit such simple concepts of reduced anti-inflammatory signaling. Such discrepancies may be linked to the complex dynamics of the inflammatory response and could reflect enhanced inflammation at other time points. We would postulate that at basal levels the LR appears to exert more influence on inflammation as a receptor with pro-inflammatory capacities than the CR while, during later stages of inflammation, the CR continues to significantly impact IL-10 production. However, since the reductions in IL-1β and TNFα were only observed in the Fat-LR KO mice, it is unclear to what extent the receptor plays a role in this process as no effect was observed in Norm mice genetically enriched with n-3 PUFAs (Fat background).

Interestingly, SPMs, namely the E series Rv, have also been shown to reduce levels of IL-1β in lung tissue during the combination of ALI and bacterial pneumonia in part by reducing neutrophil accumulation [40]. Moreover, the regulation of inflammation through TNFα and NFκB prevented transgenic *fat-1* mice from developing streptozocin-induced diabetes during which elevated levels of the RvE1 precursor 18-HEPE were detected [92]. As activated neutrophils are key producers of IL-1β and TNFα during ARDS [93,94], it is possible that RvE1 reduced levels of neutrophils in the lung, leading to a reduction in pro-inflammatory cytokines reported in our present study. RvE1-dependent generation of reactive oxygen species (ROS) and depression of anti-apoptotic signals via LRs increases the apoptosis and clearance of neutrophils during ALI [95]. Moreover, induced phagocytosis of apoptotic neutrophils by macrophages will also increase the production of pro-resolving anti-inflammatory mediators such as IL-10 [95]. Herová and colleagues (2015) found that treating human primary macrophages that expressed the CR in vitro with RvE1 (10 nM) increased IL-10 production [96]. Altogether, the previously described therapeutic effects of

RvE1 by neutrophil regulation during ARDS fit well with what we have seen in the lungs of mice in the present study.

Heavy recruitment and infiltration of neutrophils to the lung play a major role in pathogenesis during ARDS and can be monitored using markers associated with neutrophil recruitment and activation [97–104]. IL-17 and GM-CSF in particular have been identified as key modulators of inflammation in the lung whose production can be inhibited by n-3 PUFAs [105]. Consistent with the previous studies, we also saw a reduction in both IL-17 and GM-CSF in the lungs of mice, dependent on genetic n-3 PUFA enrichment. However, although the effects of n-3 PUFAs are apparent, it seems to only be the case in combination with the absence of the LR. As we hypothesized for IL-1β, TNFα, and IL-10, this could also be due to more complex regulatory mechanisms involving additional signaling pathways and binding partners aside from RvE1. Nevertheless, other researchers have found that during allergic airway inflammation, Fat mice have a lower BALF concentration of pro-inflammatory cytokines, including CXCL1 and CCL5, and elevated levels of both RvE1 and PD1 in the lung tissue [106]. Research by Haworth et al. (2008) also identified that treatment with RvE1 could decrease levels of IL-17 in BALF as well as suppress IL-17-producing T helper cells during allergic airway inflammation [11]. It has even been shown that n-3 PUFAs are able to modulate the production of CXCL1 and influence infection outcomes during the early stages of lung infection [107]. However, our present results only show a moderate impact on CXCL1 during the later time point of lung inflammation.

In comparison to the lung, there were far fewer alterations to neutrophil marker expression in the liver. Here, we did still observe the n-3 PUFA-dependent reduction in IL-17 within the Norm group with intact Rv receptors consistent with what others have reported [108–110]. The differences between results in the lung and liver may again represent a delay in the inflammatory dynamics with an origin in the lung and some spillover to the circulation inducing a response in the liver. However, the lack of a strong effect in the liver served as an indicator of the systemic inflammatory response suggesting weak contributions of humoral mediators to lung–organ interactions in our model of ARDS.

Towards the later time points, i.e., 72 h p.i., a deficiency in the CR was mainly impacting pro-inflammatory cytokines in the lung including IL-17, IL-1β, and TNFα but also CXCL5 and GM-CSF. Since CR is located on a greater variety of cell types than the LR [12,27–29,31,111], it is possible that its impact lasts beyond the initial phases of inflammation once the inflammatory leukocytes have dissipated. Overall, by knocking out the RvE1 receptors in combination with genetic n-3 PUFA enrichment, the mice seemed to experience a dampened immune response. Interestingly, such differences were largely absent in WT and Fat mice with intact Rv receptors. The fact that differences were not observed in mice with the receptors may be due to several factors. In addition to the actions of LTB_4 and chemerin, the low dose of LPS or the time points selected for analyses could also contribute [50,90].

To delineate lung-to-brain communication during ARDS and the ability to cause inflammation, further analyses were performed on the brain. Indeed, we found evidence that LPS-induced ARDS was sufficient to alter mediators of inflammation in the hypothalamus as well as neutrophil trafficking to different brain structures. Key differences were found for markers of oxidative stress, mitochondrial activity, neutrophils, cytokines, and astrocyte activation while little to no changes were seen in signaling pathways, PG synthesis, and neuroplasticity. Additionally, neutrophil trafficking and NF-IL6 immunoreactivity were assessed at the OVLT, with a leaky BBB prone to detecting circulating inflammatory mediators [53,112], and the BIF with a complete BBB. By comparing these two structures, it was possible to evaluate blood-borne signaling in lung-to-brain communication.

Oxidative stress is known to play an important role in the inflammatory response during LPS-induced inflammation in the brain [113–116]. The extent to which oxidative stress occurs can be calculated by measuring antioxidants and microglial expression of NOS2 [117,118]. While enzymes are useful when carrying out certain physiological functions, exorbitant production of NOS2 or insufficient regulation by antioxidants results in oxidative stress and cellular damage [119–121]. The anti-inflammatory effects of RvE1 combined with the presence of CR on glial cells have encouraged investigations into its applications in neuroinflammation [20,47,122,123]. Indeed, Zhang and colleagues (2022) found that, in the brain, RvE1 signaling through the CR ameliorated oxidative stress in a diabetic mouse model [124]. Our own initial evaluation of the hypothalamus showed an overall impact of CR and reduced production of NOS2 in WT-CR KO mice during later time points of LPS-induced ARDS. Moreover, CR also altered the expression of TFAM, which plays a key role in mitochondrial biogenesis and maintenance along with additional antioxidant effects [125–127]. To confirm our results on hypothalamic NOS2 mRNA expression, additional analyses on protein level were added. However, neither the carbonylation of proteins nor protein levels of Cat and SOD2 were altered by n-3 PUFA enrichment or CR receptor deficiency. Potentially, discrepancies between mRNA and protein data may be explained due to cell-specific regulations that were diluted in the protein extracts. Even though activation through the LR was connected with ROS, our initial analysis of the hypothalamus showed more differences in the CR KO line, which was why we restricted the additional tests to these groups.

As in the periphery, the role of neutrophils in the brain remained a primary interest in our present study. Recruitment of neutrophils to the brain has been recently highlighted as an important immune-to-brain communication pathway during systemic inflammation [55]. Since ARDS causes heavy recruitment of neutrophils into the lungs [97,98], we investigated if these cells could also play a role in lung-to-brain communication. Surprisingly, our initial results showed that the chemoattractant CXCL1 was elevated by n-3 PUFAs but such an increase was absent in both Rv receptor-deficient animals, although n-3 PUFAs did cause CR KO mice to produce significantly less CXCL1 than the Fat-Norm. These results were contradictory to what was expected since n-3 PUFAs reduce neutrophil infiltration and increase phagocytosis [9,128], but the precise dynamics remain to be further investigated in the future.

To better visualize neutrophil recruitment to the brain, further immunohistochemical analysis allowed us to not only quantify neutrophils but provided an opportunity to assess their recruitment to specific brain structures as well. Here, our results more accurately mirrored what has been shown previously [9,128,129]. Indeed, n-3 PUFAs eliminated the LPS-induced neutrophil peak observed in the WT-Norm mice, which was recovered by RvE1 receptor deficiency, suggesting a role for RvE1 in this response. As such, RvE1 has been previously shown to exhibit a significant impact on neutrophils during models of inflammation by attenuating LTB_4s pro-inflammatory properties [21]. Moreover a combination treatment with aspirin and the RvE1 precursor 18R-HEPE into a dorsal air pouch in a mouse model of TNFα-induced local inflammation attenuated the recruitment of neutrophils to the inflammatory site [10,12]. Recently, we have even detected increased levels of 18-HEPE in supernatants of primary CVO cultures stimulated with LPS, highlighting 18-HEPE as a promising lead with the potential to treat neuroinflammation [130]. Although specific differences could not be detected between Norm groups, increased neutrophil recruitment to the BIF was observed in n-3 PUFA-enriched CR KO mice compared to Norm in our present study. Additionally, a stronger overall effect seen in mice lacking CR as opposed to the LR suggests that modulation of neutrophils during ARDS may occur primarily through this receptor. Indeed, the RvE1/CR axis is known to induce neutrophil apoptosis and phagocytosis [131,132].

Further attention was placed on the potential for lung-to-brain communication and the ability to modulate inflammation in the hypothalamus via cytokines. Interactions of cytokines with the brain during LPS-induced inflammation could contribute to the BBB's loss of integrity and activation of brain endothelial cells [133]. In particular, IL-1β can cross the BBB as well as be produced locally by microglia [87,134,135]. We found that the mRNA expression of IL-1β and its antagonist IL-1ra were affected differently by n-3 PUFAs and RvE1 receptor KOs. The initial n-3 PUFA-dependent increase in basal levels of IL-1ra in CR KO mice may be related to microglial RvE1 expression. It is uncertain why this effect was limited to the CR KO mice in our study, but an early production of IL-1ra, such as from microglia [136] could limit LPS-induced inflammation through reduced emigration of neutrophils [137] and modulation of IL-1β pro-inflammatory activity [136,138,139]. To expand on our present mRNA expression data, further analysis is required to confirm the potential role of Rv signaling for brain IL-1β/IL-1ra action on protein levels.

Aside from cytokines and neutrophil trafficking, there are several influential signaling pathways that play a role during brain inflammation including NF-IL6, NFκB, and PPARγ [52,66,140,141]. All three of these pathways can be influenced by n-3 PUFA during models of inflammation [12,72,142,143]. In our present study, only minor impacts were found for the mRNA expression of NF-IL6 and PGC1α, a transcription coactivator that regulates signaling through PPARγ [143]. However, given that we have previously shown that NF-IL6 activation and unique lipid mediator production patterns both specifically occur in the OVLT of Fat mice during LPS-induced inflammation [72], we decided to pursue an additional analysis for NF-IL6. Even though RvE1 can limit activation of NFκB [12] and it is a prominent transcription factor for pro-inflammatory cytokines production [144], we did not determine any significant differences between groups in our present results and, thus, did not perform additional analysis for this pathway.

NF-IL6 has been shown to act as an early pro-inflammatory and late anti-inflammatory transcription factor during systemic inflammation and has been broadly used as a genomic brain cell activation marker [52,66]. More detailed results were obtained by spatiotemporal immunohistochemical assessment of NF-IL6 in the OVLT and the BIF. Interestingly, we did not observe significant differences in NF-IL6 immunoreactivity between groups nor by genetic n-3 PUFA enrichment at the level of the OVLT as we expected. However, subtle alterations due to n-3 PUFAs and RvE1 receptor deficiencies were present at the BIF suggesting some impact on brain cell activity. In particular, the Fat-CR KO mice may have experienced a more prolonged state of NF-IL6 activation, but high variability has to be taken into account. In addition, unchanged mRNA expression of the terminal enzyme for PGE2 production, namely mPGES, supports a limited impact of NF-IL6 known to be involved in the regulation of mPGES expression [66,145]. Overall, while some changes in brain cell activation marker NF-IL6 confirm modest brain alterations by LPS-induced ARDS, these minor changes again suggest that humoral signals did not play a major role in lung–brain signaling.

Glia activation was also assessed given that systemic LPS stimulation has previously been demonstrated to robustly impact the function of astrocytes and microglia [133,134,146,147]. The release of cytokines, lipid mediators, or ROS by the glial cells plays an important role in regulating neuroinflammation [134,148–150]. Therapeutic techniques that modulate the activation of microglia or astrocytes could be useful in attenuating the resulting damage [151]. While we did not detect any significant changes in the microglial activation marker CD68 (Supplementary Figure S2), GFAP, a marker for astrocytic activation, showed significant changes in CR KO mice. Interestingly, genetic n-3 PUFA enrichment increased GFAP expression in the CR KO mice compared to Norm. Additional immunohistochemical assessment of the hypothalamus at the level of the PVN confirmed that the astrocyte area was indeed increased in CR KO mice. Bousquet et al. (2011) have shown that Fat mice have reduced GFAP expression in a mouse model of Parkinson's disease [152]. Since neuroinflammation can increase astrocyte expression of CR [122,153], it is possible that, through this receptor, RvE1 is able to modulate astrocyte activation, but these observations need to be further confirmed in future studies.

Important findings of the present manuscript also pertain to new insights into changes in lipid mediators in the lung and brain during LPS-induced ARDS. For lung inflammation, SPMs and especially EPA and its derivatives such as Rvs have been well studied during ARDS and the restoration of lung homeostasis [26,38–40]. It is known that the brain, on the other hand, contains high levels of PUFAs (25–30%) consisting mainly of DHA and AA [75]. In the present study, we were able to detect several lipid mediators and their derivatives. Most of these PUFAs were altered by genetic n-3 PUFA enrichment and the absence of RvE1 receptors in the lung and the brain. The knockdown of RvE1 receptors, especially LR, mainly led to an increase in PUFAs. Yet, RvE1 receptor knockdowns combined with genetic n-3 PUFA enrichment had an overall negative impact on PUFAs in the brain. In general, genetic n-3 PUFA enrichment in Norm and CR KO mice showed increased levels of PUFAs and their derivatives compared to the WT background counterparts, in particular at 0 h and 24 h p.i. This finding is supported by previous studies, which showed an increase in free n-3 and n-6 PUFAs in plasma as a general response to severe inflammation and infection [39,154,155]. Moreover, Mayer and colleagues (2003, 2009) revealed not only that DHA and EPA levels increase after infection or inflammation in human and murine plasma, but also showed further elevated levels of these PUFAs in Fat mice and patients treated with an n-3 PUFA emulsion [39,155]. Importantly, in a model of LPS-induced ARDS, Fat mice responded with lower production of inflammatory mediators and decreased neutrophil invasion to the alveoli [39]. Indeed, in our present study, we saw that the genetically enriched Fat mice showed higher amounts of several PUFAs normalized to WT mice at 0 h p.i. but further analyses remain to be performed to determine why more anti-inflammatory effects were not detectable in the lung and the hypothalamus/brain.

As expected from previous studies [38,39,106], mice with genetic n-3 PUFA enrichment showed higher levels of n-3 PUFA EPA in the lung compared to Norm mice at 0 h p.i., but these levels decreased 24 h and 72 h p.i. Since EPA is converted to 18-HEPE and further to RvE1, higher amounts of these metabolites were found in mice with genetic n-3 PUFA enrichment compared to Norm mice at 24 h and 72 h p.i. Considering EPA is present at higher levels in the lung than the brain [73,74] and in accordance with the fact that the brain exhibited a higher percentage of DHA than EPA [75–77], a low abundance of 18-HEPE and undetectable amounts of RvE1 were expected in the brain. Nonetheless, RvE1 is a potent but unstable lipid mediator, which is rapidly converted to one of the inactive products and may, therefore, be difficult to detect [132,156,157]. These findings indicate high relevance of EPA and its derivatives in the lung while DHA and its derivatives are detected at higher levels in the brain. This may be of functional significance for lung–brain communication pathways.

The amount of DHA and especially its derivatives such as 17(S)-HDHA, 14(S)-HDHA, NPD1, and Mar1 were increased in the brain after 24 h p.i. in Norm and CR KO mice with n-3 PUFA enrichment. Of note, the 17S series DHA compounds 17(S)-HDHA and 14(S)-HDHA were proven to be potent regulators of leukocytes, reducing infiltration in vivo and they may also block cytokine production in glial cells [14]. These results are in line with other studies and show that NPD1 and Mar1 exhibit regulating effects in inflammation, which could also contribute to the resolution of inflammation in the brain [12,14,158–162].

In contrast to Norm and CR KO mice, genetic n-3 PUFA enrichment in LR KO mice did not exhibit as many elevated levels of PUFAs compared to the WT in the brain. AA and LTB_4 were especially decreased in Fat-LR KO mice after 24 h p.i. Interestingly, different studies previously reported impaired recruitment of neutrophils accompanied by lowered inflammatory cytokine production in LR-deficient mice [32], suggesting a lower inflammatory response in these animals [90]. Such reduced inflammatory responses may also explain the absence of anti-inflammatory and pro-resolving lipid mediator responses in LR KO mice in the present study but further investigation to determine the precise role of PUFAs in LR KO mice is required.

Limitations of the present study include technical challenges resulting in low numbers of samples per group. Overall, we have to acknowledge the screening character of the present study, which made use of the RRR principle of Russel and Burch [163] to perform our analyses on tissue that was harvested in collaboration from ongoing experiments for a different purpose. To reduce the amount of animals used for scientific purposes, we focused on LPS-stimulated mice. In this manner, parts of the lung were used for different experiments and we were not able to quantify neutrophils in the lung. In order to maximize the number of transgenic animals, males and females were both used for the experiment, which may have altered some of our assessments, as gender-specific alterations cannot be excluded. Previous studies have found that, compared to females, male mice have increased neutrophil influx and TNFα production during LPS-induced airway inflammation using a higher dose of LPS (50 μg) [164]. Due to low n numbers, we were not able to analyze by gender to rule out these effects in the present experiments. Future studies should expand on our present findings, for example, concerning metabolic pathways, assessing neutrophils in a more dynamic manner, as well as testing additional doses of LPS and the effects of gender. Moreover, different or more severe models of lung inflammation [165] such as a multi-hit model consisting of the intratracheal application of hydrochlorid acid and LPS [80] or simulating pulmonary infection with *Streptococcus pneumoniae* [166] may be suitable approaches to expand on the results of our present study with the aim to more precisely mimic the human situation such as the time course and components/subtypes of inflammation. Nonetheless, our present data identify an essential role for RvE1, i.e., its precursors and receptors, and neutrophils in lung-to-brain communication and explore the impact of lung inflammation, genetic n-3 PUFA enrichment, and Rv receptor deficiency on the brain.

4. Materials and Methods

4.1. Animals

The animal experiments were approved by local authorities (Regierungspräsideum Giessen: GI 20/10 Nr. G46/2017) and were performed in accordance with international legislation and the German Animal Welfare Act. All mice were obtained through in-house breeding in groups and kept under specific pathogen-free conditions on a 12 h day/night cycle at 22 ± 1 °C, relative humidity at ~50% with free access to standard laboratory chow and water. WT (C57BL/6NCrl) original breeding pairs were obtained from Charles River Laboratories (Sulzfeld, Germany), Fat (B6.129P2-Tg(CAG-fat-1)1Jxk) mice [51] were originally provided by Prof. J. X. Kang (Boston, MA, USA), CR KO (B6.129-CMKLR1 tm1_MBarnes) were originally provided by Prof. M. Barns (Takeda Cambridge, Cambridge, UK), and LR KO (B6.129S4_Ltb4r1<tm1Adl>/J) were originally purchased from Jackson Laboratory (Bar Harbor, ME, USA). Crossbreeding of Fat mice with CR KO mice obtained FAT-CR KO mice (B6-Tg(CAG-fat-1)1Jxk-CMKLR1 <tm1>Mbarnes/Ott) while crossbreeding of Fat mice with LR KO mice generated FAT-LR KO mice (B6-Tg(CAG-fat-1)1Jxk-Ltb4r1<tm1Adl>/Ott), which was performed in the lab of Prof. Dr. med. K. Mayer (Giessen, Germany). Animals of either sex aged 12 to 16 weeks (weight, ~23–25 g) were used for experiments.

4.2. Experimental Protocol

Mice were anesthetized (0.25 mg/kg KGW Medetomidin/Domitor, 1 mg/mL, Elanco, Bad Homburg, Deutschland; diluted in saline, i.p.) and, in total, 50 μL (in 10–20–20 μL portions) LPS solution (10 μg Lipopolysaccharide, *Escherichia coli* O111:B4, in 50 μL normal saline/mouse) was instilled into the lung via a 22 G indwelling venous catheter through the trachea as previously described [167–169]. Intratracheal application of LPS mimics lung inflammation of the lower airways. Therefore, we have chosen such an application route to investigate lung–brain communication. Anesthesia was antagonized (0.25 mg/kg KGW Atipamezol/Antisedan, at 5 mg/mL, Zoetis, Berlin, Germany diluted in saline, s.c.) and after LPS application (0 h, 4 h, 24 h, 72 h, or 120 h), mice were sacrificed using the same anesthesia as stated above by blood withdrawal and organs were collected.

4.3. Immunohistochemistry

4.3.1. General Protocol

To detect nuclear NF-IL6 immunoreactivity and neutrophil recruitment into the mouse brain, frozen brain sections were incubated with antibodies specific for NF-IL6 or MPO. On the day of the experiment, the brain sections were fixed in 100% ethanol (EtOH; Sigma-Aldrich, Munich, Germany) at −20 °C for 15 min then briefly dried with a hair dryer for approximately 5 s followed by three successive washes with PBS (only for sections stained with rabbit MPO and intercellular adhesion molecule 1 (ICAM1)). Other slides were air-dried at room temperature (rt) for 10 min. Next, sections were immersed in

2% paraformaldehyde (Sigma-Aldrich, Munich, Germany) diluted in PBS for 10 min at rt. After three successive washes with PBS, a blocking solution consisting of PBS, 10% normal donkey serum (NDS; Biozol Diagnostica, Eching, Germany), and 0.01% Triton X-100 (Sigma-Aldrich, Munich, Germany) was applied to the sections for 1 h at rt. Details of the primary antibody concentrations and combinations are shown in Supplementary Table S6. The sections were incubated with the primary antibodies overnight at 4 °C, followed by three successive washes with PBS, and visualization was achieved with Cy3- or Alexa-488-conjugated IgG also listed in Supplementary Table S6. Sections were counterstained with the nuclear 4.6-diamidin-2-phenylindol (DAPI) stain (MoBiTec, Göttingen, Germany) for 10 min using a 1:5000 dilution in PBS to clearly visualize the surrounding tissue and to demonstrate localization of NF-IL6 immunoreactivity. After a final three washes with PBS, a glass coverslip was used to cover the sections using glycerol/PBS solution (Citifluor, London, UK). The sections were stored at 4 °C until microscopic analysis was performed.

4.3.2. NF-IL6

A rabbit antibody was used to detect nuclear NF-IL6 signals in the mouse brain in combination with von Willebrand factor (VWF) or MPO (see also Supplementary Table S6). Immunoreactivity in the BIF or OVLT was assessed per 20× field of view by semi-quantitative evaluation using a scale from 0 to 4, as previously performed [65].

4.3.3. MPO

A rabbit or goat antibody was used to detect the neutrophil marker MPO in combination with ICAM1 or NF-IL6 (see also Supplementary Table S6). A component of myeloid cell granules MPO is commonly used in neutrophil staining [170]. Neutrophil recruitment into the BIF or OVLT was assessed per 10× field of view; all the neutrophils within the image were counted.

4.3.4. GFAP

A guinea pig antibody was used to detect immunofluorescence of GFAP in the paraventricular nucleus (PVN) and evaluate the percentage of astrocytes area per image unit. To detect GFAP immunoreactivity in the mouse brain, frozen brain sections were incubated with a GFAP-specific antibody (see Supplementary Table S6). On the day of the experiment, the brain sections were immersed in 4% paraformaldehyde (Sigma-Aldrich, Munich, Germany) diluted in Tris-buffered saline (TBS) (Carl Roth GmbH + Co. KG, Karlsruhe, Germany) for 10 min at rt. After two successive washes with TBS, permeability was increased by applying 0.25% Triton X-100 (SERVA Electrophoresis GmbH, Heidelberg, Germany) diluted in PBS for 10 min at rt. Following three successive washes with TBS, a blocking solution consisting of TBS and 20% horse serum (Thermo Fisher Scientific Inc., Waltham, MA, USA) was applied to the sections for 1 h at rt. The primary GFAP antibody (see details in Supplementary Table S6) was prepared in a solution consisting of TBS, 5% horse serum, and 0.25% Triton X-100. The sections were incubated with the primary antibody overnight at 4 °C, followed by three successive washes with TBS, and visualization was achieved with Alexa 647-conjugated IgG (see details in Supplementary Table S6). Following three successive washes, the sections were counter-stained with Dapi (Carl Roth GmbH + Co. KG, Karlsruhe, Germany) for 5 min using a 1:500 dilution in TBS to visualize the surrounding tissue. After two final washes with TBS, a glass coverslip was used to cover the sections using ROTI®Mount, an aqueous mounting medium (Carl Roth GmbH+ Co. KG, Karlsruhe, Germany). The sections were stored at 4 °C until microscopic analysis was performed.

4.3.5. Microscopic Analysis

Sections stained with NF-IL6 and MPO were analyzed using a conventional light/fluorescent Olympus BX50 microscope (Olympus Optical, Hamburg, Germany) equipped with the appropriate filter sets to detect the fluorescent conjugates used: Cy3, Alexa Fluor 488, and DAPI. Images were collected with a black-and-white Spot Insight camera (Diagnostic Instruments, Visitron Systems, Puchheim, Germany) and edited using Metamorph software (Version 7.7.5.0, Visitron Systems GmbH, Puchheim, Germany) to combine individual images into the RGB color figures. The brightness and contrast of collected images were adjusted uniformly within each experiment using Adobe Photoshop (Adobe System Inc., San Jose, CA, USA) for better visualization. Between 2 and 13 sections were imaged per animal with 1–5 animals per group used for analyses. In a few instances, due to damage to the brain or an insufficient number of slides, only one section could be collected for a structure. Additionally, as multiple brain slices were present per slide, occasionally mice have >5 sections per animal as a result of trying to obtain representative structures for other animals. Overall, some of these data were only used for descriptive purposes as statistical analyses were not possible due to low n-numbers.

The immunofluorescence of GFAP was evaluated on snapshots taken with a Zeiss-Axio Observer 7 ACR immunofluorescence microscope (Carl Zeiss Microscopy Deutschland GmbH, Oberkochen, Germany). With the Apotome.2 module, several planes could be joined together in order to form a three-dimensional network, which had previously been recorded with the Colibri lights and installed incident light fluorescence. Between 1 and 3 sections were imaged per animal with 2–5 animals per group used for analyses.

4.4. Protein Carbonyl Content

The oxidation of protein results in the production of stable carbonyl groups, which are used as a measure of oxidative injury. Carbonyl content was determined with the Protein Carbonyl Content Assay Kit (Sigma MAK094, Merck KGaA, Darmstadt, Germany) according to the manufacturer's protocol. In brief, hypothalamic tissue samples (20.0 $\pm$ 7.9 mg) were dissolved in dH2O, homogenized with a T10 basic Ultra-Turrax (IKA-Werke, Staufen, Germany), and sonicated for 10 min in a water bath at rt. Afterwards, the homogenates were centrifuged for 15 min at 13,000$\times$ g and 4 °C resulting in the protein-comprising supernatants (Heraeus Fresco 17; Thermo Fisher Scientific, Darmstadt, Germany). Next, the samples were treated with 1% streptozocin followed by successive incubation with 2,4-dinitrophenylhydrazine (DNPH), TCA, acetone, and guanidine. Absorbance was measured at 355 nm (FLUOstar OPTIMA reader, BMG Labtech, Ortenberg, Germany). To determine the amount of protein per sample, a Pierce BCA Protein Assay (Thermo Fisher Scientific, Darmstadt, Germany) was performed with 2 µL of the remaining solution.

The carbonyl content was calculated as follows:

$$\mathrm{CP(nmol\ carbonyl/mg\ protein)} = \frac{\left(\frac{A_{355}}{6.364} \times 95\right)}{\mathrm{P}} \times 1000$$

where 6.364 = extinction coefficient; 95 = total volume in well (µL); P = amount of protein in 95 µL sample; 1000 = conversion factor (µg to mg).

4.5. Protein Analysis

The pellets left over from the homogenization of the hypothalamic tissue were lysed in 0.25 M D-mannitol, 0.05 M Tris base, 1 mM EDTA, 1 mM EGTA, 1 mM DTT, and 1% Triton X-100 supplemented with protease and phosphatase inhibitor cocktail tablets (Roche Diagnostics, Mannheim, Germany). Total protein amounts were determined using the Pierce BCA Protein Assay Kit (Thermo Fisher Scientific, Darmstadt, Germany). Subsequently, 50 µg of protein in sample buffer (60 mM Tris HCl, 2% SDS, 10% glycerol, 5% β-mercaptoethanol, and 0.01% bromophenol blue) were loaded on 10% SDS-polyacrylamide gels. After gel electrophoresis and protein transfer to a polyvinylidine difluoride (PVDF) membrane (Roche Diagnostics, Mannheim, Germany), the membranes were incubated with the following antibodies: Vinculin (V9505, Mouse, 1:20,000, 116 kDa; Sigma, Munich, Germany), Catalase (sc-271803, Mouse, 1:10,000, 64 kDa; Santa Cruz, Heidelberg, Germany) and SOD-2 (NBP2-20535, Rabbit, 1:500, 25 kDa; Novus, Wiesbaden, Germany). Protein detection was realized using peroxidase-labeled secondary anti-mouse and anti-rabbit antibodies (1:1000, Vector Laboratories, Burlingame, CA, USA) and luminol-based HRP-Juice Plus (PJK GmbH, Kleinblittersdorf, Germany). The resulting chemiluminescence was imaged with a ChemiDoc XRS system (Bio-Rad Laboratories, Hercules, CA, USA). Densitometric protein quantification was performed using the Bio-Rad Image Lab Software 6.1 (Bio-Rad Laboratories, Hercules, CA, USA) according to the density of the bands (Vinculin 116 kDa, Catalase 64 kDa, SOD-2 25 kDa) in relation to WT 0 h group and normalized to Vinculin (loading control).

4.6. Cytokine Measurements

Lung and liver tissue of mice were homogenized in protease inhibitor cOmplete™ Mini (Roche, Penzberg, Germany) with Precellys 24 tissue homogenizer (Bertin Technologies SAS, Montigny-le-Bretonneux, France). After centrifugation (4 °C, 10 min, 14,000 rpm), the supernatant of these samples was used to perform the assay. The amount of protein in each sample was determined using DC Protein Assay (Bio-Rad Laboratories GmbH; Feldkirchen, Germany). Custom-made mouse magnetic bead-based multiplex assay obtained from R&D (R&D Systems, Minneapolis, MN, USA) were used to analyze levels of selected inflammatory mediators in lungs (CCL5/RANTES, CXCL1, CXCL5/LIX, GM-CSF, IL-1β, IL-6, IL-10, IL-17, TNFα; L144530) and liver (CCL5/RANTES, CXCL1, IL-1β, IL-6, IL-10, IL-17, CXCL5/LIX; L140916). To produce a comprehensive profile of cytokine levels in the tissue of the lung and liver, a luminex assay was again utilized. The cytokines measured in both tissue homogenates included CCL5/RANTES, CXCL1, CXCL5/LIX, IL-1β, IL-17, IL-10, and IL-6 while GM-CSF and TNFα were only measured in the lung. Together, these immune mediators are markers for inflammation, both initiate and modulate the sickness response and contribute to neutrophil recruitment and activation [80,101–104,171–173]. Through this analysis, it was possible to compare differences in inflammation in the infected lung tissue as well as changes in the liver response to systemic inflammation between groups [174]. The mouse magnetic bead-based multiplex assays were conducted according to the manufacturer's protocol. The assay was performed with the Bio-Plex 200 instrument (Bio-Rad Laboratories GmbH; Feldkirchen, Germany) and analyzed with the Bio-Plex Manager Software 6.1 (Bio-Rad Laboratories GmbH; Feldkirchen, Germany).

4.7. Eicosanoid Extraction and LC-MS/MS-Based Mass Spectrometric Analysis

The Eicosanoids and their derivatives were extracted from the lung and brain, and lipids from the lung were analyzed as previously described by high-pressure liquid chromatography-tandem mass spectrometry (LC-MS/MS) [56]. Accordingly, for the liquid chromatography, the 1100 Series capillary LC unit (Agilent Technologies Deutschland, Waldbronn, Germany) was used for samples of the lung and brain. Subsequently, the Esquire 3000+ ion trap mass spectrometer (Bruker Corporation, Billerica, MA, USA) was used for lung samples and the Daltonik amaZon SL (Bruker Corporation, Billerica, MA, USA) was used to examine the samples of the brain (settings: Supplementary Table S7). The characteristic mass spectrometric extracted ion chromatograms (EIC) traces were used for qualification and quantification (Supplementary Table S8).

4.8. Real Time (RT)-PCR

Total RNA from the collected hypothalamic tissue was isolated with Trizol (Thermo Fisher, Waltham, MA, USA) according to the manufacturer's protocol. One microgram of total RNA was transcribed into DNA using 50 U murine leukemia virus (MULV) reverse transcriptase, 50 μM random hexamer, and 10 mM dNTP mix (Sigma-Aldrich, Munich, Germany) in a reaction volume of 20 μL. After the DNA was reverse transcribed quantitative real-time (RT) PCR was performed in duplicate using TaqMan Gene Expression Master Mix (Thermo Fisher, Waltham, MA, USA) and the respective primers. The cDNA quantities between different groups were normalized using the housekeeping gene glyceraldehyde 3- phosphate dehydrogenase (GAPDH; 4352339E-1009032, Applied Biosystems, Waltham, MA, USA) as a reference. Assay IDs for the analyzed genes are listed as follows: BDNF (Mm04230607_s1), Cat (Mm00437992_m1), CD163 (Mm00474091_m1), CD68 (Mm03047340_m1), CXCL1 (Mm04207460_m1), ELANE (Mm00469310_m1), GFAP (Mm01253033_m1), IL-1β (Mm00434228_m1), IL-1ra (Mm00446186_m1), IL-10 (Mm00439614_m1), IL-6 (Mm00446190_m1), mPGES (Mm00452105_m1), NF-IL6 (Mm00843434_s1), NfκBia (Mm00477798_m1), NOS2 (Mm00440502_m1), PGC-1α (Mm01208835_m1), SOCS3 (Mm00545913_s1), SOD1 (Mm01344232_g1), SOD2 (Mm01313000_m1), and TFAM (Mm00447485_m1). All primers were acquired from Thermo Fisher (Applied Biosystems) unless otherwise stated.

4.9. Data Analysis and Statistics

The fold change of relative quantity values normalized to WT-Norm 0 h controls was chosen for direct statistical analysis of gene mRNA expression. Parameters measured from tissue homogenate supernatants (cytokines, lipid mediators) were normalized to the protein amount of each sample before the fold change was normalized to WT-Norm 0 h and statistical analyses were used to account for differences in the protein amount caused by environmental factors between days of preparation. Cytokines and lipid mediators (normalized to tissue weight) were quantified by regression analysis of standard curve functions. In case of a not detectable cytokine or lipid mediator showing smaller quantities in the sample than the measurable range, these sample values were set to 0. In general, groups were analyzed by two-way ANOVA with the factors genetic n-3 PUFA enrichment (WT background, Fat background) and receptor deficiency (Norm, CR KO, LR KO). Main effects were reported, and Tukey post hoc testing was performed for single-group comparison. $p < 0.05$ was applied for statistical significance and mean ± SEM is displayed in the figures.

4.9.1. Evaluation of Cytokine Measurements

The concentrations of cytokines in the lung and liver tissue were determined according to the manufacturer's instructions for the assay using the standards provided. WT and Fat groups were compared to each other within each group: Norm, CR KO, and LR KO. Additionally, the WT and Fat mice from CR KO and LR KO groups were compared to the respective Norm groups but not to each other. Each cytokine was analyzed by two-way ANOVA with the factors condition (Rv receptor deficiency) and treatment (n-3 PUFA enrichment) and a Tukey post hoc test. Outlier analyses were performed using the ROUT method [175] and values were excluded (Liver-IL-10 at 24 h: 1 WT-CR KO/CXCL1 at 24 h: 1 WT-Norm, 1 Fat-CR KO; 72 h: 1 WT-Norm/IL-17 at 24 h: 1 WT-CR KO). All statistical analyses were performed using the GraphPad Prism (Version 7; GraphPad Software, San Diego, CA, USA).

4.9.2. Evaluation of RT-PCR

The relative quantification (RQ) was determined based on the cycle threshold (ct) defined when amplification exceeds that of the background signal. Once the ct has been reached, the cDNA amplification becomes exponential and will double with each cycle and can be applied for relative quantification. The RQ values were used to compare target genes from the q-RT-PCR analysis. RQ values were normalized to the lowest expression in the control group (0 h WT-Norm) for each target gene and assigned a value of 1. All other values were shown as a multiple of the expression [176]. The relative expression from the control sample (given a value of 1) from the 0 h time point for each group was transformed and reported as fold change normalized to the WT-Norm control mice from the 0 h time point. WT and Fat groups were compared to each other within each group: Norm, CR KO, and LR KO. Additionally, the WT and Fat mice from CR KO and LR KO groups were compared to the respective Norm group but not to each other. RT-PCR for target genes was analyzed for all groups in the experiment on one day to limit variation. Analyses were performed using a two-way ANOVA with the factors condition (Rv receptor deficiency) and treatment (n-3 PUFA enrichment) and a Tukey post hoc test for both. Outlier analyses were performed using the ROUT method [175] and values were excluded (SOD1 at 72 h: 1 WT-Norm/Elane at 0 h: 1 Fat-Norm; 24 h: 1 Fat-CR KO, 1 Fat-LR KO; 72 h 1 Fat-CR KO/IL-10 at 24 h: 1 Fat-CR KO/IL-1β at 0 h: 1 WT-Norm/IL-1ra at 0 h: 1 WT-Norm, 1 WT-CR KO/NF-IL6 at 0 h: 1 Fat-Norm/mPGES at 72 h: 1 Fat-Norm). All statistical analyses were performed using the GraphPad Prism (Version 7; GraphPad Software, San Diego, CA, USA). For samples where a value could not be determined they were removed from the data set (Elane at 0 h: 1 WT-Norm, 2 Fat-LR KO; 24 h: 1 Fat-Norm; 72 h: 1 WT-Norm/IL-10 at 0 h: 3 WT-Norm, 3 Fat-Norm, 1 WT-CR KO, 2 Fat-CR KO, 1 Fat-LR KO; 24 h: 4 WT-Norm, 1 Fat-Norm, 1 WT-CR KO, 2 WT-LR KO, 1 Fat-LR KO; 72 h: 1 WT-Norm, 1 Fat-Norm, 2 WT-CR KO, 1 Fat-CR KO, 2 WT-LR KO, 1 Fat-LR KO).

4.9.3. Evaluation of Western Blot Protein Analysis

Protein levels were obtained by densitometric quantification of the Western blot data and normalization to Vinculin (loading control). The data are presented as a fold of the control group (Norm WT 0h). WT and Fat groups were compared to each other within each group: Norm and CR KO (n = 4–5). Multiple comparisons were performed by ANOVA and Tukey post hoc test. The statistical analyses were carried out using GraphPad Prism (Version 7; GraphPad Software, La Jolla, CA, USA).

4.9.4. Evaluation of Immunohistochemistry

In all instances where immunofluorescence was used, brain sections were exposed to the optimal wavelength of light and an image was collected. WT and Fat groups were compared to each other within each group: Norm, CR KO, and LR KO. Additionally, the WT and Fat mice from CR KO and LR KO groups were compared to the respective Norm group but not to each other. Analyses were performed using a two-way ANOVA with the factors condition (Rv receptor deficiency) and treatment (n-3 PUFA enrichment) and a Tukey post-hoc test for both. Outlier analyses were performed using the ROUT method [175]. All statistical analyses were performed using the GraphPad Prism (Version 7; GraphPad Software, San Diego, CA, USA).

NF-IL6 and MPO

For immunofluorescence analyses of NF-IL6 and MPO, either quantitative or semi-quantitative measurements of fluorescent signals were made. When quantitative evaluation was used, i.e., for MPO analysis, the nuclear signals were counted. Groups stained for MPO had n = 3 mice unless stated otherwise here, 'n' are reported as follows as OVLT, BIF: 0 h Fat-Norm = 3, 5/4 h WT-Norm = 2, 2; Fat-Norm = 5, 5; WT-LR KO = 4, 4/24 h WT-Norm = 3, 4; Fat-Norm = 5, 5; WT-CR KO = 4, 4; Fat-CR KO = 4, 4; WT-LR KO = 5, 5; Fat-LR KO = 4, 4/72 h WT-Norm = 3, 4; Fat-Norm = 5, 5; WT-CR KO = 4, 4; Fat-CR KO = 3, 4; WT-LR KO = 5, 5; Fat-LR KO = 3, 4/120 h Fat-Norm = 4, 5; Fat-CR KO = 2, 2; WT-LR KO = 3, 4; Fat-LR KO = 4, 4.

During semi-quantitative evaluation, i.e., NF-IL6 analysis, the level of nuclear signals present was scored for each section on a scale from 0 to 4: 0 = none, 1 = single or very few, 2 = low density, 3 = moderate density, 4 = high density; $\frac{1}{4}$ scores were assigned when slightly more or less than average signals were present but not enough to move up or down an entire level, i.e., 1.25 or 1.75 where a score of 1.25 has slightly more signal than 1 and 1.75 has slightly less signal than 2. Groups stained for NF-IL6 had n = 3 mice unless stated otherwise here, 'n' are reported as follows as OVLT, BIF: 0 h Fat-Norm = 4, 5/4 h WT-Norm = 1, 2; Fat-Norm = 5, 4/24 h WT-Norm = 4, 4; Fat-Norm = 5, 4; WT-CR KO = 4, 4; Fat-CR KO = 4, 4; Fat-LR KO = 4, 4/72 h WT-Norm = 4, 4; Fat-Norm = 5, 4; WT-CR KO = 4, 4; Fat-CR KO = 3, 4/120 h Fat-Norm = 3, 4; WT-CR KO = 2, 3; Fat-CR KO = 2, 2; WT-LR KO = 3, 4.

GFAP

For immunofluorescence analysis of GFAP, the evaluation of the images was carried out using the program ZEN 3.6 (Carl Zeiss Microscopy GmbH, Oberkochen, Germany).

The measurements were carried out with the module "Intellesis Segmentation" and the percentage of astrocytes area per image unit was determined.

4.9.5. Evaluation of LC-MS/MS

The mass-spectrometric raw data were quantified as previously described [56]. All statistical analyses were performed using the GraphPad Prism (Version 7; GraphPad Software, San Diego, CA, USA).

5. Conclusions

In our hands, even with some of our present results being descriptive, we revealed that LPS-induced ARDS increased inflammatory signaling in the lung and the hypothalamus where low humoral signaling and immune cell trafficking to the brain potentially contributed to immune-to-brain communication (Figure 10). A deficiency in RvE1 receptors as well as enhanced production of n-3 PUFAs was able to sufficiently alter the profile of several inflammatory mediators, modify GFAP immunoreactivity (for CR deficiency), and alter lipid mediators in the brain. In conclusion, we have presented evidence that genetic enrichment of n-3 PUFAs and RvE1 receptor deficiencies can alter the lipid profile in the brain during lung inflammation. In particular, acting through CR, RvE1 may be able to modulate communication to some degree through the humoral pathway and neutrophil recruitment. Moreover, we suggest a role for n-3 PUFAs to modulate neutrophil recruitment to the brain during lung inflammation. Overall, n-3 PUFAs and, in particular, SPMs, i.e., NPD1/PDX or Mar1 and SPM precursors such as 18-HEPE, 17(S)-HDHA, or 14(S)-HDHA, respectively, may represent good targets for future investigations to modulate inflammation in individuals suffering from or at risk for developing brain inflammation as a result of ARDS.

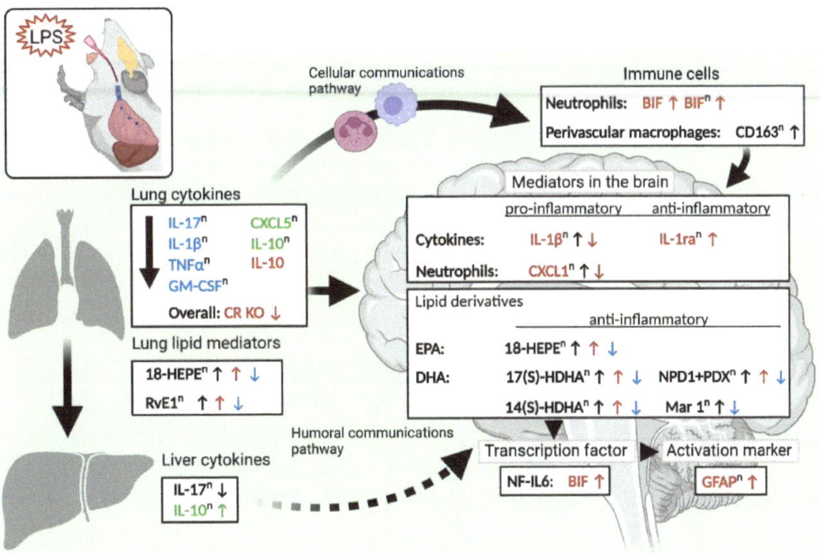

Figure 10. Influence of n-3 PUFAs and RvE1 receptor knockouts on lung and brain inflammation during ARDS. Schematic summary of the main findings: The inflammatory response in the lung, liver, and hypothalamus following intratracheal instillation with lipopolysaccharide (LPS, 10 μg) induced ARDS includes the expression of inflammatory mediators, immune cells, lipid derivatives, transcription factors, and activation markers at 0 h, 24 h, and 72 h p.i. with LPS. Further analyses of neutrophil trafficking and nuclear factor (NF)-IL6 immunoreactivity in two brain structures, namely, the bifurcation (BIF) and vascular organ of lamina terminalis (OVLT) were carried out at the same time points as well as at 4 h and 120 h p.i. Changes are shown compared to unmodified controls (Norm) bred on a wild-type

(WT) background and were color-coded for effects in resolvin (Rv) receptor-deficient groups: chemerin receptor 23 (CR) KO (red), leukotriene B4 receptor (LR) KO (blue) or both KOs (CR KO and LR KO; green). Comparisons between Norm and KO mice on a transgenic n-3 synthesizing *fat-1* (Fat) background are identified with an 'n' and include comparing Fat-Norm to WT-Norm (black). Arrows in the corresponding colors signify an increase (up arrow ↑) or decrease (down arrow ↓) for that genotype. A large arrow within the lung cytokine box indicates that all cytokines were decreased in their respective comparisons. Mediators in the brain were primarily altered in CR KO groups but black arrows also indicate effects when comparing Fat-Norm to WT-Norm. Lipid mediators are written in black due to extensive effects between groups and arrows indicate genotype-specific alterations. Larger arrows outside of text boxes show a pathway for immune-to-brain communication, when they are dashed it indicates a weak source of communication with the brain. Arrow heads (▼) were used to outline possible pathways of activation in the brain. Abbreviations: interleukin (IL), granulocyte-macrophage colony-stimulating factor (GM-CSF), chemokine (C-X-C motif) ligand (CXCL), cluster of differentiation (CD), hydroxy eicosapentaenoic acid (HEPE), hydroxy docosahexaenoic acid (HDHA), neuroprotection D1 (NPD1), protectin DX (PDX), maresin 1 (Mar1) and glial fibrillary acidic protein (GFAP).

Supplementary Materials: The following supporting information can be downloaded at: https://www.mdpi.com/article/10.3390/ijms241713524/s1.

Author Contributions: Conceptualization, C.H., J.S., M.H. and C.R.; methodology, J.S., J.H., S.R., S.K., S.M., F.J.P., J.B., C.C. and C.R.; software, J.H., J.S., S.K., J.B. and C.R.; validation, J.H., J.S., J.B. and C.R.; formal analysis, J.H. and J.S.; investigation, J.S., J.H., S.M., S.K., J.W., H.K. and C.R.; resources, M.H., J.X.K., K.M., N.S. and C.R.; data curation, J.H., J.S. and C.R.; writing—original draft preparation, J.H., J.S. and C.R.; writing—review and editing, all authors; visualization, J.H., J.S., N.S. and C.R.; supervision, N.S., M.H., C.H. and C.R.; project administration, J.S., J.H., N.S. and C.R.; funding acquisition, C.C., C.H., N.S., M.H. and C.R. All authors have read and agreed to the published version of the manuscript.

Funding: This research was partially funded by the Flexifunds "Lung–brain axis in health and disease" by the Research Campus Mid-Hessen (FCMH): 2021_3_1_2_Herden, by the EU Joint Programme—Neurodegenerative Disease Research (JPND): SOLID JPND2021-650-233; the Federal Ministry of Education and Research: 01ED2207, by the Volkswagen Foundation: Az. 9A216, and the APC was partially funded by Justus Liebig University.

Institutional Review Board Statement: The study was conducted according to the guidelines of the Declaration of Helsinki; the animal study protocol was reviewed by the Institutional Review Board (animal welfare officers) of Justus Liebig University Giessen (protocol code JLU number 863_GP and approved on 22 September 2017) by local authorities (Regierungspräsideum Giessen: GI 20/10 No. G46/2017).

Informed Consent Statement: Not applicable.

Data Availability Statement: The data presented in this study are contained within the article and supplementary material or are available upon reasonable request from the corresponding author.

Acknowledgments: We would like to thank D. Marks, J. Murgott, and B. Watzer for excellent technical assistance. Figure 10 was created with BioRender.com.

Conflicts of Interest: The authors declare no conflict of interest.

References

1. Bazinet, R.P.; Layé, S. Polyunsaturated fatty acids and their metabolites in brain function and disease. *Nat. Rev. Neurosci.* **2014**, *15*, 771–785. [CrossRef] [PubMed]
2. Layé, S.; Nadjar, A.; Joffre, C.; Bazinet, R.P. Anti-Inflammatory Effects of Omega-3 Fatty Acids in the Brain: Physiological Mechanisms and Relevance to Pharmacology. *Pharmacol. Rev.* **2018**, *70*, 12–38. [CrossRef] [PubMed]
3. Madore, C.; Leyrolle, Q.; Morel, L.; Rossitto, M.; Greenhalgh, A.D.; Delpech, J.C.; Martinat, M.; Bosch-Bouju, C.; Bourel, J.; Rani, B.; et al. Essential omega-3 fatty acids tune microglial phagocytosis of synaptic elements in the mouse developing brain. *Nat. Commun.* **2020**, *11*, 6133. [CrossRef] [PubMed]

4. Serhan, C.N.; Petasis, N.A. Resolvins and protectins in inflammation resolution. *Chem. Rev.* **2011**, *111*, 5922–5943. [CrossRef] [PubMed]
5. Serhan, C.N.; Levy, B.D. Resolvins in inflammation: Emergence of the pro-resolving superfamily of mediators. *J. Clin. Investig.* **2018**, *128*, 2657–2669. [CrossRef] [PubMed]
6. Qu, Q.; Xuan, W.; Fan, G.-H. Roles of resolvins in the resolution of acute inflammation. *Cell Biol. Int.* **2015**, *39*, 3–22. [CrossRef]
7. Mayer, K.; Sommer, N.; Hache, K.; Hecker, A.; Reiche, S.; Schneck, E.; Weissmann, N.; Seeger, W.; Hecker, M. Resolvin E1 Improves Mitochondrial Function in Human Alveolar Epithelial Cells during Severe Inflammation. *Lipids* **2019**, *54*, 53–65. [CrossRef]
8. Schwab, J.M.; Chiang, N.; Arita, M.; Serhan, C.N. Resolvin E1 and protectin D1 activate inflammation-resolution programmes. *Nature* **2007**, *447*, 869–874. [CrossRef]
9. Serhan, C.N.; Yang, R.; Martinod, K.; Kasuga, K.; Pillai, P.S.; Porter, T.F.; Oh, S.F.; Spite, M. Maresins: Novel macrophage mediators with potent antiinflammatory and proresolving actions. *J. Exp. Med.* **2009**, *206*, 15–23. [CrossRef]
10. Serhan, C.N.; Clish, C.B.; Brannon, J.; Colgan, S.P.; Chiang, N.; Gronert, K. Novel functional sets of lipid-derived mediators with antiinflammatory actions generated from omega-3 fatty acids via cyclooxygenase 2-nonsteroidal antiinflammatory drugs and transcellular processing. *J. Exp. Med.* **2000**, *192*, 1197–1204. [CrossRef]
11. Haworth, O.; Cernadas, M.; Yang, R.; Serhan, C.N.; Levy, B.D. Resolvin E1 regulates interleukin 23, interferon-γ and lipoxin A_4 to promote the resolution of allergic airway inflammation. *Nat. Immunol.* **2008**, *9*, 873–879. [CrossRef] [PubMed]
12. Arita, M.; Bianchini, F.; Aliberti, J.; Sher, A.; Chiang, N.; Hong, S.; Yang, R.; Petasis, N.A.; Serhan, C.N. Stereochemical assignment, antiinflammatory properties, and receptor for the omega-3 lipid mediator resolvin E1. *J. Exp. Med.* **2005**, *201*, 713–722. [CrossRef] [PubMed]
13. Arita, M.; Yoshida, M.; Hong, S.; Tjonahen, E.; Glickman, J.N.; Petasis, N.A.; Blumberg, R.S.; Serhan, C.N. Resolvin E1, an endogenous lipid mediator derived from omega-3 eicosapentaenoic acid, protects against 2,4,6-trinitrobenzene sulfonic acid-induced colitis. *Proc. Natl. Acad. Sci. USA* **2005**, *102*, 7671–7676. [CrossRef] [PubMed]
14. Hong, S.; Gronert, K.; Devchand, P.R.; Moussignac, R.; Serhan, C.N. Novel docosatrienes and 17S-resolvins generated from docosahexaenoic acid in murine brain, human blood, and glial cells. Autacoids in anti-inflammation. *J. Biol. Chem.* **2003**, *278*, 14677–14687. [CrossRef] [PubMed]
15. Krishnamoorthy, N.; Burkett, P.R.; Dalli, J.; Abdulnour, R.-E.E.; Colas, R.; Ramon, S.; Phipps, R.P.; Petasis, N.A.; Kuchroo, V.K.; Serhan, C.N.; et al. Cutting edge: Maresin-1 engages regulatory T cells to limit type 2 innate lymphoid cell activation and promote resolution of lung inflammation. *J. Immunol.* **2015**, *194*, 863–867. [CrossRef]
16. Serhan, C.N.; Dalli, J.; Karamnov, S.; Choi, A.; Park, C.; Xu, Z.; Ji, R.; Zhu, M.; Petasis, N.A. Macrophage proresolving mediator maresin 1 stimulates tissue regeneration and controls pain. *FASEB J.* **2012**, *26*, 1755–1765. [CrossRef]
17. Xie, W.; Wang, H.; Wang, L.; Yao, C.; Yuan, R.; Wu, Q. Resolvin D1 reduces deterioration of tight junction proteins by upregulating HO-1 in LPS-induced mice. *Lab. Investig.* **2013**, *93*, 991–1000. [CrossRef]
18. Aïd, S.; Vancassel, S.; Poumès-Ballihaut, C.; Chalon, S.; Guesnet, P.; Lavialle, M. Effect of a diet-induced n-3 PUFA depletion on cholinergic parameters in the rat hippocampus. *J. Lipid Res.* **2003**, *44*, 1545–1551. [CrossRef]
19. Serhan, C.N.; Fredman, G.; Yang, R.; Karamnov, S.; Belayev, L.S.; Bazan, N.G.; Zhu, M.; Winkler, J.W.; Petasis, N.A. Novel proresolving aspirin-triggered DHA pathway. *Chem. Biol.* **2011**, *18*, 976–987. [CrossRef]
20. Chalon, S. Omega-3 fatty acids and monoamine neurotransmission. *Prostaglandins Leukot. Essent. Fat. Acids* **2006**, *75*, 259–269. [CrossRef]
21. Arita, M.; Ohira, T.; Sun, Y.-P.; Elangovan, S.; Chiang, N.; Serhan, C.N. Resolvin E1 selectively interacts with leukotriene B4 receptor BLT1 and ChemR23 to regulate inflammation. *J. Immunol.* **2007**, *178*, 3912–3917. [CrossRef] [PubMed]
22. Campbell, E.L.; Louis, N.A.; Tomassetti, S.E.; Canny, G.O.; Arita, M.; Serhan, C.N.; Colgan, S.P. Resolvin E1 promotes mucosal surface clearance of neutrophils: A new paradigm for inflammatory resolution. *FASEB J.* **2007**, *21*, 3162–3170. [CrossRef] [PubMed]
23. Haworth, O.; Cernadas, M.; Levy, B.D. NK cells are effectors for resolvin E1 in the timely resolution of allergic airway inflammation. *J. Immunol.* **2011**, *186*, 6129–6135. [CrossRef]
24. Liu, G.; Gong, Y.; Zhang, R.; Piao, L.; Li, X.; Liu, Q.; Yan, S.; Shen, Y.; Guo, S.; Zhu, M.; et al. Resolvin E1 attenuates injury-induced vascular neointimal formation by inhibition of inflammatory responses and vascular smooth muscle cell migration. *FASEB J.* **2018**, *32*, 5413–5425. [CrossRef] [PubMed]
25. Deyama, S.; Shimoda, K.; Suzuki, H.; Ishikawa, Y.; Ishimura, K.; Fukuda, H.; Hitora-Imamura, N.; Ide, S.; Satoh, M.; Kaneda, K.; et al. Resolvin E1/E2 ameliorate lipopolysaccharide-induced depression-like behaviors via ChemR23. *Psychopharmacology* **2018**, *235*, 329–336. [CrossRef] [PubMed]
26. Luo, J.; Zhang, W.-Y.; Li, H.; Zhang, P.-H.; Tian, C.; Wu, C.-H.; Zhao, A.-N.; Chen, M.-L.; Guo, Y.-F.; Cho, Y.-C.; et al. Pro-Resolving Mediator Resolvin E1 Restores Alveolar Fluid Clearance in Acute Respiratory Distress Syndrome. *Shock* **2022**, *57*, 565–575. [CrossRef]
27. Gantz, I.; Konda, Y.; Yang, Y.K.; Miller, D.E.; Dierick, H.A.; Yamada, T. Molecular cloning of a novel receptor (CMKLR1) with homology to the chemotactic factor receptors. *Cytogenet. Cell Genet.* **1996**, *74*, 286–290. [CrossRef]
28. Samson, M.; Edinger, A.L.; Stordeur, P.; Rucker, J.; Verhasselt, V.; Sharron, M.; Govaerts, C.; Mollereau, C.; Vassart, G.; Doms, R.W.; et al. ChemR23, a putative chemoattractant receptor, is expressed in monocyte-derived dendritic cells and macrophages and is a coreceptor for SIV and some primary HIV-1 strains. *Eur. J. Immunol.* **1998**, *28*, 1689–1700. [CrossRef]

29. Goralski, K.B.; McCarthy, T.C.; Hanniman, E.A.; Zabel, B.A.; Butcher, E.C.; Parlee, S.D.; Muruganandan, S.; Sinal, C.J. Chemerin, a novel adipokine that regulates adipogenesis and adipocyte metabolism. *J. Biol. Chem.* **2007**, *282*, 28175–28188. [CrossRef]
30. Parolini, S.; Santoro, A.; Marcenaro, E.; Luini, W.; Massardi, L.; Facchetti, F.; Communi, D.; Parmentier, M.; Majorana, A.; Sironi, M.; et al. The role of chemerin in the colocalization of NK and dendritic cell subsets into inflamed tissues. *Blood* **2007**, *109*, 3625–3632. [CrossRef]
31. Kaur, J.; Adya, R.; Tan, B.K.; Chen, J.; Randeva, H.S. Identification of chemerin receptor (ChemR23) in human endothelial cells: Chemerin-induced endothelial angiogenesis. *Biochem. Biophys. Res. Commun.* **2010**, *391*, 1762–1768. [CrossRef] [PubMed]
32. Lämmermann, T.; Afonso, P.V.; Angermann, B.R.; Wang, J.M.; Kastenmüller, W.; Parent, C.A.; Germain, R.N. Neutrophil swarms require LTB4 and integrins at sites of cell death in vivo. *Nature* **2013**, *498*, 371–375. [CrossRef] [PubMed]
33. Hiram, R.; Rizcallah, E.; Marouan, S.; Sirois, C.; Sirois, M.; Morin, C.; Fortin, S.; Rousseau, E. Resolvin E1 normalizes contractility, Ca^{2+} sensitivity and smooth muscle cell migration rate in TNF-α- and IL-6-pretreated human pulmonary arteries. *Am. J. Physiol. Lung Cell. Mol. Physiol.* **2015**, *309*, L776–L788. [CrossRef] [PubMed]
34. Liu, T.; Zhang, L.; Joo, D.; Sun, S.-C. NF-κB signaling in inflammation. *Signal Transduct. Target. Ther.* **2017**, *2*, 17023. [CrossRef] [PubMed]
35. Oh, S.F.; Dona, M.; Fredman, G.; Krishnamoorthy, S.; Irimia, D.; Serhan, C.N. Resolvin E2 formation and impact in inflammation resolution. *J. Immunol.* **2012**, *188*, 4527–4534. [CrossRef]
36. Lukiw, W.J.; Cui, J.-G.; Marcheselli, V.L.; Bodker, M.; Botkjaer, A.; Gotlinger, K.; Serhan, C.N.; Bazan, N.G. A role for docosahexaenoic acid-derived neuroprotectin D1 in neural cell survival and Alzheimer disease. *J. Clin. Investig.* **2005**, *115*, 2774–2783. [CrossRef]
37. Rey, C.; Nadjar, A.; Buaud, B.; Vaysse, C.; Aubert, A.; Pallet, V.; Layé, S.; Joffre, C. Resolvin D1 and E1 promote resolution of inflammation in microglial cells in vitro. *Brain Behav. Immun.* **2016**, *55*, 249–259. [CrossRef]
38. Schaefer, M.B.; Ott, J.; Mohr, A.; Bi, M.H.; Grosz, A.; Weissmann, N.; Ishii, S.; Grimminger, F.; Seeger, W.; Mayer, K. Immunomodulation by n-3- versus n-6-rich lipid emulsions in murine acute lung injury--role of platelet-activating factor receptor. *Crit. Care Med.* **2007**, *35*, 544–554. [CrossRef]
39. Mayer, K.; Kiessling, A.; Ott, J.; Schaefer, M.B.; Hecker, M.; Henneke, I.; Schulz, R.; Günther, A.; Wang, J.; Wu, L.; et al. Acute lung injury is reduced in fat-1 mice endogenously synthesizing n-3 fatty acids. *Am. J. Respir. Crit. Care Med.* **2009**, *179*, 474–483. [CrossRef]
40. Seki, H.; Fukunaga, K.; Arita, M.; Arai, H.; Nakanishi, H.; Taguchi, R.; Miyasho, T.; Takamiya, R.; Asano, K.; Ishizaka, A.; et al. The anti-inflammatory and proresolving mediator resolvin E1 protects mice from bacterial pneumonia and acute lung injury. *J. Immunol.* **2010**, *184*, 836–843. [CrossRef]
41. Ferguson, N.D.; Fan, E.; Camporota, L.; Antonelli, M.; Anzueto, A.; Beale, R.; Brochard, L.; Brower, R.; Esteban, A.; Gattinoni, L.; et al. The Berlin definition of ARDS: An expanded rationale, justification, and supplementary material. *Intensive Care Med.* **2012**, *38*, 1573–1582. [CrossRef]
42. Heuer, J.F.; Sauter, P.; Pelosi, P.; Herrmann, P.; Brück, W.; Perske, C.; Schöndube, F.; Crozier, T.A.; Bleckmann, A.; Beißbarth, T.; et al. Effects of pulmonary acid aspiration on the lungs and extra-pulmonary organs: A randomized study in pigs. *Crit. Care* **2012**, *16*, R35. [CrossRef]
43. Steimback, P.W.; Oliveira, G.P.; Rzezinski, A.F.; Silva, P.L.; Garcia, C.S.N.B.; Rangel, G.; Morales, M.M.; Lapa E Silva, J.R.; Capelozzi, V.L.; Pelosi, P.; et al. Effects of frequency and inspiratory plateau pressure during recruitment manoeuvres on lung and distal organs in acute lung injury. *Intensive Care Med.* **2009**, *35*, 1120–1128. [CrossRef]
44. Manouchehrian, O.; Ramos, M.; Bachiller, S.; Lundgaard, I.; Deierborg, T. Acute systemic LPS-exposure impairs perivascular CSF distribution in mice. *J. Neuroinflamm.* **2021**, *18*, 34. [CrossRef] [PubMed]
45. Boitsova, E.B.; Morgun, A.V.; Osipova, E.D.; Pozhilenkova, E.A.; Martinova, G.P.; Frolova, O.V.; Olovannikova, R.Y.; Tohidpour, A.; Gorina, Y.V.; Panina, Y.A.; et al. The inhibitory effect of LPS on the expression of GPR81 lactate receptor in blood-brain barrier model in vitro. *J. Neuroinflamm.* **2018**, *15*, 196. [CrossRef]
46. Hopkins, R.O.; Suchyta, M.R.; Snow, G.L.; Jephson, A.; Weaver, L.K.; Orme, J.F. Blood glucose dysregulation and cognitive outcome in ARDS survivors. *Brain Inj.* **2010**, *24*, 1478–1484. [CrossRef] [PubMed]
47. Fries, M.; Bickenbach, J.; Henzler, D.; Beckers, S.; Dembinski, R.; Sellhaus, B.; Rossaint, R.; Kuhlen, R. S-100 protein and neurohistopathologic changes in a porcine model of acute lung injury. *Anesthesiology* **2005**, *102*, 761–767. [CrossRef]
48. Hopkins, R.O.; Weaver, L.K.; Chan, K.J.; Orme, J.F. Quality of life, emotional, and cognitive function following acute respiratory distress syndrome. *J. Int. Neuropsychol. Soc. JINS* **2004**, *10*, 1005–1017. [CrossRef]
49. Hopkins, R.O.; Gale, S.D.; Weaver, L.K. Brain atrophy and cognitive impairment in survivors of Acute Respiratory Distress Syndrome. *Brain Inj.* **2006**, *20*, 263–271. [CrossRef] [PubMed]
50. Bondue, B.; Vosters, O.; de Nadai, P.; Glineur, S.; de Henau, O.; Luangsay, S.; van Gool, F.; Communi, D.; de Vuyst, P.; Desmecht, D.; et al. ChemR23 dampens lung inflammation and enhances anti-viral immunity in a mouse model of acute viral pneumonia. *PLoS Pathog.* **2011**, *7*, e1002358. [CrossRef]
51. Kang, J.X.; Wang, J.; Wu, L.; Kang, Z.B. Transgenic mice: Fat-1 mice convert n-6 to n-3 fatty acids. *Nature* **2004**, *427*, 504. [CrossRef] [PubMed]
52. Rummel, C. Inflammatory transcription factors as activation markers and functional readouts in immune-to-brain communication. *Brain Behav. Immun.* **2016**, *54*, 1–14. [CrossRef]

53. Roth, J.; Harré, E.-M.; Rummel, C.; Gerstberger, R.; Hübschle, T. Signaling the brain in systemic inflammation: Role of sensory circumventricular organs. *Front. Biosci.(Landmark Ed.)* **2004**, *9*, 290–300. [CrossRef] [PubMed]
54. Harden, L.M.; Rummel, C.; Luheshi, G.N.; Poole, S.; Gerstberger, R.; Roth, J. Interleukin-10 modulates the synthesis of inflammatory mediators in the sensory circumventricular organs: Implications for the regulation of fever and sickness behaviors. *J. Neuroinflamm.* **2013**, *10*, 22. [CrossRef] [PubMed]
55. Pflieger, F.J.; Hernandez, J.; Schweighöfer, H.; Herden, C.; Rosengarten, B.; Rummel, C. The role of neutrophil granulocytes in immune-to-brain communication. *Temperature* **2018**, *5*, 296–307. [CrossRef] [PubMed]
56. Kiss, L.; Röder, Y.; Bier, J.; Weissmann, N.; Seeger, W.; Grimminger, F. Direct eicosanoid profiling of the hypoxic lung by comprehensive analysis via capillary liquid chromatography with dual online photodiode-array and tandem mass-spectrometric detection. *Anal. Bioanal. Chem.* **2008**, *390*, 697–714. [CrossRef]
57. Domscheit, H.; Hegeman, M.A.; Carvalho, N.; Spieth, P.M. Molecular Dynamics of Lipopolysaccharide-Induced Lung Injury in Rodents. *Front. Physiol.* **2020**, *11*, 36. [CrossRef]
58. Liu, S.; Su, X.; Pan, P.; Zhang, L.; Hu, Y.; Tan, H.; Wu, D.; Liu, B.; Li, H.; Li, H.; et al. Neutrophil extracellular traps are indirectly triggered by lipopolysaccharide and contribute to acute lung injury. *Sci. Rep.* **2016**, *6*, 37252. [CrossRef]
59. Kelly, D.P.; Scarpulla, R.C. Transcriptional regulatory circuits controlling mitochondrial biogenesis and function. *Genes Dev.* **2004**, *18*, 357–368. [CrossRef]
60. Wu, Z.; Puigserver, P.; Andersson, U.; Zhang, C.; Adelmant, G.; Mootha, V.; Troy, A.; Cinti, S.; Lowell, B.; Scarpulla, R.C.; et al. Mechanisms controlling mitochondrial biogenesis and respiration through the thermogenic coactivator PGC-1. *Cell* **1999**, *98*, 115–124. [CrossRef]
61. Voynow, J.A.; Shinbashi, M. Neutrophil Elastase and Chronic Lung Disease. *Biomolecules* **2021**, *11*, 1065. [CrossRef] [PubMed]
62. Corraliza, I. Recruiting specialized macrophages across the borders to restore brain functions. *Front. Cell. Neurosci.* **2014**, *8*, 262. [CrossRef] [PubMed]
63. Galea, I.; Felton, L.M.; Waters, S.; van Rooijen, N.; Perry, V.H.; Newman, T.A. Immune-to-brain signalling: The role of cerebral CD163-positive macrophages. *Neurosci. Lett.* **2008**, *448*, 41–46. [CrossRef] [PubMed]
64. Graeber, M.B.; Streit, W.J.; Kreutzberg, G.W. Identity of ED2-positive perivascular cells in rat brain. *J. Neurosci. Res.* **1989**, *22*, 103–106. [CrossRef] [PubMed]
65. Damm, J.; Luheshi, G.N.; Gerstberger, R.; Roth, J.; Rummel, C. Spatiotemporal nuclear factor interleukin-6 expression in the rat brain during lipopolysaccharide-induced fever is linked to sustained hypothalamic inflammatory target gene induction. *J. Comp. Neurol.* **2011**, *519*, 480–505. [CrossRef] [PubMed]
66. Schneiders, J.; Fuchs, F.; Damm, J.; Herden, C.; Gerstberger, R.; Soares, D.M.; Roth, J.; Rummel, C. The transcription factor nuclear factor interleukin 6 mediates pro- and anti-inflammatory responses during LPS-induced systemic inflammation in mice. *Brain Behav. Immun.* **2015**, *48*, 147–164. [CrossRef]
67. Nadjar, A.; Bluthé, R.-M.; May, M.J.; Dantzer, R.; Parnet, P. Inactivation of the cerebral NFκB pathway inhibits interleukin-1β-induced sickness behavior and c-Fos expression in various brain nuclei. *Neuropsychopharmacology* **2005**, *30*, 1492–1499. [CrossRef]
68. Koenig, S.; Luheshi, G.N.; Wenz, T.; Gerstberger, R.; Roth, J.; Rummel, C. Leptin is involved in age-dependent changes in response to systemic inflammation in the rat. *Brain Behav. Immun.* **2014**, *36*, 128–138. [CrossRef]
69. Rummel, C.; Matsumura, K.; Luheshi, G.N. Circulating IL-6 contributes to peripheral LPS-induced mPGES-1 expression in the rat brain. *Brain Res. Bull.* **2011**, *86*, 319–325. [CrossRef]
70. Bauer, J.; Sminia, T.; Wouterlood, F.G.; Dijkstra, C.D. Phagocytic activity of macrophages and microglial cells during the course of acute and chronic relapsing experimental autoimmune encephalomyelitis. *J. Neurosci. Res.* **1994**, *38*, 365–375. [CrossRef]
71. Damoiseaux, J.G.; Döpp, E.A.; Calame, W.; Chao, D.; MacPherson, G.G.; Dijkstra, C.D. Rat macrophage lysosomal membrane antigen recognized by monoclonal antibody ED1. *Immunology* **1994**, *83*, 140–147. [PubMed]
72. Bredehöft, J.; Bhandari, D.R.; Pflieger, F.J.; Schulz, S.; Kang, J.X.; Layé, S.; Roth, J.; Gerstberger, R.; Mayer, K.; Spengler, B.; et al. Visualizing and Profiling Lipids in the OVLT of Fat-1 and Wild Type Mouse Brains during LPS-Induced Systemic Inflammation Using AP-SMALDI MSI. *ACS Chem. Neurosci.* **2019**, *10*, 4394–4406. [CrossRef]
73. Ando, K.; Nagata, K.; Yoshida, R.; Kikugawa, K.; Suzuki, M. Effect of n-3 polyunsaturated fatty acid supplementation on lipid peroxidation of rat organs. *Lipids* **2000**, *35*, 401–407. [CrossRef]
74. Kulkarni, A.; Zhao, A.; Yang, B.; Zhang, Y.; Linderborg, K.M. Tissue-Specific Content of Polyunsaturated Fatty Acids in (n-3) Deficiency State of Rats. *Foods* **2022**, *11*, 208. [CrossRef]
75. Joffre, C.; Rey, C.; Layé, S. N-3 Polyunsaturated Fatty Acids and the Resolution of Neuroinflammation. *Front. Pharmacol.* **2019**, *10*, 1022. [CrossRef] [PubMed]
76. Ostermann, A.I.; Waindok, P.; Schmidt, M.J.; Chiu, C.-Y.; Smyl, C.; Rohwer, N.; Weylandt, K.-H.; Schebb, N.H. Modulation of the endogenous omega-3 fatty acid and oxylipin profile in vivo-A comparison of the fat-1 transgenic mouse with C57BL/6 wildtype mice on an omega-3 fatty acid enriched diet. *PLoS ONE* **2017**, *12*, e0184470. [CrossRef] [PubMed]
77. Pawlosky, R.J.; Hibbeln, J.R.; Novotny, J.A.; Salem, N. Physiological compartmental analysis of α-linolenic acid metabolism in adult humans. *J. Lipid Res.* **2001**, *42*, 1257–1265. [CrossRef]
78. Darwish, I.; Mubareka, S.; Liles, W.C. Immunomodulatory therapy for severe influenza. *Expert Rev. Anti Infect. Ther.* **2011**, *9*, 807–822. [CrossRef]

79. Gonzalvo, R.; Martí-Sistac, O.; Blanch, L.; López-Aguilar, J. Bench-to-bedside review: Brain-lung interaction in the critically ill—A pending issue revisited. *Crit. Care* **2007**, *11*, 216. [CrossRef]
80. Sahu, B.; Sandhir, R.; Naura, A.S. Two hit induced acute lung injury impairs cognitive function in mice: A potential model to study cross talk between lung and brain. *Brain Behav. Immun.* **2018**, *73*, 633–642. [CrossRef]
81. Goodman, R.B.; Pugin, J.; Lee, J.S.; Matthay, M.A. Cytokine-mediated inflammation in acute lung injury. *Cytokine Growth Factor Rev.* **2003**, *14*, 523–535. [CrossRef] [PubMed]
82. Li, C.; Chen, W.; Lin, F.; Li, W.; Wang, P.; Liao, G.; Zhang, L. Functional Two-Way Crosstalk Between Brain and Lung: The Brain-Lung Axis. *Cell. Mol. Neurobiol.* **2023**, *43*, 991–1003. [CrossRef]
83. Rummel, C.; Del Rey, A.; Bähr, L.; Krüger, K.; Peters, E. 1st European Psychoneuroimmunology Network (EPN) Autumn School: Lung-Brain Axis in Health and Disease. *Neuroimmunomodulation* **2022**, *29* (Suppl. 2), 3–8. [CrossRef] [PubMed]
84. Pelosi, P.; Rocco, P.R.M. The lung and the brain: A dangerous cross-talk. *Crit. Care* **2011**, *15*, 168. [CrossRef]
85. Nadeem, A.; Siddiqui, N.; Al-Harbi, N.O.; Attia, S.M.; AlSharari, S.D.; Ahmad, S.F. Acute lung injury leads to depression-like symptoms through upregulation of neutrophilic and neuronal NADPH oxidase signaling in a murine model. *Int. Immunopharmacol.* **2017**, *47*, 218–226. [CrossRef]
86. Meduri, G.U.; Annane, D.; Chrousos, G.P.; Marik, P.E.; Sinclair, S.E. Activation and regulation of systemic inflammation in ARDS: Rationale for prolonged glucocorticoid therapy. *Chest* **2009**, *136*, 1631–1643. [CrossRef] [PubMed]
87. Banks, W.A. The blood-brain barrier in neuroimmunology: Tales of separation and assimilation. *Brain Behav. Immun.* **2015**, *44*, 1–8. [CrossRef]
88. Carson, M.J.; Doose, J.M.; Melchior, B.; Schmid, C.D.; Ploix, C.C. CNS immune privilege: Hiding in plain sight. *Immunol. Rev.* **2006**, *213*, 48–65. [CrossRef]
89. Wang, Y.; Jin, S.; Sonobe, Y.; Cheng, Y.; Horiuchi, H.; Parajuli, B.; Kawanokuchi, J.; Mizuno, T.; Takeuchi, H.; Suzumura, A. Interleukin-1β induces blood-brain barrier disruption by downregulating Sonic hedgehog in astrocytes. *PLoS ONE* **2014**, *9*, e110024. [CrossRef]
90. Horii, Y.; Nakaya, M.; Ohara, H.; Nishihara, H.; Watari, K.; Nagasaka, A.; Nakaya, T.; Sugiura, Y.; Okuno, T.; Koga, T.; et al. Leukotriene B4 receptor 1 exacerbates inflammation following myocardial infarction. *FASEB J.* **2020**, *34*, 8749–8763. [CrossRef]
91. Harden, L.M.; Rummel, C.; Laburn, H.P.; Damm, J.; Wiegand, F.; Poole, S.; Gerstberger, R.; Roth, J. Critical role for peripherally-derived interleukin-10 in mediating the thermoregulatory manifestations of fever and hypothermia in severe forms of lipopolysaccharide-induced inflammation. *Pflug. Arch. Eur. J. Physiol.* **2014**, *466*, 1451–1466. [CrossRef] [PubMed]
92. Bellenger, J.; Bellenger, S.; Bataille, A.; Massey, K.A.; Nicolaou, A.; Rialland, M.; Tessier, C.; Kang, J.X.; Narce, M. High pancreatic n-3 fatty acids prevent STZ-induced diabetes in fat-1 mice: Inflammatory pathway inhibition. *Diabetes* **2011**, *60*, 1090–1099. [CrossRef] [PubMed]
93. Shokry, A.A.; El-Shiekh, R.A.; Kamel, G.; Bakr, A.F.; Ramadan, A. Bioactive phenolics fraction of *Hedera helix* L. (Common Ivy Leaf) standardized extract ameliorates LPS-induced acute lung injury in the mouse model through the inhibition of proinflammatory cytokines and oxidative stress. *Heliyon* **2022**, *8*, e09477. [CrossRef] [PubMed]
94. Huang, S.R.; Ma, A.Y.; Liu, Y.; Qu, Y. Effects of inflammatory factors including plasma tumor necrosis factor-α in the clinical treatment of acute respiratory distress syndrome. *Oncol. Lett.* **2017**, *13*, 5016–5020. [CrossRef]
95. El Kebir, D.; Gjorstrup, P.; Filep, J.G. Resolvin E1 promotes phagocytosis-induced neutrophil apoptosis and accelerates resolution of pulmonary inflammation. *Proc. Natl. Acad. Sci. USA* **2012**, *109*, 14983–14988. [CrossRef]
96. Herová, M.; Schmid, M.; Gemperle, C.; Hersberger, M. ChemR23, the receptor for chemerin and resolvin E1, is expressed and functional on M1 but not on M2 macrophages. *J. Immunol.* **2015**, *194*, 2330–2337. [CrossRef]
97. Ware, L.B.; Matthay, M.A. The acute respiratory distress syndrome. *N. Engl. J. Med.* **2000**, *342*, 1334–1349. [CrossRef]
98. Parsey, M.V.; Tuder, R.M.; Abraham, E. Neutrophils are major contributors to intraparenchymal lung IL-1β expression after hemorrhage and endotoxemia. *J. Immunol.* **1998**, *160*, 1007–1013. [CrossRef]
99. Mikacenic, C.; Hansen, E.E.; Radella, F.; Gharib, S.A.; Stapleton, R.D.; Wurfel, M.M. Interleukin-17A Is Associated with Alveolar Inflammation and Poor Outcomes in Acute Respiratory Distress Syndrome. *Crit. Care Med.* **2016**, *44*, 496–502. [CrossRef]
100. Bhatia, M.; Zemans, R.L.; Jeyaseelan, S. Role of chemokines in the pathogenesis of acute lung injury. *Am. J. Respir. Cell Mol. Biol.* **2012**, *46*, 566–572. [CrossRef]
101. Sawant, K.V.; Xu, R.; Cox, R.; Hawkins, H.; Sbrana, E.; Kolli, D.; Garofalo, R.P.; Rajarathnam, K. Chemokine CXCL1-Mediated Neutrophil Trafficking in the Lung: Role of CXCR2 Activation. *J. Innate Immun.* **2015**, *7*, 647–658. [CrossRef]
102. Ballinger, M.N.; Paine, R.; Serezani, C.H.C.; Aronoff, D.M.; Choi, E.S.; Standiford, T.J.; Toews, G.B.; Moore, B.B. Role of granulocyte macrophage colony-stimulating factor during gram-negative lung infection with *Pseudomonas aeruginosa*. *Am. J. Respir. Cell Mol. Biol.* **2006**, *34*, 766–774. [CrossRef]
103. Jeyaseelan, S.; Chu, H.W.; Young, S.K.; Worthen, G.S. Transcriptional profiling of lipopolysaccharide-induced acute lung injury. *Infect. Immun.* **2004**, *72*, 7247–7256. [CrossRef]
104. Matute-Bello, G.; Liles, W.C.; Radella, F.; Steinberg, K.P.; Ruzinski, J.T.; Hudson, L.D.; Martin, T.R. Modulation of neutrophil apoptosis by granulocyte colony-stimulating factor and granulocyte/macrophage colony-stimulating factor during the course of acute respiratory distress syndrome. *Crit. Care Med.* **2000**, *28*, 1–7. [CrossRef]
105. Li, C.; Yang, P.; Sun, Y.; Li, T.; Wang, C.; Wang, Z.; Zou, Z.; Yan, Y.; Wang, W.; Chen, Z.; et al. IL-17 response mediates acute lung injury induced by the 2009 pandemic influenza A (H1N1) virus. *Cell Res.* **2012**, *22*, 528–538. [CrossRef] [PubMed]

106. Bilal, S.; Haworth, O.; Wu, L.; Weylandt, K.H.; Levy, B.D.; Kang, J.X. Fat-1 transgenic mice with elevated omega-3 fatty acids are protected from allergic airway responses. *Biochim. Biophys. Acta (BBA) Mol. Basis Dis.* **2011**, *1812*, 1164–1169. [CrossRef] [PubMed]
107. Tiesset, H.; Pierre, M.; Desseyn, J.-L.; Guéry, B.; Beermann, C.; Galabert, C.; Gottrand, F.; Husson, M.-O. Dietary (n-3) polyunsaturated fatty acids affect the kinetics of pro- and antiinflammatory responses in mice with *Pseudomonas aeruginosa* lung infection. *J. Nutr.* **2009**, *139*, 82–89. [CrossRef]
108. Unoda, K.; Doi, Y.; Nakajima, H.; Yamane, K.; Hosokawa, T.; Ishida, S.; Kimura, F.; Hanafusa, T. Eicosapentaenoic acid (EPA) induces peroxisome proliferator-activated receptors and ameliorates experimental autoimmune encephalomyelitis. *J. Neuroimmunol.* **2013**, *256*, 7–12. [CrossRef]
109. Monk, J.M.; Hou, T.Y.; Turk, H.F.; Weeks, B.; Wu, C.; McMurray, D.N.; Chapkin, R.S. Dietary n-3 polyunsaturated fatty acids (PUFA) decrease obesity-associated Th17 cell-mediated inflammation during colitis. *PLoS ONE* **2012**, *7*, e49739. [CrossRef] [PubMed]
110. Qin, S.; Wen, J.; Bai, X.-C.; Chen, T.-Y.; Zheng, R.-C.; Zhou, G.-B.; Ma, J.; Feng, J.-Y.; Zhong, B.-L.; Li, Y.-M. Endogenous n-3 polyunsaturated fatty acids protect against imiquimod induced psoriasis-like inflammation via the IL-17/IL-23 axis. *Mol. Med. Rep.* **2014**, *9*, 2097–2104. [CrossRef]
111. Mastroianni, C.M.; Liuzzi, G.M.; Vullo, V.; Jirillo, E.; Delia, S.; Riccio, P. Detection of cerebrospinal fluid antibodies against myelin basic protein in patients with AIDS dementia complex. *Mol. Chem. Neuropathol.* **1991**, *14*, 227–236. [CrossRef] [PubMed]
112. Roth, J.; Blatteis, C.M. Mechanisms of fever production and lysis: Lessons from experimental LPS fever. *Compr. Physiol.* **2014**, *4*, 1563–1604. [CrossRef] [PubMed]
113. von Bernhardi, R.; Eugenín-von Bernhardi, L.; Eugenín, J. Microglial cell dysregulation in brain aging and neurodegeneration. *Front. Aging Neurosci.* **2015**, *7*, 124. [CrossRef] [PubMed]
114. Hritcu, L.; Ciobica, A.; Stefan, M.; Mihasan, M.; Palamiuc, L.; Nabeshima, T. Spatial memory deficits and oxidative stress damage following exposure to lipopolysaccharide in a rodent model of Parkinson's disease. *Neurosci. Res.* **2011**, *71*, 35–43. [CrossRef]
115. Garcia, I.J.P.; Kinoshita, P.F.; de Oliveira Braga, I.; Parreira, G.M.; Mignaco, J.A.; Scavone, C.; Barbosa, L.A.; de Lima Santos, H. Ouabain attenuates the oxidative stress induced by lipopolysaccharides in the cerebellum of rats. *J. Cell. Biochem.* **2018**, *119*, 2156–2167. [CrossRef]
116. Hritcu, L.; Ciobica, A. Intranigral lipopolysaccharide administration induced behavioral deficits and oxidative stress damage in laboratory rats: Relevance for Parkinson's disease. *Behav. Brain Res.* **2013**, *253*, 25–31. [CrossRef]
117. Nandi, A.; Yan, L.-J.; Jana, C.K.; Das, N. Role of Catalase in Oxidative Stress- and Age-Associated Degenerative Diseases. *Oxid. Med. Cell. Longev.* **2019**, *2019*, 9613090. [CrossRef]
118. Younus, H. Therapeutic potentials of superoxide dismutase. *Int. J. Health Sci.* **2018**, *12*, 88–93.
119. Loh, K.P.; Huang, S.H.; de Silva, R.; Tan, B.K.H.; Zhu, Y.Z. Oxidative stress: Apoptosis in neuronal injury. *Curr. Alzheimer Res.* **2006**, *3*, 327–337. [CrossRef]
120. Mariani, E.; Polidori, M.C.; Cherubini, A.; Mecocci, P. Oxidative stress in brain aging, neurodegenerative and vascular diseases: An overview. *J. Chromatogr. B Anal. Technol. Biomed. Life Sci.* **2005**, *827*, 65–75. [CrossRef]
121. Moreira, P.I.; Siedlak, S.L.; Aliev, G.; Zhu, X.; Cash, A.D.; Smith, M.A.; Perry, G. Oxidative stress mechanisms and potential therapeutics in Alzheimer disease. *J. Neural Transm.* **2005**, *112*, 921–932. [CrossRef] [PubMed]
122. Wang, X.; Zhu, M.; Hjorth, E.; Cortés-Toro, V.; Eyjolfsdottir, H.; Graff, C.; Nennesmo, I.; Palmblad, J.; Eriksdotter, M.; Sambamurti, K.; et al. Resolution of inflammation is altered in Alzheimer's disease. *Alzheimer's Dement.* **2015**, *11*, 40–50.e2. [CrossRef]
123. Zhang, Y.; Xu, N.; Ding, Y.; Zhang, Y.; Li, Q.; Flores, J.; Haghighiabyaneh, M.; Doycheva, D.; Tang, J.; Zhang, J.H. Chemerin suppresses neuroinflammation and improves neurological recovery via CaMKK2/AMPK/Nrf2 pathway after germinal matrix hemorrhage in neonatal rats. *Brain Behav. Immun.* **2018**, *70*, 179–193. [CrossRef] [PubMed]
124. Zhang, J.; Liu, L.; Zhang, Y.; Yuan, Y.; Miao, Z.; Lu, K.; Zhang, X.; Ni, R.; Zhang, H.; Zhao, Y.; et al. ChemR23 signaling ameliorates cognitive impairments in diabetic mice via dampening oxidative stress and NLRP3 inflammasome activation. *Redox Biol.* **2022**, *58*, 102554. [CrossRef] [PubMed]
125. Zhang, R.; Wang, J. HuR stabilizes TFAM mRNA in an ATM/p38-dependent manner in ionizing irradiated cancer cells. *Cancer Sci.* **2018**, *109*, 2446–2457. [CrossRef]
126. Kang, D.; Kim, S.H.; Hamasaki, N. Mitochondrial transcription factor A (TFAM): Roles in maintenance of mtDNA and cellular functions. *Mitochondrion* **2007**, *7*, 39–44. [CrossRef]
127. Alawdi, S.H.; El-Denshary, E.S.; Safar, M.M.; Eidi, H.; David, M.-O.; Abdel-Wahhab, M.A. Neuroprotective Effect of Nanodiamond in Alzheimer's Disease Rat Model: A Pivotal Role for Modulating NF-κB and STAT3 Signaling. *Mol. Neurobiol.* **2017**, *54*, 1906–1918. [CrossRef] [PubMed]
128. Marcheselli, V.L.; Hong, S.; Lukiw, W.J.; Tian, X.H.; Gronert, K.; Musto, A.; Hardy, M.; Gimenez, J.M.; Chiang, N.; Serhan, C.N.; et al. Novel docosanoids inhibit brain ischemia-reperfusion-mediated leukocyte infiltration and pro-inflammatory gene expression. *J. Biol. Chem.* **2003**, *278*, 43807–43817. [CrossRef]
129. Caputo, M.P.; Radlowski, E.C.; Lawson, M.A.; Antonson, A.M.; Watson, J.E.; Matt, S.M.; Leyshon, B.J.; Das, A.; Johnson, R.W. Herring roe oil supplementation alters microglial cell gene expression and reduces peripheral inflammation after immune activation in a neonatal piglet model. *Brain Behav. Immun.* **2019**, *81*, 455–469. [CrossRef]

130. Pflieger, F.J.; Wolf, J.; Feldotto, M.; Nockher, A.; Wenderoth, T.; Hernandez, J.; Roth, J.; Ott, D.; Rummel, C. Norepinephrine Inhibits Lipopolysaccharide-Stimulated TNF-α but Not Oxylipin Induction in n-3/n-6 PUFA-Enriched Cultures of Circumventricular Organs. *Int. J. Mol. Sci.* **2022**, *23*, 8745. [CrossRef]
131. Flesher, R.P.; Herbert, C.; Kumar, R.K. Resolvin E1 promotes resolution of inflammation in a mouse model of an acute exacerbation of allergic asthma. *Clin. Sci.* **2014**, *126*, 805–814. [CrossRef]
132. Serhan, C.N. Pro-resolving lipid mediators are leads for resolution physiology. *Nature* **2014**, *510*, 92–101. [CrossRef]
133. Biesmans, S.; Meert, T.F.; Bouwknecht, J.A.; Acton, P.D.; Davoodi, N.; de Haes, P.; Kuijlaars, J.; Langlois, X.; Matthews, L.J.R.; Ver Donck, L.; et al. Systemic immune activation leads to neuroinflammation and sickness behavior in mice. *Mediat. Inflamm.* **2013**, *2013*, 271359. [CrossRef]
134. Hoogland, I.C.M.; Houbolt, C.; van Westerloo, D.J.; van Gool, W.A.; van de Beek, D. Systemic inflammation and microglial activation: Systematic review of animal experiments. *J. Neuroinflamm.* **2015**, *12*, 114. [CrossRef]
135. Prinz, M.; Priller, J. Microglia and brain macrophages in the molecular age: From origin to neuropsychiatric disease. *Nat. Rev. Neurosci.* **2014**, *15*, 300–312. [CrossRef] [PubMed]
136. Pinteaux, E.; Rothwell, N.J.; Boutin, H. Neuroprotective actions of endogenous interleukin-1 receptor antagonist (IL-1ra) are mediated by glia. *Glia* **2006**, *53*, 551–556. [CrossRef] [PubMed]
137. Ulich, T.R.; Yin, S.M.; Guo, K.Z.; Del Castillo, J.; Eisenberg, S.P.; Thompson, R.C. The intratracheal administration of endotoxin and cytokines. III. The interleukin-1 (IL-1) receptor antagonist inhibits endotoxin- and IL-1-induced acute inflammation. *Am. J. Pathol.* **1991**, *138*, 521–524. [PubMed]
138. Plata-Salamán, C.R.; ffrench-Mullen, J.M.H. Interleukin-1β inhibits Ca^{2+} channel currents in hippocampal neurons through protein kinase C. *Eur. J. Pharmacol. Mol. Pharmacol.* **1994**, *266*, 1–10. [CrossRef]
139. Seckinger, P.; Williamson, K.; Balavoine, J.F.; Mach, B.; Mazzei, G.; Shaw, A.; Dayer, J.M. A urine inhibitor of interleukin 1 activity affects both interleukin 1 alpha and 1 beta but not tumor necrosis factor alpha. *J. Immunol.* **1987**, *139*, 1541–1545. [CrossRef]
140. Shih, R.-H.; Wang, C.-Y.; Yang, C.-M. NF-κB Signaling Pathways in Neurological Inflammation: A Mini Review. *Front. Mol. Neurosci.* **2015**, *8*, 77. [CrossRef]
141. Villapol, S. Roles of Peroxisome Proliferator-Activated Receptor Gamma on Brain and Peripheral Inflammation. *Cell. Mol. Neurobiol.* **2018**, *38*, 121–132. [CrossRef]
142. Draper, E.; Reynolds, C.M.; Canavan, M.; Mills, K.H.; Loscher, C.E.; Roche, H.M. Omega-3 fatty acids attenuate dendritic cell function via NF-κB independent of PPARγ. *J. Nutr. Biochem.* **2011**, *22*, 784–790. [CrossRef] [PubMed]
143. Liang, H.; Ward, W.F. PGC-1α: A key regulator of energy metabolism. *Adv. Physiol. Educ.* **2006**, *30*, 145–151. [CrossRef] [PubMed]
144. Tak, P.P.; Firestein, G.S. NF-κB: A key role in inflammatory diseases. *J. Clin. Investig.* **2001**, *107*, 7–11. [CrossRef] [PubMed]
145. Uematsu, S.; Matsumoto, M.; Takeda, K.; Akira, S. Lipopolysaccharide-dependent prostaglandin E_2 production is regulated by the glutathione-dependent prostaglandin E_2 synthase gene induced by the Toll-like receptor 4/MyD88/NF-IL6 pathway. *J. Immunol.* **2002**, *168*, 5811–5816. [CrossRef] [PubMed]
146. Hoogland, I.C.M.; Westhoff, D.; Engelen-Lee, J.-Y.; Melief, J.; Valls Serón, M.; Houben-Weerts, J.H.M.P.; Huitinga, I.; van Westerloo, D.J.; van der Poll, T.; van Gool, W.A.; et al. Microglial Activation After Systemic Stimulation with Lipopolysaccharide and *Escherichia coli*. *Front. Cell. Neurosci.* **2018**, *12*, 110. [CrossRef]
147. Biesmans, S.; Acton, P.D.; Cotto, C.; Langlois, X.; Ver Donck, L.; Bouwknecht, J.A.; Aelvoet, S.-A.; Hellings, N.; Meert, T.F.; Nuydens, R. Effect of stress and peripheral immune activation on astrocyte activation in transgenic bioluminescent Gfap-luc mice. *Glia* **2015**, *63*, 1126–1137. [CrossRef]
148. Xie, Z.; Wei, M.; Morgan, T.E.; Fabrizio, P.; Han, D.; Finch, C.E.; Longo, V.D. Peroxynitrite mediates neurotoxicity of amyloid β-peptide$_{1-42}$- and lipopolysaccharide-activated microglia. *J. Neurosci.* **2002**, *22*, 3484–3492. [CrossRef]
149. Chhor, V.; Le Charpentier, T.; Lebon, S.; Oré, M.-V.; Celador, I.L.; Josserand, J.; Degos, V.; Jacotot, E.; Hagberg, H.; Sävman, K.; et al. Characterization of phenotype markers and neuronotoxic potential of polarised primary microglia in vitro. *Brain Behav. Immun.* **2013**, *32*, 70–85. [CrossRef]
150. Cherry, J.D.; Olschowka, J.A.; O'Banion, M.K. Neuroinflammation and M2 microglia: The good, the bad, and the inflamed. *J. Neuroinflamm.* **2014**, *11*, 98. [CrossRef]
151. Michaud, J.-P.; Hallé, M.; Lampron, A.; Thériault, P.; Préfontaine, P.; Filali, M.; Tribout-Jover, P.; Lanteigne, A.-M.; Jodoin, R.; Cluff, C.; et al. Toll-like receptor 4 stimulation with the detoxified ligand monophosphoryl lipid A improves Alzheimer's disease-related pathology. *Proc. Natl. Acad. Sci. USA* **2013**, *110*, 1941–1946. [CrossRef] [PubMed]
152. Bousquet, M.; Gue, K.; Emond, V.; Julien, P.; Kang, J.X.; Cicchetti, F.; Calon, F. Transgenic conversion of omega-6 into omega-3 fatty acids in a mouse model of Parkinson's disease. *J. Lipid Res.* **2011**, *52*, 263–271. [CrossRef] [PubMed]
153. Emre, C.; Hjorth, E.; Bharani, K.; Carroll, S.; Granholm, A.-C.; Schultzberg, M. Receptors for pro-resolving mediators are increased in Alzheimer's disease brain. *Brain Pathol.* **2020**, *30*, 614–640. [CrossRef] [PubMed]
154. Weissman, C. The metabolic response to stress: An overview and update. *Anesthesiology* **1990**, *73*, 308–327. [CrossRef]
155. Mayer, K.; Fegbeutel, C.; Hattar, K.; Sibelius, U.; Krämer, H.-J.; Heuer, K.-U.; Temmesfeld-Wollbrück, B.; Gokorsch, S.; Grimminger, F.; Seeger, W. Omega-3 vs. omega-6 lipid emulsions exert differential influence on neutrophils in septic shock patients: Impact on plasma fatty acids and lipid mediator generation. *Intensive Care Med.* **2003**, *29*, 1472–1481. [CrossRef]
156. Hong, S.; Porter, T.F.; Lu, Y.; Oh, S.F.; Pillai, P.S.; Serhan, C.N. Resolvin E1 metabolome in local inactivation during inflammation-resolution. *J. Immunol.* **2008**, *180*, 3512–3519. [CrossRef]

157. Schebb, N.H.; Kühn, H.; Kahnt, A.S.; Rund, K.M.; O'Donnell, V.B.; Flamand, N.; Peters-Golden, M.; Jakobsson, P.-J.; Weylandt, K.H.; Rohwer, N.; et al. Formation, Signaling and Occurrence of Specialized Pro-Resolving Lipid Mediators-What is the Evidence so far? *Front. Pharmacol.* **2022**, *13*, 838782. [CrossRef]
158. Orr, S.K.; Palumbo, S.; Bosetti, F.; Mount, H.T.; Kang, J.X.; Greenwood, C.E.; Ma, D.W.L.; Serhan, C.N.; Bazinet, R.P. Unesterified docosahexaenoic acid is protective in neuroinflammation. *J. Neurochem.* **2013**, *127*, 378–393. [CrossRef]
159. Ariel, A.; Li, P.; Wang, W.; Tang, W.; Fredman, G.; Hong, S.; Gotlinger, K.H.; Serhan, C.N. The docosatriene protectin D1 is produced by TH2 skewing and promotes human T cell apoptosis via lipid raft clustering. *J. Biol. Chem.* **2005**, *280*, 43079–43086. [CrossRef]
160. Serhan, C.N.; Gotlinger, K.; Hong, S.; Lu, Y.; Siegelman, J.; Baer, T.; Yang, R.; Colgan, S.P.; Petasis, N.A. Anti-inflammatory actions of neuroprotectin D1/protectin D1 and its natural stereoisomers: Assignments of dihydroxy-containing docosatrienes. *J. Immunol.* **2006**, *176*, 1848–1859. [CrossRef]
161. Chiang, N.; Fredman, G.; Bäckhed, F.; Oh, S.F.; Vickery, T.; Schmidt, B.A.; Serhan, C.N. Infection regulates pro-resolving mediators that lower antibiotic requirements. *Nature* **2012**, *484*, 524–528. [CrossRef] [PubMed]
162. Zhu, M.; Wang, X.; Hjorth, E.; Colas, R.A.; Schroeder, L.; Granholm, A.-C.; Serhan, C.N.; Schultzberg, M. Pro-Resolving Lipid Mediators Improve Neuronal Survival and Increase Aβ42 Phagocytosis. *Mol. Neurobiol.* **2016**, *53*, 2733–2749. [CrossRef] [PubMed]
163. Russell, W.; Burch, R.L. The Principles of Humane Experimental Technique. *Med. J. Aust.* **1960**, *1*, 500. [CrossRef]
164. Card, J.W.; Carey, M.A.; Bradbury, J.A.; DeGraff, L.M.; Morgan, D.L.; Moorman, M.P.; Flake, G.P.; Zeldin, D.C. Gender differences in murine airway responsiveness and lipopolysaccharide-induced inflammation. *J. Immunol.* **2006**, *177*, 621–630. [CrossRef]
165. Matute-Bello, G.; Frevert, C.W.; Martin, T.R. Animal models of acute lung injury. *Am. J. Physiol. Lung Cell. Mol. Physiol.* **2008**, *295*, L379–L399. [CrossRef]
166. Gotts, J.E.; Bernard, O.; Chun, L.; Croze, R.H.; Ross, J.T.; Nesseler, N.; Wu, X.; Abbott, J.; Fang, X.; Calfee, C.S.; et al. Clinically relevant model of pneumococcal pneumonia, ARDS, and nonpulmonary organ dysfunction in mice. *Am. J. Physiol. Lung Cell. Mol. Physiol.* **2019**, *317*, L717–L736. [CrossRef] [PubMed]
167. Bi, M.H.; Ott, J.; Fischer, T.; Hecker, M.; Dietrich, H.; Schaefer, M.B.; Markart, P.; Wang, B.E.; Seeger, W.; Mayer, K. Induction of lymphocyte apoptosis in a murine model of acute lung injury—Modulation by lipid emulsions. *Shock* **2010**, *33*, 179–188. [CrossRef]
168. Hecker, M.; Ott, J.; Sondermann, C.; Schaefer, M.; Obert, M.; Hecker, A.; Morty, R.E.; Vadasz, I.; Herold, S.; Rosengarten, B.; et al. Immunomodulation by fish-oil containing lipid emulsions in murine acute respiratory distress syndrome. *Crit. Care* **2014**, *18*, R85. [CrossRef]
169. Schaefer, M.B.; Pose, A.; Ott, J.; Hecker, M.; Behnk, A.; Schulz, R.; Weissmann, N.; Günther, A.; Seeger, W.; Mayer, K. Peroxisome proliferator-activated receptor-α reduces inflammation and vascular leakage in a murine model of acute lung injury. *Eur. Respir. J.* **2008**, *32*, 1344–1353. [CrossRef]
170. Haqqani, A.S.; Sandhu, J.K.; Birnboim, H.C. A myeloperoxidase-specific assay based upon bromide-dependent chemiluminescence of luminol. *Anal. Biochem.* **1999**, *273*, 126–132. [CrossRef]
171. Donnelly, S.C.; Strieter, R.M.; Reid, P.T.; Kunkel, S.L.; Burdick, M.D.; Armstrong, I.; Mackenzie, A.; Haslett, C. The association between mortality rates and decreased concentrations of interleukin-10 and interleukin-1 receptor antagonist in the lung fluids of patients with the adult respiratory distress syndrome. *Ann. Intern. Med.* **1996**, *125*, 191–196. [CrossRef]
172. Ye, P.; Garvey, P.B.; Zhang, P.; Nelson, S.; Bagby, G.; Summer, W.R.; Schwarzenberger, P.; Shellito, J.E.; Kolls, J.K. Interleukin-17 and lung host defense against Klebsiella pneumoniae infection. *Am. J. Respir. Cell Mol. Biol.* **2001**, *25*, 335–340. [CrossRef]
173. Law, A.H.Y.; Lee, D.C.W.; Cheung, B.K.W.; Yim, H.C.H.; Lau, A.S.Y. Role for nonstructural protein 1 of severe acute respiratory syndrome coronavirus in chemokine dysregulation. *J. Virol.* **2007**, *81*, 416–422. [CrossRef] [PubMed]
174. Crispe, I.N. The liver as a lymphoid organ. *Annu. Rev. Immunol.* **2009**, *27*, 147–163. [CrossRef] [PubMed]
175. Motulsky, H.J.; Brown, R.E. Detecting outliers when fitting data with nonlinear regression—A new method based on robust nonlinear regression and the false discovery rate. *BMC Bioinform.* **2006**, *7*, 123. [CrossRef]
176. Yuan, J.S.; Reed, A.; Chen, F.; Stewart, C.N. Statistical analysis of real-time PCR data. *BMC Bioinform.* **2006**, *7*, 85. [CrossRef]

International Journal of
Molecular Sciences

MDPI

Article

Regulation of Cathelicidin Antimicrobial Peptide (CAMP) Gene Expression by TNFα and cfDNA in Adipocytes

Alexandra Höpfinger *, Andreas Schmid, Leonie Schweitzer, Marissa Patz, Anja Weber, Andreas Schäffler and Thomas Karrasch

Department of Internal Medicine III, University of Giessen, 35392 Giessen, Germany;
andreas.schaeffler@innere.med.uni-giessen.de (A.S.); thomas.karrasch@innere.med.uni-giessen.de (T.K.)
* Correspondence: alexandra.hoepfinger@innere.med.uni-giessen.de

Abstract: Understanding the complex interactions between metabolism and the immune system ("metaflammation") is crucial for the identification of key immunomodulatory factors as potential therapeutic targets in obesity and in cardiovascular diseases. Cathelicidin antimicrobial peptide (CAMP) is an important factor of innate immunity and is expressed in adipocytes. CAMP, therefore, might play a role as an adipokine in metaflammation and adipose inflammation. TNFα, cell-free nucleic acids (cfDNA), and toll-like receptor (TLR) 9 are components of the innate immune system and are functionally active in adipose tissue. The aim of the present study was to investigate the impact of TNFα and cfDNA on CAMP expression in adipocytes. Since cfDNA acts as a physiological TLR9 agonist, we additionally investigated TLR9-mediated CAMP regulation in adipocytes and adipose tissue. CAMP gene expression in murine 3T3-L1 and human SGBS adipocytes and in murine and human adipose tissues was quantified by real-time PCR. Adipocyte inflammation was induced in vitro by TNFα and cfDNA stimulation. Serum CAMP concentrations in TLR9 knockout (KO) and in wildtype mice were quantified by ELISA. In primary adipocytes of wildtype and TLR9 KO mice, CAMP gene expression was quantified by real-time PCR. CAMP gene expression was considerably increased in 3T3-L1 and SGBS adipocytes during differentiation. TNFα significantly induced CAMP gene expression in mature adipocytes, which was effectively antagonized by inhibition of PI3K signaling. Cell-free nucleic acids (cfDNA) significantly impaired CAMP gene expression, whereas synthetic agonistic and antagonistic TLR9 ligands had no effect. CAMP and TLR9 gene expression were correlated positively in murine and human subcutaneous but not in intra-abdominal/visceral adipose tissues. Male TLR9 knockout mice exhibited lower systemic CAMP concentrations than wildtype mice. CAMP gene expression levels in primary adipocytes did not significantly differ between wildtype and TLR9 KO mice. These findings suggest a regulatory role of inflammatory mediators, such as TNFα and cfDNA, in adipocytic CAMP expression as a novel putative molecular mechanism in adipose tissue innate immunity.

Keywords: TNFα; cfDNA; TLR (toll-like receptor) 9; cathelicidin anti-microbial peptide; adipocyte; adipose tissue; innate immunity

Citation: Höpfinger, A.; Schmid, A.; Schweitzer, L.; Patz, M.; Weber, A.; Schäffler, A.; Karrasch, T. Regulation of Cathelicidin Antimicrobial Peptide (CAMP) Gene Expression by TNFα and cfDNA in Adipocytes. *Int. J. Mol. Sci.* **2023**, *24*, 15820. https://doi.org/10.3390/ijms242115820

Academic Editors: Ralf Kircheis and Oliver Planz

Received: 15 May 2023
Revised: 16 October 2023
Accepted: 21 October 2023
Published: 31 October 2023

1. Introduction

The worldwide prevalence of obesity and metabolic syndrome is continuously increasing [1–3]. Metabolic syndrome is one of the main risk factors for cardiovascular diseases [4] that are the number one cause of death in public health statistics [5]. Its underlying complex interactions have not been fully elucidated until now. In particular, chronic low-grade inflammatory state, adipose inflammation, and metaflammation in obesity are assumed to play a crucial role in the development of local and systemic insulin resistance, cardiovascular diseases, and impaired immune responses [4]. Recently published data argue for a role of adipose tissue innate immunity in regulating the molecular interface of metabolism and inflammation [6,7].

Cathelicidin antimicrobial peptide (CAMP) represents an immunomodulatory peptide mainly secreted by immune cells such as monocytes, macrophages, and lymphocytes [8] with a crucial role in the innate immune system. CAMP enhances monocyte [9] and granulocyte [10] infiltration at sites of local inflammation by chemoattractant action and alters the immune response in the course of infection via toll-like receptor (TLR) modulation [11]. In bacteria, CAMP causes a loss of membrane integrity by increased permeability, leading to the death of microbial cells [12].

In 2015, for the first time, Zhang et al. described CAMP as being produced by adipocytes [13], thereby introducing CAMP as a novel adipokine. During subcutaneous Gram-positive infection with Staphylococcus aureus, CAMP secreted by adipocytes inhibits bacterial growth and is an important protective component of subdermal adipose tissue against bacterial infection [13]. In mice, diet-induced obesity results in the loss of a dermal pool of preadipocytes and thus inhibits the capacity to initiate reactive adipogenesis and to express CAMP [14]. In humans, CAMP gene expression is higher in subcutaneous than in visceral adipose tissues [15]. However, further data on the precise function and regulation of CAMP in adipocytes are sparse.

In 2021, we elucidated regulatory mechanisms of CAMP expression in murine 3T3-L1 adipocytes [16]. Activation of TLR2 and TLR4 by specific ligands (MALP2, lipopolysaccharides (LPS)) up-regulates adipocyte CAMP expression involving classical signal transduction elements such as NF-κB, PI3K and STAT3 [16]. This TLR-mediated proinflammatory activation of CAMP expression can be modified by immunomodulatory adipokines such as C1q/TNF-related protein-3 (CTRP-3) [17]. Furthermore, immuno-metabolic factors, such as bile acids, glucose, insulin, and incretins, are able to modulate CAMP expression in adipocytes in vitro [15].

In obesity, the death of adipocytes leads to an increase of cell-free nucleic acids (cfDNA) in adipose tissue and plasma [18], and adipose tissue-resident macrophages are involved in the recognition of these elevated cfDNA levels via TLR9 [18]. TLR9 is a well-characterized intracellular receptor [19] recognizing non-methylated CpG DNA motifs derived from bacteria [20], viral double-stranded DNA (dsDNA) [21], and cfDNA as a signal for severe tissue damage inside the organism [18]. Activation of TLR9 leads to an increased production of proinflammatory cytokines [22], and TLR9 signaling in macrophages is involved in obesity-induced insulin resistance [23]. Of note, Nakagawa et al. demonstrated that CAMP is required for normal TLR9 function in dendritic cells and macrophages [24]. Interestingly, adipocytes themselves express functionally active TLR9 [25]. Whether CAMP plays an additional role in TLR9 modulation in adipocytes has not been elucidated so far. A possible interaction between CAMP expression, TNFα, and cfDNA in adipocytes and adipose tissues might provide a novel molecular interface of obesity-related inflammation with adipocyte function.

Therefore, the primary aim of the present study was to investigate:

1. The underlying mechanisms involved in TNFα-mediated regulation of CAMP expression in adipocytes with respect to components of established proinflammatory signaling pathways;
2. The potential effects of cfDNA, representing an inflammatory mediator in adipose tissue and a physiological TLR9-agonist, on CAMP gene expression in murine 3T3-L1 adipocytes;
3. Circulating CAMP concentrations as well as local CAMP gene expression in adipose tissue and primary adipocytes of TLR9 knockout mice compared to wildtype mice.

2. Results

2.1. CAMP Gene Expression Is Induced during Differentiation in Murine 3T3-L1 Adipocytes and in Human SGBS Cells

During the differentiation process, 3T3-L1 adipocytes phenotypically turn from fibroblastic preadipocytes into mature adipocytes with multiple lipid vacuoles within 8 days. The pattern of adipokine expression and secretion varies depending on the stage of differentiation. In early stage 3T3-L1 preadipocytes, at days 0 and 3 of differentiation, no significant CAMP gene expression was detected. In later stages, at day 6 and in mature adipocytes (day 8), a strong and significant increase of CAMP expression was observed (Figure 1C). Human SGBS preadipocytes were differentiated over 14 days into mature adipocytes. In early preadipocytes, there was no significant CAMP expression. During differentiation, CAMP expression increased strongly until day 8, remaining at an elevated level during final adipocyte maturation (Figure 1F).

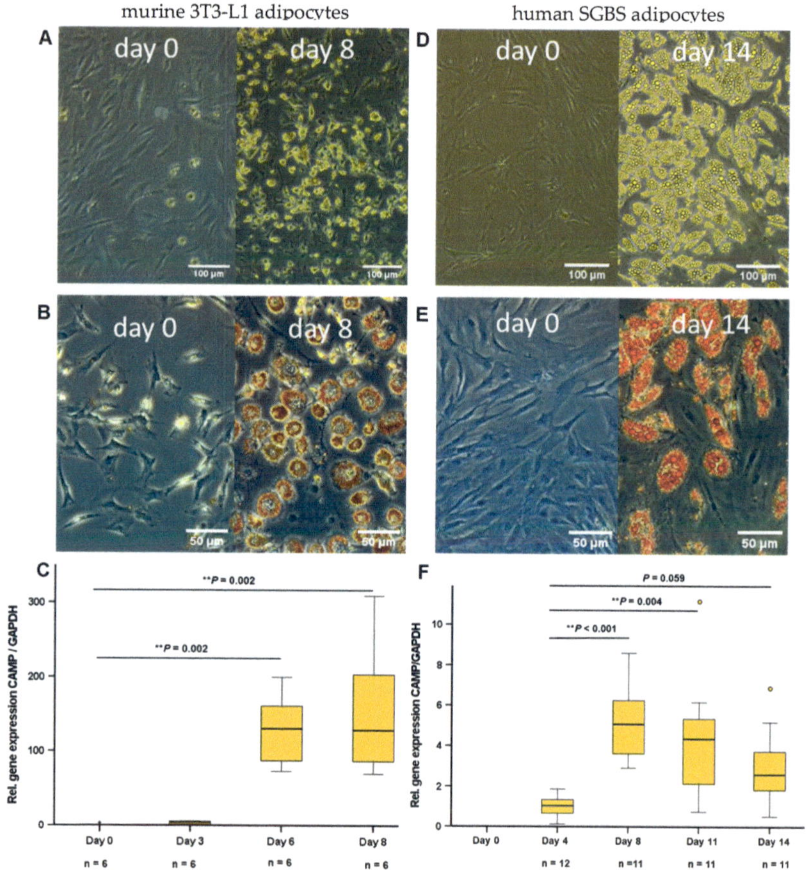

Figure 1. CAMP gene expression is induced during differentiation in murine 3T3-L1 adipocytes (**C**) and in human SGBS cells (**F**). In 3T3-L1 preadipocytes, no significant CAMP expression is detectable. At day 6 and in mature adipocytes, a considerable increase of CAMP mRNA levels is observed (**C**). CAMP expression significantly increases during differentiation of SGBS cells (**F**). Light microscopy of naïve (**A**,**D**) and Oil Red O stained (**B**,**E**) adipocytes is shown. During differentiation, 3T3-L1 fibroblasts on day 0 mature to adipocytes on day 8 (**A**,**B**). SGSB preadipocytes on day 0 mature to adipocytes on day 14 (**D**,**E**). CAMP expression was investigated by quantitative real-time PCR. The Kruskal–Wallis test was applied for calculation of p values and statistical significance ($p < 0.05$). N = 6–12 wells were investigated per experimental setting. Gene expression levels are given in relative gene expression as compared to day 0 in 3T3-L1 and day 4 in SGBS cells. (** $p < 0.01$).

2.2. TNFα-Induced CAMP Gene Expression Can Be Antagonized by Inhibitors of Intracellular Signal Transduction Pathways

In 3T3-L1 adipocytes, a state of inflammation was induced by stimulation with TNFα. Treatment with doses of 2 ng/mL TNFα ($p = 0.005$) and 10 ng/mL TNFα ($p < 0.001$) significantly increased CAMP gene expression (Figure 2A). Inhibition of the PI3K (by 5 μM LY294002) signal transduction pathway effectively antagonized TNFα-induced CAMP gene expression ($p = 0.032$) (Figure 2B), whereas inhibition of NF-κB (5 μM BAY11-7085), STAT3 (50 μM S3I-201), MAPK (5 μM SB239063), or MEK-1/2 (5 μM U0126) signal transduction pathways did not antagonize TNFα-induced CAMP expression (Figure 2B,C). In contrast, CAMP mRNA levels were considerably elevated by 50 μM S3I-201 and 5 μM U0126 treatment, in addition to the TNFα-induced effect (Figure 2C). Basal CAMP mRNA levels were not significantly affected by inhibition of any of these signal transduction pathways (Figure 2D).

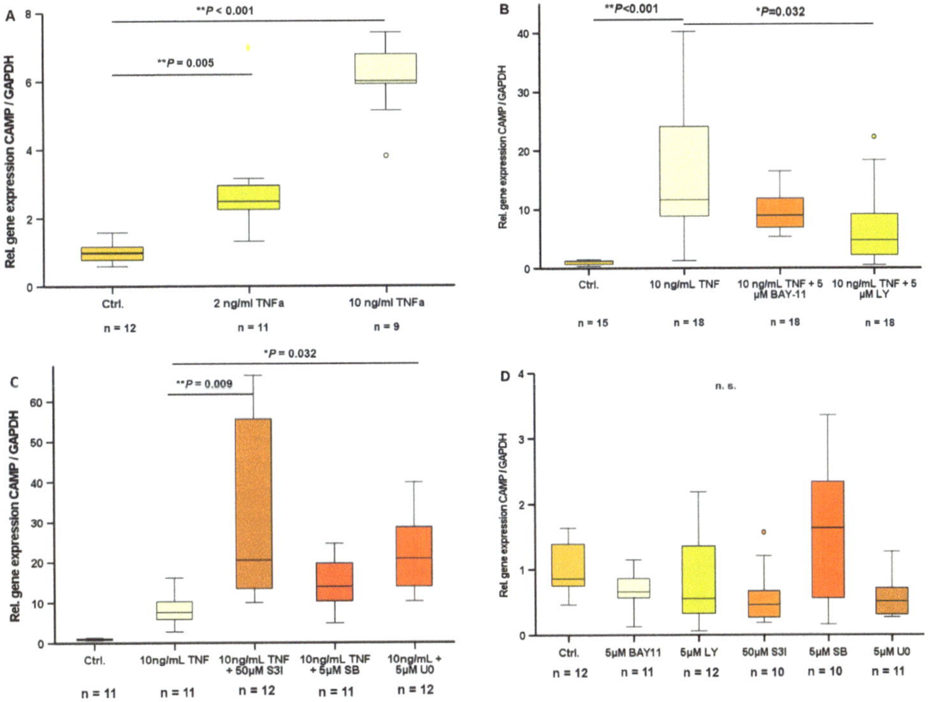

Figure 2. TNFα-induced CAMP gene expression is antagonized by inhibitor of PI3K signaling: TNFα induces CAMP expression in 3T3-L1 adipocytes (**A**). Inhibition of the PI3K pathway by 5 μM LY294002 antagonizes TNFα-induced CAMP expression, whereas inhibition of the NF-κB (5 μM BAY11-7085) (**B**), STAT3 (50 μM S3I-201), MAPK (5 μM SB239063), and MEK-1/-2 (5 μM U0126) pathways does not inhibit TNFα-induced CAMP expression (**C**). Basal CAMP expression is not significantly modified by inhibition of classical signal transduction pathways (**D**) (* $p < 0.05$, ** $p < 0.01$). N = 9–18 samples were investigated. Gene expression levels are given in relative gene expression as compared to control.

2.3. Cell-Free Nucleic Acids (cfDNA) Impair CAMP Expression in 3T3-L1 Adipocytes

Adipocytes release cfDNA as a response to inflammatory stress [18] and cfDNA acts as an endogenous ligand for TLR9 in immune cells [26]. In the present study, we investigated whether cfDNA from adipocytes has an impact on CAMP expression in adipocytes. In

3T3-L1 adipocytes, cfDNA reduced CAMP expression with an effective dosage of 1 μg/mL cfDNA (p = 0.002), whereas 100 ng/mL proved to be ineffective (Figure 3).

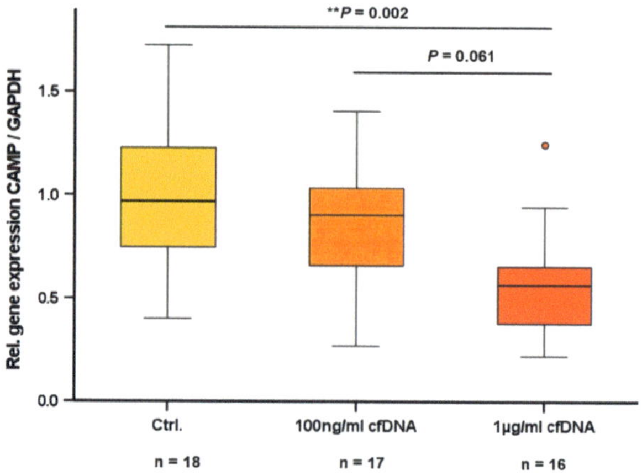

Figure 3. Adipocytic cell-free nucleic acids (cfDNA) impair CAMP expression in 3T3-L1 adipocytes: cfDNA (1 μg/mL) reduces CAMP mRNA levels in 3T3-L1 adipocytes. CAMP expression was investigated by quantitative real-time PCR. Kruskal–Wallis test was applied for calculation of p values and statistical significance ($p < 0.05$). N = 16–18 wells were investigated per experimental setting. Gene expression levels are given in relative gene expression as compared to control. (** $p < 0.01$).

2.4. CAMP and TLR9 mRNA Levels Are Positively Correlated in Human Subcutaneous Adipose Tissue

cfDNA is increased in the adipose tissue and plasma of obese patients [18]. Since cfDNA represents a physiological TLR9 agonist [27] and cfDNA inhibited CAMP gene expression in adipocytes in vitro (Figure 3), gene expression levels of CAMP and TLR9 were analyzed using quantitative real-time PCR in subcutaneous and visceral adipose tissue specimens obtained from obese individuals undergoing bariatric surgery. Interestingly, TLR9 and CAMP gene expression levels in subcutaneous adipose tissue were positively correlated (*rho* = +0.282; p = 0.008; n = 88) (Figure 4A). No correlation of TLR9 and CAMP expression was detected in human visceral adipose tissue (*rho* = +0.034; p = 0.747; n = 90). Furthermore, CAMP mRNA quantities in human visceral and subcutaneous adipose tissue compartments correlated positively (*rho* = +0.224; p = 0.005; n = 157) (Figure 4B).

2.5. CAMP and TLR9 Gene Expression Levels Are Positively Correlated in Murine Subcutaneous Adipose Tissue

It is well known that subcutaneous and intra-abdominal adipose tissues differ in the numerical ratio of adipocytes to stromal vascular cells as well as in functional characteristics [28]. Therefore, we additionally investigated both adipose tissue compartments separately in murine samples. Subcutaneous and intra-abdominal adipose tissue specimens were resected from wildtype C57BL/6J mice. Similar to human adipose tissue samples, CAMP and TLR9 gene expression levels in subcutaneous adipose tissue were found to be positively correlated (rho = +0.657; p = 0.011; n = 14) (Figure 5). In contrast, there was no significant correlation of TLR9 and CAMP mRNA levels in intra-abdominal adipose tissue (rho = −0.178; p = 0.543; n = 14).

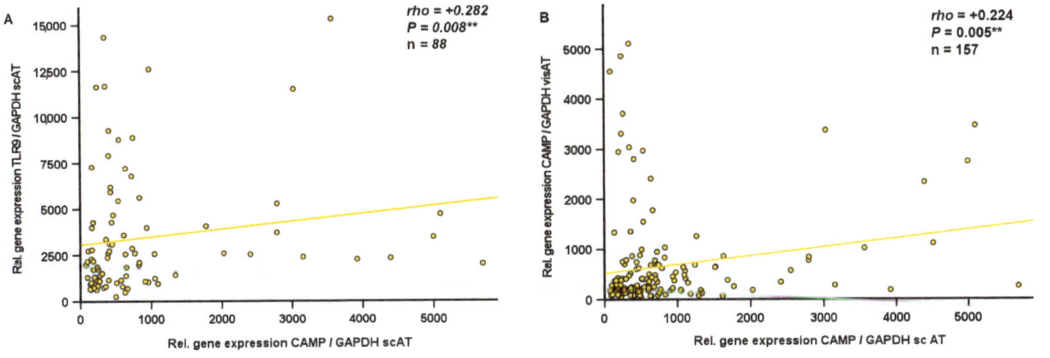

Figure 4. CAMP and TLR9 mRNA levels are positively correlated in human subcutaneous adipose tissue (n = 88) (**A**). CAMP gene expression levels in subcutaneous and visceral adipose tissue are positively correlated (n = 157) (**B**). Specimens of human subcutaneous adipose tissue were collected during bariatric surgery. CAMP and TLR9 expression were investigated by quantitative real-time PCR and Spearman-rho test was applied for calculation of correlation coefficient and statistical significance (** $p < 0.01$).

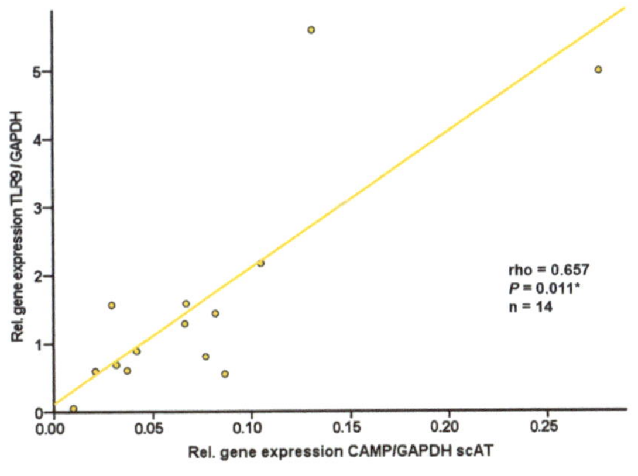

Figure 5. CAMP and TLR9 gene expression levels are positively correlated in murine subcutaneous adipose tissue: TLR9 expression in subcutaneous adipose tissue is positively correlated with CAMP expression in subcutaneous adipose tissue of wildtype mice. Samples of subcutaneous adipose tissue from wildtype mice were collected at the age of 12 months. CAMP and TLR9 expression were investigated by quantitative real-time PCR. The Spearman-rho test was applied for calculation of correlation coefficient and statistical significance ($p < 0.05$). N = 14 samples were investigated. (* $p < 0.05$).

2.6. TLR9 Deficiency Is Associated with Reduced Systemic CAMP Serum Concentrations in Male Mice

To further investigate a possible interrelation between CAMP and TLR9 expression levels, we analyzed blood serum samples from wildtype mice and TLR9 KO mice. Serum samples were collected at the age of 5–8 months and circulating CAMP levels were quantified via ELISA. Of note, CAMP serum concentrations were significantly lower ($p = 0.011$) in male TLR9 KO mice when compared to wildtype mice (WT: 6122.70 ± 1962.79 pg/mL vs. TLR9 KO: 4445.01 ± 826.42 pg/mL) (Figure 6A). This difference between genotypes was not observed in female mice (WT: 3735.38 ± 973.51 pg/mL vs. TLR9 KO: 3937.66 ± 1847.37 pg/mL) (Figure 6B).

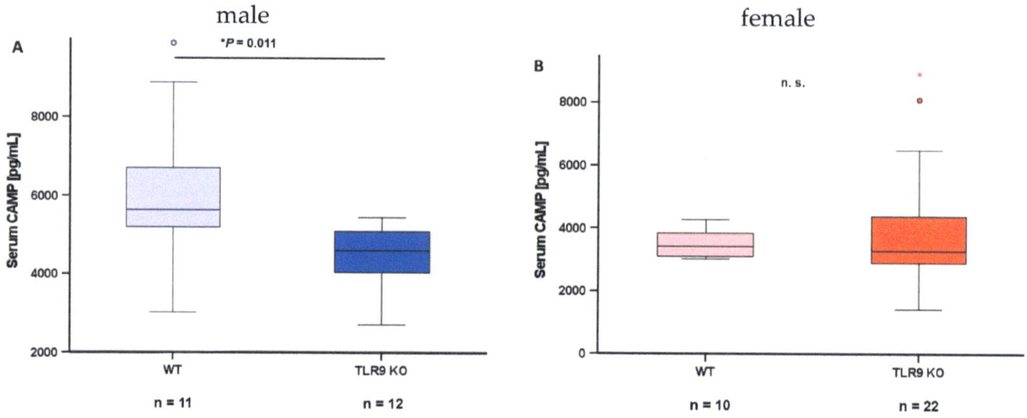

Figure 6. TLR9 KO is associated with reduced systemic CAMP concentrations in male mice: CAMP concentration in blood serum is significantly lower ($p = 0.011$) in male TLR9 KO mice than in wildtype mice (**A**), whereas this difference was not observed in female animals (**B**). Blood serum samples from wildtype and TLR9 KO mice were collected at the age of 5–8 months. Blood serum CAMP concentrations were measured by ELISA and Mann–Whitney U-test was applied for calculation of statistical significance ($p < 0.05$). Samples from n = 10–22 animals per group were investigated. n.s.: not significant, * $p < 0.05$; *: statistical outlier >2.5 interquartile range.

2.7. *Synthetic TLR9 Ligands Do Not Modify CAMP Gene Expression in Murine Adipocytes*

In immune cells, the activation of TLR9 generally leads to an increased production of proinflammatory cytokines [22]. Oligodeoxynucleotides (ODNs) are established synthetic ligands of TLR9 [29]: ODN2087 is a commonly used TLR7/TLR9 inhibitor, whereas ODNA, ODNB, and ODNC represent agonists inducing TLR9 activity [25]. Therefore, we investigated whether ODNA, ODNB, ODNC, and ODN2087 modify basal CAMP gene expression in murine adipocytes in vitro (Figure 7). Doses of 5 µg/mL and 20 µg/mL ODNA, 5 µg/mL, 20 µg/mL ODNB, 10 µg/mL and 20 µg/mL ODN C, and 20 µg/mL ODN2087 were applied in murine 3T3L-1 adipocytes. However, none of the tested ODNs significantly affected adipocyte CAMP gene expression (Figure 7).

2.8. *Knockout of TLR9 Does Not Affect CAMP Gene Expression in Murine Primary Subcutaneous and Intra-Abdominal Adipocytes*

We detected significant correlations of CAMP and TLR9 gene expression in human and murine subcutaneous adipose tissue, and TLR9 knockout in male mice resulted in reduced CAMP serum levels. However, synthetic TLR9 ligands—as opposed to cfDNA—did not modify CAMP gene expression in murine adipocytes in vitro. In order to further investigate the impact of TLR9 on CAMP expression in adipocytes, we tested the cellular effect of TLR9 knockout on CAMP gene expression in primary adipocytes isolated from wildtype mice and TLR9 KO mice. Due to the diverging constitution of subcutaneous and intra-abdominal adipose tissue, we evaluated adipocytes from both compartments separately (Figure 8A,B). Of note, we detected no difference in CAMP gene expression in subcutaneous (Figure 8A) and intra-abdominal (Figure 8B) adipocytes of wildtype and TLR9 knockout mice.

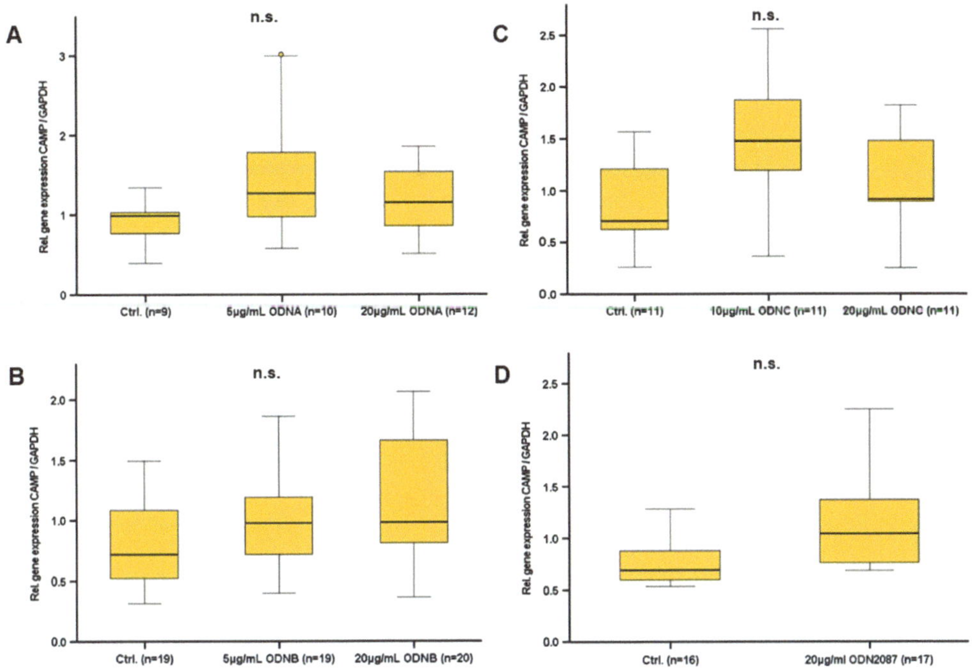

Figure 7. Synthetic TLR9 ligands do not modify CAMP gene expression in murine adipocytes: Stimulation of murine 3T3-L1 adipocytes with various doses of ODNA (**A**), ODNB (**B**), ODNC (**C**), and ODN2087 (**D**) does not significantly affect CAMP gene expression. CAMP expression was investigated by quantitative real-time PCR. The Kruskal–Wallis test (**A**–**C**) and Mann–Whitney U-test (**D**) were applied for calculation of p values and statistical significance ($p < 0.05$). N = 11–20 wells were investigated per experimental setting. Gene expression levels are given in relative gene expression as compared to control. n.s.: not significant.

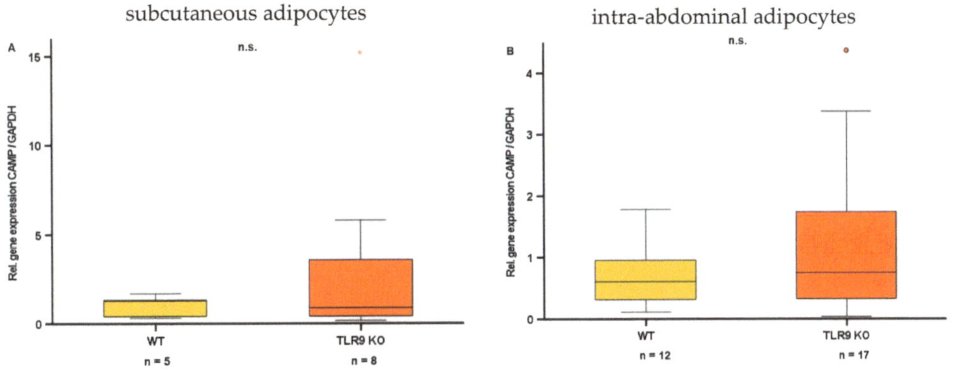

Figure 8. CAMP gene expression in primary adipocytes from subcutaneous (**A**) and intra-abdominal adipose tissue (**B**) does not differ between wildtype (WT) and TLR9 knockout (KO) mice. CAMP expression was investigated by quantitative real-time PCR, and Mann–Whitney U-test was applied for calculation of statistical significance ($p < 0.05$). Samples from n = 5–17 animals per group were analyzed. n.s.: not significant; *: statistical outlier > 2.5 interquartile range.

3. Discussion

During adipocyte differentiation, the expression and secretion of various adipokines increase [30]. Zhang et al. reported that CAMP gene expression in mature murine 3T3-L1 adipocytes is increased when compared to preadipocytes [13], which was confirmed by our group [31]. In the experiments presented here, we could further demonstrate a strong and persistent induction of CAMP gene expression during differentiation of human SGBS preadipocytes into mature adipocytes, a finding comparable to the expression of CAMP during in vitro differentiation of primary human preadipocytes [13]. SGBS adipocytes [32] therefore might represent an applicable cell culture model for the analysis of CAMP regulation in human adipocytes, improving the translational potential of in vitro findings for future clinical and therapeutical applications.

The capacity of host defense in subcutaneous bacterial infection depends on metabolic factors, which is demonstrated by impaired host defense in obesity [14]. Importantly, adipocyte-derived CAMP plays a crucial role in the local host defense of subcutaneous adipose tissue [13]. As an antimicrobial peptide, CAMP is part of the innate immune system. The regulation of CAMP in adipocytes by other factors of the innate immune system is unknown so far. The proinflammatory cytokine TNFα is expressed in adipose tissue and affects adipose tissue-resident macrophage activation [33]. In obesity, increased levels of TNFα are observed, and TNFα levels correlate with the extent of adiposity and associated insulin resistance [33]. Remarkably, we found that TNFα significantly induces CAMP gene expression in 3T3-L1 adipocytes. This inflammatory elevation of CAMP expression was antagonized by the inhibition of PI3K signaling. Of note, pharmacological antagonists of other classical signal transduction pathways exerted either no (NF-κB, MAPK) or even contrary effects (STAT3, MEK-1/2). Basal, non-stimulated CAMP gene expression levels were not affected by the inhibition of these signal transduction pathways, suggesting a specific TNFα-/inflammation-related mechanism.

In obesity, an increased quantity of visceral adipose tissue and adipocyte hypertrophy are associated with systemic low-grade inflammation, representing a key aspect of the metabolic syndrome [6]. Macrophage infiltration into visceral adipose tissue induces local inflammatory stress in adipocytes [18]. Subsequent adipocyte degeneration results in the release of nucleic acids (cell-free DNA, cfDNA) [18], presumably exerting paracrine effects on adipocytes [34]. In obesity, levels of cfDNA in circulating blood are increased and positively correlated with the quantity of visceral adipose tissue [18]. cfDNA has even been suggested as a biomarker for obesity [35]. Local cfDNA acts as an endogenous ligand of TLR9 on immune cells leading to an increased recruitment of these cells to the site of inflammation [26]. Of note, CAMP is required for normal TLR9 function in dendritic cells and macrophages [24]. Our group demonstrated previously that there is significant and functional TLR9 expression in adipocytes [25]. The activation or inhibition of TLR9 activity affects the pattern of adipokine secretion in 3T3-L1 adipocytes in vitro [25]. We therefore aimed to investigate whether adipocyte CAMP gene expression is modulated by cfDNA and if TLR9 is involved in this context.

Interestingly, stimulation with cfDNA significantly reduced CAMP gene expression, suggesting that cfDNA represents a yet unknown paracrine factor in adipose tissue inhibiting CAMP expression in adipocytes. As CAMP gene expression was regulated by the physiological TLR9 agonist cfDNA, we investigated a possible association between CAMP and TLR9 expression in vivo. Considering the differing cellular composition and function of subcutaneous and intra-abdominal adipose tissues, we comparatively investigated these compartments in mice and in human individuals. In subcutaneous adipose tissue of wildtype mice, TLR9 and CAMP mRNA levels were positively correlated, whereas no significant correlation of these parameters was observed in intra-abdominal adipose tissue. In accordance with our findings in mice, TLR9 and CAMP expression were positively correlated in human subcutaneous but not in visceral adipose tissue. Thus, our data indicate a putatively compartment-specific association of TLR9 and CAMP expression in subcutaneous fat. Future studies elaborating on these data should focus on primary adipocytes and immune cells residing in adipose tissue—such as macrophages and T cells—in order to elucidate which cell types predominantly contribute to the observed correlation, and also whether the differences between subcutaneous and intra-abdominal/visceral adipose tissue might depend on the diverging cellular composition of both compartments.

Additionally, CAMP serum levels were measured in C57BL/6J wildtype and TLR9 knockout mice. Systemic CAMP concentrations were significantly lower in TLR9-deficient mice when compared to wildtype animals. Of note, this difference between genotypes was exclusively observed in male but not in female mice. This unexpected sexual dimorphism needs to be further elucidated, bearing in mind that testosterone or estradiol did not exert a significant impact on CAMP expression in adipocytes in previous experiments in vitro [15]. In accordance with the present findings in mice, male patients in this previous study exhibited higher serum concentrations and subcutaneous adipose tissue mRNA levels of CAMP when compared to women [15]. Taken together, these data strongly argue for a sexual dimorphism in adipose tissue CAMP gene expression that might also affect circulating CAMP levels. With the underlying regulatory processes remaining to be elucidated, future studies should address this important issue at the molecular level of sexual hormones, potentially affecting the transcriptional and/or secretory regulation of CAMP in more detail. In particular, future experiments should investigate CAMP gene expression in primary murine adipocytes isolated from males versus females to further elaborate on a potential sexual dimorphism and on a potential impact of impaired adipocyte differentiation.

Since we were able to demonstrate a strong association between the gene expression of CAMP and TLR9 in (subcutaneous) adipose tissue and on systemic levels in vivo, we asked if this association might be mediated by TLR9-dependent CAMP regulation in adipocytes. Thus, we investigated the effect of synthetic TLR9 ligands on CAMP gene expression in adipocytes in vitro. Representing well-established TLR9 ligands, the oligodeoxynucleotides ODN1585 (ODNA), ODN1826 (ODNB), ODN2395 (ODNC), and ODN2087 are known to modify immune responses by modulating TLR9 activity [36]. Of note, CAMP gene expression was affected neither by ODNA, ODNB, ODNC, nor by ODN2087 in adipocytes. Additionally, we analyzed CAMP gene expression in primary adipocytes from wildtype and TLR9 knockout mice. Importantly, we observed no significant difference between CAMP gene expression levels in primary adipocytes from wildtype mice as compared to TLR9 knockout mice, neither for subcutaneous nor intra-abdominal adipose tissue. Together, these findings suggest that—although cfDNA impairs CAMP gene expression in adipocytes in vitro—the association of CAMP and TLR9 expression levels in adipose tissues presumably mainly depends on other cell types that are different from adipocytes, like monocytes or macrophages in the stromal vascular fraction of adipose tissue, which are a known source of CAMP [37]. Furthermore, the observed inhibitory effect of cfDNA on CAMP gene expression in adipocytes might be mediated via additional receptors, for example TLR7, as cfDNA is a putative TLR7 agonist as well [34]. Future studies on the putative regulatory interactions of CAMP and TLR7 in adipocytes are necessary to verify this hypothesis and to further elucidate the mechanisms underlying CAMP expression in adipose tissue.

Overall, we focused on a gene expression analysis of CAMP in adipocytes and adipose tissue in order to gain a basic insight into the regulatory relations between the inflammatory factors of innate immunity and CAMP. The predominant analysis on the level of transcriptional regulation represents a limitation concerning the physiological conclusions drawn from the present work. Therefore, future research elaborating on the provided data should apply immunological techniques such as Western blot and immunocytochemistry in order to analyze CAMP regulation in adipocytes and adipose tissue at the protein level and to confirm and further elucidate the biological relevance of these processes.

4. Materials and Methods

4.1. 3T3-L1 Cell Culture and Stimulation Experiments

Murine 3T3-L1 fibroblasts [38] were cultured and differentiated into mature adipocytes as described previously [39]. Briefly, cells were cultured at 37 °C and 5% CO_2 in Dulbecco's Modified Eagle Medium (Biochrom AG, Berlin, Germany) supplemented with 10% newborn calf serum (Sigma-Aldrich, Deisenhofen, Germany) and were differentiated into adipocytes in DMEM/F12/glutamate medium (Lonza, Basel, Switzerland) supplemented with 20 μM 3-isobutyl-methyl-xanthine (Serva, Heidelberg, Germany), 1 μM corticosterone, 100 nM insulin, 200 μM ascorbate, 2 μg/mL transferrin, 5% fetal calf serum (FCS, Sigma-Aldrich, Deisenhofen, Germany), 1 μM biotin, 17 μM pantothenic acid (all from Sigma Aldrich, Deisenhofen Germany), and 300 μg/mL Pedersen-fetuin (MP Biomedicals, Illkirch, France) [40,41]. A differentiation protocol reported in the literature [38,42–45] was used with slight modifications. Visual control of the cellular phenotype and of lipid accumulation by light-microscopy was performed during all stages of the differentiation process.

Mature adipocytes were incubated under serum-free conditions prior to stimulation experiments. TLR9 agonistic oligodeoxynucleotide (ODN) ODN1585 (referred to as ODNA), ODN1826 (ODNB), ODN2395 (ODNC), and inhibitory ODN2087 were purchased from Invivogen (San Diego, CA, USA) and were dissolved in H2O under sterile conditions. ODNA (5 μg/mL, 20 μg/mL), ODNB (5 μg/mL, 20 μg/mL), ODNC (10 μg/mL, 20 μg/mL), and ODN2087 (20 μg/mL) were applied in two separate overnight (18 h) stimulation experiments. All applied doses were determined following the manufacturer's recommendations for effective dosage as well as by previous preliminary testing for potential cytotoxicity in adipocyte culture. Cell-free nucleic acids (referred to as cfDNA) were isolated from 3T3-L1 adipocytes as described previously [34] and below, and were applied in doses of 100 ng/mL and 1 μg/mL. As an inflammatory stimulus, TNFα was applied at a dosage of 2 ng/mL and 10 ng/mL. Furthermore, co-stimulation experiments were performed with 10 ng/mL TNFα and inhibitors of different signal transduction pathways (NF-κB inhibitor BAY-11 (5 mM), STAT3 inhibitor S3I-201 (50 mM), selective MAPK inhibitor SB239063 (5 mM), MEK-1/-2 inhibitor U0126 (5 mM), and phosphatidylinositol 3-kinase (PI3K) inhibitor LY294002 (5 mM), all purchased from Merck). The dosage was applied as described in our previous study [16].

All applied stimulatory doses were within the concentration range recommended by the manufacturer and had been determined either by preliminary tests or previous experiments in adipocyte culture with respect to dose effects and cell viability. Furthermore, LDH (lactate dehydrogenase) concentration was measured in the cell supernatants (Cytotoxicity Detection Kit, Roche, Mannheim, Germany) of all experiments in order to exclude any unintended cytotoxic effects.

4.2. SGBS Cell Culture

The Simpson–Golabi–Behmel syndrome (SGBS) preadipocyte cell strain represents primary human cells originated from adipose tissue specimens of a patient suffering from SGBS [32]. They were kindly provided by Prof. Martin Wabitsch (University of Ulm, Germany). The cells were differentiated into mature adipocytes within 14 days of culture following the provider's established protocol [32]. SGBS preadipocytes were cultured in DMEM/F12 (1:1) (Invitrogen, Darmstadt, Germany) supplemented with 10% FCS (all purchased from Invitrogen, Darmstadt, Germany). Differentiation into mature adipocytes was induced at confluence. After 3 washing steps with phosphate-buffered saline (PBS), cells were cultured in serum-free medium supplemented with 0.01 g/mL transferrin, 20 nM insulin, 0.2 nM triiodothyronine, and 100 nM cortisol (all purchased from Sigma-Aldrich, Deisenhofen, Germany). During the initial 4 days of differentiation, 2 μM rosiglitazone (BRL 49653) (Cayman, Tallinn, Estonia), 250 μM isobutylmethylxanthine (IBMX), and 25 nM dexamethasone (all purchased from Sigma-Aldrich, Deisenhofen, Germany) were also added to the medium. The culture medium was replaced every third or fourth day. Adipocyte differentiation was completed after 14 days. The characteristic adipocyte morphology was visually controlled by light microscopy.

4.3. Preparation of Cell-Free Nucleic Acids from 3T3-L1 Adipocytes

Prior to isolation of cell-free nucleic acids (cfDNA) using the QIAamp® DNA Micro Kit according to the manufacturer's instructions (Qiagen, Hilden, Germany), murine 3T3-L1 adipocytes were exposed to inflammatory stress induced by stimulation with 50 ng/mL tumor necrosis factor α (TNFα) (Biomol, Hamburg, Germany). After 18 h, adipocytes were harvested and cfDNA was isolated from cell lysates as described earlier [34]. Differing doses of cfDNA (100 ng/mL; 1 μg/mL) were applied for overnight (18 h) stimulation experiments in 3T3-L1, while treatment with solvent control containing no cfDNA served as a control setting.

4.4. Human Visceral and Subcutaneous Adipose Tissues for Gene Expression Analysis

Human adipose tissue specimens were obtained from the ROBS (Research in Obesity and Bariatric Surgery) study [46], an open-label, non-randomized, monocentric, prospective, and observational study of patients undergoing either bariatric surgery or dietary intervention at the University Hospital of Giessen, Germany. Subcutaneous and visceral adipose tissues were obtained intra-surgically from patients receiving bariatric surgery (gastric sleeve or Roux-en-Y gastric bypass). Detailed information about the ROBS study cohort was published previously and can be drawn from the literature [46]. The study was approved by the local ethical committee at the University of Giessen, Germany (file: AZ 101/14). All patients gave informed consent for their participation in the study. Data anonymization and privacy policy were accurately applied.

4.5. Quantification of CAMP Gene Expression in Murine Intra-Abdominal and Subcutaneous Adipose Tissues and Primary Adipocytes

Intra-abdominal and subcutaneous adipose tissue compartments were resected from C57BL/6J-Tlr9^{M7Btlr}/Mmjax mice (TLR9-KO) [25] and wildtype C57BL/6J mice (WT) that were bred under standard conditions and fed normal chow (age 5–8 months). Animals were euthanized for the sampling of serum, subcutaneous and intra-abdominal adipose tissue, and primary adipocytes conformably to the German animal protection law (§4 Abs. 3 *Tierschutzgesetz*). A specific announcement was made at the local ethical committee (*Regierungspraesidium* Giessen; internal registration number: 710_M) that was subsequently approved.

Primary adipocytes from intra-abdominal and subcutaneous adipose tissue were isolated as described earlier [34]. Briefly, small portions of the intra-abdominal and subcutaneous adipose tissue from C57BL/6J-Tlr9^{M7Btlr}/Mmjax mice (TLR9-KO) [25] and wildtype C57BL/6J mice (WT) were digested with 0.225 U/mL collagenase NB 6 (#17458, SERVA Electrophoresis; Heidelberg, Germany) for 30–60 min at 37 °C. In a subsequent centrifugation step (200 rcf, 10 min, 4 °C), the adipocytes were separated from the stromal vascular cells (SVC) and later transferred into TRIzol®-Reagent (Life Technologies GmbH, Darmstadt, Germany) for mRNA isolation, as described below (Section 4.6). The purity of the adipocytes versus SVC after our isolation procedure was assessed by real-time PCR, applying adiponectin as a biomarker for the adipocyte fraction and CD45 (indicating leukocyte lineage) for the stroma-vascular cell fraction [34].

4.6. Isolation of mRNA and Quantitative Real-Time PCR Analysis of CAMP Gene Expression in Murine and Human Adipose Tissues and Adipocytes

Total mRNA was isolated from frozen human and murine total adipose tissues and from 3T3-L1, SGBS, and primary adipocytes following a protocol described previously [39]. Tissues were homogenized in TRIzol®-Reagent (Life Technologies GmbH, Darmstadt, Germany) in combination with gentleMACS dissociator and M-tubes (Miltenyi Biotec GmbH, Bergisch Gladbach, Germany) for dissociation and RNA was isolated applying

RNeasy® Mini Kit (Qiagen, Hilden, Germany) including DNase (RNase-Free DNase Set, Qiagen, Hilden, Germany).

For the gene expression analysis, reverse transcription of RNA (QuantiTect Reverse Transcription Kit from Qiagen, Hilden, Germany) was performed in order to generate corresponding cDNA for real-time PCR (RT-PCR) (iTaq Universal SYBR Green Supermix, CFX Connect RT-PCR system; Bio-Rad, Munich, Germany). Expression levels of the target gene CAMP were normalized to the gene expression of glyceraldehyde-3-phosphate dehydrogenase (GAPDH) as a house-keeping gene applying the $\Delta\Delta$CT method. The following primer-pairs were used:

Murine CAMP:
5′-CCCAAGTCTGTGAGGTTCCG-3′/5′-GTGCACCAGGCTCGTTACA-3′
Murine GAPDH:
5′-TGTCCGTCGTGGATCTGAC-3′/5′-AGGGAGATGCTCAGTGTTGG-3′
Murine TLR9:
5′-CATCTCCCAACATGGTTCTCC-3′/5′-GCAGAGAAACGGGGTACAGA-3′
Human CAMP:
5′-TAGATGGCATCAACCAGCGG-3′/5′-CTGGGTCCCCATCCATCGT-3′
Human GAPDH:
5′-GAGTCCACTGGCGTCTTCAC-3′/5′-CCAGGGGTGCTAAGCAGTT-3′
Human TLR9:
5′-CCCCCAGCATGGGTTTCT-3′/5′-TGGAGCTCACAGGGTAGGAA-3′

All oligonucleotides used were purchased from Metabion, Martinsried, Germany.

4.7. Quantification of CAMP Concentrations in Murine Blood Serum

Healthy C57BL/6J (WT) and C57BL/6J-Tlr9^{M7Btlr}/Mmjax mice (TLR9-KO) were bred under standard conditions being fed a chow diet *ad libitum* until euthanasia for organ and tissue resection and blood serum collection. Concentrations of CAMP in blood serum from C57BL/6J-Tlr9^{M7Btlr}/Mmjax mice (TLR9-KO) and wildtype C57BL/6J mice (WT) (age 5–8 months) were measured in technical duplicates by enzyme-linked immunosorbent assay (ELISA) (kit purchased from Abbexa Ltd., Cambridge, UK) and are expressed as means ± standard deviation in the text and displayed as box plots in the figures. The test range of the applied ELISA kit was 1.56–100 ng/mL. All measurements exceeding an intra-duplicate variance of 20% were repeated. Animal studies were performed at the University of Giessen, Germany, and were approved by the local government agency.

4.8. Statistical Analysis

For explorative data analysis, a statistical software package (SPSS 28.0) was used. Non-parametric numerical parameters were analyzed using the Mann–Whitney U-test (for 2 unrelated samples) or the Kruskal–Wallis test (>2 unrelated samples). A correlation analysis was performed by using the Spearman-rho test. p values below 0.05 (two-tailed) were considered statistically significant. In the figures, the data are presented as box plots, with the box representing the second and third quartile of values, the whiskers giving the inter quartile range, and statistical outliers outside inter quartile range being indicated by dots and stars.

5. Conclusions

As a part of the innate immune system, adipocytes, as well as adipose tissue as a whole organ, play a crucial role in immune-metabolism ("metaflammation", "adipose inflammation") [47]. Here, we investigate and discuss a novel aspect of the inflammation-mediated regulation of CAMP in adipocytes and adipose tissues. As visualized in Figure 9, our results implicate that CAMP expression in adipocytes is induced by TNFα involving the PI3K pathway. Furthermore, the physiological TLR7/TLR9 ligand cell-free nucleic acids (cfDNA) impairs CAMP gene expression in adipocytes. Further studies are necessary in order to investigate the molecular mechanisms and functional consequences of this novel relation within innate immunity and metaflammation in more detail. Most importantly, TNFα, CAMP, and adipose tissue-derived cfDNA might provide potential molecular targets in future anti-inflammatory drug therapies addressing metabolic inflammation and concomitant morbidities.

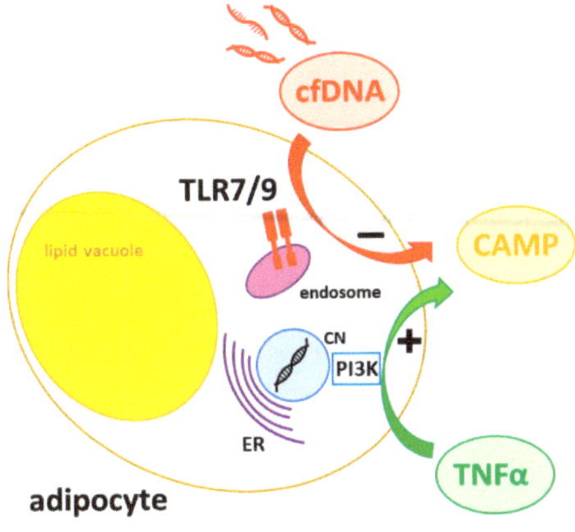

Figure 9. Putative mechanisms of CAMP regulation in adipocytes. Adipocyte CAMP gene expression is induced by TNFα via the PI3K pathway. The impaired CAMP gene expression by cfDNA might be modulated via endosomal receptors, e.g., TLR7/9. ER: endoplasmatic reticulum; CN: cell nucleus.

Author Contributions: Conceptualization, A.H., T.K., A.S. (Andreas Schmid) and A.S. (Andreas Schäffler); methodology, A.H., T.K. and A.S. (Andreas Schmid); validation, A.H. and A.S. (Andreas Schmid); formal analysis, A.H., T.K. and A.S. (Andreas Schmid); investigation, L.S., M.P., A.W. and A.S. (Andreas Schmid); resources, A.H. and T.K.; data curation, A.H. and A.S. (Andreas Schmid); writing—original draft preparation, A.H. and A.S. (Andreas Schmid); writing—review and editing, A.H., T.K., L.S., M.P., A.W., A.S. (Andreas Schmid) and A.S. (Andreas Schäffler); visualization, A.H. and A.S. (Andreas Schmid); supervision, T.K. and A.S. (Andreas Schäffler); project administration, T.K. and A.S. (Andreas Schäffler); funding acquisition, A.H. and T.K. All authors have read and agreed to the published version of the manuscript.

Funding: The research was funded by DEUTSCHE FORSCHUNGSGEMEINSCHAFT (to A. Höpfinger, grant number HO 6929/2-1, and to T. Karrasch, grant number KA 1846/4-2) and from UKGM GIESSEN (to A. Höpfinger, grant number 3/2022 GI).

Institutional Review Board Statement: The study was conducted in accordance with the Declaration of Helsinki, and approved by the local Institutional Ethics Committee of Giessen University Hospital, Germany (Ethik-Kommission am Fachbereich Medizin, 12.06.2014, identification code: 101/14). Animal experiments were performed at the University of Giessen, Germany, and all sampling of animal tissue was conducted according to an announcement made at the local Ethical Committee (Regierungspräsidium Giessen) conformable to §4 Abs. 3 Tierschutzgesetz on 12 March 2019. Internal project identification code 710_M was assigned to the project at the University of Giessen. The study did not comprise any animal experiments subject to approval by the Ethics Committee.

Informed Consent Statement: Informed consent was obtained from all subjects involved in the study.

Data Availability Statement: The data presented in this study are available on reasonable request from the corresponding author.

Acknowledgments: The excellent administrative and technical support of the ROBS study consortium is highly appreciated. We thank Kathrin Ebeling, Lisa Knüpfer, Nicole Odenthal, Melina Gwiazdowski, and Veronika Eckert for excellent technical assistance during in vitro experiments and sample acquisition.

Conflicts of Interest: The authors declare no conflict of interest. The funders had no role in the design of the study; in the collection, analyses, or interpretation of data; in the writing of the manuscript; or in the decision to publish the results.

References

1. Engin, A. The Definition and Prevalence of Obesity and Metabolic Syndrome. *Adv. Exp. Med. Biol.* **2017**, *960*, 1–17. [CrossRef] [PubMed]
2. Übergewicht und Adipositas | Statista. Available online: https://de.statista.com/statistik/studie/id/14790/dokument/uebergewicht-und-adipositas-statista-dossier/ (accessed on 24 July 2022).
3. Jokinen, E. Obesity and cardiovascular disease. *Minerva Pediatr.* **2015**, *67*, 25–32. [PubMed]
4. Silveira Rossi, J.L.; Barbalho, S.M.; Reverete de Araujo, R.; Bechara, M.D.; Sloan, K.P.; Sloan, L.A. Metabolic syndrome and cardiovascular diseases: Going beyond traditional risk factors. *Diabetes. Metab. Res. Rev.* **2022**, *38*, e3502. [CrossRef] [PubMed]
5. Mathers, C.D.; Loncar, D. Projections of global mortality and burden of disease from 2002 to 2030. *PLoS Med.* **2006**, *3*, e442. [CrossRef]
6. Kawai, T.; Autieri, M.V.; Scalia, R. Adipose tissue inflammation and metabolic dysfunction in obesity. *Am. J. Physiol. Cell Physiol.* **2021**, *320*, C375–C391. [CrossRef]
7. Suganami, T.; Ogawa, Y. Role of chronic inflammation in adipose tissue in the pathophysiology of obesity. *Nihon Rinsho* **2013**, *71*, 225–230.
8. Agerberth, B.; Charo, J.; Werr, J.; Olsson, B.; Idali, F.; Lindbom, L.; Kiessling, R.; Jörnvall, H.; Wigzell, H.; Gudmundsson, G.H. The human antimicrobial and chemotactic peptides LL-37 and alpha-defensins are expressed by specific lymphocyte and monocyte populations. *Blood* **2000**, *96*, 3086–3093. [CrossRef]
9. de Yang, D.; Chen, Q.; Schmidt, A.P.; Anderson, G.M.; Wang, J.M.; Wooters, J.; Oppenheim, J.J.; Chertov, O. LL-37, the neutrophil granule- and epithelial cell-derived cathelicidin, utilizes formyl peptide receptor-like 1 (FPRL1) as a receptor to chemoattract human peripheral blood neutrophils, monocytes, and T cells. *J. Exp. Med.* **2000**, *192*, 1069–1074. [CrossRef]
10. Tjabringa, G.S.; Ninaber, D.K.; Drijfhout, J.W.; Rabe, K.F.; Hiemstra, P.S. Human cathelicidin LL-37 is a chemoattractant for eosinophils and neutrophils that acts via formyl-peptide receptors. *Int. Arch. Allergy Immunol.* **2006**, *140*, 103–112. [CrossRef]
11. Lee, E.Y.; Lee, M.W.; Wong, G.C.L. Modulation of toll-like receptor signaling by antimicrobial peptides. *Semin. Cell Dev. Biol.* **2019**, *88*, 173–184. [CrossRef]
12. Gao, W.; Xing, L.; Qu, P.; Tan, T.; Yang, N.; Li, D.; Chen, H.; Feng, X. Identification of a novel cathelicidin antimicrobial peptide from ducks and determination of its functional activity and antibacterial mechanism. *Sci. Rep.* **2015**, *5*, 17260. [CrossRef] [PubMed]
13. Zhang, L.-J.; Guerrero-Juarez, C.F.; Hata, T.; Bapat, S.P.; Ramos, R.; Plikus, M.V.; Gallo, R.L. Innate immunity. Dermal adipocytes protect against invasive Staphylococcus aureus skin infection. *Science* **2015**, *347*, 67–71. [CrossRef]
14. Zhang, L.-J.; Guerrero-Juarez, C.F.; Chen, S.X.; Zhang, X.; Yin, M.; Li, F.; Wu, S.; Chen, J.; Li, M.; Liu, Y.; et al. Diet-induced obesity promotes infection by impairment of the innate antimicrobial defense function of dermal adipocyte progenitors. *Sci. Transl. Med.* **2021**, *13*, 577. [CrossRef] [PubMed]
15. Hochberg, A.; Patz, M.; Karrasch, T.; Schäffler, A.; Schmid, A. Serum Levels and Adipose Tissue Gene Expression of Cathelicidin Antimicrobial Peptide (CAMP) in Obesity and During Weight Loss. *Horm. Metab. Res.* **2021**, *53*, 169–177. [CrossRef] [PubMed]
16. Höpfinger, A.; Karrasch, T.; Schäffler, A.; Schmid, A. Regulation of CAMP (cathelicidin antimicrobial peptide) expression in adipocytes by TLR 2 and 4. *Innate Immun.* **2021**, *27*, 184–191. [CrossRef] [PubMed]
17. Karrasch, T.; Höpfinger, A.; Schäffler, A.; Schmid, A. The adipokine C1q/TNF-related protein-3 (CTRP-3) inhibits Toll-like receptor (TLR)-induced expression of Cathelicidin antimicrobial peptide (CAMP) in adipocytes. *Cytokine* **2021**, *148*, 155663. [CrossRef]
18. Nishimoto, S.; Fukuda, D.; Higashikuni, Y.; Tanaka, K.; Hirata, Y.; Murata, C.; Kim-Kaneyama, J.-R.; Sato, F.; Bando, M.; Yagi, S.; et al. Obesity-induced DNA released from adipocytes stimulates chronic adipose tissue inflammation and insulin resistance. *Sci. Adv.* **2016**, *2*, e1501332. [CrossRef]
19. Blasius, A.L.; Beutler, B. Intracellular toll-like receptors. *Immunity* **2010**, *32*, 305–315. [CrossRef]
20. Hemmi, H.; Takeuchi, O.; Kawai, T.; Kaisho, T.; Sato, S.; Sanjo, H.; Matsumoto, M.; Hoshino, K.; Wagner, H.; Takeda, K.; et al. A Toll-like receptor recognizes bacterial DNA. *Nature* **2000**, *408*, 740–745. [CrossRef]
21. Krug, A.; French, A.R.; Barchet, W.; Fischer, J.A.A.; Dzionek, A.; Pingel, J.T.; Orihuela, M.M.; Akira, S.; Yokoyama, W.M.; Colonna, M. TLR9-dependent recognition of MCMV by IPC and DC generates coordinated cytokine responses that activate antiviral NK cell function. *Immunity* **2004**, *21*, 107–119. [CrossRef]
22. Kawai, T.; Akira, S. Toll-like receptors and their crosstalk with other innate receptors in infection and immunity. *Immunity* **2011**, *34*, 637–650. [CrossRef] [PubMed]
23. Revelo, X.S.; Ghazarian, M.; Chng, M.H.Y.; Luck, H.; Kim, J.H.; Zeng, K.; Shi, S.Y.; Tsai, S.; Lei, H.; Kenkel, J.; et al. Nucleic Acid-Targeting Pathways Promote Inflammation in Obesity-Related Insulin Resistance. *Cell Rep.* **2016**, *16*, 717–730. [CrossRef] [PubMed]
24. Nakagawa, Y.; Gallo, R.L. Endogenous intracellular cathelicidin enhances TLR9 activation in dendritic cells and macrophages. *J. Immunol.* **2015**, *194*, 1274–1284. [CrossRef] [PubMed]

25. Thomalla, M.; Schmid, A.; Neumann, E.; Pfefferle, P.I.; Müller-Ladner, U.; Schäffler, A.; Karrasch, T. Evidence of an anti-inflammatory toll-like receptor 9 (TLR 9) pathway in adipocytes. *J. Endocrinol.* **2019**, *240*, 325–343. [CrossRef]
26. Guillerey, C.; Mouriès, J.; Polo, G.; Doyen, N.; Law, H.K.W.; Chan, S.; Kastner, P.; Leclerc, C.; Dadaglio, G. Pivotal role of plasmacytoid dendritic cells in inflammation and NK-cell responses after TLR9 triggering in mice. *Blood* **2012**, *120*, 90–99. [CrossRef]
27. Pohar, J.; Lainšček, D.; Ivičak-Kocjan, K.; Cajnko, M.-M.; Jerala, R.; Benčina, M. Short single-stranded DNA degradation products augment the activation of Toll-like receptor 9. *Nat. Commun.* **2017**, *8*, 15363. [CrossRef]
28. Ibrahim, M.M. Subcutaneous and visceral adipose tissue: Structural and functional differences. *Obes. Rev.* **2010**, *11*, 11–18. [CrossRef]
29. Takeshita, F.; Gursel, I.; Ishii, K.J.; Suzuki, K.; Gursel, M.; Klinman, D.M. Signal transduction pathways mediated by the interaction of CpG DNA with Toll-like receptor 9. *Semin. Immunol.* **2004**, *16*, 17–22. [CrossRef]
30. Sheng, X.; Tucci, J.; Malvar, J.; Mittelman, S.D. Adipocyte differentiation is affected by media height above the cell layer. *Int. J. Obes.* **2014**, *38*, 315–320. [CrossRef]
31. Schmid, A.; Karrasch, T.; Thomalla, M.; Schlegel, J.; Salzberger, B.; Schäffler, A.; Hanses, F. Innate Immunity of Adipose Tissue in Rodent Models of Local and Systemic Staphylococcus aureus Infection. *Mediators Inflamm.* **2017**, *2017*, 5315602. [CrossRef]
32. Wabitsch, M.; Brenner, R.E.; Melzner, I.; Braun, M.; Möller, P.; Heinze, E.; Debatin, K.M.; Hauner, H. Characterization of a human preadipocyte cell strain with high capacity for adipose differentiation. *Int. J. Obes. Relat. Metab. Disord.* **2001**, *25*, 8–15. [CrossRef] [PubMed]
33. Tzanavari, T.; Giannogonas, P.; Karalis, K.P. TNF-alpha and obesity. *Curr. Dir. Autoimmun.* **2010**, *11*, 145–156. [CrossRef] [PubMed]
34. Thomalla, M.; Schmid, A.; Hehner, J.; Koehler, S.; Neumann, E.; Müller-Ladner, U.; Schäffler, A.; Karrasch, T. Toll-like Receptor 7 (TLR7) Is Expressed in Adipocytes and the Pharmacological TLR7 Agonist Imiquimod and Adipocyte-Derived Cell-Free Nucleic Acids (cfDNA) Regulate Adipocyte Function. *Int. J. Mol. Sci.* **2022**, *23*, 8475. [CrossRef] [PubMed]
35. Camuzi Zovico, P.V.; Gasparini Neto, V.H.; Venâncio, F.A.; Soares Miguel, G.P.; Graça Pedrosa, R.; Kenji Haraguchi, F.; Barauna, V.G. Cell-free DNA as an obesity biomarker. *Physiol. Res.* **2020**, *69*, 515–520. [CrossRef]
36. Becker, Y. CpG ODNs treatments of HIV-1 infected patients may cause the decline of transmission in high risk populations—A review, hypothesis and implications. *Virus Genes* **2005**, *30*, 251–266. [CrossRef]
37. Kościuczuk, E.M.; Lisowski, P.; Jarczak, J.; Strzałkowska, N.; Jóźwik, A.; Horbańczuk, J.; Krzyżewski, J.; Zwierzchowski, L.; Bagnicka, E. Cathelicidins: Family of antimicrobial peptides. A review. *Mol. Biol. Rep.* **2012**, *39*, 10957–10970. [CrossRef]
38. Green, H.; Kehinde, O. An established preadipose cell line and its differentiation in culture II. Factors affecting the adipose conversion. *Cell* **1975**, *5*, 19–27. [CrossRef]
39. Schmid, A.; Hochberg, A.; Kreiß, A.F.; Gehl, J.; Patz, M.; Thomalla, M.; Hanses, F.; Karrasch, T.; Schäffler, A. Role of progranulin in adipose tissue innate immunity. *Cytokine* **2019**, *125*, 154796. [CrossRef]
40. Zaitsu, H.; Serrero, G. Pedersen fetuin contains three adipogenic factors with distinct biochemical characteristics. *J. Cell. Physiol.* **1990**, *144*, 485–491. [CrossRef]
41. Bachmeier, M.; Löffler, G. Adipogenic activities in commercial preparations of fetuin. *Horm. Metab. Res.* **1994**, *26*, 92–96. [CrossRef]
42. Green, H.; Kehinde, O. Formation of normally differentiated subcutaneous fat pads by an established preadipose cell line. *J. Cell. Physiol.* **1979**, *101*, 169–171. [CrossRef] [PubMed]
43. Green, H.; Meuth, M. An established pre-adipose cell line and its differentiation in culture. *Cell* **1974**, *3*, 127–133. [CrossRef] [PubMed]
44. MacDougald, O.A. Transcriptional Regulation of Gene Expression During Adipocyte Differentiation. *Annu. Rev. Biochem.* **1995**, *64*, 345–373. [CrossRef] [PubMed]
45. Cornelius, P. Regulation of Adipocyte Development. *Annu. Rev. Nutr.* **1994**, *14*, 99–129. [CrossRef]
46. Brock, J.; Schmid, A.; Karrasch, T.; Pfefferle, P.; Schlegel, J.; Busse, I.; Hauenschild, A.; Schmidt, B.; Koukou, M.; Arapogianni, E.; et al. Progranulin serum levels and gene expression in subcutaneous vs visceral adipose tissue of severely obese patients undergoing bariatric surgery. *Clin. Endocrinol.* **2019**, *91*, 400–410. [CrossRef] [PubMed]
47. Hotamisligil, G.S. Inflammation, metaflammation and immunometabolic disorders. *Nature* **2017**, *542*, 177–185. [CrossRef]

International Journal of
Molecular Sciences

MDPI

Communication

MYD88 and Proinflammatory Chemokines in Aortic Atheromatosis: Exploring Novel Statin Effects

Konstantinos S. Mylonas [1,*], Michail Peroulis [2], Dimitrios Schizas [3] and Alkistis Kapelouzou [4]

1 Department of Cardiac Surgery, Onassis Cardiac Surgery Center, 176 71 Athens, Greece
2 Vascular Surgery Unit, Department of Surgery, Faculty of Medicine, University of Ioannina, 451 10 Ioannina, Greece; mperoulis@gmail.com
3 First Department of Cardiac Surgery, Laikon General Hospital, National and Kapodistrian University of Athens, 176 71 Athens, Greece; schizasad@gmail.com
4 Clinical, Experimental Surgery & Translational Research, Biomedical Research Foundation Academy of Athens, 176 71 Athens, Greece; akapel@bioacademy.gr
* Correspondence: ksmylonas@gmail.com; Tel.: +306-987-024653

Abstract: Atherosclerosis is driven by a diverse range of cellular and molecular processes. In the present study, we sought to better understand how statins mitigate proatherogenic inflammation. 48 male New Zealand rabbits were divided into eight groups, each including 6 animals. The control groups received normal chow for 90 and 120 days. Three groups underwent a hypercholesterolemic diet (HCD) for 30, 60, and 90 days. Another three groups underwent HCD for 3 months, followed by normal chow for one month, with or without rosuvastatin or fluvastatin. The cytokine and chemokine expressions were assessed in the samples of thoracic and abdominal aorta. Rosuvastatin significantly reduced MYD88, CCL4, CCL20, CCR2, TNF-α, IFN-β, IL-1b, IL-2, IL-4, IL-8, and IL-10, both in the thoracic and abdominal aorta. Fluvastatin also downregulated MYD88, CCR2, IFN-β, IFN-γ, IL-1b, IL-2, IL-4, and IL-10 in both aortic segments. Rosuvastatin curtailed the expression of CCL4, IFN-β, IL-2, IL-4, and IL-10 more effectively than fluvastatin in both types of tissue. MYD88, TNF-α, IL-1b, and IL-8 showed a stronger downregulation with rosuvastatin compared to fluvastatin only in the thoracic aorta. The CCL20 and CCR2 levels reduced more extensively with rosuvastatin treatment only in abdominal aortic tissue. In conclusion, statin therapy can halt proatherogenic inflammation in hyperlipidemic animals. Rosuvastatin may be more effective in downregulating MYD88 in atherosclerotic thoracic aortas.

Keywords: MYD88; CCL4; CCL20; CCR2; atherosclerosis; statin

Citation: Mylonas, K.S.; Peroulis, M.; Schizas, D.; Kapelouzou, A. MYD88 and Proinflammatory Chemokines in Aortic Atheromatosis: Exploring Novel Statin Effects. *Int. J. Mol. Sci.* **2023**, *24*, 9248. https://doi.org/10.3390/ijms24119248

Academic Editors: Oliver Planz and Ralf Kircheis

Received: 22 April 2023
Revised: 15 May 2023
Accepted: 24 May 2023
Published: 25 May 2023

1. Introduction

Atherosclerotic cardiovascular disease (CVD) is the primary cause of death worldwide. In 2015, more than 17 million people died from CVD, accounting for over 30% of all deaths globally [1]. Among these deaths, coronary heart disease and stroke were responsible for an estimated 7.4 million and 6.7 million, respectively. In the United States, the cost of treating CVD exceeds USD 300 billion per year, with predictions suggesting that the total cost, both direct and indirect, could reach almost USD trillion by 2030 [2].

From a pathophysiological standpoint, atherosclerosis constitutes a complex disease involving a wide array of cellular and molecular mechanisms [3,4]. Indeed, both the innate and the adaptive arms of the immune system participate in various steps of atherogenesis [5]. Within this context, toll-like receptors (TLRs) have emerged as the key players in inflammatory phenomena that drive the development of atherosclerosis [6,7]. An extensive body of literature has actually shown that a hyperlipidemic diet can induce an overexpression of proinflammatory cytokines and TLRs both in rabbit and human aortas [8–10]. According to recent data, highly immunogenic oxidized low-density lipoproteins (LDL) are primarily associated with the upregulation of TLR2/4 and others [11,12].

TLR signaling pathways can be categorized into two categories: a MYD88 (myeloid differentiation primary response protein 88)-dependent pathway, which triggers NF-κB (nuclear factor kappa-light-chain-enhancer of activated B cells) activation, and a TRIF (TIR domain-containing adapter-inducing IFN-β)-dependent pathway [13,14]. MYD88 is crucial for the activation of all TLRs (excluding TLR3) and some members of the IL-1 receptor family. It stimulates the MAPK (mitogen-activated protein kinase) and NF-κB signaling pathways, leading to the production and release of proatherogenic cytokines [15].

In the present animal model, we assessed the expression of MYD88 and its correlation with inflammatory markers in the setting of aortic atherosclerosis. We also explored the effect of commonly used statins on these factors.

2. Results

2.1. Effects of Atherogenesis and Statin Therapy on Aortic Cytokine mRNA Expression

The expression patterns of MYD88, NF-kB, CCL4, CCL20, CCR2, IFN-β, IFN-γ, TNF-α, IL-1b, IL-2, IL-4, IL-8, IL-10, and tIL-18 were measured against β-actin during the course of atherogenesis (G30, G60, G90, G120) and treatment with statins (GF and GR). In the control groups (C90 and C120), the mRNA expression levels of all biomarkers were very low both in the thoracic and abdominal aortas. The cytokine expression increased during the process of atherosclerosis and decreased with statin treatment. The mRNA production of all biomarkers in the thoracic aorta were higher compared to the abdominal aorta (Tables 1 and 2).

Table 1. Statistical analysis of the thoracic aorta mRNA expression for each biomarker across all study groups.

Thoracic								
Groups	**C90**	**G30**	**G60**	**G90**	**C120**	**G120**	**GF120**	**GR 120**
N	6	6	6	6	6	6	6	6
	MYD88							
Mean	0.19	0.84	1.99	2.28	0.49	4.81	3.94	1.59
SD	0.05	0.17	0.40	0.21	0.25	0.46	0.49	0.23
	NF-kB							
Mean	0.18	1.70	4.21	6.49	0.17	11.44	10.15	9.18
SD	0.02	0.21	0.81	0.74	0.04	2.25	1.80	1.63
	CCL4							
Mean	0.18	6.78	8.92	12.06	0.18	16.03	12.58	6.11
SD	0.03	0.92	0.93	2.05	0.05	1.80	1.42	0.99
	CCL20							
Mean	0.18	9.20	12.45	16.23	0.20	20.75	18.26	15.84
SD	0.03	0.65	1.80	1.87	0.03	1.81	2.68	3.06
	CCR2							
Mean	0.19	9.02	11.85	17.14	0.18	20.29	3.51	2.36
SD	0.02	0.74	1.17	1.76	0.06	2.60	1.55	1.09
	IFNβ							
Mean	0.16	2.87	3.19	3.21	0.18	7.16	3.25	1.55
SD	0.04	0.70	0.56	0.43	0.03	0.99	0.11	0.24

Table 1. *Cont.*

Thoracic								
Groups	**C90**	**G30**	**G60**	**G90**	**C120**	**G120**	**GF120**	**GR 120**
	IFNγ							
Mean	0.15	1.37	2.38	2.66	0.17	2.84	1.58	1.33
SD	0.04	0.15	0.31	0.41	0.04	0.35	0.47	0.34
	TNFα							
Mean	0.25	5.25	7.39	12.98	0.19	15.76	11.49	8.55
SD	0.17	1.02	0.45	0.88	0.04	0.90	1.75	0.87
	Ilb							
Mean	0.14	4.48	6.73	7.43	0.13	11.46	5.05	3.43
SD	0.04	0.50	0.51	0.99	0.06	1.69	1.00	0.74
	IL-2							
Mean	0.18	13.91	15.27	20.04	0.17	24.25	13.82	7.62
SD	0.01	1.69	1.72	2.47	0.05	1.83	0.99	1.61
	IL-4							
Mean	0.13	10.50	14.27	21.83	0.15	30.30	12.39	10.78
SD	0.02	2.42	2.59	2.04	0.03	3.53	3.09	1.92
	IL-8							
Mean	0.17	2.86	5.38	8.25	0.22	10.10	8.76	7.55
SD	0.05	1.14	1.12	1.74	0.05	1.62	1.90	1.74
	IL-10							
Mean	0.17	3.14	4.41	5.91	0.18	6.63	4.88	2.71
SD	0.01	1.16	0.70	0.99	0.02	1.20	1.53	0.73
	IL-18							
Mean	0.20	1.20	1.66	2.65	0.19	3.15	2.10	1.91
SD	0.01	0.63	0.57	1.11	0.03	1.05	0.12	0.12

Myeloid differentiation primary response 88 (MYD88); nuclear factor kappa-light-chain-enhancer of activated B cells (NF-κB); C–C chemokine ligand (CCL4); chemokine (C–C motif) ligand 20 (CCL20); C–C chemokine receptor type 2 (CCR2); interferon beta (IFN-β); interferon gamma (IFN-γ); tumor necrosis factor alpha (TNF-α); interleukin 1b (IL-1b); interleukin 2 (IL-2); interleukin 4 (IL-4); interleukin 8 (IL-8); interleukin 10 (IL-10); interleukin 18 (IL-18).

Table 2. Statistical analysis of the abdominal aorta mRNA expression for each biomarker across all study groups.

Abdominal								
Groups	**C90**	**G30**	**G60**	**G90**	**C120**	**G120**	**GF120**	**GR 120**
N	6	6	6	6	6	6	6	6
	MYD88							
Mean	0.16	0.55	0.44	1.27	0.18	2.61	2.01	1.76
SD	0.03	0.25	0.22	0.35	0.06	0.44	0.40	0.28
	NF-kB							
Mean	0.18	1.25	2.57	3.88	0.16	9.82	8.32	8.26
SD	0.02	0.21	0.32	0.25	0.05	1.89	2.05	1.35

Table 2. *Cont.*

Abdominal								
Groups	**C90**	**G30**	**G60**	**G90**	**C120**	**G120**	**GF120**	**GR 120**
	CCL4							
Mean	0.17	5.37	7.21	8.55	0.17	9.66	8.39	7.52
SD	0.02	1.09	0.77	0.86	0.05	1.63	1.60	1.42
	CCL20							
Mean	0.18	7.06	10.4	13.83	0.22	17.33	12.70	6.21
SD	0.03	2.48	1.00	2.24	0.07	2.32	5.72	1.15
	CCR2							
Mean	0.19	5.88	7.45	12.16	0.18	16.84	4.31	1.98
SD	0.02	1.56	0.91	1.57	0.04	1.36	1.53	0.6
	IFNβ							
Mean	0.16	1.96	2.36	2.40	0.18	5.37	2.61	1.35
SD	0.02	0.44	0.33	0.50	0.03	0.48	0.45	0.20
	IFNγ							
Mean	0.16	0.88	1.50	1.60	0.17	5.91	1.72	1.57
SD	0.03	0.13	0.32	0.33	0.03	0.34	0.23	0.23
	TNFα							
Mean	0.30	347	514	9.88	0.27	12.98	11.58	10.58
SD	0.19	0.47	0.76	1.06	0.16	1.40	1.34	1.22
	Ilb							
Mean	0.12	2.11	2.60	4.36	0.17	5.69	2.85	2.22
SD	0.04	0.54	0.44	0.69	0.05	0.46	0.82	0.44
	IL-2							
Mean	0.15	10.35	12.50	14.50	0.22	15.67	13.22	6.31
SD	0.03	1.30	1.37	1.51	0.07	2.25	2.16	1.66
	IL-4							
Mean	0.12	9.59	12.27	14.42	0.15	15.92	13.65	8.96
SD	0.03	0.67	1.36	2.48	0.04	1.87	2.48	1.45
	IL-8							
Mean	0.173	2.08	3.38	5.13	0.19	7.01	7.06	5.08
SD	0.02	0.89	0.72	0.67	0.04	1.72	0.70	0.83
	IL-10							
Mean	0.19	2.33	3.10	4.1	0.18	4.71	3.18	1.53
SD	0.01	1.15	1.16	0.83	0.02	0.71	0.89	0.79
	IL-18							
Mean	0.18	0.54	1.44	1.90	0.17	2.33	1.83	1.45
SD	0.03	0.46	0.45	0.76	0.03	0.76	0.65	0.89

Myeloid differentiation primary response 88 (MYD88); nuclear factor kappa-light-chain-enhancer of activated B cells (NF-κB); C–C chemokine ligand (CCL4); chemokine (C–C motif) ligand 20 (CCL20); C-C chemokine receptor type 2 (CCR2); interferon beta (IFN-β); interferon gamma (IFN-γ); tumor necrosis factor alpha (TNF-α); interleukin 1b (IL-1b); interleukin 2 (IL-2); interleukin 4 (IL-4); interleukin 8 (IL-8); interleukin 10 (IL-10); interleukin 18 (IL-18).

2.1.1. Thoracic Aorta

After the resumption of a normal diet (G120), MYD88, NF-kB, CCL4, IFN-β, TNF-α, IL-1b, IL-2, and IL-4 mRNA expression statistically increased compared with the G30, G60, and G90 groups in the thoracic aorta. On the other hand, G120 levels for IFN-γ, IL-8, IL-10, and IL-18 mRNA were significantly higher compared to the G30 and G60 subgroups.

Moreover, we found that the mRNA gene expression of MYD88 and CCR2 statistically decreased in the GF group compared to the G30, G60, G90, and G120 thoracic groups. In addition, NF-kB (vs. G30, G60, G90), CCL4 (vs. G30, G60, G120), CCL20 (vs. G30, G60), IFN-β (vs. G30, G120), IFN-γ (G60, G90, G120), TNF-α (vs. G30, G60, G120), IL-1b (vs. G60, G90, G120), IL-2 (vs. G90, G120), IL-4 (vs. G90, G120), IL-8 (vs. G60, G90), and IL-10 (vs. G120) were also downregulated in the thoracic GF subanalyses.

For the first time, we found that rosuvastatin treatment is more efficient for MYD88, CCL4, IFN-β TNF- α, IL-1b, IL-2, and IL-10 downregulation in the thoracic aorta (7 out of 14), while fluvastatin treatment appeared to exert more potent effects on the rest of the biomarkers (7 out of 14) (Figure 1). In detail, the following biomarkers were significantly reduced in the GR cohorts: MYD88 (vs. G30, G90, G120, GF); NF-kB (vs. G30, G60, G90); CCL4 (vs. G60, G90, G120, GF); CCL20 (vs. G30, G120); CCR2 (vs. G30, G60, G90, G120); IFN-β (vs. G30, G60, G90, G120, GF); IFN-γ (vs. G60, G90, G120); TNF-α (vs. G30, G90, G120, GF); IL-1b (vs. G30, G60, G90, G120, GF); IL-2 (vs. G30, G60, G90, GF); IL-4 (vs. G90, G120); IL-8 (vs. G30, G120); IL-10 (vs. (G90, G120, GF); and IL-18 (vs. G120).

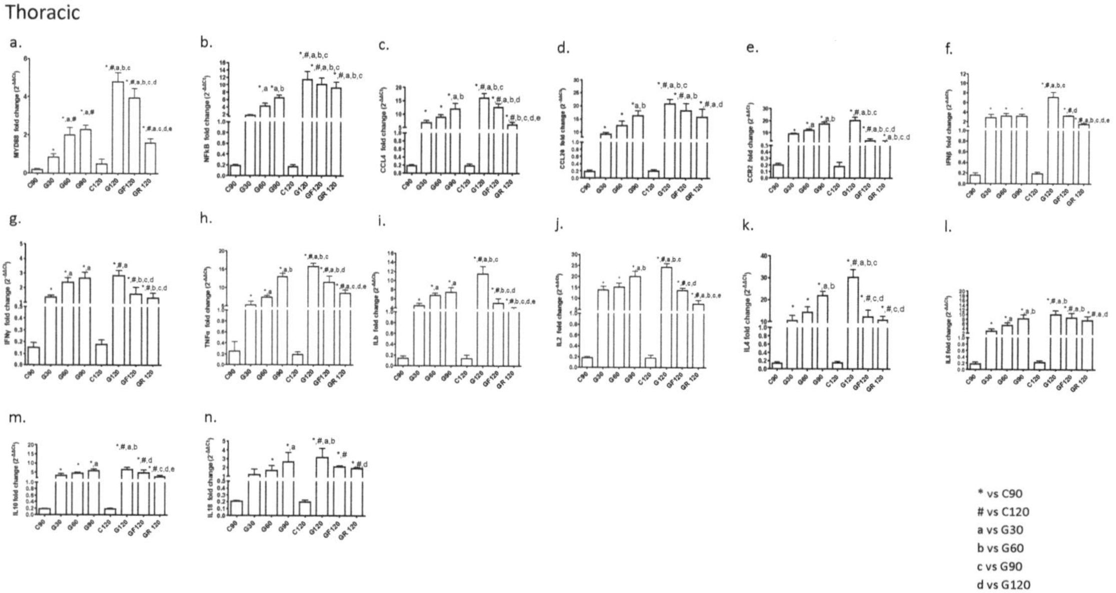

Figure 1. Effects of a high-fat diet and statin treatment on the thoracic rabbit aorta. (**a**) MYD88; (**b**) NF-kB; (**c**) CCL4; (**d**) CCL20; (**e**) CCR2; (**f**) IFNβ; (**g**) IFNγ; (**h**) TNFα; (**i**) IL-1b; (**j**) IL-2; (**k**) IL-4; (**l**) IL-8; (**m**) IL-10; (**n**) IL-18. Significant difference ($p < 0.05$) versus C90 (*), C120 (#), G30 (a), G60 (b), G90 (c), G120 (d), and GF120 (e).

2.1.2. Abdominal Aorta

In the G120 group of the abdominal aorta, the following cytokines were significantly upregulated: MYD88 (vs. G30, G60, G90); NF-kB (vs. G30, G60, G90); CCL4 (vs. G30, G60); CCL20 (vs. G30, G60); CCR2 (vs. G30, G60, G90); IFN-β (vs. G30, G60, G90); IFN-γ (vs. G30, G60, G90); TNF-α (vs. G30, G60, G90); IL-1b(vs. G30, G60, G90); IL-2(vs. G30, G60); IL-4 (vs. G30, G60); IL-8 (vs. G30, G60, G90); IL-10 (vs. G30, G60); and IL-18 (vs. G30, G90).

Similarly, rosuvastatin treatment (GR) significantly downregulated MYD88 (vs. G30, G60, G120), NF-kB (vs. G30, G60, G90), CCL4 (vs. G30, G90, G120), CCL20 (vs. G90, G120, GF), CCR2 (vs. G60, G90, G120, GF), IFN-β (vs. G60, G90, G120, GF), IFN-γ (vs. G30, GF), TNF-α (vs. G30, G60, G120); IL-1b (vs. G60, G90, G120), IL-2 (vs. G30, G60, G90, G120, GF), IL-4 (vs. G60, G90, G120, GF), IL-8 (vs. G30, G120), and IL-10 (vs. G90, G120, GF). As described herein, rosuvastatin treatment was more efficient for CCL20, CCR2, IFN-β, IFN-γ, IL-2, and IL-10 downregulation in the abdominal aortic segments.

Nevertheless, fluvastatin treatment (GF) also led to a significant attenuation in the proatherogenic biomarkers, including MYD88 (vs. G30, G60, G90, G120), NF-kB (vs. G30, G60, G90), CCL4 (vs. G30), CCL20 (vs. G30, G60), CCR2 (vs. G60, G90, G120), IFN-β (vs. G120), IFN-γ (vs. G120), TNF-α (vs. G30, G60), IL-1b (vs. G90, G120), IL-2 (vs. G30), IL-4 (vs. G30), IL-8 (vs. G30, G60, G90), IL-10 (vs. G120), and IL-18 (vs. G30). The graphical representation of these analyses can be found in Figure 2.

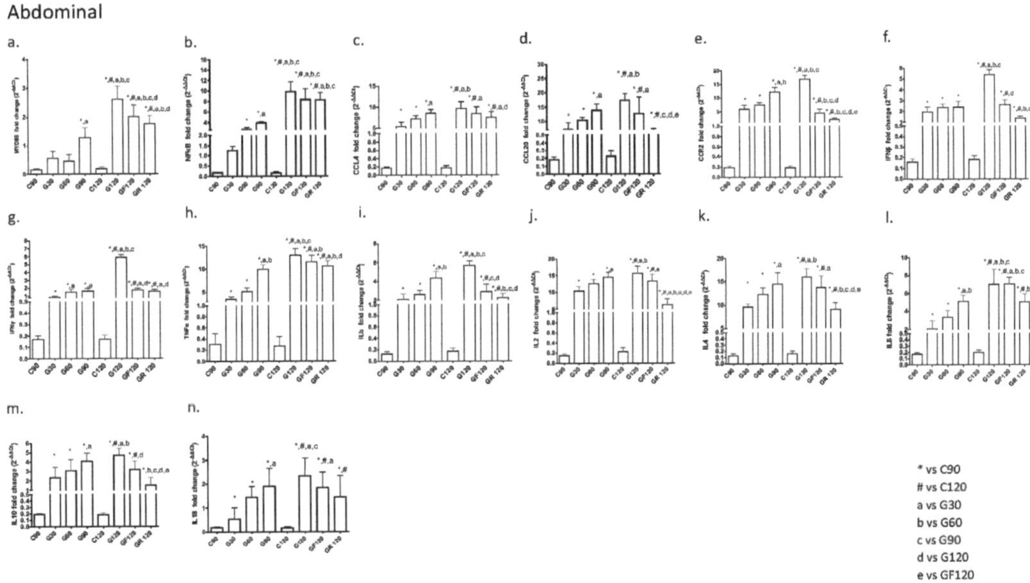

Figure 2. Effects of a high-fat diet and statin treatment on the abdominal rabbit aorta. (**a**) MYD88; (**b**) NF-kB; (**c**) CCL4; (**d**) CCL20; (**e**) CCR2; (**f**) IFNβ; (**g**) IFNγ; (**h**) TNFα; (**i**) IL-1b; (**j**) IL-2; (**k**) IL-4; (**l**) IL-8; (**m**) IL-10; (**n**) IL-18. Significant difference ($p < 0.05$) versus C90 (*), C120 (#), G30 (a), G60 (b), G90 (c), G120 (d), and GF120 (e).

2.2. Correlations between MYD88 and Other Biomarkers during Aortic Atherogenesis and Statin Treatment

In order to assess whether MYD88-mediated signaling pathways are responsible for the progression of atherosclerosis and inflammation, we performed an analysis of the correlation coefficients between MYD88 and all other biomarkers in both aortas.

We noted a significant upregulation between MYD88 and CCL20 (G60); IFNβ (G90); NF-kB, IL18 (G120); NF-kB, CCR2, IL1b, IL18 (GF120); and CCR2 (GR120) in the thoracic aorta during the inflammatory process of atherosclerosis (Table 3). Secondly, we noticed that the MYD88 downregulation was significantly correlated with NF-kB (G90); IL4 (G120); CCL4, CCL20 (GF120); and NF-kB, IL1b (GR120) in the thoracic aorta.

Table 3. Correlations between MYD88 and other biomarkers in the thoracic aorta.

	MYD88	MYD88	MYD88	MYD88	MYD88	MYD88
Group	**G30**	**G60**	**G90**	**G120**	**GF120**	**GR120**
Biomarker						
NFKB			−0.04 (−0.82)	+0.001 (0.45)	+0.04 (0.81)	−0.09 (−0.64)
CCL4					−0.02 (−0.881)	
CCL20		+0.02 (0.86)			0.01 (0.89)	
CCR2					+0.03 (−0.85)	+0.04 (0.82)
IFNβ			+0.01 (0.89)			
IFNγ						
TNFa						
ILb					+0.02 (0.86)	−0.01 (−0.91)
IL2						
IL4				−0.02 (−0.86)		
IL8						
IL10						
IL18				+0.02 (0.86)	+0.03 (0.83)	

Correlations between MYD88 and other biomarkers in the thoracic aorta. Green boxes are statistically significant; red boxes are not statistically significant. Upregulation symbol "+"; downregulation symbol "−". $p < 0.005$, parenthesis indicates the correlation coefficient.

We also observed that MYD88 is significantly upregulated with IL18 (G30); and TNFa and IL4 (G90) in the abdominal aorta (Table 4). Finally, we documented a significant downregulation of MYD88 with IL4 (G60); and CCL20, IL4 (GR120) in the abdominal aorta.

Table 4. Correlations between MYD88 and other biomarkers in the abdominal aorta.

	MYD88	MYD88	MYD88	MYD88	MYD88	MYD88
Group	**G30**	**G60**	**G90**	**G120**	**GF120**	**GR120**
Biomarker						
NFKB						
CCL4						
CCL20						0.001 (0.97)
CCR2						
IFNβ						
IFNγ						
TNFa						
ILb			+0.02 (0.50)			
IL2						
IL4		−0.008 (−0.92)	+0.04 (0.81)			0.04 (−0.82)
IL8						
IL10						
IL18	+0.01 (0.89)					

Correlations between MYD88 and other biomarkers in the thoracic aorta. Green boxes are statistically significant; red boxes are not statistically significant. Upregulation symbol "+"; downregulation symbol "−". $p < 0.005$, parenthesis indicates the correlation coefficient.

3. Discussion

Many popular cardiovascular medications seem to exhibit off-target anti-inflammatory properties [16]. Statins afford cardiovascular protection primarily by reducing LDL levels through the inhibition of HMG-CoA (3-hydroxy-3-methylglutaryl-Coenzyme A) reductase. Nevertheless, a plethora of data suggest that statins decrease cardiovascular inflammation as well. First, statins inhibit prenylated protein production and the mevalonate pathway, while inducing the Krüppel-like factor and nitric oxide synthase expression [17]. Statins also reduce endothelial cell activation and inhibit the interferon-mediated induction of major histocompatibility complex Class II expression, thereby decreasing T-cell activation [16,18]. Our study found that cytokine expression increased during the progression of atherosclerosis, but was reduced with statin treatment.

Rosuvastatin treatment significantly reduced the production of several biomarkers, including MYD88, CCL4, CCL20, CCR2, TNF-α, IFN-β, IL-1b, IL-2, IL-4, IL-8, and IL-10, in both the thoracic and abdominal aorta. Rosuvastatin reduced CCL4, IFN-β, IL-2, IL-4, and IL-10 more effectively than fluvastatin in both aortic segments. MYD88, TNF-α, IL-1b, and IL-8 showed a stronger downregulation with rosuvastatin compared to fluvastatin only in the thoracic aorta, while CCL20 and CCR2 levels reduced more extensively with rosuvastatin treatment only in abdominal aortic tissue. The IL18 production curtailed with the administration of rosuvastatin as well, but only in thoracic aortas. Fluvastatin treatment also reduced proatherogenic inflammation compared to normal diet alone, with the downregulation of MYD88, CCR2, IFN-β, IFN-γ, IL-1b, IL-2, IL-4, and IL-10 in both aortic segments. Interestingly, the CCL4 and TNF-α levels diminished with fluvastatin only in the thoracic aorta.

To explore the role of MYD88 in atherosclerosis, many groups have used MYd88−/− mice in the setting of ApoE or LDL receptor deficiency. The global deletion of MYD88 or TLR4 in these models resulted in smaller atherosclerotic plaques, at least partly due to an impaired macrophage recruitment and the lower levels of chemokines (i.e., CCL2 and CCL4) [19]. MYD88 also seems to influence macrophage function within plaques. Of note, macrophages isolated from MYD88−/− mice exhibited reduced activation, lipid accumulation, and foam cell formation in response to oxidized LDL exposure. The production of reactive oxygen species (ROS) was also reduced [20,21]. Interestingly, injecting MYd88−/− B1a cells into mice diminishes the secretion of atheroprotective IgM, which deposits in lesions and blocks T-cell infiltration and IL-1β/TNFα production to halt atherogenesis [22,23].

Work on MYD88 deletion using the Tie2-Cre driver resulted in decreased levels of IL-1β, IL-6, and CXCL1, as well as the downregulation of key adhesion molecules such as MCP1, ICAM1, and VCAM1. There are also data suggesting that the MYD88-dependent production of GM-CSF can cause monocytes to differentiate into inflammatory macrophages, thus promoting atherosclerosis [24]. The activation of MYD88-dependent pathways in aortic endothelial cells, following exposure to endotoxins (i.e., LPS and oxidized LDL), induces the expression of proprotein convertase subtilisin/kexin type 9 (PCSK9). Considering that PCSK9 is a negative regulator of the LDL receptor, these data propose an additional proatherogenic effect of MYD88 that predicates upon endothelial lipid processing [25].

Although the versatile roles of interleukins, interferons, and TNF are well established in atherogenesis, many other chemokines partake in this multi-step process. Among others, CCL20 upregulation has been documented in atherosclerotic carotid arteries and coronary plaques. Serum CCL20 levels are also increased in patients with hyperlipidemia [26]. LDL seems to induce the expression of CCL20 in VSMC, in a dose-dependent fashion through the MAPK and NF-kB-mediated pathways. The increased levels of CCL20 significantly accelerate lymphocyte and monocyte migration, whilst promoting cellular adhesion [27,28]. CCR2 appears to exert similar effects on immune cell mobilization and activation [29]. On the other hand, CCL4 triggers the release of pro-inflammatory cytokines, including IL1, IL6, and TNF-α. It also stimulates the production of ROS, and metalloproteinases 2 and 9 [30].

The current study has some limitations that need to be acknowledged. First, we included a finite number of test subjects in each group to adhere to the 3R principle (replacement, reduction, refinement) [31]. Second, our study was not designed to determine the specific mechanisms by which statins modify cytokine and chemokine production. In the future, our group will be granularly investigating the biomechanics of aortic atheromatosis, and its effects on biomarker production. Unique variations in terms of length, diameter, curvature, and branch network of the thoracic and abdominal aorta certainly affect the intrinsic hemodynamics and shear stress forces [32,33]. These variations likely account for the differences in proatherogenic cytokine expression that were observed.

4. Methods

4.1. Animal Experimental Protocol

The experimental protocol was approved by the Animal Care and Use Committee of the Athens Prefecture Veterinarian Service, Greece (K/3319/4-5-2009). All experiments took place in the animal facilities of the Center of Experimental Surgery, Biomedical Research Foundation Academy of Athens (BRFAA), according to the guidelines set by the National Research Council's Guide for Care and Use of Laboratory Animals.

The protocol of the study has been previously published [34,35]. In this experiment, we analyzed thoracic and abdominal aortic tissue samples from 48 male New Zealand rabbits (Trompetas Breeding Laboratories; Attiki, Greece) with an average body weight of 2.8 ± 0.2 kg. The subjects were randomly divided into eight groups including six animals each, and underwent one week of acclimatization. The control groups were fed normal chow for 90 (C90) and 120 (C120) days. Three groups underwent hyperlipidemic feeds with HCD2RB19 (Mucedola, Milano, Italy) for 1 (G30), 2 (G60) and 3 (G90) months. Another three groups were fed HCD for 3 months, followed by (a) normal chow for 1 month (G120) and (b) the administration of fluvastatin (GF120) or rosuvastatin (GR120) for 1 month. Rosuvastatin (0.7 mg/kg BW) and fluvastatin (2 mg/kg BW) were administered daily per os. The rabbits were sacrificed through an intravenous overdose of sodium pentobarbital (100–120 mg/kg). The thoracic and abdominal aortas were removed and rinsed with distilled (DEPC-treated) water and shock-frozen to −140 °C for mRNA analysis.

4.2. mRNA Analysis of Rabbit Aortic Tissue

Total RNA was isolated from thoracic and abdominal aortas using the Trizol kit (Invitrogen, Life Technologies, New York, NY, USA), according to the manufacturer's protocol [36]. The purity of the extracted RNA was measured with a spectrophotometer. RNA was reverse-transcribed, and the resulting cDNA was synthesized by RT (M-MLV Reverse transcriptase, Sigma), and a real-time quantitative polymerase chain reaction was performed using SYBR Green (Invitrogen, Life Technologies, NY, USA), according to our institutional protocol [37]. All experiments were repeated at least three times and the amplification signals from each target gene were normalized to b-actin (housekeeping gene). All primers for β-actin, MYD88, NF-kB, CCL4, CCL20, CCR2, IFN-β, IFN-γ, TNF-a, IL-1b, IL-2, IL-4, IL-8, IL10, and IL-18 were synthesized by TIB Molbiol (Syntheselabor GmbH, Berlin, Germany). The sequences of the primer pairs were designed with the Beacon Designer V7.0 software (Premier Biosoft International, Palo Alto, CA, USA; Table 1). The primer sequences are provided in Supplementary Table S1.

4.3. Statistical Analysis

All statistical calculations were performed using a one-way ANOVA for comparisons between the groups. Differences were considered significant when a two-tailed $p < 0.05$ was calculated (power analysis = 0.95, alpha = 0.05, beta = 0.05). The correlation between the measured variables was assessed through Pearson analysis. Data are presented as the mean ± standard deviation (mean ± SD). All statistical calculations were performed using GraphPad Prism version 4.03 (GraphPad Software, San Diego, CA, USA).

5. Conclusions

The present animal model further delineates the association between inflammatory phenomena and atherosclerosis. Several temporo-spatial differences were identified with regards to cytokine expression in the aorta. Our findings suggest that even one month of combination treatment with statins and a normal diet can halt the progression of atherosclerosis, both in the thoracic and the abdominal aorta. Furthermore, rosuvastatin may be more potent than fluvastatin in downregulating many key proatherogenic cascades, including the MYD88 pathway in the thoracic aorta. Overall, our study provides an opportunity to better understand at which stage of atherogenesis we can intervene by providing the most effective statin treatment.

Supplementary Materials: The supporting information can be downloaded at: https://www.mdpi.com/article/10.3390/ijms24119248/s1.

Author Contributions: Conceptualization, A.K., K.S.M. and M.P.; Data curation, M.P., A.K. and K.S.M.; Formal analysis, A.K.; Funding acquisition, None; Investigation, K.S.M., M.P., D.S. and A.K.; Methodology, A.K., K.S.M. and M.P.; Project administration, K.S.M. and A.K. Resources, K.S.M., M.P., D.S. and A.K.; Software, A.K.; Supervision, A.K.; Validation, K.S.M., M.P., D.S. and A.K.; Visualization, A.K. and K.S.M.; Writing—original draft, K.S.M. and A.K.; Writing—review and editing, K.S.M., M.P., D.S. and A.K. All authors have read and agreed to the published version of the manuscript.

Funding: This research received no external funding.

Institutional Review Board Statement: The animal study protocol was approved by the Veterinary Directorate of the Prefecture of Athens and the Institutional Board of Review of the School of Medicine of the National and Kapodistrian University of Athens (no.: 3231/26 June 2018).

Informed Consent Statement: Not applicable.

Data Availability Statement: Raw data may be provided by the authors upon request.

Conflicts of Interest: The authors declare no conflict of interest.

References

1. Benjamin, E.J.; Blaha, M.J.; Chiuve, S.E.; Cushman, M.; Das, S.R.; Deo, R.; de Ferranti, S.D.; Floyd, J.; Fornage, M.; Gillespie, C.; et al. Heart Disease and Stroke Statistics-2017 Update: A Report From the American Heart Association. *Circulation* **2017**, *135*, e146–e603. [CrossRef] [PubMed]
2. Heidenreich, P.A.; Trogdon, J.G.; Khavjou, O.A.; Butler, J.; Dracup, K.; Ezekowitz, M.D.; Finkelstein, E.A.; Hong, Y.; Johnston, S.C.; Khera, A.; et al. Forecasting the future of cardiovascular disease in the United States: A policy statement from the American Heart Association. *Circulation* **2011**, *123*, 933–944. [CrossRef] [PubMed]
3. Getz, G.S.; Reardon, C.A. Atherogenic lipids and macrophage subsets. *Curr. Opin. Lipidol.* **2015**, *26*, 357–361. [CrossRef] [PubMed]
4. Gomez, D.; Owens, G.K. Smooth muscle cell phenotypic switching in atherosclerosis. *Cardiovasc. Res.* **2012**, *95*, 156–164. [CrossRef]
5. Hansson, G.K.; Hermansson, A. The immune system in atherosclerosis. *Nat. Immunol.* **2011**, *12*, 204–212. [CrossRef]
6. Quillard, T.; Araújo, H.A.; Franck, G.; Shvartz, E.; Sukhova, G.; Libby, P. TLR2 and neutrophils potentiate endothelial stress, apoptosis and detachment: Implications for superficial erosion. *Eur. Heart J.* **2015**, *36*, 1394–1404. [CrossRef]
7. Margiana, R.; Alsaikhan, F.; Al-Awsi, G.R.L.; Patra, I.; Sivaraman, R.; Fadhil, A.A.; Al-Baghdady, H.F.A.; Qasim, M.T.; Hameed, N.M.; Mustafa, Y.F.; et al. Functions and therapeutic interventions of non-coding RNAs associated with TLR signaling pathway in atherosclerosis. *Cell. Signal.* **2022**, *100*, 110471. [CrossRef]
8. Mylonas, K.S.; Sarantis, P.; Kapelouzou, A.; Karamouzis, M.V.; Kapetanakis, E.I.; Kontzoglou, K.; Iliopoulos, D.C.; Nikiteas, N.; Schizas, D. Mechanosensitive Stem-Cell Genes and Klotho in Atherosclerotic Aortas: Regulating Spatially Deranged Expression Patterns Using Colchicine Regimens. *J. Clin. Med.* **2022**, *11*, 6465. [CrossRef]
9. Mylonas, K.S.; Karangelis, D.; Androutsopoulou, V.; Chalikias, G.; Tziakas, D.; Mikroulis, D.; Iliopoulos, D.C.; Nikiteas, N.; Schizas, D. Stem cell genes in atheromatosis: The role of Klotho, HIF1α, OCT4, and BMP4. *IUBMB Life* **2022**, *74*, 1003–1011. [CrossRef]
10. Mylonas, K.S.; Kapelouzou, A.; Spartalis, M.; Mastrogeorgiou, M.; Spartalis, E.; Bakoyiannis, C.; Liakakos, T.; Schizas, D.; Iliopoulos, D.; Nikiteas, N. KLF4 Upregulation in Atherosclerotic Thoracic Aortas: Exploring the Protective Effect of Colchicine-based Regimens in a Hyperlipidemic Rabbit Model. *Ann. Vasc. Surg.* **2022**, *78*, 328–335. [CrossRef]

11. Miller, Y.I.; Viriyakosol, S.; Worrall, D.S.; Boullier, A.; Butler, S.; Witztum, J.L. Toll-like receptor 4-dependent and -independent cytokine secretion induced by minimally oxidized low-density lipoprotein in macrophages. *Arterioscler. Thromb. Vasc. Biol.* **2005**, *25*, 1213–1219. [CrossRef]
12. Lee, J.Y.; Hwang, D.H. The modulation of inflammatory gene expression by lipids: Mediation through Toll-like receptors. *Mol. Cells* **2006**, *21*, 174–185.
13. O'Neill, L.A.; Dunne, A.; Edjeback, M.; Gray, P.; Jefferies, C.; Wietek, C. Mal and MyD88: Adapter proteins involved in signal transduction by Toll-like receptors. *J. Endotoxin Res.* **2003**, *9*, 55–59. [CrossRef]
14. Muendlein, H.I.; Connolly, W.M.; Cameron, J.; Jetton, D.; Magri, Z.; Smirnova, I.; Vannier, E.; Li, X.; Martinot, A.J.; Batorsky, R.; et al. Neutrophils and macrophages drive TNF-induced lethality via TRIF/CD14-mediated responses. *Sci. Immunol.* **2022**, *7*, eadd0665. [CrossRef]
15. Kong, P.; Cui, Z.Y.; Huang, X.F.; Zhang, D.D.; Guo, R.J.; Han, M. Inflammation and atherosclerosis: Signaling pathways and therapeutic intervention. *Signal Transduct. Target. Ther.* **2022**, *7*, 131. [CrossRef]
16. Mylonas, K.S.; Iliopoulos, D.; Nikiteas, N.; Schizas, D. Looking for the Achilles heel of atheromatosis: Could it be immunotherapy? *Immunotherapy* **2021**, *13*, 557–560. [CrossRef]
17. Jain, M.K.; Ridker, P.M. Anti-inflammatory effects of statins: Clinical evidence and basic mechanisms. *Nat. Rev. Drug Discov.* **2005**, *4*, 977–987. [CrossRef]
18. Ii, M.; Losordo, D.W. Statins and the endothelium. *Vasc. Pharmacol.* **2007**, *46*, 1–9. [CrossRef]
19. Michelsen, K.S.; Wong, M.H.; Shah, P.K.; Zhang, W.; Yano, J.; Doherty, T.M.; Akira, S.; Rajavashisth, T.B.; Arditi, M. Lack of Toll-like receptor 4 or myeloid differentiation factor 88 reduces atherosclerosis and alters plaque phenotype in mice deficient in apolipoprotein E. *Proc. Natl. Acad. Sci. USA* **2004**, *101*, 10679–10684. [CrossRef]
20. Howell, K.W.; Meng, X.; Fullerton, D.A.; Jin, C.; Reece, T.B.; Cleveland, J.C., Jr. Toll-like receptor 4 mediates oxidized LDL-induced macrophage differentiation to foam cells. *J. Surg. Res.* **2011**, *171*, e27–e31. [CrossRef]
21. Singh, R.K.; Haka, A.S.; Asmal, A.; Barbosa-Lorenzi, V.C.; Grosheva, I.; Chin, H.F.; Xiong, Y.; Hla, T.; Maxfield, F.R. TLR4 (Toll-Like Receptor 4)-Dependent Signaling Drives Extracellular Catabolism of LDL (Low-Density Lipoprotein) Aggregates. *Arterioscler. Thromb. Vasc. Biol.* **2020**, *40*, 86–102. [CrossRef] [PubMed]
22. Subramanian, M.; Thorp, E.; Hansson, G.K.; Tabas, I. Treg-mediated suppression of atherosclerosis requires MYD88 signaling in DCs. *J. Clin. Investig.* **2013**, *123*, 179–188. [CrossRef] [PubMed]
23. Engelbertsen, D.; Rattik, S.; Wigren, M.; Vallejo, J.; Marinkovic, G.; Schiopu, A.; Björkbacka, H.; Nilsson, J.; Bengtsson, E. IL-1R and MyD88 signalling in CD4+ T cells promote Th17 immunity and atherosclerosis. *Cardiovasc. Res.* **2018**, *114*, 180–187. [CrossRef] [PubMed]
24. Yu, M.; Zhou, H.; Zhao, J.; Xiao, N.; Roychowdhury, S.; Schmitt, D.; Hu, B.; Ransohoff, R.M.; Harding, C.V.; Hise, A.G.; et al. MyD88-dependent interplay between myeloid and endothelial cells in the initiation and progression of obesity-associated inflammatory diseases. *J. Exp. Med.* **2014**, *211*, 887–907. [CrossRef] [PubMed]
25. Liu, S.; Deng, X.; Zhang, P.; Wang, X.; Fan, Y.; Zhou, S.; Mu, S.; Mehta, J.L.; Ding, Z. Blood flow patterns regulate PCSK9 secretion via MyD88-mediated pro-inflammatory cytokines. *Cardiovasc. Res.* **2020**, *116*, 1721–1732. [CrossRef]
26. Wan, W.; Murphy, P.M. Regulation of atherogenesis by chemokine receptor CCR6. *Trends Cardiovasc. Med.* **2011**, *21*, 140–144. [CrossRef]
27. Calvayrac, O.; Rodríguez-Calvo, R.; Alonso, J.; Orbe, J.; Martín-Ventura, J.L.; Guadall, A.; Gentile, M.; Juan-Babot, O.; Egido, J.; Beloqui, O.; et al. CCL20 is increased in hypercholesterolemic subjects and is upregulated by LDL in vascular smooth muscle cells: Role of NF-κB. *Arterioscler. Thromb. Vasc. Biol.* **2011**, *31*, 2733–2741. [CrossRef]
28. Manthey, H.D.; Cochain, C.; Barnsteiner, S.; Karshovska, E.; Pelisek, J.; Koch, M.; Chaudhari, S.M.; Busch, M.; Eckstein, H.H.; Weber, C.; et al. CCR6 selectively promotes monocyte mediated inflammation and atherogenesis in mice. *Thromb. Haemost.* **2013**, *110*, 1267–1277.
29. Zernecke, A.; Weber, C. Chemokines in atherosclerosis: Proceedings resumed. *Arterioscler. Thromb. Vasc. Biol.* **2014**, *34*, 742–750. [CrossRef]
30. Chang, T.T.; Yang, H.Y.; Chen, C.; Chen, J.W. CCL4 Inhibition in Atherosclerosis: Effects on Plaque Stability, Endothelial Cell Adhesiveness, and Macrophages Activation. *Int. J. Mol. Sci.* **2020**, *21*, 6567. [CrossRef]
31. Schechtman, L.M. Implementation of the 3Rs (refinement, reduction, and replacement): Validation and regulatory acceptance considerations for alternative toxicological test methods. *ILAR J.* **2002**, *43* (Suppl. S1), S85–S94. [CrossRef]
32. Bassiouny, H.S.; Zarins, C.K.; Kadowaki, M.H.; Glagov, S. Hemodynamic stress and experimental aortoiliac atherosclerosis. *J. Vasc. Surg.* **1994**, *19*, 426–434. [CrossRef]
33. Padilla, J.; Jenkins, N.T.; Vieira-Potter, V.J.; Laughlin, M.H. Divergent phenotype of rat thoracic and abdominal perivascular adipose tissues. *Am. J. Physiol. Regul. Integr. Comp. Physiol.* **2013**, *304*, R543–R552. [CrossRef]
34. Kapelouzou, A.; Giaglis, S.; Peroulis, M.; Katsimpoulas, M.; Moustardas, P.; Aravanis, C.V.; Kostakis, A.; Karayannakos, P.E.; Cokkinos, D.V. Overexpression of Toll-Like Receptors 2, 3, 4, and 8 Is Correlated to the Vascular Atherosclerotic Process in the Hyperlipidemic Rabbit Model: The Effect of Statin Treatment. *J. Vasc. Res.* **2017**, *54*, 156–169. [CrossRef]
35. Tziakas, D.; Chalikias, G.; Kapelouzou, A.; Tentes, I.; Schäfer, K.; Karayannakos, P.; Kostakis, A.; Boudoulas, H.; Konstantinides, S. Erythrocyte membrane cholesterol and lipid core growth in a rabbit model of atherosclerosis: Modulatory effects of rosuvastatin. *Int. J. Cardiol.* **2013**, *170*, 173–181. [CrossRef]

36. Chomczynski, P.; Sacchi, N. Single-step method of RNA isolation by acid guanidinium thiocyanate-phenol-chloroform extraction. *Anal. Biochem.* **1987**, *162*, 156–159. [CrossRef]
37. Kapelouzou, A.; Katsimpoulas, M.; Kontogiannis, C.; Lidoriki, I.; Georgiopoulos, G.; Kourek, C.; Papageorgiou, C.; Mylonas, K.S.; Dritsas, S.; Charalabopoulos, A.; et al. A High-Cholesterol Diet Increases Toll-like Receptors and Other Harmful Factors in the Rabbit Myocardium: The Beneficial Effect of Statins. *Curr. Issues Mol. Biol.* **2021**, *43*, 818–830. [CrossRef]

International Journal of *Molecular Sciences*

MDPI

Article

The Impact of Extracellular Histones and Absence of Toll-like Receptors on Cardiac Functional and Electrical Disturbances in Mouse Hearts

Randall Loaiza [1,2,†], Fatemeh Fattahi [3,4,†], Miriam Kalbitz [4,5,6], Jamison J. Grailer [4,7], Mark W. Russell [8], Jose Jalife [1,9], Hector H. Valdivia [1,10], Firas S. Zetoune [4] and Peter A. Ward [4,*]

1 Center for Arrhythmia Research, University of Michigan, Ann Arbor, MI 48109, USA; r.loaiza@gmail.com (R.L.); jjalife@umich.edu (J.J.); hhvaldiv@medicine.wisc.edu (H.H.V.)
2 CENIBiot Laboratory, The National Center of High Technology (CeNAT-CONARE), San José 10109, Costa Rica
3 Division of Allergy and Clinical Immunology, Department of Internal Medicine, University of Michigan Medical School, Ann Arbor, MI 48109, USA; ffattahi@umich.edu
4 Department of Pathology, University of Michigan Medical School, Ann Arbor, MI 48109, USA; m.kalbitz@googlemail.com (M.K.); jamison.grailer@promega.com (J.J.G.); zetoune@umich.edu (F.S.Z.)
5 Department of Orthopaedic Trauma, Hand, Plastic and Reconstructive Surgery, University Hospital of Ulm, 89081 Ulm, Germany
6 Military Medical City Hospital, Doha 486441, Qatar
7 Integrated Biology R&D, Bioassay Development, Promega Corporation, Madison, WI 53711, USA
8 Department of Pediatrics, University of Michigan Medical School, Ann Arbor, MI 48109, USA; mruss@umich.edu
9 Centro Nacional de Investigaciones Cardiovasculares (CNIC), 28029 Madrid, Spain
10 Department of Medicine, Cardiovascular Research Center, University of Wisconsin School of Medicine and Public Health, Madison, WI 53705, USA
* Correspondence: pward@umich.edu; Tel.: +1-734-647-2921
† These authors contributed equally to this work.

Citation: Loaiza, R.; Fattahi, F.; Kalbitz, M.; Grailer, J.J.; Russell, M.W.; Jalife, J.; Valdivia, H.H.; Zetoune, F.S.; Ward, P.A. The Impact of Extracellular Histones and Absence of Toll-like Receptors on Cardiac Functional and Electrical Disturbances in Mouse Hearts. *Int. J. Mol. Sci.* **2024**, *25*, 8653. https://doi.org/10.3390/ijms25168653

Academic Editors: Oliver Planz and Ralf Kircheis

Received: 29 March 2024
Revised: 2 August 2024
Accepted: 4 August 2024
Published: 8 August 2024

Abstract: In polymicrobial sepsis, the extracellular histones, mainly released from activated neutrophils, significantly contribute to cardiac dysfunction (septic cardiomyopathy), as demonstrated in our previous studies using Echo-Doppler measurements. This study aims to elucidate the roles of extracellular histones and their interactions with Toll-like receptors (TLRs) in cardiac dysfunction. Through ex vivo assessments of ECG, left ventricle (LV) function parameters, and in vivo Echo-Doppler studies in mice perfused with extracellular histones, we aim to provide comprehensive insights into the mechanisms underlying sepsis-induced cardiac dysfunction. Langendorff-perfused hearts from both wild-type and TLR2, TLR3, or TLR4 knockout (KO) mice were examined. Paced mouse hearts were perfused with histones to assess contractility and relaxation. Echo-Doppler studies evaluated cardiac dysfunction after intravenous histone injection. Histone perfusion caused defects in contractility and relaxation, with TLR2 and TLR3 KO mice being partially protected. Specifically, TLR2 KO mice exhibited the greatest reduction in Echo-Doppler abnormalities, while TLR4 KO exacerbated cardiac dysfunction. Among individual histones, H1 induced the most pronounced abnormalities in cardiac function, apoptosis of cardiomyocytes, and LDH release. Our data highlight significant interactions between histones and TLRs, providing insights into histones especially H1 as potential therapeutic targets for septic cardiomyopathy. Further studies are needed to explore specific histone–TLR interactions and their mechanisms.

Keywords: complement; C5a; sepsis; heart dysfunction; arrhythmia; cardiomyocytes; neutrophils

1. Introduction

One of the critical organ systems profoundly impacted during severe sepsis is the cardiovascular system, notably the heart. Organ dysfunction arising from sepsis has been extensively researched [1]. Cardiac dysfunction emerges as a frequent complication in septic patients and septic mice, often culminating in a substantial mortality rate. Sepsis in humans is frequently associated with mortality rates ranging from 70% to 90% with cardiovascular impairment compared to 20% in those without impairment, highlighting the critical role of cardiovascular dysfunction in sepsis-related mortality [2]. These dysfunctions are distinct from ischemia-reperfusion injury and are reversible if the patient survives sepsis [1]. Surprisingly, in humans with infectious sepsis, coronary arterial blood flow measurements have demonstrated unimpeded flow and normal tissue pO2 levels [3,4].

More than two decades ago, evidence suggested depressed myocardial performance during sepsis, indicating a role of factors contributing to cardiac dysfunction [5]. Significant elevations in serum cytokines, such as IL-1β, IL-6, TNF-α, IL-8, and the IL-7 family of cytokines, have been noted during sepsis [6–10]. The inflammatory responses initiated by these cytokines likely contribute to widespread organ system failure in sepsis. We have shown that sepsis in mice activates the NLRP3 inflammasome, leading to the release of IL-1β, IL-6, and other cytokines and chemokines, causing cardiomyocyte (CM) dysfunction [11,12]. Subsequent research has highlighted the role of extracellular histones, released into the extracellular compartment during sepsis, as an important driver inducing myocardial dysfunction during sepsis [13]. These histones are the chief component of chromatin that can act as damage-associated molecular patterns (DAMPs) when released to the extracellular compartment and have been postulated to be a major cause of cardiac failure along with other organ failures during sepsis [11,14–17]. Both C5a receptors (C5aR1 and C5aR2) are known to be required for histone appearance in plasma during polymicrobial sepsis [11,18,19]. Histones are likely released from neutrophil extracellular traps (NETs) and macrophage extracellular traps (METs) after activation of phagocytes by C5a at the acute phase of sepsis [11,20]. The histones are postulated to play a significant role not only in sepsis but also in trauma-related organ dysfunction [21,22] and non-infectious conditions that clinically mimic sepsis such as chemical injury or ischemic injury [23,24]. In humans with infectious sepsis, plasma histones (H2B, H3) have been found at levels similar to those observed in septic mice, further emphasizing their potential significance [25]. Notably, deaths occurring during septic shock are often linked to cardiac dysfunction caused by extracellular histones [13]. Using a neutralizing antibody to histones, immunofluorescence studies revealed histones appearance in CMs and in plasma after cecal ligation and puncture (CLP), involving heart dysfunction [11]. Additionally, observations in COVID-19 infections have suggested evidence of extracellular histone release and CM dysfunction, further highlighting the relevance of our findings [26,27]. Our current study provides evidence that septic cardiac abnormalities occur due to histones.

Electrical and mechanical dysfunctions in the heart often develop in septic patients and are frequently associated with a poor prognosis. Ventricular arrhythmias are a serious complication of sepsis, often associated with increased mortality risk in ICU-admitted patients [28]. Other cardiac complications include decreased ventricular contraction, defective relaxation kinetics, ventricular dilatation, and electrical conduction abnormalities [29–31]. Septic patients may exhibit prolongation of QRS complexes, as defined by electrocardiography (ECG), which usually disappears during recovery from septic shock, indicating reversibility [32]. In the septic mice, these disturbances in functional responses of hearts were defined by Echo-Doppler parameters [11]. In polymicrobial sepsis in mice, early cardiac defects in CM functions have been linked to reduced levels of homeostatic proteins, notably Na+/K+-ATPase [33]. Prolongation of the action potentials is another CM defect observed during sepsis [33]. Developing and validating a clinical risk score for QT prolongation has allowed for the prediction of which septic patients are at the highest risk for developing sepsis-associated cardiomyopathy, aiding in monitoring and treatment decisions [34,35].

Toll-like receptors (TLRs) notably TLR2 and TLR4, are proposed as receptors for extracellular histones [14,23]. Our previous research also suggests TLR3 and TLR9 as other potential histone receptors, supported by echocardiographic studies showing the protective effect in the absence of these TLRs [36]. As stated above, extracellular histones serve as DAMPs, released mainly from NETs during sepsis. Increasing evidence links TLR activation, especially TLR2, TLR3, and TLR4, to responses observed during septic cardiomyopathy in mice [37]. Histones may bind to several TLRs, suggesting their potential role in cardiac dysfunction [23,24,36,38]. While it is established that sepsis causes cardiac dysfunction [1,11,29,39,40], the specific contribution of extracellular histones to cardiac dysfunction and ECG changes needs further investigation. The current study design investigates the direct effects of histones on cardiac function and assesses the impact of various TLRs, including TLR2, TLR3, and TLR4 knockout (KO) on changes in electrical and functional responses in the mouse heart ex vivo and in vivo. Our study also directly examines the effects of individual histones (H1, H2A, H2B, H3, H4) on cardiac function and the defective electrical responses of hearts exposed to these individual histones. We explore the role of each individual histone on perfused hearts concerning electrical and functional activities in hearts from wild-type (Wt) mice. These investigations aim to identify histones that are pathogenic for CMs while considering the impact of different TLRs and each individual histone on cardiac abnormalities.

2. Results

2.1. Functional Abnormalities in Wt Mouse Hearts Perfused Ex Vivo with Histone Mix

For these studies, Wt mouse hearts electrically paced were perfused for up to 20 min at 37 °C with control buffer (PBS), with the histone mix (Figure 1A–F). For each condition, the concentration of the histone mix varied from 5–20 µg/mL as indicated and involved the use of 11 hearts for each condition. As indicated, several different functional parameters were measured. In Figure 1A, either buffer or the histone mix (5, 10, or 20 µg/mL) was used as the perfusate. Tracings for both the buffer and different doses of histone mix are shown in the expanded view on the left portion of the tracing for 300 milliseconds (ms), while the right side (black compact tracings) represents 15 s of perfusion. Perfusion of hearts with buffer resulted in tracings shown in the first left panel of Figure 1A. Subsequent perfusion with three different concentrations of the histone mix (5, 10, 20 µg/mL) in the remaining panels showed progressively reduced Left Ventricular (LV) Pressures (LVP) amplitudes, with reductions and widening of the amplitudes. At the 20 µg/mL concentration of the histone mix, LVP amplitudes collapsed, and bradycardia developed. In Figure 1B, the temporal progression is depicted following the perfusion of a histone mix (5 µg/mL), as shown in the expanded view in Figure 1C. Within 1 min of perfusion, observable widening of amplitudes and the onset of bradycardia became apparent. By the 8th minute, LVP had collapsed, and severe bradycardia had developed. In Figure 1D–F, perfused Wt hearts were assessed for LVP (Figure 1D), end-diastolic pressures (Figure 1E), or relaxation pressures (Figure 1F) over a 20 min perfusion time. The horizontal dashed lines indicate the point of heart failure, where tracings revealed $\geq$80% loss of pressure values, denoting heart failure. Black bars represent responses to buffer perfusion (Wt Control), while white bars represent pressures in hearts perfused with the histone mix (Wt histones) at different doses (5–20 µg/mL). As evident in Figure 1D, LVP and end-diastolic pressures (Figure 1E) began to decline with the perfusion containing 10 µg/mL of the histone mix. With the histone mix of 20 µg/mL, there were statistically significant reductions in all three pressure frames (Figure 1D–F). In Figure 1F, relaxation pressures were measured. Starting with the histone mix of 5 µg/mL, there was a decline in relaxation pressures, which became more pronounced as the histone mix concentrations of 10 µg/mL and 20 µg/mL were used.

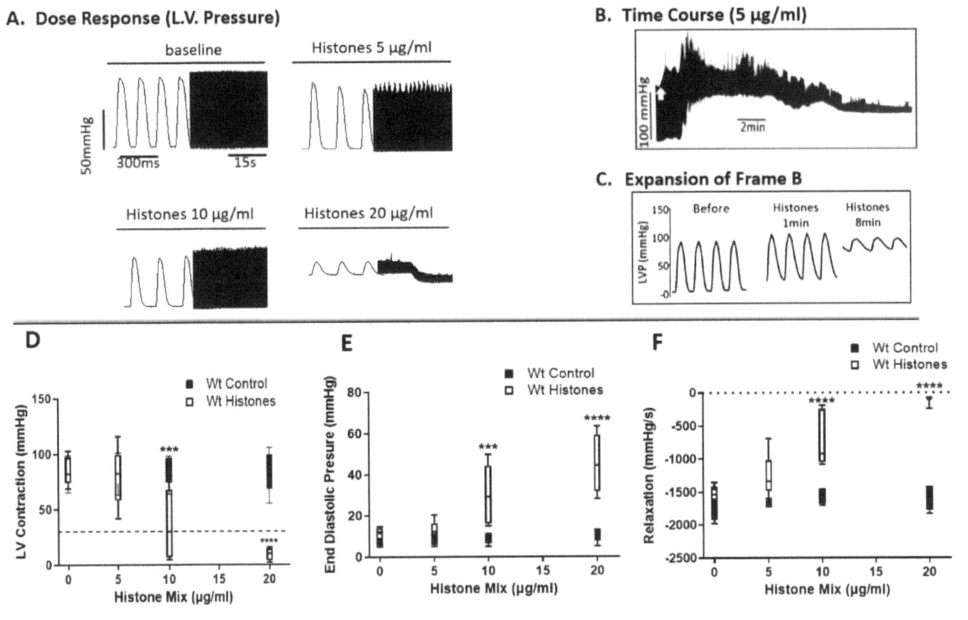

Figure 1. Functional abnormalities in Wt mouse hearts perfused ex vivo with histone mix. Wt mouse hearts, electrically paced, were perfused ex vivo with control buffer or histone mix (5, 10, and 20 µg/mL) for up to 20 min at 37 °C (**A–F**). Each condition involved 11 hearts. (**A**) Tracings for buffer and various histone mix concentrations showed progressive reduction and widening of left ventricular pressure (LVP) amplitudes. At 20 µg/mL, LVP amplitudes collapsed, inducing bradycardia. (**B**) Temporal progression after perfusing with a 5 µg/mL histone mix, revealing widening amplitudes and bradycardia within 1 min. (**C**) LVP collapsed by the 8th minute. (**D–F**) Displays LVP, end-diastolic pressures, and relaxation pressures, respectively, over 20 min. Dashed lines mark heart failure (≥80% pressure loss). Black bars represent buffer (Wt Control), while white bars represent histone mix (Wt Histones) at different doses. (**D**) Declining LVP with 10 µg/mL histone mix, and 20 µg/mL significantly reduces all pressures (**D–F**). (**F**) Relaxation pressures declined with histone mix, accentuated at 10 and 20 µg/mL concentrations. In (**D–F**), data are shown as max–min box plots and analyzed by two−way ANOVA followed by Sidak's multiple comparison test. Significance is indicated as follows: *** $p < 0.001$ and **** $p < 0.0001$. The dashed line in (**D**) indicates the heart failure threshold, and in (**F**), shows 0 mmHg/s as the baseline.

2.2. ECG Changes in Wt Hearts Perfused with Histones or After Induction of Polymicrobial Sepsis by CLP

In Figure 2A, Wt hearts ($n \geq 5$ hearts for each experiment) were perfused with three different concentrations of the histone mix (5, 10, or 20 µg/mL) for up to 12 min at 37 °C. By 12 min of perfusion there was a distinct widening of the QT interval (Figure 2A). The baseline duration of the QTc interval of the heart was 67.4 ± 3.1 ms, which then increased to 71.8 ± 3.6 ms, and 87.2 ± 5 ms only 6 and 12 min after histone perfusion, respectively (Figure 2A). At the two higher concentrations of the histone mix, (10, 20 µg/mL) there was collapse of ECG tracings. In Figure 2B, hearts were perfused with the histone mix (5 µg/mL) mix over a 12 min period of time. Figure 2B-1 shows a significantly increased QT interval by 6 min of perfusion (approximately 15 ms increase), peaking at 8 min (22 ms increase in QT interval), followed by a gradual decline (all changes became significant at 6 min and remained so through 12 min). In Figure 2B-2, there was a steady increase in the QRS interval starting at 6 min of perfusion and continuing at 12 min. In Figure 2B-3, there was no increase in the PR interval. ECG recordings were also obtained in Wt mice before CLP (time 0) and 6 and 12 h after polymicrobial sepsis (Figure 2C). Sham controls and CLP mice (n − 5 mice in each group) were used 6 and 12 h after CLP. The following intervals (ms) were measured: QT, QRS, and PR. The QT intervals in the septic mice were significantly elevated at 6 hr and further elevated at 12 after CLP. The QRS interval was also significantly elevated at 6 and 12 h in CLP hearts. Consistent with the studies of hearts perfused with histone mix (Figure 2B), sepsis did not cause statistically significant changes in the PR interval (Figure 2C).

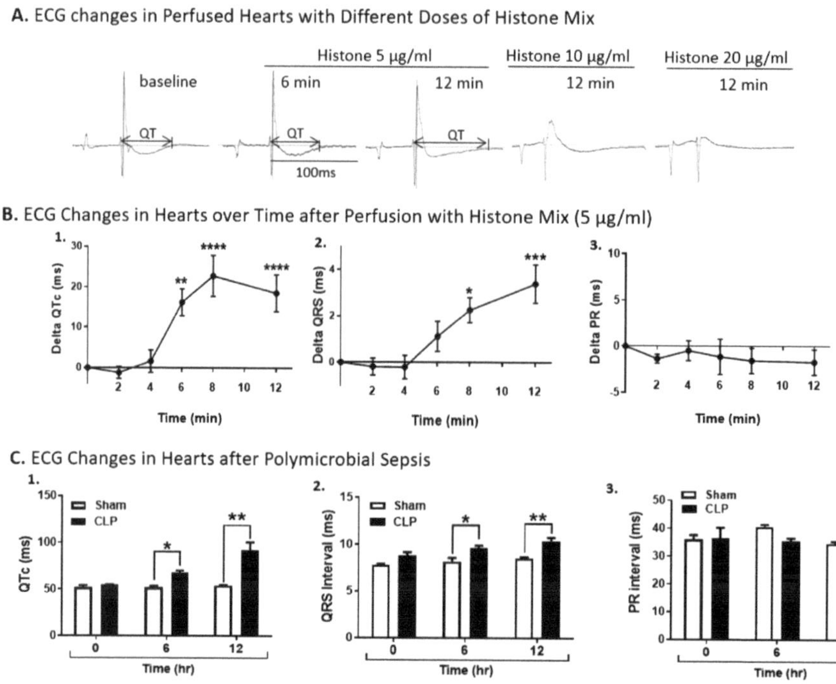

Figure 2. Electrocardiography (ECG) changes in Wt hearts perfused with histones or after induction of polymicrobial sepsis by cecal ligation and puncture (CLP). (**A**) Wt hearts (n ≥ 5) were perfused with histone mix (5, 10, and 20 μg/mL) for 12 min at 37 °C. By 12 min, there was a distinct QT interval widening, and at higher concentrations (10, 20 μg/mL), ECG tracings collapsed. (**B**) Hearts perfused with a 5 μg/mL histone mix for 12 min. (**B-1**) Increased QT interval peaking at 8 min, (**B-2**) steady increase in QRS interval, and (**B-3**) no change in PR interval. (**C**) ECG recordings before and after CLP in Wt mice (n = 5) at 6 and 12 h. Septic mice displayed significantly elevated QTc (Bazett's formula) and QRS intervals at both time points, while PR interval remained unchanged. Sham controls showed no alterations. All values in (**B**,**C**) are expressed as means ± SEM. Data sets were analyzed by one-way ANOVA followed by Dunnett's multiple comparison test for (**B**), and two-way ANOVA followed by Sidak's multiple comparison test for (**C**). Significance is indicated as follows: * $p < 0.05$, ** $p < 0.01$, *** $p < 0.001$, and **** $p < 0.0001$.

2.3. Ability of Histone Mix to Cause Cessation of Beating Mouse Hearts as a Function of Histone Concentration and Effect of TLR Absence

As shown in Figure 3A, Wt non-paced spontaneously beating mouse hearts were perfused with buffer or with the histone mix at increasing concentrations (5–20 μg/mL). The cessation of beating hearts was determined by visual inspection. In the presence of buffer, all hearts over 12 min of exposure (n ≥ 10 hearts) per group continued beating during perfusion with buffer. In the presence of 5 μg/mL of the histone mix, 63% of the hearts were still beating after 12 min. As the concentrations of the histone mix were increased to 10, 15, and 20 μg/mL, there was a progressive increase in hearts caused by progressively reduced beating or cessation of beating. At the histone mix of 20 μg/mL, all 11 hearts had ceased beating over the interval of 12 min. Figure 3B contrasts these responses with non-paced spontaneously beating mouse hearts from various TLR KOs including TLR2, TLR3, and TLR4 KO compared to Wt, as described in Figure 3A. With 10 μg/mL of the histone mix, the percentage of functional (beating) Wt hearts decreased to 20% in Wt hearts and this number decreased to 0% in the presence of 20 μg/mL histones in Wt or TLR4 KO hearts. At all concentrations, the detrimental effects of the histones were evident 5 min after the start of perfusion. TLR2 KO hearts show more protection from the adverse effects of the histone mix. LV dysfunction (reduced contractility) occurred in only one out of eleven hearts perfused with 5 μg/mL histones, while in 80%, the function was maintained. TLR3 KO hearts were also more resistant to injury, with nearly 30% of TLR3 KO hearts still beating 20 min after perfusion with the highest concentration of histones. In contrast, the absence of TLR4 provided no protection for the beating hearts at any concentration of the histone mix. Accordingly, the absence of TLR4 surprisingly did not afford any protective effects during ex vivo perfusion with histones, in marked contrast to the protective effects in TLR2 and TLR3 KO hearts. These findings highlight the protective effects of TLR2 KO, and to a lesser extent TLR3 KO, in maintaining heart function during histone mix perfusion, while such protective effects were absent in TLR4 KO hearts.

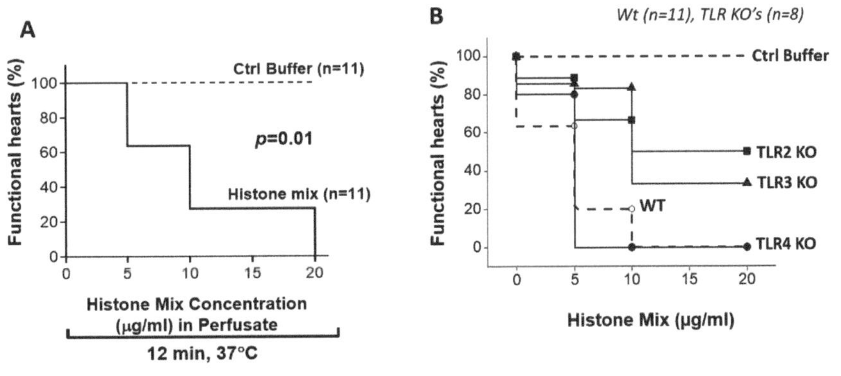

Figure 3. Ability of histone mix to cause cessation of beating mouse hearts as a function of histone concentration and effect of TLR absence. (**A**) Non-paced beating mouse hearts (Wt) were perfused with histone mix (5, 10, and 20 μg/mL) for 12 min at 37 °C. Cessation of beating hearts increased with histone concentration, reaching 100% at 20 μg/mL. (**B**) Comparison of Wt and TLRs KO hearts' responses to 10 μg/mL histones. Wt hearts showed 20% beating, decreasing to 0% at 20 μg/mL. TLR2 KO exhibited better preservation, while TLR4 KO hearts ceased beating at all concentrations. LV dysfunction was observed in TLR2 KO (5 μg/mL), sparing 80% function. TLR3 KO provided partial protection. TLR4 absence offered no protection. Perfusing hearts with the control buffer (PBS) does not cause any functional defects, as shown by the dashed line. Comparisons of survival curves were analyzed using the log-rank (Mantel-Cox) test. $p < 0.05$ is considered as significant.

2.4. *Functional Abnormalities in Wt Mouse Hearts Perfused with Histone Mix and Effects of TLR Absence on Functional Responses of Mouse Hearts to Histone Mix*

In our investigation of functional abnormalities, electrically paced Wt mouse hearts were compared with hearts from TLR2, TLR3, and TLR4 KO mice, all perfused for up to 20 min at 37 °C with buffer or the histone mix. The mechanical function was evaluated in Langendorff-perfused hearts (n = 11 Wt and n ≥ 8 for each TLR KO mice). Perfusing hearts with three different concentrations of the histone mix (5, 10, 20 μg/mL) led to progressive reductions in LVP generation, characterized by both reduced and widened LVP amplitudes in Wt hearts, as also described in Figure 4A. At the 20 μg/mL concentration of the histone mix, LVP amplitudes in Wt hearts plummeted to the point of collapse, accompanied by the development of bradycardia.

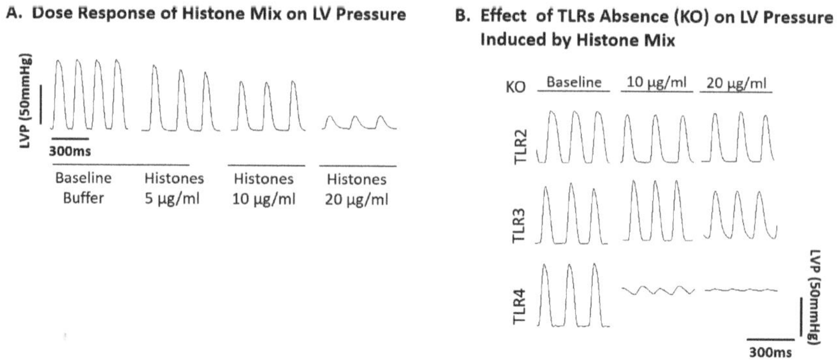

Figure 4. Functional abnormalities in Wt mouse hearts perfused with histone mix and effects of TLR absence on functional responses of mouse hearts to histone mix. Wt mouse hearts (n = 11) and hearts from TLR2, TLR3, and TLR4 KO mice (n ≥ 8 each) were electrically paced and perfused for 20 min at 37 °C with buffer (recorded as the baseline) or histone mix (5, 10, 20 µg/mL). (**A**) Wt hearts' LVP amplitude reduction, collapsing at 20 µg/mL histones. (**B**) Illustrates TLR2, TLR3, and TLR4 KO baseline LVP (n = 8). TLR2 KO exhibited widened LV pressures with histones, TLR3 KO had a modest drop (20%) at 20 µg/mL, while TLR4 KO had virtually ablated LV pressures at both histone concentrations. TLR2 KO and TLR3 KO displayed protective effects with minimal changes or modest decline at higher histone doses. In contrast, TLR4 KO hearts were vulnerable, showing dysfunction even at lower histone concentrations.

As shown in Figure 4B (n = 8 hearts), baseline (ctrl buffer perfused) LVP were similar in TLR2, TLR3, and TLR4 KO hearts, with pressures reaching nearly 100 mmHg. In TLR2 KO hearts, the areas between the LV pressures began to widen with the two concentrations of histones. In TLR3 KO hearts, there was a slight (20%) drop in pressures at the higher concentration (20 µg/mL) of histones. In the case of TLR4 KO hearts, the responses were dramatically different. At both concentrations of histones (10, 20 µg/mL), the LV pressures were virtually ablated. The reasons why LVP was abolished in hearts from TLR4 KO mice exposed to the histone mix as exposed to responses of TLR2 and 3 KO hearts are not currently known.

In summary, TLR2 KO hearts demonstrated a protective effect, exhibiting minimal changes in LV function exposed to histone mix doses (20 µg/mL). TLR3 KO hearts, to a lesser extent, showed protection with changes in LV function initiating at the 20 µg/mL dose. In stark contrast, TLR4 KO hearts displayed no protection with LV function changes evident at a lower histone mix dose (10 µg/mL), indicating a lack of resilience compared to the Wt mouse heart. These findings highlight the distinct protective effects of TLR2 KO and, to a lesser extent, TLR3 KO while underscoring the vulnerability of TLR4 KO hearts in the face of histone mix-induced LV functional alterations.

2.5. Effect of TLR Absence on Delta QTc Prolongation in Murine Hearts Perfused with Histone Mix

In Figure 5, Wt paced hearts were perfused with buffer (baseline) or the histone mix (5, 10, and 20 µg/mL), and ECG recordings were measured for 12 min with 2 min intervals. Due to marked abnormalities in hearts perfused with 10 or 20 µg/mL of the histone mix, the focus was placed on changes in hearts exposed to the 5 µg/mL histone mix at the 12 min perfusion interval. At this point, the QT interval (QTc) was clearly prolonged, as described in Figure 5A. Figure 5A demonstrates the ECG results over 12 min, representing hearts perfused with the histone mix (5 µg/mL), revealing marked and rapid changes in QTc over time and genotypes. Wt hearts displayed progressive QTc prolongation during the first 8 min, followed by a plateau. Concurrently, QTc interval of TLR2 and TLR3 KO mice remained consistent with the baseline (Figure 5A). In sharp contrast, TLR4

KO hearts displayed the largest average increase in the QTc interval among the three groups (Figure 5A). Thus, TLR2 and TLR3 KO mice showed minimal changes, suggesting protection. In contrast, TLR4 KO mice showed exaggerated QTc prolongation at all time points, accentuating the effects of histone mix exposure. Figure 5B compares the ECG pattern of TLR4 KO (worse) and TLR2 KO (protective), evidently showing that the most significant variations in ECG were observed in the duration of the QTc interval.

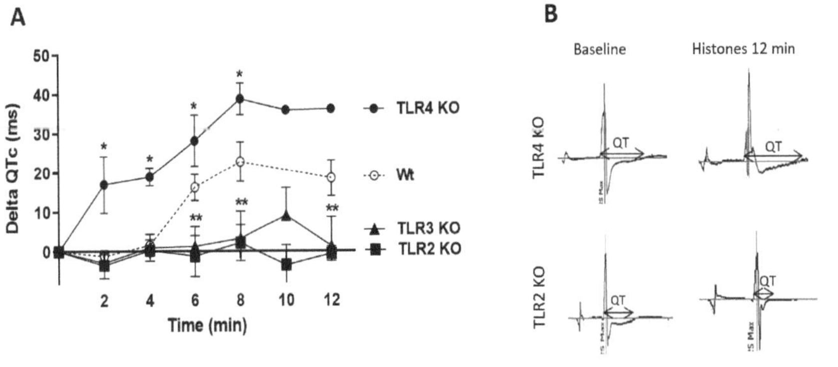

Figure 5. Effect of TLR absence on delta QTc prolongation in murine hearts perfused with histone mix. Wt paced hearts were perfused with buffer or histone mix (5, 10, 20 μg/mL) for 12 min, and ECG recordings were measured at 2 min intervals. (**A**) 5 μg/mL histone mix-induced QTc prolongation over 12 min in Wt hearts; TLR2, and TLR3 KO mice displayed minimal changes, while TLR4 KO mice exhibited exaggerated prolongation. (**B**) ECG patterns of TLR4 KO (worse) and TLR2 KO (protective). The values in 5A are expressed as means ± SEM. Data sets were analyzed by one-way ANOVA followed by Dunnett's multiple comparison test. Significance between groups in each time point is indicated as * $p < 0.05$ and ** $p < 0.01$.

Comparatively, the figure with detailed values showcased the impact of TLR absence on corrected QTc duration. Wt mice exhibited increasing ΔQTc values over time, reaching 18.5 ms at 12 min. TLR2 KO mice showed a protective effect with significantly lower ΔQTc values (slightly increased by 2 ms) while TLR3 KO mice displayed reduced prolongation (increased by 9 ms), indicating robust protection at 12 min. Conversely, TLR4 KO mice exhibited significantly higher ΔQTc values (increased by almost 40 ms from 68 ± 5 ms at baseline only 8 min after histones perfusion), suggesting a pro-arrhythmic effect. In summary, TLR2 and TLR3 KO provided protection against QTc prolongation, with TLR2 KO showing the most significant reduction. In contrast, TLR4 KO led to pronounced pro-arrhythmic effects, highlighting the dynamic responses of murine hearts to TLR KO.

2.6. Echo-Doppler Parameters Measurement after Intravenous Infusion of Histone Mix

As a follow-up to the data in previous figures, which indicated that the absence of TLR2 or TLR3 protected the heart from defects developing after ex vivo perfusion of hearts with the histone mix, while the absence of TLR4 caused an intensification of functional defects, complementary in vivo experiments were conducted in Wt, TLR2 KO (as the most protective), and TLR4 KO (as the least protective) mice in terms of heart electrical and functional effects following histone mix perfusion. Mice were intravenously injected with the histone mix (65 mg/kg body weight), simulating levels of plasma histones appearing after CLP [11]. This amount of histone mixes employed mimics the plasma levels of histones appearing in the plasma of Wt mice after CLP [11]. As evident in Figure 6, both systolic and diastolic parameters were evaluated in Wt, TLR2 KO, and TLR4 KO mice at various time points (including 10, 60, 90 min, and 24 h as indicated) after i.v. histones injection. In Wt mice, a significant decrease in heart rate was noted 90 min after histones injection (Figure 6A) compared to values before injection. By 24 h, values had returned to the pre-infusion (time 0) values. A more intense drop in heart rate after histones infusion was observed in TLR4 KO but not in TLR2 KO animals during the first 90 min post-infusion. Stroke volumes also declined acutely after histones infusion in Wt mice (Figure 6B). However, stroke volumes were not significantly affected by histones infusion in TLR2 KO nor TLR4 KO mice. The lower heart rate and stroke volume resulted in a lower calculated cardiac output in Wt and TLR4 KO mice but not in TLR2 KO mice after histone infusion compared to pre-infusion values (Figure 6C). Heart rate, stroke volume, and cardiac output had essentially returned to baseline in all animals by the 24 h time point. Left ventricular ejection fraction was slightly increased in Wt and TLR4 KO animals after histone infusion (Figure 6D). This may represent an adaptation to loading conditions and heart rate rather than a primary effect of histone infusion to increase contractility. Again, TLR2 KO mice did not show a significantly sustained change in ejection fraction after histone infusion. There was also a modest reduction in other parameters (Figure 6E–H) after histone infusion into Wt animals, and to a greater extent in TLR4 KO animals. Isovolumic relaxation time was significantly prolonged after histone infusion into Wt and TLR4 KO mice, but this was not found in TLR2 KO mice (Figure 6E). The effect of histone infusion on other measures, including the mitral E/A ratio, was more variable, potentially due to differences in preload and secondary compensatory mechanisms between the pre- and post-infusion states (Figure 6F–H). Together, these data indicate an intensification of defective Echo-Doppler parameters in TLR4 KO mice compared to Wt mice and nearly no defective parameters developing in TLR2 KO mice following i.v. injection of the histone mix.

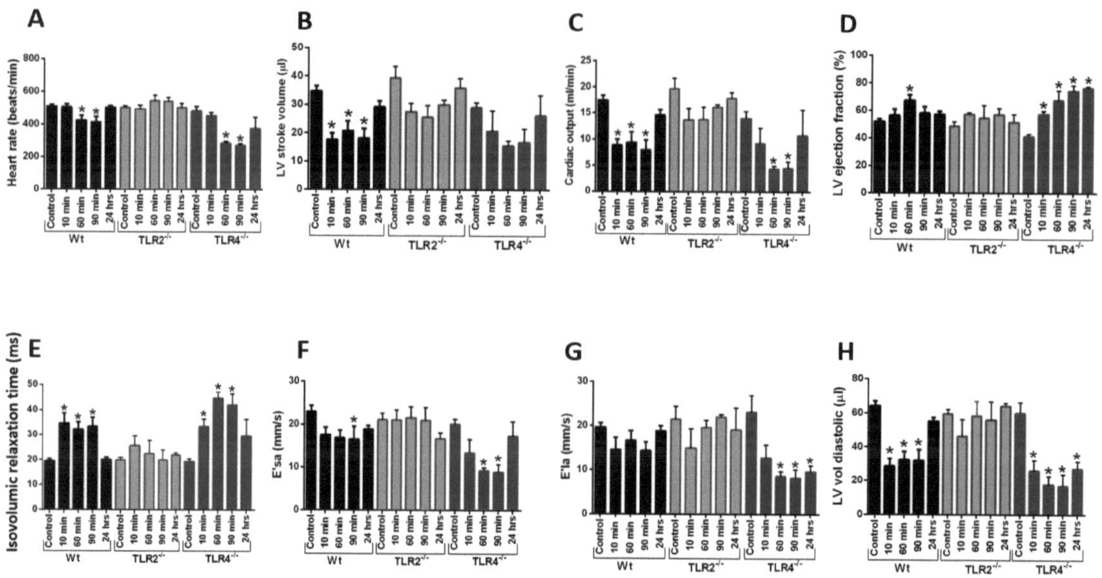

Figure 6. Echo−doppler parameters in Wt, TLR2 KO, and TLR4 KO hearts after intravenous infusion of histone mix. Effects of TLR2 or TLR4 KO on responses of mouse hearts after i.v. infusion of histone mix. Wt, TLR2 KO (protective), and TLR4 KO (least protective) mice were intravenously injected with histone mix (65 mg/kg). Echo−Doppler parameters were evaluated at various time points post−injection. (**A**) Significant heart rate decreases in Wt and TLR4 KO but not TLR2 KO groups 90 min post−injection, returning to baseline by 24 h. TLR4 KO exhibited a more intense drop compared to Wt. (**B**) Stroke volumes decreased significantly in Wt but not in TLR2 KO or TLR4

KO mice. (**C**) Cardiac output dropped significantly in Wt and TLR4 KO but not in TLR2 KO mice. (**D**) Left ventricular ejection fraction slightly increased in Wt and to a bigger extent in TLR4 KO after infusion. TLR2 KO showed no sustained change. (**E**) Isovolumic relaxation time increased significantly in Wt and TLR4 KO but not in TLR2 KO mice. (**F–H**) Other parameters displayed variable responses especially in TLR4 KO. Each bar represents Echo–Doppler measurements from 5 mice. Values are expressed as means ± SEM. Differences were assessed with one–way ANOVA, and group differences were determined using Tukey's post–hoc test. Differences were considered significant if $p < 0.05$ as shown by one star.

2.7. Effects of Individual Histones on LV Function, ECG Changes, and Cytokine Release on Perfused Hearts

Since we found the effect of histone mix in our previous figures, we extended our studies to elucidate the effect of each individual histone (H1, H2A, H2B, H3, and H4) present in the histone mix. Our earlier in vivo and in vitro studies showed the various biological effects of different individual histones [41]. For these studies, we used only the Wt hearts. The studies described in Figures 1 and 2 were carried out similar to histone mix protocols.

In Figure 7A, hearts were perfused with buffer or individual histones (5 μg/mL) over a time frame of 0–150 min. H1 reduced LVP most significantly followed closely by H2A and H2B. H4 had little effect on LVP, while H3 induced slightly increased pressures above those pressures. When the perfusate was buffer, minimal effects developed. Based on these data, it is not surprising that H1 followed by H2A and H2B had the most inhibitory effects. As will be shown in Figure 8, these histones induced the most pronounced indications of CM lysis [lactate dehydrogenase (LDH) release] and apoptosis (TUNEL staining), techniques previously detailed [18,36,41].

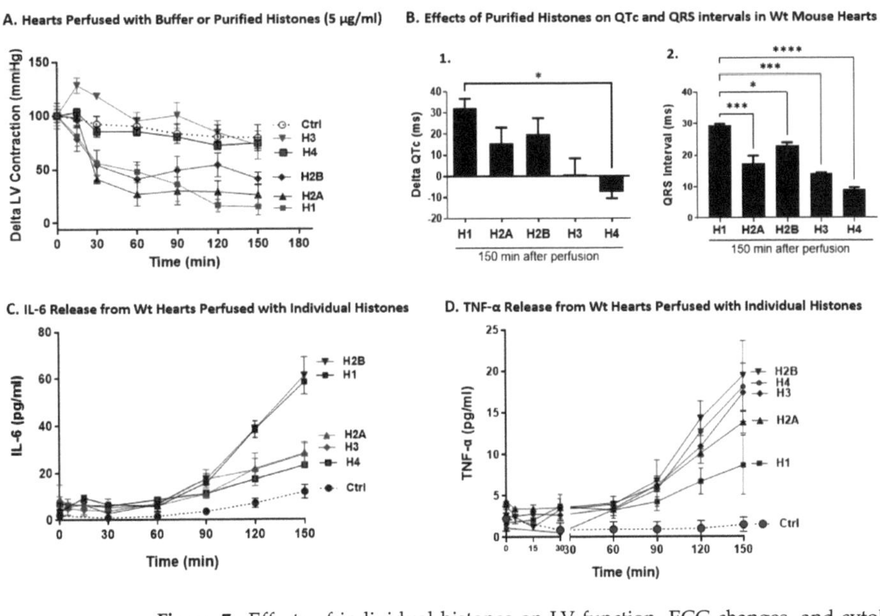

Figure 7. Effects of individual histones on LV function, ECG changes, and cytokine release on perfused hearts. (**A**) Significant reduction in LVP was observed with H1, followed by H2A and H2B. Minimal effects were observed with H4, while H3 slightly increased pressures. (**B**) QT interval pro–

longation was noted with H1 (33 ms) and to a lesser extent with H2A and H2B (18–22 ms) but not with H3 and H4. (**B-1**) QT interval was more prolonged by perfusion with H1 compared to other individual histones. (**B-2**) QRS interval prolongation was caused by H1, H2A, and H2B. Overall, H1 emerges as the primary histone-inducing functional and electrical dysfunction. (**C**) Cytokine release was highest for IL-6 with H1 and H2B, followed by H2A, H3, and H4. (**D**) TNF-α responses were most pronounced with H2B, followed by H4, H3, H2A, and H1. Statistical analyses included one–way ANOVA with Tukey's post–hoc analysis for comparisons among histones in (**A**) and with H1 in (**B**). Significant differences in (**A**) are as follows: Control vs. H1 $p < 0.05$), Control vs. H2A ($p < 0.05$), H1 vs. H3 ($p < 0.01$), H2A vs. H3 ($p < 0.01$), and H2B vs. H3 ($p < 0.05$). In (**A**), these p–values are not shown due to space constraints. In (**B**), * $p < 0.05$. *** $p < 0.001$, **** $p < 0.0001$.

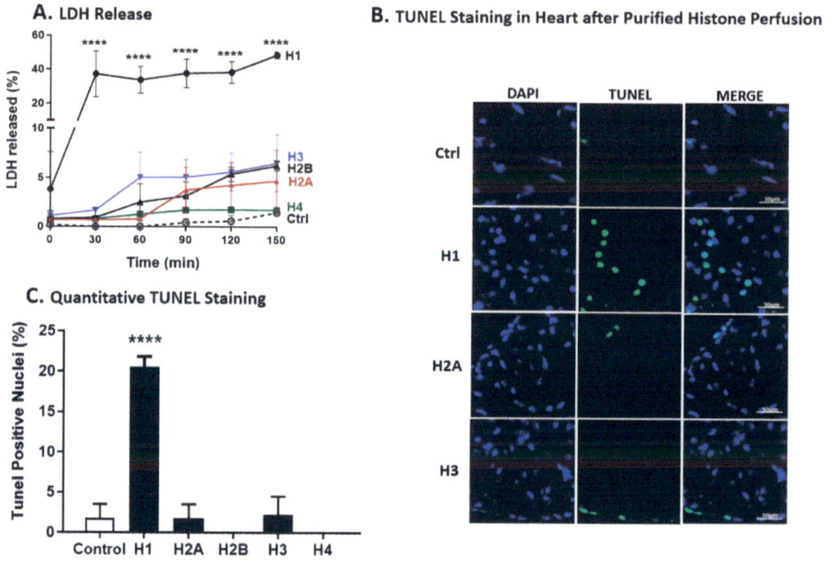

Figure 8. LDH Release and TUNEL Staining of Wt Hearts Atria after Perfusion with Individual Histones. Wt mouse hearts were perfused with individual histones (5 µg/mL) or buffer at 37 °C for specified durations. (**A**) Within 30 to 150 min, H1 induced approximately 40% LDH release, primarily in the initial 30 min while other histones caused only 2–5% release. (**B**) TUNEL staining demonstrates that H1 induced maximal apoptosis (around 20%). Blue: DAPI staining; Green: TUNEL staining. (**C**) Other histones showed minimal staining. H2B and H4 exhibited no staining. Scale bars = 50 µm. Values are expressed as means ± SEM. Differences were assessed with one-way ANOVA, and group differences were determined with Tukey's procedure. Differences were considered significant if $p < 0.05$. In (**A**,**C**), **** $p < 0.0001$.

In Figure 7B, we calculated the extent to which individual histones would alter QT and QRS intervals using perfusion of Wt hearts with each 5 of the individual histones (5 µg/mL). For each bar, n $\geq$ 5 mice. In Figure 7B-1, the QT interval was prolonged by perfusion with H1 by approximately 33 ms while H2A and H2B caused some prolongation (18–22 ms) of this interval. However, H3 and H4 did not cause prolongation of this interval. When the effects of individual histones on ECG parameters were examined, the same three histones (H1, H2A, and H2B) caused prolongation of the QRS interval (Figure 7B-2). In summary, these data indicate H1 as the most predominant individual histone causing functional and electrical dysfunction in the perfused hearts compared to other individual histones.

Figure 7C,D show the levels of cytokine release from Wt hearts perfused with individual histones. It is known that polymicrobial sepsis is often associated with cytokine release from various organs as well as from phagocytic cells [42–49]. Wt hearts were perfused with buffer or with individual histones (each at 5 μg/mL) for the times indicated (6–150 min), at 37 °C. Cytokine levels were measured in perfusion fluids by ELISA. Generally, IL-6, H1, and H2B caused the highest levels of cytokine release followed by H2A, H3, H4 (Figure 7C). In the case of TNF-α, the most robust responses occurred with H2B > H4 > H3, followed by H2A and H1 (Figure 7D). Different cell types in the heart where TNF-α and IL-6 originate might explain the difference. Interestingly, concentrations of TNF-α and IL-6 do not directly correlate with cardiac dysfunction or cell death. This observation suggests a complex pathophysiology that extends beyond simple cytokine release. Factors such as exposure duration, cytokine concentration, and model dependency must be considered, as evidenced by the contradictory effects of TNF-α observed post-myocardial infarction and in heart failure cases in humans and non-human models [50].

2.8. LDH Release and TUNEL Staining of Wt Hearts after Perfusion with Individual Histones

For the experiments in Figure 8A, Wt mouse hearts were perfused with buffer or with individual histones (5 μg/mL) for the indicated periods of time at 37 °C. The dose of individual histones (5 μg/mL) was the same as employed for most of the other experiments. As shown in Figure 8A, between 30 and 150 min of perfusion with H1, approximately 40% of LDH release occurred, virtually all of it during the first 30 min of perfusion, followed by a plateau, whereas, with the presence of the other individual histones, only 2–5% LDH release occurred.

To determine which individual histone causes the highest apoptosis in the heart LV, the TUNEL staining was employed as described before [18]. Figure 8B shows the TUNEL staining of LV frozen sections of perfused hearts with individual histones while Figure 8C shows the quantitative of the TUNEL staining. In accordance with the release of LDH from the perfused heart (Figure 8A), TUNEL staining of LV frozen sections of perfused hearts showed maximal TUNEL staining (approximately 20%) in hearts perfused with H1. Perfusion with the other histones resulted in little or no evidence of TUNEL-positive cells (Figure 8B). H2B and H4 showed no staining. In summary, these data indicate that H1 as the most predominant individual histones causing apoptosis in the CMs compared to other individual histones.

3. Discussion

Extracellular histones are key emerging proinflammatory and prothrombic factors, particularly in polymicrobial sepsis [14,16,19,51]. Histone release, predominantly from PMNs via complement activation products such as the C5a anaphylatoxin, leads to the formation of NETs, facilitating bacterial killing. Those NETs become the main source of circulating plasma histones in the polymicrobial sepsis induced by CLP [11]. Our investigations indicated that histones are linked to adverse outcomes in sepsis, including organ injury and early development of cardiac dysfunction (septic cardiomyopathy) [11]. Specifically, histone-blocking antibodies decreased the cardiac dysfunction in septic mice induced by CLP. Interfering with the C5a and NLRP3 inflammasome complex can mitigate histone appearance in polymicrobial sepsis models [11,12,33]. Accordingly, a recent study has confirmed the presence of plasma histones (H2B and H3) in sepsis patients, with concentrations reaching 10–20 μg/mL (as measured by ELISA) which closely align with the levels we observed in the plasma of septic mice [11].

In this report, we present findings that direct perfusion of a histone mixture into mouse hearts leads to functional and electrical abnormalities. Our data also provide evidence that some, but not all, individual histones have adverse effects on the heart, with H1 showing the largest overall alterations relative to the other individual histones. H1 can induce cell death (as shown by TUNEL staining and LDH) of LV CMs, similar to acute lung injury and phagocytes from earlier studies [41]. Also, certain histones can cause proinflammatory cytokines release from the heart, some of which have cardio-suppressive effects while others induce proinflammatory responses potentially exacerbating cardiac dysfunction and leading to worse outcomes [12,50]. Quite similar events occur in neutrophils and macrophages [41].

Although the molecular targets of histones have not been completely elucidated, their interaction with TLRs has attracted considerable attention [23,24,36,38]. Our findings from TLR KO hearts perfused with a histone mixture support the hypothesis that this pathway may contribute to their harmful effects in polymicrobial sepsis.

The impact of TLR stimulation on ventricular myocyte contractility has been previously reported, underscoring TLRs' critical role in modulating cardiac function [52]. Numerous reports on sepsis highlight the importance of TLR2, TLR3, and TLR4 in cardiac dysfunction and cardiomyopathy, although their precise roles remain incompletely understood. Our study explores histones as potential mediators of cardiomyopathy through interactions with TLRs. Consistent with findings in sepsis and other cardiomyopathy models, TLR2 KO and TLR3 KO are shown to be protective, whereas TLR4 KO exacerbates cardiac dysfunction and cardiomyopathy. For instance, Ma et al. reported that TLR2 blockade attenuated cardiac dysfunction and inhibited cardiac fibrosis in a chronic cardiomyopathy model induced by doxorubicin injection, while TLR4 blockade exacerbated cardiac dysfunction and fibrosis by amplifying myocardial inflammation [53]. Similarly, Mathur et al. found that TLR2 deficiency reduced CM inflammatory responses, whereas TLR4 deficiency had minimal impact [54]. In line, TLR2 KO mice showed better cardiac function and survival in a doxorubicin-induced cardiomyopathy model compared to Wt mice in another study [55]. Zhu et al. found that activating TLR4 offers a protective advantage to CMs, shielding them from stress-induced damage through MyD88- and NOS2-dependent pathways [56]. In terms of TLR3, in line with our results, Gao et al. showed that TLR3 deficiency attenuated cardiac dysfunction and increased survival outcomes following CLP-induced sepsis in mice [57].

Collectively, our research is consistent with the protective effects of TLR2 and TLR3 KO, and not TLR4 KO [53,54,56–60]. One limitation of our study is that we did not investigate the effect of each individual histone on cardiac function in TLR2 KO or TLR4 KO models in vivo. Further research is needed to determine whether specific histones, acting through individual TLRs, contribute to the observed mechanical and electrical changes in cardiac function. This could identify potential mechanistic targets unique to each histone subtype.

Furthermore, our grasp of how individual histones and their epigenetic modifications (acetylation, methylation, phosphorylation, ubiquitination, SUMOylation) influence biological responses remains limited. For example, the implications of converting arginine to citrulline in histones are not well understood. Previous studies using a neutralizing histone antibody (clone BWA3) that targets H2A and H4 [11,61], have shown reductions in the tissue-damaging effects of histones across various models, including polymicrobial sepsis [11,14], acute lung injury [19,62], and chemical-induced organ injury [23]. Furthermore, citrullination by peptidyl arginine deiminase 4 (PAD4) not only modifies histones but may also trigger the formation of autoantibodies, exacerbating autoimmune conditions such as rheumatoid arthritis and systemic lupus erythematosus [63–65]. Development of these auto antibodies may intensify the underlying autoimmune disease [66]. Understanding the intricate relationships between histone modifications and their biological outcomes continues to be a significant challenge, as evidenced by the variable effects of histone methylation on biological responses, highlighting the complex nature of histone-mediated signaling pathways.

Although the literature suggests that histones bind to TLRs releasing proinflammatory products [23,24,36,38,67], it is premature to define the specific functions of histones related to these interactions. With the presence of both native and chemically modified histones, there is still much to understand about their biological roles before we can effectively target them in acute conditions like sepsis, trauma, and acute lung injury. Most studies have focused on 28-day survival rates, leaving uncertainties about the immediate impacts on heart function in sepsis. Furthermore, clinical trials have not been structured to evaluate the developments following polymicrobial sepsis. Clinical trials from 2010 to 2021 investigated the role of endotoxin (LPS) and its interaction with TLR4 antagonists, such as eritoran and TAK-242 but yielded mixed results. Despite initial expectations, these studies failed to provide conclusive evidence of efficacy [58], leading the pharmaceutical industry to halt further research in septic patients. To improve outcomes, it is important to understand the molecular events unfolding during the early stages of sepsis in order to develop innovative therapies.

In conclusion, our findings underscore the intricate roles of histones and TLRs in inflammatory responses and cardiac dysfunction. Elucidating specific interactions between individual histones and TLRs is crucial for developing targeted therapies in acute inflammatory conditions such as sepsis. The need for new strategies to increase our knowledge of how histones adversely affect biological systems and cause cell and organ dysfunction, like cardiomyopathy, is a pressing problem. Specifically, focusing on H1 might offer an interesting prospect.

4. Materials and Methods

4.1. Animals

All animal procedures were performed in accordance with U.S. National Institutes of Health guidelines and were approved by the University of Michigan Committee on the Use and Care of Animals (protocol PRO00005749). Male age-matched C57BL/6 (wild type), TLR2-KO ($TLR2^{-/-}$), TLR3-KO ($TLR3^{-/-}$), and TLR4-KO ($TLR4^{-/-}$) were purchased from the Jackson Laboratories (Bar Harbor, ME). All animals were housed under specific pathogen-free conditions with free access to food and water.

4.2. Whole Heart Perfusions, Functional LVP and Electrical (ECG) Measurements

Male and female adult C57BL/6 mice aged three to four months underwent treatment involving a peritoneal injection of 0.075 mL heparin followed by cervical dislocation. Subsequently, the hearts were rapidly excised and immersed in ice-cold PBS (pH 7.4). The aorta was cannulated on a Langendorff perfusion apparatus (AD Instruments, Colorado Springs, CO), and warm (37 °C), oxygenated Krebs–Henseleit Buffer (KHB) supplemented with 1.8 mM Ca^{2+} was perfused for 10–15 min to stabilize the heart. The left atrium was excised, and a water-filled balloon connected to a pressure transducer was inserted into the left ventricle to record LVP. Perfusate flow and balloon pressure were adjusted to achieve a diastolic pressure of around 8 mmHg and a systolic pressure of around 95 mmHg. Hearts unable to reach the target pressures with a maximal perfusion flow of 4 mL/min were discarded. Following an approximate 10 min stabilization period, the perfusion solution was switched to KHB (1.8 mM Ca^{2+}) containing either the histone mix or individual histones. Surface electrodes were placed at L1 and L2 levels for ECG recordings, connected to an octal bioamp and a PowerLab unit. Raw tracings were filtered using 5 Hz high pass and 450 Hz low pass digital filters, and analysis was conducted using PowerLab software (version 7, AD Instruments, Boston, MA, USA). Each data point represents the average of at least 15 superimposed tracings. Semi-automated customized macros were employed to derive functional parameters for both LVP and ECG. Perfusate samples (2 mL) were collected at 0, 30, 60, 90, 120, and 150 min of histone perfusion, flash frozen, and stored at −80 °C for cytokine quantification. ECG measurements involved recordings for right versus left volume, with continuous perfusion of histones using Langendorff perfusions.

4.3. Protocols for Wt Mouse Heart Perfusion with Histone Mixture or Individual Histones

Wt or different TLRs KO Langendorff hearts including TLR2 KO, TLR3 KO, and TLR4 KO were perfused at 37 °C for various intervals of time with the histone mix. This histone mix was already tested in our earlier studies suggesting that the biological responses induced by histones in phagocytes were not due to the contamination with LPS [41,62]. We also investigated the effects of individual histones, including H1, H2A, H2B, H3, and H4, by perfusing them in Wt Langendorff hearts. Perfusion of hearts with the histone mix or individual histones continued for up to 150 min, with data collected at 30 min intervals, unless otherwise noted. Readouts included abnormal LV functions and defective ECG tracings, cytokine release, the release of LDH, and TUNEL staining.

4.4. Cytokine ELISA

Cytokines (IL-6 and TNF-α) levels were assessed in the collected perfusate samples using the R&D DuoSet ELISA kit (R&D Systems, Minneapolis, MN, USA) following the manufacturer's guidelines.

4.5. LDH Cytotoxicity Assay and TUNEL Staining

An increase in LDH levels released from cells indicates damage to the cells and leakage resulting from membrane disruption. To assess the cardiac damage and cytotoxic effects induced by individual histones, we performed the LDH assay on the perfusate samples, following the guidelines provided by the manufacturer (Cayman Chemical, Ann Arbor, MI, USA), as previously demonstrated in our studies [36,41].

TUNEL staining on frozen heart tissue was applied according to the manufacturer's recommendations, using the special kit (Promega, Madison, WI, USA) and performed as described in our recent publication [18]. Frozen sections of left ventricles were mounted in ProLong Gold Anti-Fade reagent containing DAPI (Life Technologies, Carlsbad, CA, USA). Images were acquired on a Nikon A-1 confocal system with Nikon Elements software (version 3.2, Nikon Instruments Inc., Tokyo, Japan). The average number of TUNEL-positive cells/field was determined in a blinded fashion from $\geq$10 fields over several sections per heart. Notably, while electrical and functional measurements were conducted with hearts being electrically paced, no pacing was employed for cytokine assessment and LDH/TUNEL staining.

4.6. Echo-Doppler Functional Studies on TLR2, TLR3, and TLR4 KO Mice

Echocardiograms were performed as we described in our previous studies [11,12,33,36,68] according to the recommendations of the American Society of Echocardiography. All echocardiograms were performed by a registered echocardiographer who was blinded to genotype. TLR2 KO mice ($n \geq 7$), TLR4 KO mice (for each, $n \geq 7$), and control Wt mice ($n \geq 7$) were weighed and anesthetized with inhaled isoflurane. Imaging was performed using a Vevo 770 High-Resolution In Vivo Imaging System (Visualsonics Inc., Toronto, ON, Canada) equipped with an RMV-707 30 MHz RMV (Real-Time Visualization) (up to 45 MHz) scanhead. LV volumes were measured from the parasternal long-axis view at the level of the tips of the leaflets of the mitral valve at end-systole (VolS) and end-diastole (VolD) and used to calculate stroke volume (SV = VolD − VolS) and ejection fraction (EF % = endocardial SV/endocardial VolD × 100). Cardiac output (CO = SV × heart rate) was calculated from stroke volume and heart rate. Isovolemic relaxation time was determined from the apical 4-chamber view using pulse wave Doppler acquisition and was measured from the closure of the aortic valve to the onset of mitral inflow. Imaging was performed at 0, 10, 60, and 90 min and 24 h after histone (65 mg/kg body weight) injection (i.v.).

4.7. Reagents

Mixed calf thymus histones (purified, Type II-A) and histones H1 and H3 (purified from calf thymus) were obtained from Roche (Indianapolis, IN, USA). The remaining individual histones (H2A, H2B, and H4) were in recombinant form and were purchased from Cayman Chemical (Ann Arbor, MI, USA).

4.8. Statistical Analysis

For statistical analysis, GraphPad Prism version 10.0 (Graphpad, San Diego, CA, USA) was used. All values are expressed as mean ± SEM. Data sets were analyzed using one-way or two-way ANOVA, followed by Dunnett's, Sidak, or Tukey's multiple comparison tests as appropriate for each dataset. Survival curve comparisons were evaluated using the log-rank (Mantel-Cox) test with <0.05 considered significant.

Author Contributions: Conceptualization, R.L. and P.A.W.; methodology, R.L., M.K., J.J.G. and P.A.W.; validation, R.L., M.K., J.J.G., M.W.R. and P.A.W.; formal analysis, R.L., F.F., F.S.Z. and P.A.W.; investigation, R.L., F.F., M.K., J.J.G. and F.S.Z.; resources, P.A.W.; writing—original draft preparation, P.A.W.; writing—review and editing, F.F. and P.A.W.; supervision, M.W.R., J.J., H.H.V. and P.A.W.; project administration, P.A.W.; funding acquisition, J.J. and P.A.W. All authors have read and agreed to the published version of the manuscript.

Funding: These studies were supported by the National Institutes of Health NIGMS grants R01-GM29507 and R01-GM61656 (P.A.W.) and NHLBI grant R01-HL163943, CIBERCV (CB16/11/00458) and European Union's Horizon 2020 Research and Innovation programme (MAESTRIA) under grant agreement GA-965286 (J.J.). Additional support was provided by the Stobbe Endowment Fund, University of Michigan (P.A.W.), by the endorsement of Dr. Charles Parkos, Carl Vernon Weller Professor of Pathology, and Chair of the Department of Pathology, University of Michigan Medical School.

Institutional Review Board: All animal procedures were performed within the U.S. National Institutes of Health guidelines and were approved by the University of Michigan Committee on the Use and Care of Animals (protocol PRO00005749).

Informed Consent Statement: Not applicable.

Data Availability Statement: Data is contained within the article.

Acknowledgments: The authors would like to thank Kimber Converso-Baran, research sonographer and echocardiographic specialist, University of Michigan, Department of Molecular and Integrative Physiology Phenotyping Core, for the excellent heart function imaging services provided.

Conflicts of Interest: The authors declare no potential or competing financial interests. The funders had no role in the design of the study; in the collection, analyses, or interpretation of data; in the writing of the manuscript; or in the decision to publish the results.

References

1. Merx, M.W.; Weber, C. Sepsis and the heart. *Circulation* **2007**, *116*, 793–802. [CrossRef] [PubMed]
2. Parrillo, J.E.; Parker, M.M.; Natanson, C.; Suffredini, A.F.; Danner, R.L.; Cunnion, R.E.; Ognibene, F.P. Septic shock in humans. Advances in the understanding of pathogenesis, cardiovascular dysfunction, and therapy. *Ann. Intern. Med.* **1990**, *113*, 227–242. [CrossRef] [PubMed]
3. Cunnion, R.E.; Schaer, G.L.; Parker, M.M.; Natanson, C.; Parrillo, J.E. The coronary circulation in human septic shock. *Circulation* **1986**, *73*, 637–644. [CrossRef] [PubMed]
4. Hasanin, A.; Mukhtar, A.; Nassar, H. Perfusion indices revisited. *J. Intensive Care* **2017**, *5*, 24. [CrossRef] [PubMed]
5. Abel, F.L. Myocardial function in sepsis and endotoxin shock. *Am. J. Physiol.* **1989**, *257 Pt 2*, R1265–R1281. [CrossRef]
6. Kurt, A.N.; Aygun, A.D.; Godekmerdan, A.; Kurt, A.; Dogan, Y.; Yilmaz, E. Serum IL-1beta, IL-6, IL-8, and TNF-alpha levels in early diagnosis and management of neonatal sepsis. *Mediat. Inflamm.* **2007**, *2007*, 31397. [CrossRef] [PubMed]
7. Sato, R.; Nasu, M. A review of sepsis-induced cardiomyopathy. *J. Intensive Care* **2015**, *3*, 48. [CrossRef] [PubMed]
8. Daix, T.; Mathonnet, A.; Brakenridge, S.; Dequin, P.F.; Mira, J.P.; Berbille, F.; Morre, M.; Jeannet, R.; Blood, T.; Unsinger, J.; et al. Intravenously administered interleukin-7 to reverse lymphopenia in patients with septic shock: A double-blind, randomized, placebo-controlled trial. *Ann. Intensive Care* **2023**, *13*, 17. [CrossRef] [PubMed]
9. Deng, P.; Tang, N.; Li, L.; Zou, G.; Xu, Y.; Liu, Z. Diagnostic value of combined detection of IL-1beta, IL-6, and TNF-alpha for sepsis-induced cardiomyopathy. *Med. Clin.* **2022**, *158*, 413–417. [CrossRef]
10. Casey, L.C.; Balk, R.A.; Bone, R.C. Plasma cytokine and endotoxin levels correlate with survival in patients with the sepsis syndrome. *Ann. Intern. Med.* **1993**, *119*, 771–778. [CrossRef]
11. Kalbitz, M.; Grailer, J.J.; Fattahi, F.; Jajou, L.; Herron, T.J.; Campbell, K.F.; Zetoune, F.S.; Bosmann, M.; Sarma, J.V.; Huber-Lang, M.; et al. Role of extracellular histones in the cardiomyopathy of sepsis. *FASEB J.* **2015**, *29*, 2185–2193. [CrossRef]
12. Kalbitz, M.; Fattahi, F.; Grailer, J.J.; Jajou, L.; Malan, E.A.; Zetoune, F.S.; Huber-Lang, M.; Russell, M.W.; Ward, P.A. Complement-induced activation of the cardiac NLRP3 inflammasome in sepsis. *FASEB J.* **2016**, *30*, 3997–4006. [CrossRef] [PubMed]
13. Kakihana, Y.; Ito, T.; Nakahara, M.; Yamaguchi, K.; Yasuda, T. Sepsis-induced myocardial dysfunction: Pathophysiology and management. *J. Intensive Care* **2016**, *4*, 22. [CrossRef] [PubMed]
14. Xu, J.; Zhang, X.; Pelayo, R.; Monestier, M.; Ammollo, C.T.; Semeraro, F.; Taylor, F.B.; Esmon, N.L.; Lupu, F.; Esmon, C.T. Extracellular histones are major mediators of death in sepsis. *Nat. Med.* **2009**, *15*, 1318–1321. [CrossRef] [PubMed]

15. Chaput, C.; Zychlinsky, A. Sepsis: The dark side of histones. *Nat. Med.* **2009**, *15*, 1245–1246. [CrossRef] [PubMed]
16. Nakahara, M.; Ito, T.; Kawahara, K.; Yamamoto, M.; Nagasato, T.; Shrestha, B.; Yamada, S.; Miyauchi, T.; Higuchi, K.; Takenaka, T.; et al. Recombinant thrombomodulin protects mice against histone-induced lethal thromboembolism. *PLoS ONE* **2013**, *8*, e75961. [CrossRef] [PubMed]
17. Ekaney, M.L.; Otto, G.P.; Sossdorf, M.; Sponholz, C.; Boehringer, M.; Loesche, W.; Rittirsch, D.; Wilharm, A.; Kurzai, O.; Bauer, M.; et al. Impact of plasma histones in human sepsis and their contribution to cellular injury and inflammation. *Crit. Care* **2014**, *18*, 543. [CrossRef]
18. Grailer, J.J.; Fattahi, F.; Dick, R.S.; Zetoune, F.S.; Ward, P.A. Cutting edge: Critical role for C5aRs in the development of septic lymphopenia in mice. *J. Immunol.* **2015**, *194*, 868–872. [CrossRef]
19. Bosmann, M.; Grailer, J.J.; Ruemmler, R.; Russkamp, N.F.; Zetoune, F.S.; Sarma, J.V.; Standiford, T.J.; Ward, P.A. Extracellular histones are essential effectors of C5aR- and C5L2-mediated tissue damage and inflammation in acute lung injury. *FASEB J.* **2013**, *27*, 5010–5021. [CrossRef]
20. Fattahi, F.; Grailer, J.J.; Jajou, L.; Zetoune, F.S.; Andjelkovic, A.V.; Ward, P.A. Organ distribution of histones after intravenous infusion of FITC histones or after sepsis. *Immunol. Res.* **2015**, *61*, 177–186. [CrossRef]
21. Abrams, S.T.; Zhang, N.; Manson, J.; Liu, T.; Dart, C.; Baluwa, F.; Wang, S.S.; Brohi, K.; Kipar, A.; Yu, W.; et al. Circulating histones are mediators of trauma-associated lung injury. *Am. J. Respir. Crit. Care Med.* **2013**, *187*, 160–169. [CrossRef] [PubMed]
22. Huber-Lang, M.; Lambris, J.D.; Ward, P.A. Innate immune responses to trauma. *Nat. Immunol.* **2018**, *19*, 327–341. [CrossRef] [PubMed]
23. Xu, J.; Zhang, X.; Monestier, M.; Esmon, N.L.; Esmon, C.T. Extracellular histones are mediators of death through TLR2 and TLR4 in mouse fatal liver injury. *J. Immunol.* **2011**, *187*, 2626–2631. [CrossRef] [PubMed]
24. Allam, R.; Scherbaum, C.R.; Darisipudi, M.N.; Mulay, S.R.; Hagele, H.; Lichtnekert, J.; Hagemann, J.H.; Rupanagudi, K.V.; Ryu, M.; Schwarzenberger, C.; et al. Histones from dying renal cells aggravate kidney injury via TLR2 and TLR4. *J. Am. Soc. Nephrol.* **2012**, *23*, 1375–1388. [CrossRef] [PubMed]
25. Garcia-Gimenez, J.L.; Roma-Mateo, C.; Carbonell, N.; Palacios, L.; Peiro-Chova, L.; Garcia-Lopez, E.; Garcia-Simon, M.; Lahuerta, R.; Gimenez-Garzo, C.; Berenguer-Pascual, E.; et al. A new mass spectrometry-based method for the quantification of histones in plasma from septic shock patients. *Sci. Rep.* **2017**, *7*, 10643. [CrossRef] [PubMed]
26. Sarubbo, S.; Annicchiarico, L.; Corsini, F.; Zigiotto, L.; Herbet, G.; Moritz-Gasser, S.; Dalpiaz, C.; Vitali, L.; Tate, M.; De Benedictis, A.; et al. Planning Brain Tumor Resection Using a Probabilistic Atlas of Cortical and Subcortical Structures Critical for Functional Processing: A Proof of Concept. *Oper. Neurosurg.* **2021**, *20*, E175–E183. [CrossRef] [PubMed]
27. Lagedal, R.; Eriksson, O.; Sorman, A.; Huckriede, J.B.; Kristensen, B.; Franzen, S.; Larsson, A.; Bergqvist, A.; Alving, K.; Forslund, A.; et al. Impaired Antibody Response Is Associated with Histone-Release, Organ Dysfunction and Mortality in Critically Ill COVID-19 Patients. *J. Clin. Med.* **2022**, *11*, 3419. [CrossRef]
28. Annane, D.; Sebille, V.; Duboc, D.; Le Heuzey, J.Y.; Sadoul, N.; Bouvier, E.; Bellissant, E. Incidence and prognosis of sustained arrhythmias in critically ill patients. *Am. J. Respir. Crit. Care Med.* **2008**, *178*, 20–25. [CrossRef]
29. Hunter, J.D.; Doddi, M. Sepsis and the heart. *Br. J. Anaesth.* **2010**, *104*, 3–11. [CrossRef]
30. Vieillard-Baron, A.; Caille, V.; Charron, C.; Belliard, G.; Page, B.; Jardin, F. Actual incidence of global left ventricular hypokinesia in adult septic shock. *Crit. Care Med.* **2008**, *36*, 1701–1706. [CrossRef] [PubMed]
31. Hoover, D.B.; Ozment, T.R.; Wondergem, R.; Li, C.; Williams, D.L. Impaired heart rate regulation and depression of cardiac chronotropic and dromotropic function in polymicrobial sepsis. *Shock* **2015**, *43*, 185–191. [CrossRef]
32. Rich, M.M.; McGarvey, M.L.; Teener, J.W.; Frame, L.H. ECG changes during septic shock. *Cardiology* **2002**, *97*, 187–196. [CrossRef]
33. Kalbitz, M.; Fattahi, F.; Herron, T.J.; Grailer, J.J.; Jajou, L.; Lu, H.; Huber-Lang, M.; Zetoune, F.S.; Sarma, J.V.; Day, S.M.; et al. Complement Destabilizes Cardiomyocyte Function In Vivo after Polymicrobial Sepsis and In Vitro. *J. Immunol.* **2016**, *197*, 2353–2361. [CrossRef]
34. Varriale, P.; Ramaprasad, S. Septic cardiomyopathy as a cause of long QT syndrome. *J. Electrocardiol.* **1995**, *28*, 327–329. [CrossRef]
35. Tisdale, J.E.; Jaynes, H.A.; Kingery, J.R.; Mourad, N.A.; Trujillo, T.N.; Overholser, B.R.; Kovacs, R.J. Development and validation of a risk score to predict QT interval prolongation in hospitalized patients. *Circ. Cardiovasc. Qual. Outcomes* **2013**, *6*, 479–487. [CrossRef]
36. Fattahi, F.; Russell, M.W.; Malan, E.A.; Parlett, M.; Abe, E.; Zetoune, F.S.; Ward, P.A. Harmful Roles of TLR3 and TLR9 in Cardiac Dysfunction Developing during Polymicrobial Sepsis. *Biomed. Res. Int.* **2018**, *2018*, 4302726. [CrossRef]
37. Liu, Y.C.; Yu, M.M.; Shou, S.T.; Chai, Y.F. Sepsis-Induced Cardiomyopathy: Mechanisms and Treatments. *Front. Immunol.* **2017**, *8*, 1021. [CrossRef]
38. Semeraro, F.; Ammollo, C.T.; Morrissey, J.H.; Dale, G.L.; Friese, P.; Esmon, N.L.; Esmon, C.T. Extracellular histones promote thrombin generation through platelet-dependent mechanisms: Involvement of platelet TLR2 and TLR4. *Blood* **2011**, *118*, 1952–1961. [CrossRef]
39. Maeder, M.; Fehr, T.; Rickli, H.; Ammann, P. Sepsis-associated myocardial dysfunction: Diagnostic and prognostic impact of cardiac troponins and natriuretic peptides. *Chest* **2006**, *129*, 1349–1366. [CrossRef]
40. Rudiger, A.; Singer, M. Mechanisms of sepsis-induced cardiac dysfunction. *Crit. Care Med.* **2007**, *35*, 1599–1608. [CrossRef]
41. Fattahi, F.; Grailer, J.J.; Lu, H.; Dick, R.S.; Parlett, M.; Zetoune, F.S.; Nunez, G.; Ward, P.A. Selective Biological Responses of Phagocytes and Lungs to Purified Histones. *J. Innate Immun.* **2017**, *9*, 300–317. [CrossRef] [PubMed]

42. Joulin, O.; Petillot, P.; Labalette, M.; Lancel, S.; Neviere, R. Cytokine profile of human septic shock serum inducing cardiomyocyte contractile dysfunction. *Physiol. Res.* **2007**, *56*, 291–297. [CrossRef] [PubMed]
43. Janssen, S.P.; Gayan-Ramirez, G.; Van den Bergh, A.; Herijgers, P.; Maes, K.; Verbeken, E.; Decramer, M. Interleukin-6 causes myocardial failure and skeletal muscle atrophy in rats. *Circulation* **2005**, *111*, 996–1005. [CrossRef] [PubMed]
44. Cain, B.S.; Meldrum, D.R.; Dinarello, C.A.; Meng, X.; Joo, K.S.; Banerjee, A.; Harken, A.H. Tumor necrosis factor-alpha and interleukin-1beta synergistically depress human myocardial function. *Crit. Care Med.* **1999**, *27*, 1309–1318. [CrossRef] [PubMed]
45. Parrillo, J.E.; Burch, C.; Shelhamer, J.H.; Parker, M.M.; Natanson, C.; Schuette, W. A circulating myocardial depressant substance in humans with septic shock. Septic shock patients with a reduced ejection fraction have a circulating factor that depresses in vitro myocardial cell performance. *J. Clin. Investig.* **1985**, *76*, 1539–1553. [CrossRef] [PubMed]
46. Kumar, A.; Thota, V.; Dee, L.; Olson, J.; Uretz, E.; Parrillo, J.E. Tumor necrosis factor alpha and interleukin 1beta are responsible for in vitro myocardial cell depression induced by human septic shock serum. *J. Exp. Med.* **1996**, *183*, 949–958. [CrossRef]
47. Standiford, T.J.; Strieter, R.M. TNF and IL-1 in sepsis: Good cytokines gone bad. *J. Lab. Clin. Med.* **1992**, *120*, 179–180.
48. Pierrakos, C.; Vincent, J.L. Sepsis biomarkers: A review. *Crit. Care* **2010**, *14*, R15. [CrossRef] [PubMed]
49. Prucha, M.; Bellingan, G.; Zazula, R. Sepsis biomarkers. *Clin. Chim. Acta* **2015**, *440*, 97–103. [CrossRef] [PubMed]
50. Atefi, G.; Zetoune, F.S.; Herron, T.J.; Jalife, J.; Bosmann, M.; Al-Aref, R.; Sarma, J.V.; Ward, P.A. Complement dependency of cardiomyocyte release of mediators during sepsis. *FASEB J.* **2011**, *25*, 2500–2508. [CrossRef]
51. Gillrie, M.R.; Lee, K.; Gowda, D.C.; Davis, S.P.; Monestier, M.; Cui, L.; Hien, T.T.; Day, N.P.; Ho, M. Plasmodium falciparum histones induce endothelial proinflammatory response and barrier dysfunction. *Am. J. Pathol.* **2012**, *180*, 1028–1039. [CrossRef] [PubMed]
52. Boyd, J.H.; Mathur, S.; Wang, Y.; Bateman, R.M.; Walley, K.R. Toll-like receptor stimulation in cardiomyoctes decreases contractility and initiates an NF-kappaB dependent inflammatory response. *Cardiovasc. Res.* **2006**, *72*, 384–393. [CrossRef] [PubMed]
53. Ma, Y.; Zhang, X.; Bao, H.; Mi, S.; Cai, W.; Yan, H.; Wang, Q.; Wang, Z.; Yan, J.; Fan, G.C.; et al. Toll-like receptor (TLR) 2 and TLR4 differentially regulate doxorubicin induced cardiomyopathy in mice. *PLoS ONE* **2012**, *7*, e40763. [CrossRef]
54. Mathur, S.; Walley, K.R.; Wang, Y.; Indrambarya, T.; Boyd, J.H. Extracellular heat shock protein 70 induces cardiomyocyte inflammation and contractile dysfunction via TLR2. *Circ. J.* **2011**, *75*, 2445–2452. [CrossRef] [PubMed]
55. Nozaki, N.; Shishido, T.; Takeishi, Y.; Kubota, I. Modulation of doxorubicin-induced cardiac dysfunction in toll-like receptor-2-knockout mice. *Circulation* **2004**, *110*, 2869–2874. [CrossRef] [PubMed]
56. Zhu, X.; Zhao, H.; Graveline, A.R.; Buys, E.S.; Schmidt, U.; Bloch, K.D.; Rosenzweig, A.; Chao, W. MyD88 and NOS2 are essential for toll-like receptor 4-mediated survival effect in cardiomyocytes. *Am. J. Physiol. Heart Circ. Physiol.* **2006**, *291*, H1900–H1909. [CrossRef] [PubMed]
57. Gao, M.; Ha, T.; Zhang, X.; Liu, L.; Wang, X.; Kelley, J.; Singh, K.; Kao, R.; Gao, X.; Williams, D.; et al. Toll-like receptor 3 plays a central role in cardiac dysfunction during polymicrobial sepsis. *Crit. Care Med.* **2012**, *40*, 2390–2399. [CrossRef] [PubMed]
58. Opal, S.M.; Laterre, P.F.; Francois, B.; LaRosa, S.P.; Angus, D.C.; Mira, J.P.; Wittebole, X.; Dugernier, T.; Perrotin, D.; Tidswell, M.; et al. Effect of eritoran, an antagonist of MD2-TLR4, on mortality in patients with severe sepsis: The ACCESS randomized trial. *JAMA* **2013**, *309*, 1154–1162. [CrossRef] [PubMed]
59. Tidswell, M.; Tillis, W.; Larosa, S.P.; Lynn, M.; Wittek, A.E.; Kao, R.; Wheeler, J.; Gogate, J.; Opal, S.M.; Eritoran Sepsis Study, G. Phase 2 trial of eritoran tetrasodium (E5564), a toll-like receptor 4 antagonist, in patients with severe sepsis. *Crit. Care Med.* **2010**, *38*, 72–83. [CrossRef]
60. Rice, T.W.; Wheeler, A.P.; Bernard, G.R.; Vincent, J.L.; Angus, D.C.; Aikawa, N.; Demeyer, I.; Sainati, S.; Amlot, N.; Cao, C.; et al. A randomized, double-blind, placebo-controlled trial of TAK-242 for the treatment of severe sepsis. *Crit. Care Med.* **2010**, *38*, 1685–1694. [CrossRef]
61. Monestier, M.; Fasy, T.M.; Losman, M.J.; Novick, K.E.; Muller, S. Structure and binding properties of monoclonal antibodies to core histones from autoimmune mice. *Mol. Immunol.* **1993**, *30*, 1069–1075. [CrossRef] [PubMed]
62. Grailer, J.J.; Canning, B.A.; Kalbitz, M.; Haggadone, M.D.; Dhond, R.M.; Andjelkovic, A.V.; Zetoune, F.S.; Ward, P.A. Critical role for the NLRP3 inflammasome during acute lung injury. *J. Immunol.* **2014**, *192*, 5974–5983. [CrossRef] [PubMed]
63. Sur Chowdhury, C.; Giaglis, S.; Walker, U.A.; Buser, A.; Hahn, S.; Hasler, P. Enhanced neutrophil extracellular trap generation in rheumatoid arthritis: Analysis of underlying signal transduction pathways and potential diagnostic utility. *Arthritis Res. Ther.* **2014**, *16*, R122. [CrossRef] [PubMed]
64. Sokolove, J.; Bromberg, R.; Deane, K.D.; Lahey, L.J.; Derber, L.A.; Chandra, P.E.; Edison, J.D.; Gilliland, W.R.; Tibshirani, R.J.; Norris, J.M.; et al. Autoantibody epitope spreading in the pre-clinical phase predicts progression to rheumatoid arthritis. *PLoS ONE* **2012**, *7*, e35296. [CrossRef]
65. Costa, O.; Monier, J.C. Antihistone antibodies detected by ELISA and immunoblotting in systemic lupus erythematosus and rheumatoid arthritis. *J. Rheumatol.* **1986**, *13*, 722–725.
66. Bruschi, M.; Petretto, A.; Bertelli, R.; Galetti, M.; Bonanni, A.; Pratesi, F.; Migliorini, P.; Candiano, G.; Vaglio, A.; Ghiggeri, G.M. Post-translational modified proteins are biomarkers of autoimmune-processes: NETosis and the inflammatory-autoimmunity connection. *Clin. Chim. Acta* **2017**, *464*, 12–16. [CrossRef] [PubMed]

67. Darwiche, S.S.; Ruan, X.; Hoffman, M.K.; Zettel, K.R.; Tracy, A.P.; Schroeder, L.M.; Cai, C.; Hoffman, R.A.; Scott, M.J.; Pape, H.C.; et al. Selective roles for toll-like receptors 2, 4, and 9 in systemic inflammation and immune dysfunction following peripheral tissue injury. *J. Trauma. Acute Care Surg.* **2013**, *74*, 1454–1461. [CrossRef]
68. Fattahi, F.; Grailer, J.J.; Parlett, M.; Lu, H.; Malan, E.A.; Abe, E.; Russell, M.W.; Frydrych, L.M.; Delano, M.J.; Zetoune, F.S.; et al. Requirement of Complement C6 for Intact Innate Immune Responses in Mice. *J. Immunol.* **2020**, *205*, 251–260. [CrossRef]

MDPI AG
Grosspeteranlage 5
4052 Basel
Switzerland
Tel.: +41 61 683 77 34

International Journal of Molecular Sciences Editorial Office
E-mail: ijms@mdpi.com
www.mdpi.com/journal/ijms

MDPI

www.ingramcontent.com/pod-product-compliance
Lightning Source LLC
Chambersburg PA
CBHW041416210926
51850CB00013B/3283
* 9 7 8 3 7 2 5 8 2 3 4 4 4 *